Multi-Agent-Based Simulations Applied to Biological and Environmental Systems

Diana Francisca Adamatti
Universidade Federal do Rio Grande, Brazil

A volume in the Advances in Computational Intelligence and Robotics (ACIR) Book Series

www.igi-global.com

Published in the United States of America by
IGI Global
Information Science Reference (an imprint of IGI Global)
701 E. Chocolate Avenue
Hershey PA, USA 17033
Tel: 717-533-8845
Fax: 717-533-8661
E-mail: cust@igi-global.com
Web site: http://www.igi-global.com

Library of Congress Cataloging-in-Publication Data

Names: Adamatti, Diana Francisca, editor.
Title: Multi-agent-based simulations applied to biological and environmental systems / Diana Francisca Adamatti, editor.
Description: Hershey PA : Information Science Reference, 2017.
Identifiers: LCCN 2016044784| ISBN 9781522517566 (hardcover) | ISBN 9781522517573 (ebook)
Subjects: LCSH: Biological systems--Computer simulation. | Multiagent systems.
Classification: LCC QH324.2 .M853 2017 | DDC 570.1/13--dc23 LC record available at https://lccn.loc.gov/2016044784

This book is published in the IGI Global book series Advances in Computational Intelligence and Robotics (ACIR) (ISSN: 2327-0411; eISSN: 2327-042X)

British Cataloguing in Publication Data
A Cataloguing in Publication record for this book is available from the British Library.

Advances in Computational Intelligence and Robotics (ACIR) Book Series

Ivan Giannoccaro
University of Salento, Italy

ISSN:2327-0411
EISSN:2327-042X

Mission

While intelligence is traditionally a term applied to humans and human cognition, technology has progressed in such a way to allow for the development of intelligent systems able to simulate many human traits. With this new era of simulated and artificial intelligence, much research is needed in order to continue to advance the field and also to evaluate the ethical and societal concerns of the existence of artificial life and machine learning.

The **Advances in Computational Intelligence and Robotics (ACIR) Book Series** encourages scholarly discourse on all topics pertaining to evolutionary computing, artificial life, computational intelligence, machine learning, and robotics. ACIR presents the latest research being conducted on diverse topics in intelligence technologies with the goal of advancing knowledge and applications in this rapidly evolving field.

Coverage

- Robotics
- Cognitive Informatics
- Synthetic Emotions
- Fuzzy systems
- Adaptive and Complex Systems
- Computational Logic
- Cyborgs
- Pattern Recognition
- Artificial Intelligence
- Heuristics

IGI Global is currently accepting manuscripts for publication within this series. To submit a proposal for a volume in this series, please contact our Acquisition Editors at Acquisitions@igi-global.com or visit: http://www.igi-global.com/publish/.

The Advances in Computational Intelligence and Robotics (ACIR) Book Series (ISSN 2327-0411) is published by IGI Global, 701 E. Chocolate Avenue, Hershey, PA 17033-1240, USA, www.igi-global.com. This series is composed of titles available for purchase individually; each title is edited to be contextually exclusive from any other title within the series. For pricing and ordering information please visit http://www.igi-global.com/book-series/advances-computational-intelligence-robotics/73674. Postmaster: Send all address changes to above address.

Titles in this Series

For a list of additional titles in this series, please visit: www.igi-global.com

Multi-Core Computer Vision and Image Processing for Intelligent Applications
Mohan S. (Al Yamamah University, Saudi Arabia) and Vani V. (Al Yamamah University, Saudi Arabia)
Information Science Reference • copyright 2017 • 292pp • H/C (ISBN: 9781522508892) • US $210.00 (our price)

Developing and Applying Optoelectronics in Machine Vision
Oleg Sergiyenko (Autonomous University of Baja California, Mexico) and Julio C. Rodriguez-Quiñonez (Autonomous University of Baja California, Mexico)
Information Science Reference • copyright 2017 • 341pp • H/C (ISBN: 9781522506324) • US $205.00 (our price)

Pattern Recognition and Classification in Time Series Data
Eva Volna (University of Ostrava, Czech Republic) Martin Kotyrba (University of Ostrava, Czech Republic) and Michal Janosek (University of Ostrava, Czech Republic)
Information Science Reference • copyright 2017 • 282pp • H/C (ISBN: 9781522505655) • US $185.00 (our price)

Integrating Cognitive Architectures into Virtual Character Design
Jeremy Owen Turner (Simon Fraser University, Canada) Michael Nixon (Simon Fraser University, Canada) Ulysses Bernardet (Simon Fraser University, Canada) and Steve DiPaola (Simon Fraser University, Canada)
Information Science Reference • copyright 2016 • 346pp • H/C (ISBN: 9781522504542) • US $185.00 (our price)

Handbook of Research on Natural Computing for Optimization Problems
Jyotsna Kumar Mandal (University of Kalyani, India) Somnath Mukhopadhyay (Calcutta Business School, India) and Tandra Pal (National Institute of Technology Durgapur, India)
Information Science Reference • copyright 2016 • 1015pp • H/C (ISBN: 9781522500582) • US $465.00 (our price)

Applied Artificial Higher Order Neural Networks for Control and Recognition
Ming Zhang (Christopher Newport University, USA)
Information Science Reference • copyright 2016 • 511pp • H/C (ISBN: 9781522500636) • US $215.00 (our price)

Handbook of Research on Generalized and Hybrid Set Structures and Applications for Soft Computing
Sunil Jacob John (National Institute of Technology Calicut, India)
Information Science Reference • copyright 2016 • 607pp • H/C (ISBN: 9781466697980) • US $375.00 (our price)

www.igi-global.com

701 E. Chocolate Ave., Hershey, PA 17033
Order online at www.igi-global.com or call 717-533-8845 x100
To place a standing order for titles released in this series, contact: cust@igi-global.com
Mon-Fri 8:00 am - 5:00 pm (est) or fax 24 hours a day 717-533-8661

Table of Contents

Detailed Table of Contents

computer-aided landscape ecology. More specifically, we introduce the "ecosystems as agent societies" and "landscapes as multi-societal agent systems" approaches to ecosystems and landscapes, together with the core elements of the agent-based architectural models that support such approaches. The elements of those architectural models are then used to formally capture the main organizational and functional aspects of ecosystems and landscapes.

Thomas Portegys, Dialectek, USA
Gabriel Pascualy, University of Michigan, USA
Richard Gordon, Gulf Specimen Marine Laboratory and Wayne State University, USA
Stephen P McGrew, New Light Industries, USA
Bradly J. Alicea, OpenWorm, USA

A cellular automaton model, Morphozoic, is presented. Morphozoic may be used to investigate the computational power of morphogenetic fields to foster the development of structures and cell differentiation. The term morphogenetic field is used here to describe a generalized abstraction: a cell signals information about its state to its environment and is able to sense and act on signals from nested neighborhood of cells that can represent local to global morphogenetic effects. Neighborhood signals are compacted into aggregated quantities, capping the amount of information exchanged: signals from smaller, more local neighborhoods are thus more finely discriminated, while those from larger, more global neighborhoods are less so. An assembly of cells can thus cooperate to generate spatial and temporal structure. Morphozoic was found to be robust and noise tolerant. Applications of Morphozoic presented here include: 1) Conway's Game of Life, 2) Cell regeneration, 3) Evolution of a gastrulation-like sequence, 4) Neuron pathfinding, and 5) Turing's reaction-diffusion morphogenesis.

Vladimir Rocha, University of São Paulo, Brazil
Anarosa Alves Franco Brandão, University of São Paulo, Brazil

Recently, there has been an explosive growth in the use of wireless devices, mainly due to the decrease in cost, size, and energy consumption. Researches into Internet of Things have focused on how to continuously monitor these devices in different scenarios, such as environmental and biodiversity tracking, considering both scalability and efficiency while searching and updating the devices information. For this, a combination of an efficient distributed structure and data aggregation method is used, allowing a device to manage a group of devices, minimizing the number of transmissions and saving energy. However, scalability is still a key challenge when the group is composed of a large number of devices. In this chapter, the authors propose a scalable architecture that distributes the data aggregation responsibility to the devices of the boundary of the group, and creates agents to manage groups and the interaction among them, such as merging and splitting. Experimental results showed the viability of adopting this architecture if compared with the most widely used approaches.

models, methodologies, and frameworks. In particular, the adoption of agent-based modelling (ABM) in the field of multicellular systems biology is explored, focussing on the challenging scenarios of developmental biology. After motivating why agent-based abstractions are critical in representing multicellular systems behaviour, MABS is discussed as the source of the most natural and appropriate mechanism for analysing the self-organising behaviour of systems of cells. As a case study, an application of MABS to the development of Drosophila Melanogaster is finally presented, which exploits the ALCHEMIST platform for agent-based simulation.

Section 2
Applications in Biological and Environmental Systems

Risk assessment for human health and ecosystems by exposure to chemicals is an important process to aid in the mitigation of affected areas. Generally, this process is carried out in isolated spots and therefore may be ineffective in mitigating. This chapter describes an architecture of a multi-agent system for environmental risk assessment in areas contaminated as often occur in mining, oil exploration, intensive agriculture and others. Plan multiple points in space-time matrix where each agent carries out exposure assessment and the exchange of information on toxicity, to characterize and classify risk in real time. Therefore, it is an architecture model with multi-agent that integrates ontology by semantic representation, classifies risks by decision rules by support vectors machines with multidimensional data. The result is an environment to exchange information that provides knowledge about the chemical contamination, which can assist in the planning and management of mitigation of the affected area.

Microbial Fuel Cells (MFC) could generate electrical energy combined with the wastewater treatment and they can be a promising technological opportunity. This chapter presents an agent-based model and simulation of MFC comparing it with analytical models, to show that this approach could model and simulate these problems with more abstraction and with excellent results.

Nowadays, urban mobility and air quality issues are prominent, due to the heavy traffic of vehicles and the emission of pollutants dissipated in the atmosphere. In the literature, a model of optimal control of traffic lights using Genetic Algorithms (GA) has been proposed. These algorithms have been introduced in the context of control traffic. In order to search for possible solutions to the problems of traffic lights in major urban centers. Thus, the study of the dispersion of pollutants and Genetic Algorithms with simulations performed in Urban Mobility Simulator SUMO (Simulation of Urban Mobility), seek satisfactory solutions to such problems. The AG uses the crossing of chromosomes, in this case the times of the traffic lights, featuring the finest green light times and the sum of each of the pollutants each simulation cycle. The simulations were performed and the results compared analyzes showed that the use of the genetic algorithm is very promising in this context.

Tertiary protein structure prediction in silico is currently a challenging problem in Structural Bioinformatics and can be classified according to the computational complexity theory as an NP-hard problem. Determining the 3-D structure of a protein is both experimentally expensive, and time-consuming. The agent-based paradigm has been shown a useful technique for the applications that have repetitive and time-consuming activities, knowledge share and management, such as integration of different knowledge sources and modeling of complex systems, supporting a great variety of domains. This chapter provides an integrated view and insights about the protein structure prediction area concerned to the usage, application and implementation of multi-agent systems to predict the protein structures or to support and coordinate the existing predictors, as well as it is advantages, issues, needs, and demands. It is noteworthy that there is a great need for works related to multi-agent and agent-based paradigms applied to the problem due to their excellent suitability to the problem.

The phytoplanktons are organisms that have limited locomotion about the current being drift in aquatic environment. Another characteristic of phytoplankton their growth and energy are result about photosynthetic process. It is important to emphasize that the phytoplankton is the main primary producer of aquatic environment, it means that, it is the base the aquatic food chain . The organic material produced by phytoplankton is responsible in provide the material and energy which sustains the growth of fish, crustaceans and mollusks, in marine ecosystems. Because of this, it is important to know the factors that interfere with their accumulation in environments mainly in fishing regions. In this way, this study tries to demonstrate the importance of retention time, often caused by hydrological issues, in the variation of phytoplankton biomass in the estuary of the Patos Lagoon (ELP), in Rio Grande/RS. To do that, we created one model that simulates this environment, using techniques of multi-agent-based simulation and its implementation was done with the NetLogo tool.

Foreword

Simulation is a third way of doing science (Axelrod, 1997).

By building on a computer artificial societies using and sharing resources in a virtual landscape, Epstein and Axtell (1996) promoted a new way of thinking for "Social science from the bottom up". Their purpose was to demonstrate that it is possible to explain many concepts from the Social Sciences perspective via computer simulations based on relatively simple models. By designing step-by-step populations of simple agents located in a *Sugarscape*, they contributed, with others, to explain "How does the heterogeneous micro-world of individual behaviors generate the global macroscopic regularities of the society?" (Epstein & Axtell, 1996). *Sugarscape*, the virtual world they have implemented, allows them to experiment various hypothesis on the emergence of social structures such as migratory phenomena, trade exchanges, crisis and wars. In other words, they seek to explain complex social phenomena from simple but dynamic representations.

Thus from the origin, the Agent-Based Model (ABM) paradigm aims at explaining global emergent patterns from individual behaviors of agents interacting with their environment (including other agents). So, to claim such bottom-up approach, it is necessary to conceive our model by designing individual behaviors and by formulizing how the agents interact. Then, by running the model, the simulation time will let the entities evolve and interact. The simulation lets the model to express itself: by activating the agents, the time animates the model (from Lat. *animare*: to give life) and lets us see how global phenomena may emerge. Distinguishing model and simulation is thus of importance. The modeler should conceive agents with restricted skills (local perception, no or partial control on the others, etc.) and similarly, he shouldn't specify how a population of entities must evolve. Therefore, instead of trying to predefine the states of the agents and to fix on how the system must evolve, the modeler should try to find the basic rules that conduct the actors' strategies and may reflect activities observed in reality.

In the frame of ABM applied to Biological and Environmental Systems, it is sometimes tempting for a modeler to design his agents in a very descriptive way. This work is of interest to describe practices observed on the field. But these descriptions can be seen as frozen: a sequential set of facts without explanation. On the contrary, the agents need a kind of liberty in their choices: an emulated autonomy that allows them to take decisions and finally to confer them some adaptation capabilities. Consequently, after having described the practices of the actors, we should gain in abstraction by trying to extract the basic mechanisms and decision points that conduct their behaviors. Hence, when running a simulation (after having translated these schemes in a computer language), the agents are able to act in a more freely way, without a strong control on the sequences of their actions. Then, if coherent patterns are observed at individual and global levels, we can estimate that we were able to acquire a more essential level of understanding of the studied system. From a very descriptive representation, the model has improved its generic nature: with our understanding of the system, we are able to explain the reasons why some social or ecological behaviors occur.

Models, however, are not crystal balls and the resulting simulation is not a prediction (Bradbury, 2002). It is an artifact, which by nature is highly unrealistic. It is nonetheless essential because, it helps us to assess the logical consequences of multiple and interdependent mechanisms. It is a crutch for understanding because our brain is simply not able to anticipate how several joined dynamics produce a global behavior. In this sense, having designed a model doesn't mean to be able to anticipate all its outputs. After having checked the consistency of it functioning (Balci, 1998), we should explore it parameters space. The exploration phase gives us a better understanding of the model significance. This knowledge allows us to better anticipate its reactions, to better explain its results and to provide answers to the questions at the origin of the modeling process. Therefore, the role played by the model is essentially to learn and understand (Grimm & Railsback, 2005).

One advantage of ABM (which should rather be considered as a weakness) is the ability to endlessly complicate a model to work towards a perfect and realistic representation of the world. While mathematical models require condensing knowledge, conciseness isn't a constraint for the design of ABM. One can incorporate more and more elements and details.

Anyway, simplistic versus sophisticated models is always a topical issue. In this debate, some argue it is essential to build complicated models to address social issues. They criticize the excessive simplicity of some ABM with too poor reasoning agents. This approach is also consistent with the ideas of some sociologists, who (when they do not reject any idea of modeling) recommend using cognitive agents to generate social dynamics. In this regard, the BDI architecture often used to model cognitive agents does not rely on robust scientific evidences: it is not derived from precepts of neuroscience nor psychology nor philosophy. In the paper entitled "From KISS to KIDS", Edmonds and Moss (2004) propose an "anti-simplistic" approach for modeling. They criticize the usual KISS (Keep It Simple Stupid) approach, which requires the modeler to make preliminary choices and to eliminate elements that seem unimportant a priori. The risk is to eliminate information that could be fundamental to correctly describe the structure and dynamics of the studied system. In contrast, they claim the KIDS approach (Keep It Descriptive Stupid), which aims to incorporate into a model all available information on the system.

However, even for very simple models, the probability to unintentionally introduce simulation bias is not zero. Given the sensitivity of ABMs with respect to time and interaction management, but also with respect to the initial conditions, and even to the quality of the pseudo-random number generator, not to mention errors related to floating point calculation, there are significant chances of finding results which may be the consequences of biases. Today, we cannot demonstrate the characteristics of an ABM. Then one can legitimately question the reliability of its results since many biases may occur, especially from complicated ABM. Yet these problems are generally underestimated: we usually prefer to improve the agent decision-making or the realism of it behavior at the expense of a clear and unbiased management of their activation or interactions.

Besides the problem of reliability, designing a very sophisticated model means running the risk of not being able to explain its results. If the decisions of the agents are too complex and the processes too intricately linked, it quickly becomes impossible to provide a comprehensible explanation about the simulated outputs. Yet the main purpose of modeling is not to mimic reality nor to simplify complexity, but rather to try to understand it. Even if it looks wonderful, an ABM remains essentially unrealistic and its results remain not provable. Without any explanatory dimension, it becomes useless. Running after the fantasy of the perfect model seems doomed to failure. Consequently, if it is useful to criticize a model, the lack of sophistication should not be blamed.

Between extreme simplicity and endless sophistication, it is obviously necessary to find a balance so that the model can be useful for understanding or decision. However, given the current knowledge on ABM simulation, throwaway models should be preferred to cathedrals models. Instead of seeking complex models (in their structures and mechanisms), it seems preferable to regain a form of complexity through simulation (as explained earlier in this text). At least, studying simplifications of an elaborated model is a necessary stage. "It allows the researchers to establish robust results and stylized facts, which constitute references for the study of more complicated dynamics" (Deffuant, Weisbuch, Amblard, & Faure, 2003).

ABM modeling, like any modeling, should obey the principle of parsimony. This is not to say that simplicity is a guarantee of truth, but rather that looking for concision requires identifying and understanding the basic mechanisms at work in the phenomenon under study. In other words, it is important to say as simply as possible the most complex things as would say William Ockham, the medieval philosopher.

Pierre Bommel
CIRAD, France

Pierre Bommel *is a modeler scientist at CIRAD (the French Agricultural Research Center for International Development). As member of the GREEN Research Unit, he contributes to promote the Companion Modeling approach (http://www.commod.org) dedicated to participatory modeling and simulation involving local stakeholders. Through the development of CORMAS, a simulation plateform for Agent-Based Models (ABM, http://cormas.cirad.fr), he has been focusing on the development and the use of ABM for renewable resources management. As associated professor in Brazil, at the university of Brasilia (UnB) then at Rio de Janeiro (PUC), he developed models related to environmental management, such as breeding adaptation to drought in the Uruguay or as breeding and deforestation in the Amazon. He is currently based at the University of Costa Rica working on adaptation of agriculture and livestock to Climate Changes.*

REFERENCES

Axelrod, R. (1997). Advancing the art of simulation in the social sciences. In R. Conte, R. Hegselmann, & P. Terna (Eds.), *Simulating social phenomena*. Berlin: Springer-Verlag.

Balci, O. (1998). Verification, validation, and accreditation. In *Proceedings of the 30th Conference on Winter Simulation*. IEEE Computer Society Press. doi:10.1109/WSC.1998.744897

Bradbury, R. H. (2002). Futures, prediction and other foolishness. In M. A. Janssen (Ed.), *Complexity and ecosystem management: The theory and practice of multi-agent systems* (pp. 48–62). Cheltenham, UK: Edward Elgar.

Deffuant, G., Weisbuch, G., Amblard, F., & Faure, T. (2003). Simple is beautiful . . . and necessary. *Journal of Artificial Societies and Social Simulation*, *6*(1).

Edmonds, B., & Moss, S. (2004). *From KISS to KIDS - An "anti-simplistic" modelling approach.Joint Workshop on Multi-Agent and Multi-Agent-Based Simulation*, New York, NY.

Epstein, J. M., & Axtell, R. L. (1996). *Growing Artificial Societies. Social science from the bottom up*. Washington, DC: Brookings Institution Press.

Gary, P. J., Izquierdo, L. R., & Gotts, N. M. (2005). The ghost in the model (and other effects of floating point arithmetic). *Journal of Artificial Societies and Social Simulation*, *8*(1). Retrieved from http://jasss.soc.surrey.ac.uk/8/1/5.html

Grimm, V., & Railsback, S. F. (2005). *Individual-based Modeling and Ecology*. Princeton, NJ: Princeton University Press. doi:10.1515/9781400850624

Preface

The main topic of this book is the Multi-Agent-Based Simulations (MABS) applied to biological and environmental domains. The interdisciplinary character of MABS is an important challenge faced by all researchers, while demanding a difficult interlacement of different theories, methodologies, terminologies and points of view. MABS has provided architectures and platforms for the implementation and simulation of relatively autonomous agents and it has contributed to the establishment of the agent-based computer simulation paradigm. The agent-based approach enhances the potentialities of computer simulation as a tool for theorizing about social scientific issues. In particular, the notion of an extended computational agent, implementing cognitive capabilities, is giving encouragement to the construction and exploration of artificial societies, since it facilitates the modeling of artificial societies of autonomous intelligent agents.

Although it is inherently an interdisciplinary discipline, MABS is a difficult activity. This book brings new insights on the discipline, looking through its challenges. The book presents some current discussions on agent-based simulation and modeling, addressing theoretical, methodological, technical and instrumental issues concerning the areas of Biological and Environmental Systems, and focusing on applications, so offering different kinds of models and tools that can help the reader to face complex developments. The book is divided into 2 sections.

Section 1, "Theoretical Models and Tools," provides some new agent-based models or tools to advance current research in the theme, to help a better understanding of the area and also assist in the development of new simulations.

Chapter 1, "Ignition of Algorithm Mind: The Role of Energy in Neuronal Assemblies," by Magessi and Antunes, presents a simple multi-agent model, where we explored the role of energy and respective limits on neuronal assemblies.

Chapter 2, "Ecosystems as Agent Societies: Landscapes as Multi-Societal Agent Systems," by Rocha Costa, introduces the ecosystems as agent societies and landscapes as multi-societal agent systems approaches to ecosystems and landscapes, together with the core elements of the agent-based architectural models that support such approaches.

Chapter 3, "Morphozoic, Cellular Automata with Nested Neighborhoods as a Metamorphic Representation of Morphogenesis," by Portegys, Pascualy, Gordon, McGrew and Alicea, presents a cellular automaton model, Morphozoic that may be used to investigate the computational power of morphogenetic fields to foster the development of structures and cell differentiation.

Chapter 4, "A Scalable Multiagent Architecture for Monitoring Biodiversity Scenarios," by Rocha and Brandão, proposes a scalable architecture that distributes the data aggregation responsibility to the devices of the boundary of the group, and creates agents to manage groups and the interaction among them, such as merging and splitting.

Chapter 5, "MASE: A Multi-Agent-Based Environmental Simulator," by Ralha and Abreu, presents research carried out under the MASE project, including the definition of a conceptual model to characterize the behavior of individuals that interact in the dynamics of land-use and cover change. A computational tool for analyzing environmental scenarios of land change was developed, called MASE – Multi-Agent System for Environmental Simulation.

Chapter 6, "Modelling and Simulating Complex Systems in Biology: Introducing NetBioDyn – A Pedagogical and Intuitive Agent-Based Software," by Ballet, Rivière, Pothet, Theron, Pichavant, Abautret, Fronville and Rodin, presents NetBioDyn, an original software aimed at biologists (students, teachers, researchers) to easily build and simulate complex biological mechanisms observed in multicellular and molecular systems. Thanks to its specific graphical user interface guided by the multi-agent paradigm, this software does not need any prerequisite in computer programming.

Chapter 7, "Agent-Based Modelling in Multicellular Systems Biology," by Montagna and Omicini, discusses the content of multi-agent based simulation (MABS) applied to computational biology, i.e, to modelling and simulating biological systems by means of computational models, methodologies, and frameworks. MABS is discussed as the source of the most natural and appropriate mechanism for analysing the self-organising behaviour of systems of cells.

Section 2, "Applications in Biological and Environmental Systems," is the core of the book, presenting applications in the research focus fields, such as chemical contamination, three-dimensional protein structure prediction, air pollution, biodiversity conservation and social inclusion or tuberculosis bacillus growth curve.

Chapter 8, "Architecture with Multi-Agent for Environmental Risk Assessment by Chemical Contamination," by Andrade and Modesto, describes an architecture of a multi-agent system for environmental risk assessment in areas contaminated as often occur in mining, oil exploration, intensive agriculture and others. Plan multiple points in space-time matrix where each agent carries out exposure assessment and the exchange of information on toxicity, to characterize and classify risk in real time.

Chapter 9, "Microbial Fuel Cells Using Agent-Based Simulation: Review and Basic Modeling," by Machado, Adamatti and Gonçalves, presents an agent-based model and simulation of Microbial Fuel Cells comparing it with analytical models, to show that this approach could model and simulate these problems with more abstraction

Chapter 10, "Use SUMO Simulator for the Determination of Light Times in Order to Reduce Pollution: A Case Study in the City Center of Rio Grande, Brazil," by Born, Souza de Aguiar and Adamatti, proposes a study of the dispersion of pollutants and Genetic Algorithms with simulations performed in Urban Mobility Simulator SUMO (Simulation of Urban Mobility) to calibrate the times of the traffic lights, featuring the finest green light times and the sum of each of the pollutants each simulation cycle.

Chapter 11, "Multi-Agent Systems in Three-Dimensional Protein Structure Prediction," by Corrêa and Dorn, provides an integrated view and insights about the protein structure prediction area concerned to the usage, application and implementation of multi-agent systems to predict the protein structures or to support and coordinate the existing predictors, as well as it is advantages, issues, needs, and demands

Chapter 12, "Biomass Variation of Phytoplankton Using Agent-Based Simulation: A Case Study to Estuary of the Patos Lagoon," by Porcellis, Abreu and Adamatti, simulates the phytoplankton, the main primary producer of aquatic environment - the base the aquatic food chain and tries to demonstrate the importance of retention time, often caused by hydrological issues, in the variation of phytoplankton biomass in the estuary of the Patos Lagoon (ELP), in Rio Grande/RS.

Chapter 13, "Participatory Management of Protected Areas for Biodiversity Conservation and Social Inclusion: Experience of the SimParc Multi-Agent Based Serious Game," by Briot, Irving, Vasconcelos Filho, de Melo, Alvarez, Sordoni and Lucena, present their experience in a serious game research project, named SimParc, about multi-agent support for participatory management of protected areas for biodiversity conservation and social inclusion.

Chapter 14, "Using Probability Distributions in Parameters of Variables at Agent-Based Simulations: A Case Study for the TB Bacillus Growth Curve," by Moraes, Borba, Werhli, von Groll and Adamatti, models the growth curve of Mycobacterium tuberculosis using MABS, aiming to simulate the curve with the minimum possible error when compared to in vitro results.

Finally, I hope this book provides a comprehensive and integrated view of the current discussions and investigations on MABS in the biological and environmental domains.

Diana F. Adamatti
Universidade Federal do Rio Grande, Brazil

Section 1
Theoretical Models and Tools

Chapter 1
Ignition of Algorithm Mind:
The Role of Energy in Neuronal Assemblies

Nuno Trindade Magessi
FCUL, Portugal

Luis Antunes
FCUL, Portugal

ABSTRACT

The ignition of the algorithmic mind is a fascinating phenomenon that occurs in our brains. The algorithm mind is related to our reasoning. When we use it, we consume a lot of resources from our brains like energy. The ignition process is triggered by reflective mind and it works through neuronal assemblies. Specific neurons are ignited and then it begins a recruitment process for other neurons in order to assemble a complex structure. To understand these mechanisms, we have developed a simple multi-agent model, where we explored the role of energy and respective limits on neuronal assemblies. The available and consumed energy are the keystones to ignite the algorithm mind and to find out the limit that interrupts our reasoning's. The connections between incumbent and new neurons are at the same level as the connections established only between the new neurons in the case of algorithmic mind. Unlike, the autonomous mind established more connections, only between new neurons. Finally, the algorithmic mind consumes more energy than autonomous mind, which has a clearly declining trend.

INTRODUCTION

The Algorithm Mind was a concept launched and defined by Keith Stanovich (2008). It was a work where he brought a new approach with an intuit of giving a step forward from the two systems of thinking (Kahneman, 2011). Basically, Daniel Kahneman defended the existence of two systems of thought inside our brains: system I and II. System I, is considered as being fast, automatic, frequent, emotional and subconscious. Instead System II is slower, requiring additional efforts, is less frequent, logical, calculating and conscious (Kahneman, 2011). System I and System II debate the use of reasoning or lack of it regarding the decision making process. Instead, Stanovich disagreed with this simplistic approach of dividing the way that we think. Proceeding with his logical reasoning he suggested a mind with tri

DOI: 10.4018/978-1-5225-1756-6.ch001

process theory. According to this theory, Stanovich suggested the division of the mind in three distinct minds: reflective, algorithmic and autonomous. The three process approach was an advance on the discussion of great rationality debate in cognitive science and a step forward in trying to work out the impacts of dual process theory (Stanovich, 2008). The author has focused his work on debating add-ons and issues belonging to the dual-process theory. Then, he worked on the effects of these ideas with the intuit of building his perspective on human rationality.

Rationality, in terms of physiological perspective works under the prospect of neuronal assemblies (Greenfield, 2002). Neuronal assemblies are structures of transiently synchronized neurons belonging to a variety of neural systems and are thought to encode sensory information or store short-term memories (Galan, 2006). This phenomenon is reminiscent of the formation of clusters in models of coupled phase oscillators (Golomb et al 2002; Galan, 2006). In reality, all the dynamic involving neural networks is normally described by systems of coupled phase oscillators as long as networks have weak connections and the neurons' firing frequencies are roughly constant.

Now, if rationality is sustained on the neuronal assemblies, we should question:

- Why do our brains are considered lazy giving more use to system I in deterrence of system II?
- Or according to the tri-process of Stanovich why people use more the autonomous mind for solving problems instead of using more the algorithmic mind?

We have investigated the mechanisms leading to the emergence of these neural assemblies with models of coupled oscillators (Galan, 2006). We took in consideration that the formation of synchronized assemblies is a rather general phenomenon. And we used as assumption the mathematical analysis done by Galan among others to support the necessary conditions for the occurrence of synchronized neuronal assemblies.

If we analyse the subjacent mathematical formulas, we can verify that energy represents a specific and important role in the way that neuronal assemblies are structured on the tri process of mind. The brain, besides other elements, works under a spectrum of energy (Kahneman, 2011). Typically, these elements are combined in a certain way, with the objective to produce energy that puts all necessary synapsis working. It is also interesting to see human brains emitting waves, like when a person focuses his attention and tries to perceive some risk or remembering something important or even trying to solve a complex problem. This activity fires thousands of neurons simultaneously at the same frequency generating a wave, at a rate closer 10 to 100 cycles per second (Kahneman, 2011). There are several types of brain waves and the algorithmic mind occurs when the brain waves become beta and gamma, depending on the implicit reasoning needs. Waves that are the fastest ones comparing to the waves used by automatic mind. Automatic mind usually works with out more by using alpha waves which are considerable slower. So the question arises: if the autonomous mind is associated to system I which is faster, instantaneous and the subconscious works with waves in a slower frequency mode, why does the algorithmic mind which takes more time, is infrequent and conscious works with waves in a higher frequency? Isn't supposed to work in an opposite way? In other words, does it make sense to have a lower energy frequency for a mind working intensively and a higher energy frequency for a mind working less intensively? Why the systems work inversely and are asymmetric?

To tackle this problem, we decided to explore the role of energy in the ignition of the algorithmic mind and try to explain why exists this negative correlation between the characteristics of each type of mind and the frequency of energy. For that we built a simulator under multi agent based system methodology where neurons establish synapses in order to build neuronal assemblies. We will compare neuronal assemblies that occur with automatic mind and algorithmic mind.

In this sense the chapter is organised as follows.

- In section 2, we review the most relevant literature on this matter and give an explanation why this item is so important to sustain our approach.
- In section 3, we outline the specificities of the problem and its related concepts.
- Section 4 describes the proposed model, and the thinking process behind the simulations of neuronal assemblies.
- In section 5 we analyse our results and discuss them in the light of the tri process theory of mind.
- Finally, on section 6, we draw out our conclusions.

REVISITING THE LITERATURE OF RACIONALITY: DUAL AND TRI PROCESS

In several works, (Stanovich, 1999, 2004; Stanovich & West, 2000, 2003) Stanovich proposes the splitting of human thinking structure moving a step forward from dual-process theory to a tri-process. He argued that this split would help to better understand the nature of the disputes about the great rationality debate inside cognitive science. The proposal establishes differences between System I and System II structures since they don't have the same baseline and these differences have tremendous impact on individual's fulfilment. The analytical system fits more with individual's needs, integrated as a homogeneous body than it is the System I. A system that is more closely tuned to the old reproductive goals of "sub-personal replicators" (Stanovich, 2004). The system I is more non-reflectively related.

When the results of the two systems enter in conflict, people will probably be in a best position if they follow the analytical system which overrides the triggered result of System I (Stanovich, 2008). In cases where exists a response to a conflict overriding could be the best strategy to be taken. The instrumental rationality is normally undervalued by evolutionary psychology and adaptive modelling. This happens because evolutionary psychology sustain that exists many non-normative responses in reasoning experiments. It is clearly important to enhance that evolutionarily adaptive behaviour is far from the rational behaviour concept. (Stanovich 2004) Evolutionary psychology avoids this, since it considers the assumption that an adaptive behaviour is rational, which is not true most of the times, according to Stanovich. This merger represents a fundamental error of extreme importance to us humans. Meanings of rationality must remain strong and reliable with the element where improvement is stills in doubt. Keeping in mind that the end goal to keep up this consistency is the distinctive "interests" of interveners that must be expressively perceived.

Stanovich and other researchers have demonstrated that the reaction is in conformity with many evolutionary analyses. It is also the modal response on many heuristics and biases tasks, where the most cognitively capable subjects give the reaction that is instrumentally rational (Stanovich, 2008)

According with the patterns identified on data Stanovich agreed that the evolutionary psychologists are probably correct when they affirm that most of System I reactions are evolutionarily adaptive and developmentally versatile. However, the author also highlight that evolutionary interpretations are compatible with the positions of heuristics and biases defenders. The alternative reaction given by the minority of people is rational. Individuals with higher analytic intelligence have more propensity to override System I in order to generate reactions that are epistemically and instrumentally rational. These findings are consistent with a comparative union of rationality (Evans, 2007). For sure, such a combination could be said to be prone as a two process hypothesis (Kahneman, 2000). Many theorists consider System I and the Heuristic System as the same. A kind of Singular System. (Evans 2006; Carruthers 2006). Meanwhile, nowadays this system should not be considered as singular, but in fact as plural. Stanovich advocates that both systems are in reality a group of systems in the brain that operate autonomously in reaction to the unleashing of their own stimuli (Stanovich, 2008). It means that, he admits the existence of heterogeneous systems which are not under the control of the analytic processing system. He has suggested the acronym of "The Autonomous Set of Systems" to separate both systems. Unfortunately, the author couldn't finish his conclusions in how the system works at a physiological level. He had the perspective that TASS processes would be considered as being modular but not restricted to modular sub processes that that fits the classic Fodorian criteria (Fodor, 1983). At this stage, it is proposed that TASS should be in line with the Darwinian mind of quasi-modules defended by evolutionary psychologists. This happens because TASS is assumed to have its domain sustained on unconscious processes with implicit learning and conditioning. TASS is also considered as having many rules discriminating stimuli and principles of decision which are done automatically (Shiffrin & Schneider, 1977). Finally, some authors considered that emotions as regulators of the processes associated to behaviours are inside in the context of TASS (Evans, 2003)

Thus, TASS processes are interconnected on the basis of autonomy and not in terms of modularity. They answer automatically to a sequence of stimuli and their execution is independent from the origin of input. Besides TASS is not under the control of the analytic processing system: System II. Finally, Stanovich argue that TASS could execute and provide outputs that are in conflict with the results of a simultaneous computation being carried out by System II. The attrition between dual-process models could be compared to the modules resulting from evolutionary adaptation, if we look to the learned information in TASS. This learned information can be just as much as a threat to rational behaviour. It means that, only in cases of need the System II will override it, like the evolutionary modules that fire inappropriately in a modern environment. The automatic rules who were learned can be overgeneralized. These rules have the independency to trigger behaviour when the situation is an exception to the class of events they are meant to cover (Hsee & Hastie, 2006; Stanovich 2008).

The conceptualization of System II has also launched some concerns to Stanovich. He defended that System II must be understood in terms of two levels of processing. It means that he split off System II in a algorithmic level and a reflective level. He based his arguments by taking into account the logic behind TASS override. TASS will actualize its short-chained objectives unless overridden by the algorithmic mechanisms executing the long-rope objectives of the analytic system. Be that as it may, override itself is started by higher level of control.

So, the algorithmic level belongs to the analytic system and subordinated to the "higher-level goal states and epistemic thinking dispositions" (Cacioppo et al., 1996; Stanovich & West, 1997, 2007). These states of mind and epistemic dispositions belong to what is considered by Stanovich as the reflective mind of processing. The reflective mind is associated to the control states which moderate behaviour at a high level.

We can identify the high-level goal states in the intelligent agents built by artificial intelligence researchers (Franklin, 1995; Pollock, 1995; Sloman, 1993; Sloman & Chrisley, 2003).

At this moment, based on the works of Stanovich, literature have split the concept of mind in three types or levels. First TASS or System I is actually called as autonomous mind. The algorithmic level of System II is called the algorithmic mind and the reflective level of system II as, the reflective mind. This tri process theory is set against to what Spinoza defended in seventeenth century the three conceptual functions of mind (Spinoza,1677). However, both authors discovered in individuals, different behaviours for reasoning and reaction. For us exists a parallelism between the reaction function of Spinoza and the autonomous mind of Stanovich. The same happens if we compare the reasoning function of Spinoza with the algorithm mind of Stanovich. The reflect function and reflective mind are quite different, since the last on assumes itself as the high level of control charges but they use the same part of the brain.

These concepts are not new to artificial intelligence researchers. Artificial Intelligence took seriously and with a lot of interest the tri-process or structure of mind (Sloman & Chrisley, 2003; Samuels, 2005). The gap between algorithmic and reflective mind will be no way nicely as those that have traditionally differentiated System I and II just because the algorithmic and reflective mind will both share properties like capacity-limited serial processing that differentiate them from the autonomous mind. This limit in capacity is in fact the energy limit and available on the individual's brains. The limit of energy is the maximum capacity to store energy in our brain and remain inside it. The available energy depends from the capacity of the body to generate this energy. Therefore, two concepts interrelated but different.

In any case, there are some explanations that can give the algorithmic and reflective differences some thought. Individual differences in some imperative thinking pitfalls, for example, the inclination toward the side deduction and the propensity toward uneven intuition are moderately autonomous of insight and intelligence (Stanovich & West, 2007). The imperative thinking abilities are important to dodge individuals' side bias and one-side bias are instantiated at the reflective level of mind instead of the algorithmic level. Second, over an assortment of undertakings from the heuristics and biases literature, it has reliably found that rational thinking miens will anticipate volatility in these tasks after the impacts of general knowledge have been controlled (Bruine de Bruin et al., 2007; Klaczynski & Lavallee, 2005; Kokis et al., 2002; Parker & Fischhoff, 2005; Stanovich & West, 1997, 2000; Toplak & Stanovich, 2003).

Thinking miens is normally associated to the individual's goals and epistemic values. It is additionally related to expansive tendencies of pragmatic and epistemic self-regulation at a deliberate level of analysis.

The empirical studies described above show that these distinctive sorts of cognitive indicators are tapping divisible fluctuation, and the reason this is not out of the ordinary is because of the subjective limit measures. For example, insight and thinking miens map on to various levels of analysis in cognitive theory.

Figure 1 represents the theoretical approach of Stanovich about the individual differences in tri-process structure. This theoretical approach suggests that oscillations in fluid intelligence derives from the variations of efficiency when processing algorithm mind (Carroll, 1993). Instead, thinking miens is associated to the individual variations at an intentional level, which means at a reflective mind. Meanwhile, the empirical studies described above, rational thinking miens have also considered the assessments of epistemic regulation, for instance the active open-minded thinking and dogmatism or cognitive regulation such as need for cognition (Cacioppo et al., 1996).

We considered that system I is completely disseminated with TASS and Stanovich with his pairs have solved the problem of splitting system II in what concerns the different characteristics of reflective and algorithmic minds.

One of the motives to consider is the tripartite mental structure that derived from the failures detected in cognitive functioning and distinct manifestation of each type of mind. For example, interruptions in the operation of the algorithmic mind apparently results from deficiencies in general in the intellectual abilities of subjects with mental disorders (Anderson, 1998). And these deficiencies are varying constantly. On the other hand, there are few deficiencies varying continuously in the autonomous mind. The major

Figure 1. Individual differences in tripartite structure
Source: Stanovich, 2008

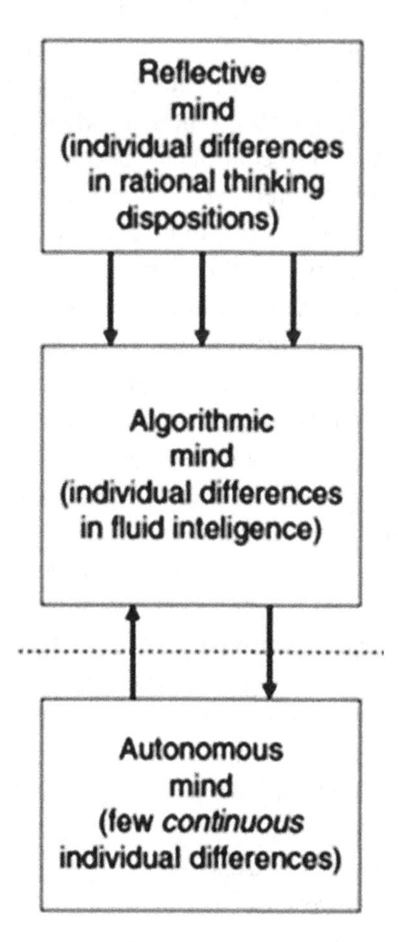

existing differences are related with the damage of cognitive modules resulting in inconstant cognitive dysfunctionalities like autism. (Bermudez, 2001). In complete contrast are many psychiatric disorders, particularly delusions which implicate intentional-level functioning of the reflective mind (Bermudez 2001). "Impairments in which they manifest themselves are of the sort that would standardly be explained at the personal level, rather than at the subpersonal level." Many of the symptoms of psychiatric disorders involve impairments of rationality.

Thus, there is an important sense on literature, in which rationality is a more immersive and complete construction than intelligence. At least considers both aspects of System II. Rationality is an "organismic-level concept" (Stanovich, 2008). It concerns the actions of an entity in its environment that serve its goals. Rationality is achieved when we have a good articulation between a reflective level and an algorithmic-level that enables it to carry out the actions and to process the environment in a way that enables the correct beliefs to be fixed and the correct actions to be taken. Thus, individual differences in rational thought and action can arise because of individual differences in intelligence or because of individual differences in thinking miens (Stanovich, 2008). Simply put, the concept of rationality encompasses well the thinking dispositions and algorithmic-level capacity, whereas the concept of intelligence, at least as it is commonly operationalized. Basically, is largely confined to algorithmic-level capacity.

NEURONAL ASSEMBLIES STRUCTURES

One of the obstacles for neuroscience remains without doubt, the lack of an integrated system that reproduces in full the connection of micro events at a macro level, after neuronal events that generate our behaviour and that occur in our brain (Greenfield, 2005). However, the concept of 'neuronal assemblies' is gradually gaining recognition: large scale (tens of millions) neuronal coalitions that are highly transient (sub-second) and which thereby could provide the essential missing link between molecular/cellular neurobiology and the cognitive neuroscience of non-invasive imaging of active brain regions. There is no clear ontogenetic line that is crossed as the brain grows in the womb - no single event, nor change in brain physiology, and certainly not at birth, when consciousness might be generated in an all-or-none fashion.

A neuronal assembly is a group of neurons that maintain strengthened synaptic connections among each other, maintaining all together active at the same time. The neurons in assemblies are not necessarily all physically close to one another. They can be distributed across various parts of the brain. Moreover, a single neuron can belong to several different assemblies and can be recruited into new assemblies at any time. Thus neuronal assemblies are not stable, but dynamic, and not necessarily localized, but often distributed.

Neuronal assemblies have other characteristics that should be taken in consideration for modelling consciousness. For instance, researchers have already known that independently of a neuronal assembly is activated partially at the beginning, this ignition when achieve a certain threshold, may then be disseminated across the entire assembly. Thus, if memory is successfully activated partially in brains, that may cause the turn off of the entire neuronal assembly. Consequently, it enters in a mechanism of filling the memory lapses or gaps and allowing us to recover it, in its plenitude. Some neuronal assemblies are known that at the beginning of ignition are not related with one another. Instead they are ignited in a repeatedly way, at the same time, which make them strongly connected. This is in fact the fundamental principle of learning at a cellular level, the mechanism by which our brains let us store the world's regularities in our memories, which of course has a definite adaptive value. Neuronal assemblies become more complex and great with individuals' age. Besides the dimension of neuronal assemblies is dependent from the need of diverse regions of the brains to process a decision, an action or a determined behaviour. The extension of transient assembly of neurons interferes directly with the extension of consciousness. Or consciousness might vary in degree from one moment to the next one depending of neuronal assembly. Voltage-sensitive dye images gave us the capability to characterize this fundamental brain mechanism since it was poorly understood. Once these assemblies are completely characterized physiologically, we are in conditions to generate and experiment, falsifiable hypotheses that link subjective experience with physico-chemical events in the brain (see Figure 2).

In terms of modelling neuronal assemblies, it's common to be assumed that they are on the circumstances of philosophical materialism. It is also assumed that any change in our mental states necessarily involves a change in the states of our neurons. Most of these models are designed with the agreement of the fact, that our sensory perceptions as well as our more abstract thoughts correspond to the activity of vast networks or assemblies of neurons. An activity that is subject to a complex set of dynamics.

Most of these models also recognize that at any given time, the brain is processing far more information than we are aware of. To cite only one example, regarding our sense of vision, we have the false impression that we are consciously aware of everything that lies in our field of vision, but actually, we consciously experience only a small portion of this scene.

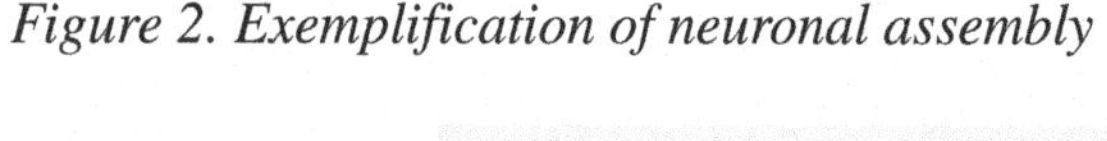

Figure 2. Exemplification of neuronal assembly

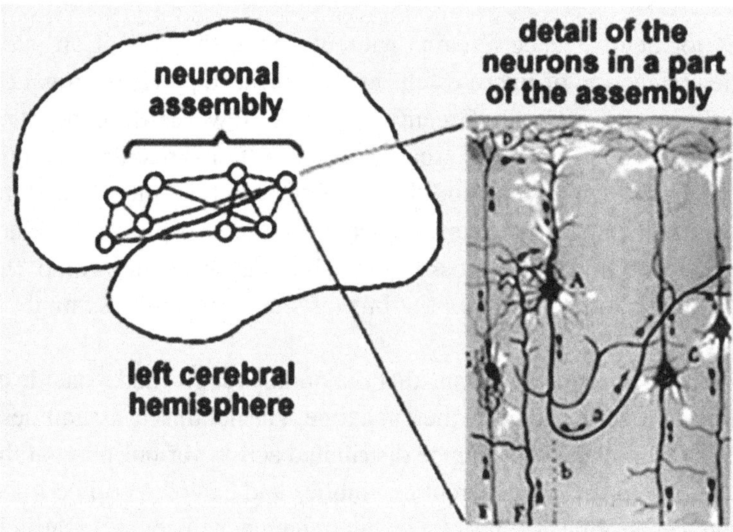

The frequency of nerve impulses cannot simultaneously indicate the intensity of a stimulus and whether or not it is a conscious one. Hence, there must be some other mechanism by which a piece of conscious content is selected. How then does the brain manage to distinguish a stimulus that is intense but unconscious and a stimulus that is less intense but conscious? The brain must have some other mechanism that lets it assess the objective importance of a stimulus and at the same time distinguish between conscious and unconscious mental representations. This mechanism must also account for another problem that doesn't seem like one if you don't consider how sensory information is processed in the brain, but that becomes a real puzzle if you do. This problem arises from the fact that the brain uses numerous specialized circuits to process the various properties of perceived objects in parallel.

Neuronal ensembles encode information works in a way of multiple edits by many participants. Neuroscientists have discovered that individual neurons are very noisy. For example, by examining the activity of only a single neuron in the visual cortex, it is very difficult to reconstruct the visual scene that the owner of the brain is looking at. An individual neuron does not 'know' everything and is likely to make mistakes. This problem is solved by the brain having billions of neurons. Information processing by the brain could consider all the population of neurons and it is also distributed. Most of the time each neuron knows a little bit about everything, and the more neurons participate in a specific work, the more precise is the information encoding. In the distributed processing scheme, individual neurons may exhibit neuronal noise, but the neuronal population, on average cancels all that noise.

An alternative hypothesis to the assembly is the theory that there exist highly specialized neurons that serve as the mechanism for neuronal encoding. In the visual system, such cells are often referred to as grandmother cells because they would respond in a very specific circumstances. Neuroscientists have indeed found that some neurons provide better information than others, and a population of such expert neurons has an improved signal to noise ratio. However, the basic principle of encoding holds: large neuronal populations do better than single neurons.

The emergence of specific neural assemblies is thought to provide the functional elements of brain activity that execute the basic operations of informational processing (Fingelkurts & Fingelkurts, 2004; Fingelkurts et al., 2005). Neuronal code or the 'language' that neuronal assemblies speak is very far from being understood. Currently, there are two main theories about neuronal code. The rate encoding theory states that individual neurons encode behaviourally significant parameters by their average firing rates, and the precise time of the occurrences of neuronal spikes is not important. The temporal encoding theory, on the contrary, states that precise timing of neuronal spikes is an important encoding mechanism. Neuronal oscillations that synchronize activity of the neurons in an assembly appear to be an important encoding mechanism. For example, oscillations have been suggested to underlie visual feature binding (Gray, Singer and others). In additions, sleep stages and waking are associated with distinct oscillatory patterns.

DISTINCT BRAINWAVES

Brains that work out need a powerful combination of many elements, where one of them is energy (Lopes da Silva, 1991). The energy produced by a brain is dependent from its use. If it is an automatic decision that we do with high frequency using the automatic mind it is very likely that a brain will use some part of its own capacity. On the other hand, if it is a problem-solving and complex decision using the algorithmic mind we expect a higher use of energy. The energy is spread out through brainwaves following the neuronal assemblies in which are building up during its activity (Galambos et al. 1981). Brainwaves are the connectors among neurons within our brains when we need to think or to act for example. Brainwaves establish the communications between neurons and are produced by synchronised electrical pulses (Lopes da Silva, 1991). Normally the detection of brainwaves is done by sensors placed on the scalp (Galambos et al., 1981). There are several types of brainwaves, according to their bandwidths, to describe their functions like a thought of a continuous spectrum of consciousness (Lopes da Silva, 1991). Their frequency could be slow, loud and functional or fast, subtle and complex. Brainwaves can modify according to what we intend to do or according to what we are feeling (Klausberger et al., 2003). When we are dominated by slower brainwaves it means that perhaps we are tired, with slow but frequent thoughts or simply dreaming. The higher frequencies are present when we are wired or hyper-alert depending of an absolute need. Brainwaves are produced from a high complex process reflecting different aspects when they occur in different locations in the brain (Lopes da Silva, 1991). Their speed is measured in cycles per second (Hertz). The first case is infra low brainwave (0.5Hz) normally known as slow cortical potential. This type is normally associated to the basic cortical rhythms that sustain our high brain functions, like for example in brain timing and network function. They are like a motor of a car when it is stopped but still running, in a standby situation. Infra-low waves are very difficult to be detected and we need more studies to understand better their nature.

The next waves in terms of speed are delta waves. Delta waves range from 0.5 Hz to 3 Hz and are also slower but faster and loudest than infra-low. Delta brainwaves are associated to a deepest meditation or a dreamless sleep. Delta waves suspend external awareness and are the source of empathy. Healing and regeneration are stimulated in this state and that is why deep restorative sleep is so essential to the healing process.

The next group of waves are Theta waves varying their speed from 3Hz to 8 Hz. These waves occur when we are sleeping and are also present in deep meditation like Delta waves. Its main goal is associated to the learning process and memory. With theta waves, our senses extract information from the external world centring our attention on signals existing inside us. These waves are associated to imagination, intuition, fears and information beyond the normal awareness processes. These type waves fall in automatic characterization of mind or into system I according to the dual process theory.

Alpha waves are other type of brain waves varying from 8 to 12 Hz and happen during the flowing of thoughts. There are used in present situations aiding overall coordination, tranquillity and integration of self. These waves are in consonance to autonomous mind of Stanovich.

Beta brainwaves are present in our normal waking state of consciousness. This type of waves ranges from 12 to 38Hz. Beta waves dominate our normal waking state of consciousness when attention is directed towards cognitive tasks and the outside world. Beta is a fast activity, present when we are alert, attentive, engaged in a problem solving, judgment, decision making, and engaged in focused mental activity. Beta waves are subdivided in terms of intensity in order to discriminate different stages. Normally, they are associated to highly complex thoughts, integrating new experiences, resulting in high anxiety or excitement. They are waves processed by high frequency, however is not very efficient to run the brain using a tremendous amount of energy. These waves fall in the description of algorithmic mind of Stanovich.

Gamma brainwaves are the fastest of brain waves and related to simultaneous processing of information from different brain areas. These waves range from 38 to 42Hz to transpose information rapidly as the subtlest of the brainwave frequencies. The mind has to be calm to access it. Gamma brainwaves are also above the frequency of neuronal firing, still how it is generated remains unknown from scientific community.

The movements and the interchangeability of brainwaves on the daily bases is dependent from our experiences and the constant changes that occur in the world surround us. Experiences and outside world are in continuous friction among each other. When our brainwaves are out of balance, there will be corresponding problems in our emotional or neuro-physical health. Any process that changes our perception changes our brainwaves.

TRAVELLING THROUGH NEURONAL ASSEMBLY COMPUTING

Neural assembly computing (NAC) is a framework for researching computational operations realized by spiking nervous cell assemblies and for designing spiking neural machines. The core of NAC is to understand the way that these assemblies interact and how it transforms in information processing with causal and hierarchical relations (Ranhel, 2013). Consequently, NAC tries to mismatch how assemblies reproduce states of the world. Or how the neuronal assemblies impose their control on information carried by spike streaming (Ranhel, 2013). Another important feature of this framework is to perceive how neuronal assemblies generated parallel and threaded processes by boning out and deconstructing other assemblies. The framework is also focused in create memory loops and where neurons coalitions interact by following logical functions.

Literature is not abundant in what concerns to the use of multi-agent systems in studying neuronal assemblies and the effects of energy on the tri-process of mind. Most of the studies centred their research on neural networks frameworks. Some of the works explore specific issues or adaptations of multi agent systems in studying neuronal networks. For example, (Wang & Cai, 2015) elaborated an article about the reinforcement ideas into a multi-agent system. They analysed two different multi-agent systems. The first one is the fully-connected neural network and consists of multiple single neurons. The second one is the simplified mechanical arm system which is controlled by multiple neurons (Wang & Cai, 2015). They assumed that each neuron is similar to an agent and it can be implemented Gibbs sampling of the posterior probability of stimulus features. They incorporated physics into the model constraints. The simulation results revealed inconclusive ideas where authors couldn't overpass the issues of computational complexity and consequently they recognize the need to better understand the problem.

We must not forget that labelling a certain group of neurons for building an assembly is a challenging task that only recently has been progressing through the use of more advanced recording techniques and analysis tools (Buzsaki, 2010; Lopes dos Santos et al., 2011; Canolty et al., 2012). However, it remains unclear how neuronal structures form, organize, cooperate and interact over time (Kopell et al., 2011).

Intelligent multi agent systems have great potentials to study the dynamic of neuronal structures. One of the important issues to apply intelligent multi agent systems in neuronal assemblies is to develop and reproduce the mechanisms behind them that support a model to reflect the whole complexity of tri-process mind.

An interesting study about computational neuroscience provides many examples in which cell networks (brains) store memories of geometrical states and coordinate their activity towards proximal and distant goals (Pezzulo & Levin, 2015). In their work they propose a perspective by programming large-scale morphogenesis that requires the exploration of information processing where cellular structures work toward specific shapes. In non-neural cells, as in the brain, bioelectric signaling implements information processing, decision-making, and memory in regulating pattern and its remodeling. Thus, approaches used in computational neuroscience to understand goal-seeking neural systems offer a toolbox of techniques to model and control regenerative pattern formation (Pezzulo & Levin, 2015). They proposed the developmental bioelectricity as a regulator of patterning, and propose that target morphology could be encoded within tissues as a kind of memory, using the same molecular mechanisms and algorithms so successfully exploited by the brain (Pezzulo & Levin, 2015). They suggest that the evident similarities between concepts in memory/decision-making and regenerative patterning are "not merely anthropomorphic ways of speaking about the remarkable robustness of shape control, but underlie real homologies of molecular mechanisms and underlying control logic". Bioelectrical signalling in non-neural cells was considered as a major regulator or large-scale anatomy. It also demonstrates that the differences between neural and non-neural cells are not fundamental, because all cells have the capacity of establishing networks with highly and tunable electric synapses which propagate signals via voltage dynamics and neurotransmitter signalling. These are some of the few examples existing in literature, which brought us the incentive of developing this field of neuro computing by using multi-agent systems.

NEURONAL ASSEMBLIES MODEL

As it was demonstrated above, there is a difference between the automatic and the algorithmic mind. One of the principal differences is the frequency that individuals use their mind and the frequency of the flux of energy. It is an inverse relation between both frequencies. Neuronal assemblies arise from the interconnection among different type of neurons. Their main goal is to provide all the necessary connections among the specific cortical regions of the brain in order to generate a thought or a behaviour.

Model Description

In this sense, we developed a model that we called Energy of Neuronal Assemblies I (ENA I), which brings together the capacity of neurons in establishing connections with other neurons, where energy assumes the key point in all process. The main objective of ENA I is to understand what is the principal role of energy in influencing the ignition of algorithmic mind. By doing that, we can formulate a strong hypothesis: why automatic and algorithmic mind are different? Exposing the reason why, in most of the times, we prefer to avoid algorithm mind and use more the automatic mind.

This model of collaboration networks illustrates how the behaviour of neurons in assembling together in small groups for short and long-term reasoning's can give an increase to a variety of large-scale network structures over time. The rules of the model are to draw upon observations of neuroscience imaging and the dual or tri-process theories of rationality. Many of the general features found in the networks of neuronal structures can be captured by the ENA-I model with two simple parameters: the proportion of new neurons participating in the neuronal assemblies' structures and the energy consumption to build these structures.

Methodology

A Multi Agent Based System is the most adequate method to build a model in reproducing neuronal assemblies' phenomenon and inherent behaviour of neurons(agents). It includes parameters like the number of neurons connected at the same time to establish a neuronal assembly or the consumption of energy to build them up. The advantage of this methodology gives us the opportunity to add attributes to neurons. Even other features that may generate complex relationships dynamics. In our case, the method is the more appropriated since allows us to bring heterogeneity to neurons in terms of relation with energy. Those factors have not been integrated on neuro computing models in such a way to bring an explanation for the distinction between the dual process or the algorithmic mind and the automatic mind.

Parameters

The ENA-I model presented here is empirically calibrated using the basic features of neuronal system as well as the indicators of typically described in literature, which give us the capacity to reproduce in a simplistic way the behaviour of neurons in what energy concerns. Some relevant setup parameters include the frequency of energy (ω_i) which defines the type of waves that we are in front of. The energy frequency varies between 0 and 42HZ incorporating all types of waves. The other parameter is the neuron working on H_i. This parameter tells the number of steps a neuron will remain in the neuronal structures collaborating before it´s activity disappears. It is the periodic function. Then we have the probability of a neuron to be chosen and become a member of new structure depending of its capacity in transmitting energy. Finally, we have the parameter which gives us the probability that the group of neurons being assembled will include a previous neuron of an incumbent on the structure, given that the group has at least one incumbent.

Agents

Agents represent neurons, programmed as possessing certain common attributes. Neurons are divided in two categories:

1. New and
2. Assembled.

While new have the incumbency to join in order to form the structure, the assembled ones serve as anchors for the new neurons that were called to assembled so that our brain could give an answer to a determined thought and consequently behaviour. Neurons receive and transmit energy to other neurons in order to assembled neuronal structures. Neurons have the capacity to regenerate themselves as nerve cells. They also have a time scheduling for their activity, which mean that they can be activated and suddenly deactivated. Everything depends from the type of neuron, the region of where it belongs to and the behaviour from what is required. This model is very simple and it is only focus on the flux of energy used for automatic and for algorithmic mind. Neurons are assumed as being completely synchronised, which means they go into the path that is commonly followed up.

Decisional Algorithm

The decisional algorithm for each neuron to be assembled is only based on the energy transmition. These neural assemblies are thought to encode sensory information. This phenomenon is reminiscent to the formation of clusters in models of coupled-phase oscillators (Golomb et al., 1992). In fact, the dynamics of neural networks can be described by systems of coupled-phase oscillators as longs networks have weak connections and the neurons' firing frequencies are roughly constant. For that reason, we use as start point the algorithm described by (Hoppensteadt & Izhikevich, 1997; Rinzel & Ermentrout, 1998):

$$\frac{d\varphi_i}{dt} = \omega_i + \sum_{j \neq i}^{N} J_{ij} H_i \left(\omega_i - \omega_j \right) \tag{1}$$

Where φ_i is the oscillator's phase, ω_i is the angular firing frequency, H_i is the 2π-periodic interaction function, J_{ij} represents the strength of the synaptic connections between neurons i and j and can be positive (excitatory) or negative (inhibitory). When the interactions between neurons are much faster than the firing frequency, as is commonly the case, H_i approximates the infinitesimal phase-resetting curve of neuron i. The PRC quantifies how much the period of the oscillator increases or decreases in response to a perturbation occurring at any phase.

System Dynamics

Our model simulates the dynamics in a simplified virtual environment in which neurons, each time that turn, proceeds to collect new neurons and implements a neuronal assembly structure. The simulated environment has the sufficient flexibility to add other relevant attributes to neurons and different capabilities to tackle neuronal assembly phenomenon. At this moment it is out of scope, in this stage.

First of all, we define the global variables at the beginning of the code program, before any function definitions. This group of variables are created with the incumbency of being accessible to all agents, in such case the neurons and can be used anywhere in a model. In this model we have the following variables: new neuron, neuronal structure size, major structure size and neuronal assemblies. New neuron is the agent who has never been yet assembled and it is prompt to be recruited for a new neuronal assembly. The neuronal structure size establishes the current running size of neuronal assembly being explored. The major structure size gives the dimension of the largest possible structure of a neuronal assembly. Finally, the neuronal assemblies list the assembled structures of neurons (See Figure 3).

After defining the global variables is time to add the characteristics to the neurons. Neurons could be incumbent if they were recruited before or assembled if belongs to the neuronal assembly structure that is being created in determined moment. The downtime refers to the number of time steps passed since the last recruitment of the neuron to an assembled structure. And last the explored, which refers to the neuronal assemblies that were computed as output results and used as inputs in designed graphs. The next step is the setup procedures where we generate the environment and the neurons recruited and assembled. The second process is the definition of the main procedures. In these procedures, we contemplate the possibility of neurons being incumbents, assemble a new structure where the neurons are chosen based on the decisional algorithm referenced above. It is also built the procedure to establish the dynamic of the energy flow among neurons linking each other. (See Figure 4)

Figure 3. System dynamics of ENA-I

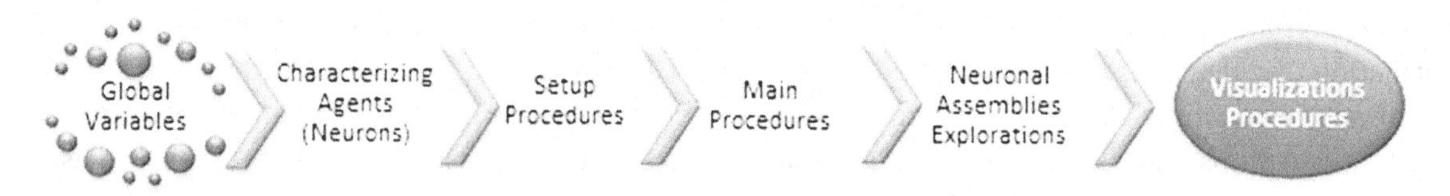

Figure 4. Characterizing neurons of ENA-I

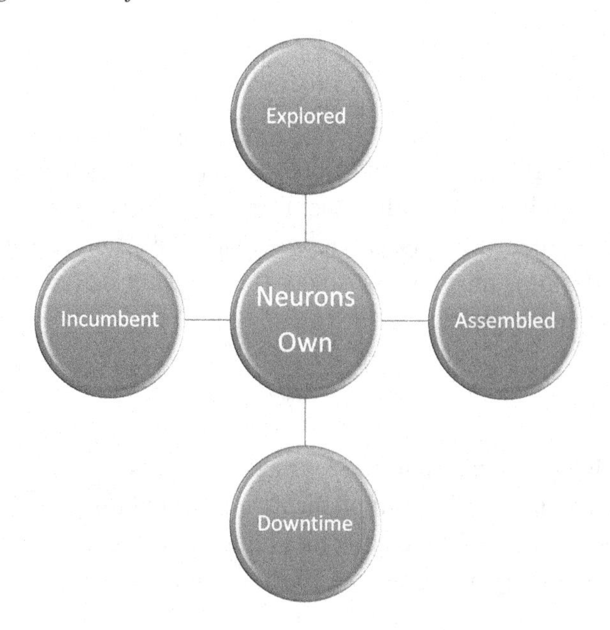

The third process of this multi-agent model is the one related to the network exploration, where neurons are assembled and generating the respective structures according the specific needs of each type of mind. The structures are flexible and heterogeneous, reflecting the dynamics produce by the decisional algorithm. Finally, the model has a program specifically created to ensure the procedures related to the visualization of the obtain neuronal assemblies and the graphs representing the output results.

Results

The ENA-I model allows us to observe how different mechanisms may yield, from given initial conditions to different outputs in terms of neuronal assemblies. Our first goal is to check the difference between the autonomous and algorithmic minds in terms of percentage of available neurons used in neuronal assemblies. The second goal is to count the number of connections among different type of intervenient neurons. Finally, our third and main goal is to do the assessment of the consumption of energy comparing the autonomous mind and the algorithmic mind.

As we can see, after one hundred runs and doing sensitive analyses through variations on parameters, that algorithmic mind uses 100% of the amount, of available neurons (see Figure 5). Instead, autonomous mind presents a huge use of neurons at the firsts cycles ($10m/s^2$). Then it starts to decrease after 100 cycles as it is demonstrated by the tendency expressed through a logarithmic function (see Figure 6). There is a considerable difference between algorithmic and autonomous mind, in terms of used neurons. The autonomous mind, after several cycles and with a lot variance, only use at maximum 40% of available neurons. A case where it is very clear that neuronal assemblies are more volatile in terms of neurons used in the case of autonomous mind and where isn't necessary a large quantity of neurons to think fast and automatically.

Results also demonstrate a substantive difference in terms of the number of neuron connections and respective types. Once more, there are more connections between the different types of neurons in the algorithmic mind than in the autonomous mind. A very comprehensible result since algorithmic mind is supposed to need more brain resources and contemplates plus brain areas than autonomous mind. Algorithmic mind uses substantially more neurons that were used previously in the past by other reasoning's, than autonomous mind. The difference is so great that new neurons by using autonomous mind reaches a maximum of 40 connections which is less than the neurons used previously. This simulation indicates that algorithm mind needs considerably the action of neurons previously used and consequently repeated to achieve the purposed target, demonstrating the high complexity of a neuronal assembly system and the turn backs that must be done to ignite the algorithmic mind (see Figure 7 and 8). These results suggest that the algorithm mind use neurons who have a more stable and frequent relation during their activity. In other words, the neurons used on ignition of algorithm mind have preference for recruiting other neurons who were incumbent and fruitfully in their relation considering the success of the past connections. It means that it is generated a kind of repeated and historical relationship among connected neurons in past. Perhaps it is a reflex of a learning process inside of each neuron that guides the neurons on their recruiting process. This induces a more propensity to recruit from brain areas which have more density of neurons and where they present a widely strengthened knowledge among them, about what is going to happen.

Figure 5. Percentage of neurons in neuronal assemblies: Algorithmic mind

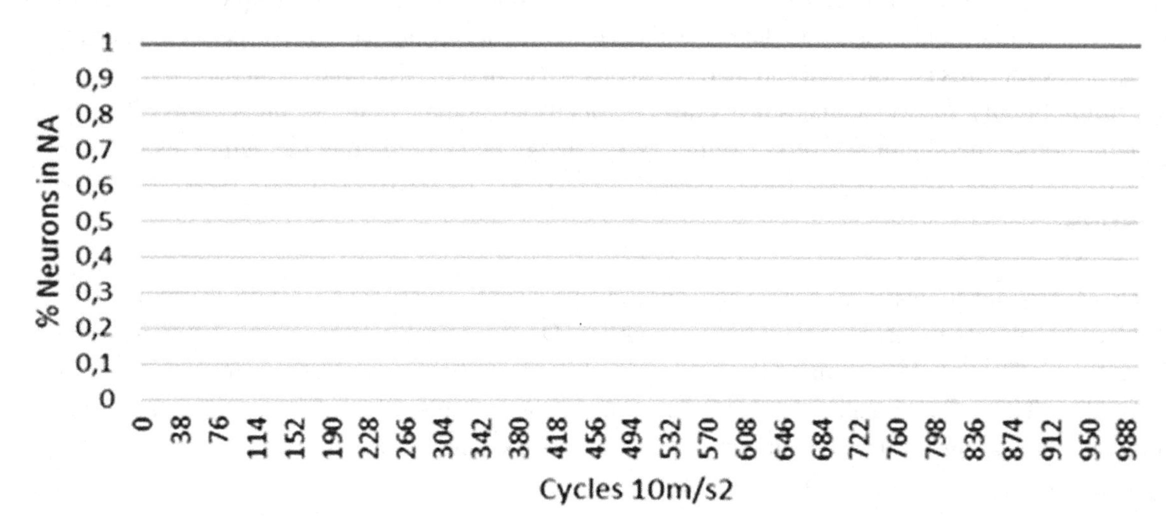

Figure 6. Percentage of neurons in neuronal assemblies: Automomous mind

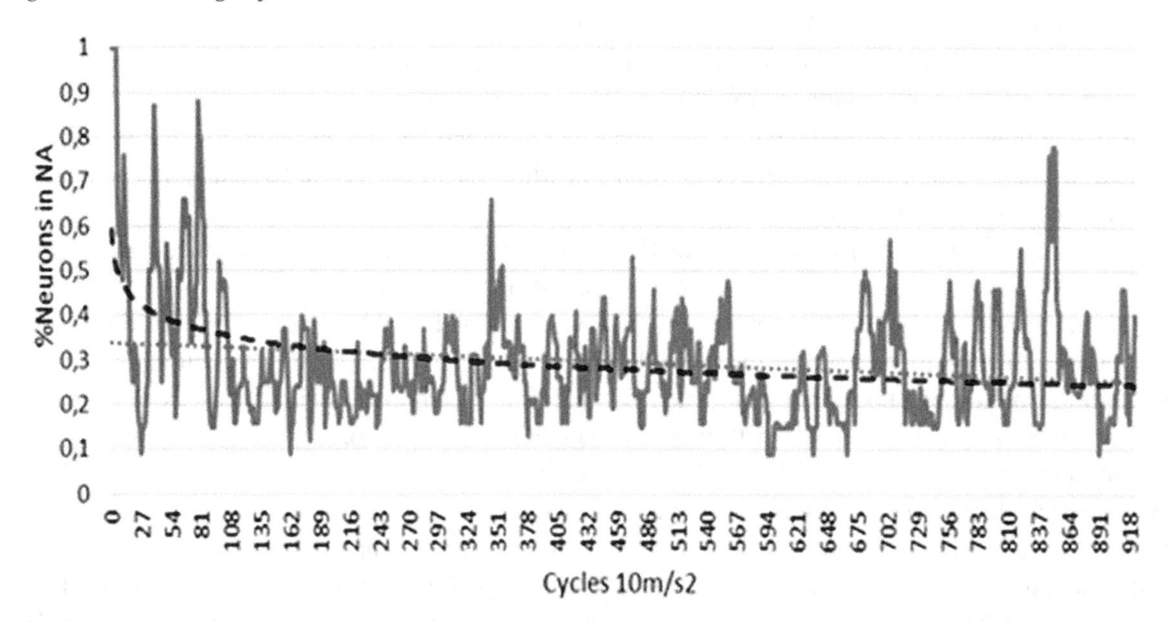

It is quite interesting to verify the importance and the greatness of the gap existent on the autonomous mind between the new neurons that are recruited to proceed with the mind task and the incumbent neurons who belong, in that moment, into the structure. The case is completely different with algorithmic mind where the connections of new neurons to the structure and incumbent neurons are practically at the same baseline. Connections between these neuron types vary very close of each other. This reflects that neuronal assemblies of algorithmic mind are continuously building up between incumbent and new neurons and the relations established among them. There is a strong need to recruit more new and incumbent neurons available in our brain system, when we are in front of a complex task.

Figure 7. Number of neurons connections in neuronal assemblies: Algorithmic mind

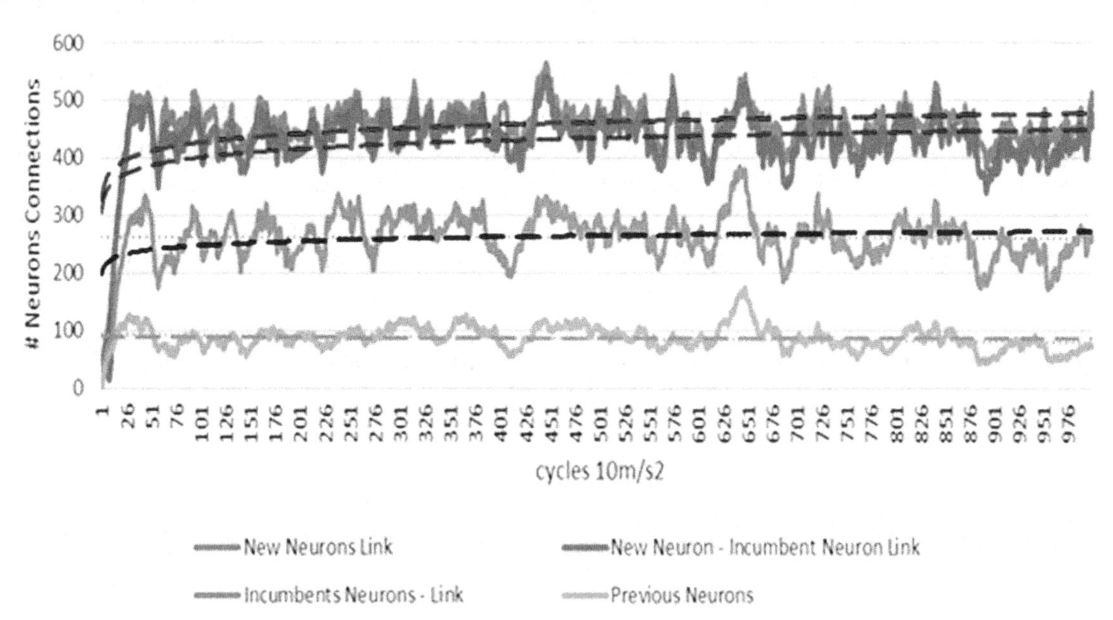

Figure 8. Number of neurons connections in neuronal assemblies: Automomous mind

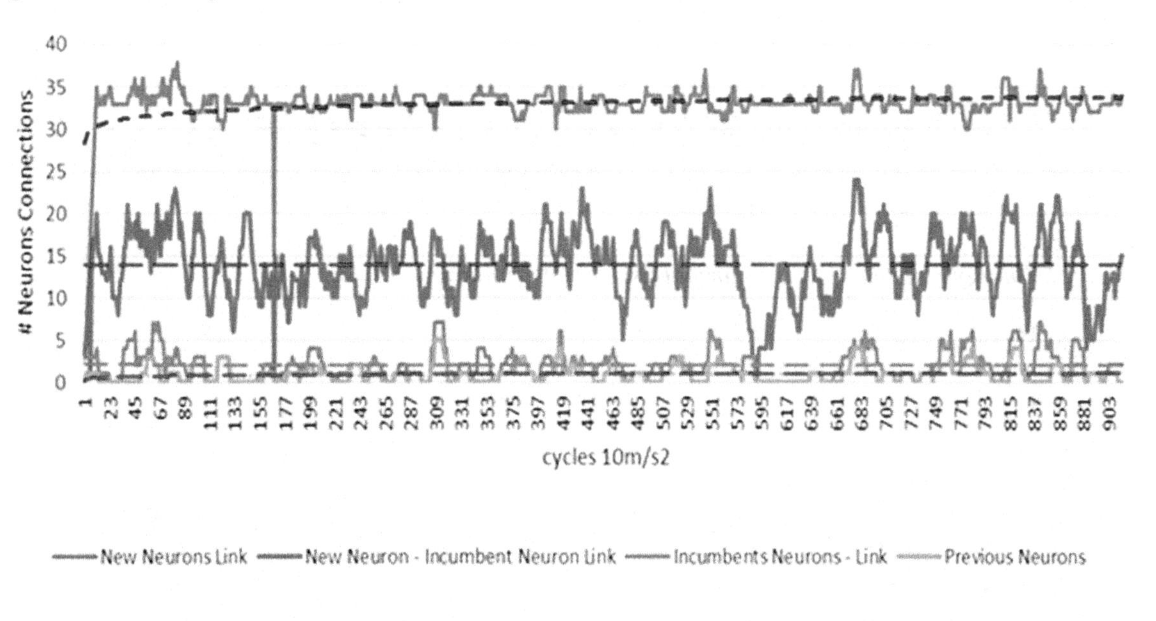

In terms of energy consumption, the obtained results have demonstrated that the energy consumption is significantly high for algorithmic mind with values ranging from 50 and 70 watts. On the other hand, autonomous mind ranges from 4 to 6 watts. Both minds are significantly different in their tendency. If for one side, the algorithm mind increases exponentially the consumption of energy at the beginning of its use and then stabilizes oscillating among the values described above. The autonomous mind has the opposite tendency it has a high consumption of energy at the beginning of its use and starts to decrease smoothly in a continuous way (See Figure 9 and 10).

An important fact that normally is dismissed by literature that covers the tri process theory of rationality is the role of energy. As it was described before alpha waves are normally associated to autonomous mind. They are typically slower than beta waves which are more associated to the algorithmic mind. Alpha waves have high dimension than beta waves which are shorter reflecting bigger cycle duration. So, if the autonomous mind is considered faster than the algorithmic mind and permanently continuous the expectation is to have a higher frequency, in terms of brain waves. However, what happens it is completely the opposite of our expectations. Algorithmic mind is faster in terms of brain waves frequency which results in a substantial consumption of energy to achieve its goals. While the algorithmic mind

Figure 9. Energy consumption in neuronal assemblies: Algorithmic mind

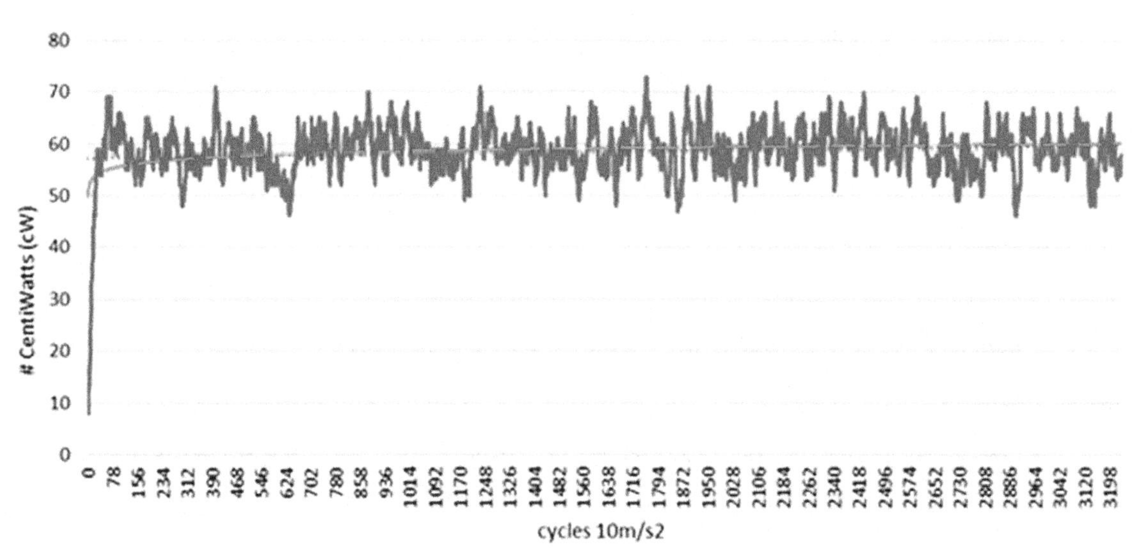

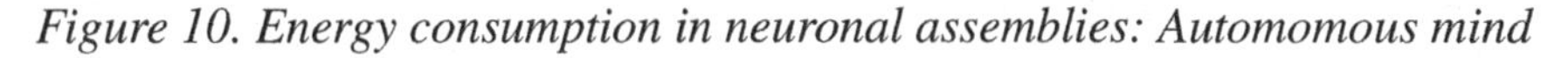

Figure 10. Energy consumption in neuronal assemblies: Automomous mind

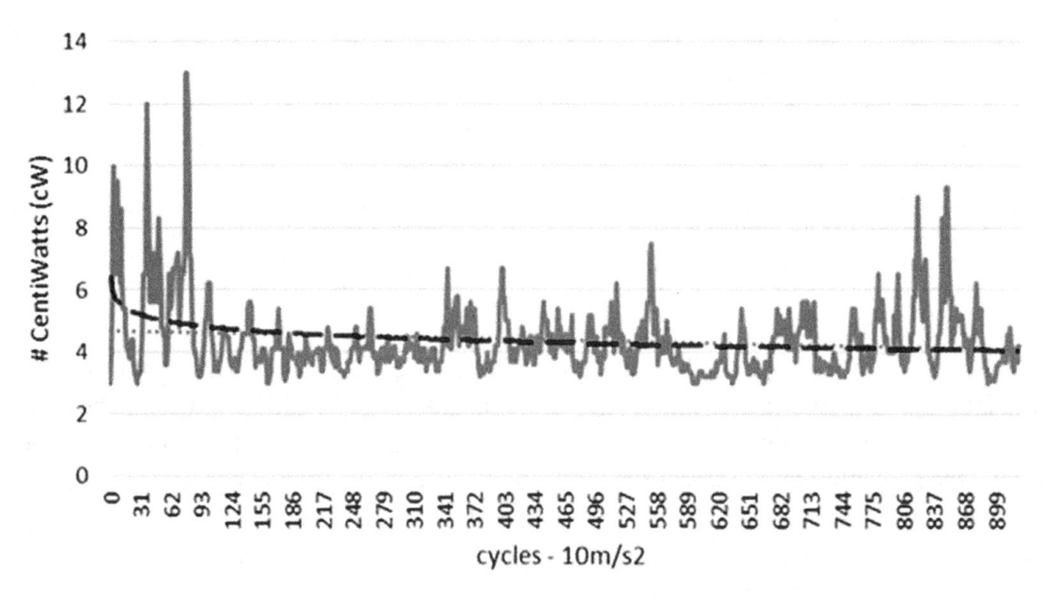

has a greater energy consumption than the autonomous mind with a slightly upward trend, the energy on autonomous mind decreases with the increasing of cycles, until stabilize.

So, we are in front of an inverse effect where the autonomous mind is faster than the algorithmic mind because it needs a lower electromagnetic frequency. Since it doesn't need to assembly a lot of neurons to reproduce an action or behaviour and consequently doesn't need many watts to work out. The algorithmic mind needs more time to ignite and reproduce a reasoning because it uses a higher electromagnetic frequency to put it working, consuming a substantial energy to build huge structures of neurons in order to process different types of information from different areas of the brain. The brains are clearly economistic because they affect their resources according to their needs.

FUTURE RESEARCH DIRECTIONS

As a result of this study, many questions arise for further research. The first one is to know what kind of impact there would be if we added several attributes to neurons. Please note that only the number of used neurons, the connections of new and incumbent were taken into account as also the energy consumption. However, it is important to analyse different types of neuronal assemblies with distinct neurons from multiple regions and layers of the brain. The second issue not discussed on this theoretical work is to understand more deeply, how the economistic role of the brain is regulated. In this case, it is important to clarify how the brains and respective mechanisms affect resources to fulfil the necessary needs.

CONCLUSION

The ignition of algorithmic mind assumes a crucial role on the way that the brain works out, in what concerns rationality. The transformation of the dual process proposed by Kanhneman in a tri process sustain by Kevin Stanovich is not a new concept. The literature has demonstrated that exists a consolidated tri process on rationality, where neuronal assemblies' structures performs an important mechanism in the way that brains work in terms of rationality.

So the main objective of this chapter was to analyse, through a multi-agent system, how neuronal assemblies' structure ignite the algorithmic mind and suggests a strong hypothesis for the difference between autonomous mind and algorithmic mind sustain under the role of energy inside of the brain. This is a topic normally neglected on psychology literature. According to that, we have compared the autonomous mind with the algorithmic mind, in terms of energy consumptions and the use of resources of the brain.

According to the output results, we verify that it is a substantial difference in terms of resources (neurons) used out, established connections and energy consumption. We realise that even the autonomous mind has the characteristic to be faster, simple and without complexity inducing the use of a few neuronal resources in order to give fast responses or actions that normally characterizes this type of mind. In achieving its target is in fact the less consumer of energy and neuronal resources. This type of mind has also a less frequency in terms of brains waves than the algorithmic mind which is slower but with higher frequency and a lot of energy consumption.

So, we are in front of a strong hypothesis that the differences among the tri process of rationality and the consequent ignition of neuronal assemblies is regulated by the electromagnetic resources, especially the consumption of energy. Following this idea, the brains should be viewed as economistic instruments adjusting the resources to the needs, in terms of efficiency and effectiveness.

REFERENCES

Anderson, K., & Silver, J. M. (1998). Modulation of Anger and Aggression. *Seminars in Clinical Neuropsychiatry*, *3*, 232–242. PMID:10085211

Bermúdez, J. L. (2001). Normativity and Rationality in Delusional Psychiatric Disorders. *Mind & Language*, *16*(5), 493–457. doi:10.1111/1468-0017.00179

Bruine de Bruin, W., Parker, A. M., & Fischhoff, B. (2007). Individual differences in adult decision-making competence. *Journal of Personality and Social Psychology*, *92*(5), 938–956. doi:10.1037/0022-3514.92.5.938 PMID:17484614

Buzsaki, G. (2010). Neural syntax: Cell assemblies, synapsembles, and readers. *Neuron*, 68. PMID:21040841

Canolty, R. T., Cadieu, C. F., Koepsell, K., Ganguly, K., Knight, R. T., & Carmena, J. M. (2012). Detecting event-related changes of multivariate phase coupling in dynamic brain networks. *Journal of Neurophysiology*, *107*(7), 2020–2031. doi:10.1152/jn.00610.2011 PMID:22236706

Carroll, J. B. (1993). *Human cognitive abilities: A survey of factor-analytic studies*. Cambridge, UK: Cambridge University Press. doi:10.1017/CBO9780511571312

Carruthers, P. (2006). *The Architecture of the Mind: Massive Modularity and the Flexibility of Thought*. Oxford University Press. doi:10.1093/acprof:oso/9780199207077.001.0001

Evans, J. St. B. T. (2003). In two minds: Dual process accountsof reasoning. *Trends in Cognitive Sciences*, *7*(10), 454–459. doi:10.1016/j.tics.2003.08.012 PMID:14550493

Evans, J. St. B. T. (2006). The heuristic-analytic theory of reasoning: Extension and evaluation. *Psychonomic Bulletin & Review*, *13*(3), 378–395. doi:10.3758/BF03193858 PMID:17048720

Evans, J. St. B. T. (2007). On the resolution of conflict in dual-process theories of reasoning. *Thinking & Reasoning*, *13*(4), 321–329. doi:10.1080/13546780601008825

Evans, J. St. B. T. (2008). Dual-processing accounts of reasoning, judgment and social cognition. *Annual Review of Psychology*, *59*(1), 255–278. doi:10.1146/annurev.psych.59.103006.093629 PMID:18154502

Fingelkurts, A., & Fingelkurts, A. (2004). Making complexity simpler: Multivariability and metastability in the brain. *The International Journal of Neuroscience*, *114*(7), 843–862. doi:10.1080/00207450490450046 PMID:15204050

Fingelkurts, A., Fingelkurts, A., & Kähkönen, S. (2005). Functional connectivity in the brain, is it an elusive concept? *Neuroscience and Biobehavioral Reviews*, *28*(8), 827–836. doi:10.1016/j.neubiorev.2004.10.009 PMID:15642624

Fodor, J. (1983). *The Modularity of Mind*. Cambridge, MA: MIT Press.

Franklin, S. (1995). *Artificial minds*. Cambridge, MA: MIT Press.

Galán, R., Ermentrout, G., & Urban, N. (2006). Predicting synchronized neural assemblies from experimentally estimated phase-resetting curves. *Neurocomputing*, 1–2.

Golomb, D., Hansel, D., Shraiman, B., & Sompolinsky, H. (1992). Clustering in globally coupled phase oscillators. *Physical Review A.*, *45*(6), 3516–3530. doi:10.1103/PhysRevA.45.3516 PMID:9907399

Gray, J. R., Chabris, C. F., & Braver, T. S. (2003). Neural mechanisms of general fluid intelligence. *Nature Neuroscience*, *6*(3), 316–322. doi:10.1038/nn1014 PMID:12592404

Greenfield, S. A. (2002). Mind, brain and consciousness. *The British Journal of Psychiatry*, *181*, 91–93. PMID:12151275

Greenfield, S. A., & Collins, T. F. T. (2005). A neuroscientific approach to consciousness. *Progress in Brain Research Journal*, *150*, 11–23. doi:10.1016/S0079-6123(05)50002-5 PMID:16186012

Hoppensteadt, F., & Izhikevich, E. (1997). *Weakly Connected Neural Networks*. Berlin: Springer. doi:10.1007/978-1-4612-1828-9

Hsee, C. K., & Hastie, R. (2006). *Decision and Experience: Why Don't We Choose What Makes Us Happy? Trends in Cognitive Sciences*. Retrieved from http://ssrn.com/abstract=929914

Kahneman, D. (2000). A psychological point of view: Violations of rational rules as a diagnostic of mental processes[Commentary on Stanovich and West]. *Behavioral and Brain Sciences*, *23*(5), 681–683. doi:10.1017/S0140525X00403432

Kahneman, D. (2011). *Thinking, fast and slow*. New York: Farrar, Straus and Giroux.

Klaczynski, P. A., & Lavallee, K. L. (2005). Domain-specific identity, epistemic regulation, and intellectual ability as predictors of belief-based reasoning: A dual-process perspective. *Journal of Experimental Child Psychology*, *92*(1), 1–24. doi:10.1016/j.jecp.2005.05.001 PMID:16005013

Klausberger, T., Magill, P. J., Marton, L. F., Roberts, J. D., Cobden, P. M., Buzsaki, G., & Somogyi, P. (2003). Brain-state- and cell-type-specific firing of hippocampal interneurons in vivo. *Nature*, *421*(6925), 844–848. doi:10.1038/nature01374 PMID:12594513

Kokis, J., Macpherson, R., Toplak, M., West, R. F., & Stanovich, K. E. (2002). Heuristic and analytic processing: Age trends and associations with cognitive ability and cognitive styles. *Journal of Experimental Child Psychology*, *83*(1), 26–52. doi:10.1016/S0022-0965(02)00121-2 PMID:12379417

Kopell, N., Whittington, M. A., & Kramer, M. A. (2011). Neuronal assembly dynamics in the beta1 frequency range permits short-term memory. *Proceedings of the National Academy of Sciences of the United States of America*, *108*(9), 37793784. doi:10.1073/pnas.1019676108 PMID:21321198

Lopes da Silva, F. (1991). Neural mechanisms underlying brain waves: from neural membranes to networks. *Electroencephalography and Clinical Neurophysiology, 79*(2), 81-93. 10.1016/0013-4694(91)90044-5

Lopes dos Santos, V., Conde-Ocazionez, S., Nicolelis, M., Ribeiro, S. T., & Tort, A. B. L. (2011). Neuronal assembly detection and cell membership specification by principal component analysis. *PLoS ONE*, *6*(6), 20996. doi:10.1371/journal.pone.0020996 PMID:21698248

Parker, A. M., & Fischhoff, B. (2005). Decision-making competence: External validation through an individual differences approach. *Journal of Behavioral Decision Making*, *18*(1), 1–27. doi:10.1002/bdm.481

Pezzulo, G., & Levin, M. (2015). Re-Membering the Body: Applications of computational neuroscience to the top-down control of regeneration of limbs and other complex organs. *Integrative Biology : Quantitative Biosciences from Nano to Macro*, *7*(12), 1487–1517. doi:10.1039/C5IB00221D PMID:26571046

Pollock, J. L. (1995). *Cognitive carpentry: A blueprint for how to build a person*. Cambridge, MA: MIT Press.

Ranhel, J. (2013). Neural Assemblies and Finite State Automata. In *BRICS-CCI-CBIC '13 Proceedings of the 2013 BRICS Congress on Computational Intelligence and 11th Brazilian Congress on Computational Intelligence* (pp. 28-33). IEEE Computer Society. doi:10.1109/BRICS-CCI-CBIC.2013.16

Rinzel, J., & Ermentrout, B. (1998). Article. In C. Koch & I. Segev (Eds.), Methods in Neuronal Modelling (2nd ed.; p. 251). Cambridge, MA: MIT Press.

Samuels, R. (2005). The complexity of cognition: Tractability arguments for massive modularity. In P. Carruthers, S. Laurence, & S. Stich (Eds.), *The innate mind* (pp. 107–121). Oxford, UK: Oxford University Press. doi:10.1093/acprof:oso/9780195179675.003.0007

Schneider, W., & Shiffrin, R. M. (1977, January). Controlled and automatic human information processing: I. Detection, search, and attention. *Psychological Review*, *84*(1), 1–66. doi:10.1037/0033-295X.84.1.1

Sloman, A. (1993). The mind as a control system. In C. Hookway & D. Peterson (Eds.), *Philosophy and cognitive science* (pp. 69–110). Cambridge, UK: Cambridge University Press.

Sloman, A., & Chrisley, R. (2003). Virtual machines and consciousness. *Journal of Consciousness Studies*, *10*, 133–172.

Spinoza, B. (1677). Ethica ordine geometrico demonstrata. In *English translation: "The Ethics*. Radford, VA: Wilder Publications.

Stanovich, K. E. (2004). Balance in psychological research: The dual process perspective. *Behavioral and Brain Sciences*, *27*(03), 357–358. doi:10.1017/S0140525X0453008X

Stanovich, K. E. (2008). Higher-order preferences and the Master Rationality Motive. *Thinking & Reasoning*, *14*(1), 111–127. doi:10.1080/13546780701384621

Stanovich, K. E., & West, R. F. (1997). Reasoning independently of prior belief and individual differences in actively open-minded thinking. *Journal of Educational Psychology*, *89*(2), 342–357. doi:10.1037/0022-0663.89.2.342

Stanovich, K. E., & West, R. F. (1999). Discrepancies between normative and descriptive models of decision making and the understanding/acceptance principle. *Cognitive Psychology*, *38*(3), 349–385. doi:10.1006/cogp.1998.0700 PMID:10328857

Stanovich, K. E., & West, R. F. (2000). Individual differences in reasoning: Implications for the rationality debate? *Behavioral and Brain Sciences*, *23*(5), 645–665. doi:10.1017/S0140525X00003435 PMID:11301544

Stanovich, K. E., & West, R. F. (2003). Evolutionary versus instrumental goals: How evolutionary psychology misconceives human rationality. In D. Over (Ed.), *Evolution and the psychology of thinking: The debate* (pp. 171–230). Hove, UK: Psychology Press.

Stanovich, K. E., & West, R. F. (2007). Natural myside bias is independent of cognitive ability. *Thinking & Reasoning*, *13*(3), 225–247. doi:10.1080/13546780600780796

Stanovich, K. E., & West, R. F. (2008). On the relative independence of thinking biases and cognitive ability. *Journal of Personality and Social Psychology*, *94*(4), 672–695. doi:10.1037/0022-3514.94.4.672 PMID:18361678

Stanovich, K. E., & West, R. F. (2008). On the failure of intelligence to predict myside bias and one-sided bias. *Thinking & Reasoning*, *14*, 129–167. doi:10.1080/13546780701679764

Stanovich, K. E., West, R. F., & Toplak, M. E. (2011). The complexity of developmental predictions from dual process models. *Developmental Review*, *31*(2-3), 103–118. doi:10.1016/j.dr.2011.07.003

Toplak, M. E., & Stanovich, K. E. (2003). Associations between myside bias on an informal reasoning task and amount of post-secondary education. *Applied Cognitive Psychology*, *17*(7), 851–860. doi:10.1002/acp.915

Wang, Z., & Cai, M. (2015). Reinforcement Learning Applied to Single Neuron. In *Computer Science and Artificial Intelligence*. Cornell University.

ADDITIONAL READING

Evans, J. St. B. T., & Stanovich, K. E. (2013). Dual-process theories of higher cognition: Advancing the debate. *Perspectives on Psychological Science*, *8*(3), 223–241, 263–271. doi:10.1177/1745691612460685 PMID:26172965

Fingelkurts, A. A., & Fingelkurts, A. A. (2001). Operational architectonics of the human brain biopotential field: Towards solving the mind-brain problem. *Brain and Mind*, *2*(3), 261–296. doi:10.1023/A:1014427822738

Fingelkurts, A. A., Fingelkurts, A. A., Rytsälä, H., Suominen, K., Isometsä, E., & Kähkönen, S. (2007). Impaired functional connectivity at EEG alpha and theta frequency bands in major depression. *Human Brain Mapping*, *28*(3), 247–261. doi:10.1002/hbm.20275 PMID:16779797

Stanovich, K. E., & Stanovich, P. J. (2010). A framework for critical thinking, rational thinking, and intelligence. In D. Preiss & R. J. Sternberg (Eds.), *Innovations in educational psychology: Perspectives on learning, teaching and human development* (pp. 195–237). New York: Springer.

Stanovich, K. E., Toplak, M. E., & West, R. F. (2008). The development of rational thought: A taxonomy of heuristics and biases. *Advances in Child Development and Behavior*, *36*, 251–285. doi:10.1016/S0065-2407(08)00006-2 PMID:18808045

Stanovich, K. E., West, R. F., & Toplak, M. E. (2011). Individual differences as essential components of heuristics and biases research. In K. Manktelow, D. Over, & S. Elqayam (Eds.), *The science of reason: A festschrift for Jonathan St. B. T. Evans* (pp. 335–396). New York: Psychology Press.

Stanovich, K. E., West, R. F., & Toplak, M. E. (2014). Rationality, intelligence, and the defining features of Type 1 and Type 2 processing. In J. Sherman, B. Gawronski, & Y. Trope (Eds.), *Dual process theories of the social mind* (pp. 80–91). New York: Guilford Press.

Toplak, M., Liu, E., Macpherson, R., Toneatto, T., & Stanovich, K. E. (2007). The reasoning skills and thinking dispositions of problem gamblers: A dual-process taxonomy. *Journal of Behavioral Decision Making*, *20*(2), 103–124. doi:10.1002/bdm.544

KEY TERMS AND DEFINITIONS

Algorithmic Mind: The algorithmic mind is a level that belongs to System II. The individual differences is associated to fluid intelligence.

Autonomous Mind: The autonomous mind is a level that belongs to System I, with few continuous individual difference.

Neuronal Assembly: A population of nervous system cells (or cultured neurons) involved in a particular neural computation.

Reflective Mind: The reflective mind is a level that belongs to System II. The individual differences is in rational thinking dispositions.

Chapter 2
Ecosystems as Agent Societies, Landscapes as Multi-Societal Agent Systems

Antonio Carlos da Rocha Costa
Universidade Federal do Rio Grande, Brazil & Universidade Federal do Rio Grande do Sul, Brazil

ABSTRACT

Landscape ecology concerns the analysis, modeling and management of landscapes and their component ecosystems, mostly in view of the effects of the anthropic actions that they may suffer. As such, landscape ecology is well amenable to be supported by agent-based computational tools. In this chapter, we introduce the concept of "multi-societal agent system", a formal architectural model for distributed multi-agent systems, and we interpret it in ecological terms, to serve as an agent-based theoretical foundation for computer-aided landscape ecology. More specifically, we introduce the "ecosystems as agent societies" and "landscapes as multi-societal agent systems" approaches to ecosystems and landscapes, together with the core elements of the agent-based architectural models that support such approaches. The elements of those architectural models are then used to formally capture the main organizational and functional aspects of ecosystems and landscapes.

INTRODUCTION

This work concerns the use of agent-oriented concepts in the formal ontological account of environmental systems. In particular, it concerns the use of the concept of agent society for the formal account of the main organizational and functional aspects of ecosystems, and the concepts of import-export agent society and multi-societal agent system for the formal account of the main organizational and functional aspects of landscapes.

The concept of agent society is used here in the sense in which it has been used in our previous works, e.g., (Costa, 2014). The concept of multi-societal agent system, and its component concept of import-export agent society, are being introduced in the present work, for the first time.

DOI: 10.4018/978-1-5225-1756-6.ch002

The environmental concepts were taken from (Pidwirny, 2009), complemented with (Wu, 2012). For the most part, however, they were not taken in the literal form they appear in those publications, but in a form adapted to the agent-based modeling point of view. This was a necessary step, to make them fit the conceptual framework of the agent societies.

STRUCTURE OF THE CHAPTER

The basic ecological concepts, in their revised form, are summarized in Section 3. The main organizational and functional features of ecosystems and landscapes (as evinced by the revised form of such concepts) are presented in Sections 4 and 5, respectively.

The concepts of agent society and multi-societal agent systems are given in Section 6. The way ecosystems and landscapes can be modeled in terms of agent societies and inter-societal systems is formally shown in Section 7 and 8, respectively. Section 8 also considers how to account for biomes and the biosphere as multi-societal agent systems.

Section 9 briefly deals with the issue of anthropic action on landscapes, sketching a general way to integrate multi-societal models of landscapes and human societies into models for landscapes that are subject to anthropic action.

Section 10 is the Conclusion.

THE BASIC ECOLOGICAL CONCEPTS

We characterize ecosystems, their components and related concepts, in terms of four dimensions[1], taking as primitive the concept of individual (i.e., a particular individual organism).

- The *populational* dimension, encompassing the set of individuals and types of individuals that constitute the ecosystem;
- The *organizational* dimension, encompassing the ways individuals and sets of individuals relate to each other, in terms of their interactions;
- The *functional* dimension, encompassing the functions that individuals and sets of individuals perform for each other, and the ways the interactions among organisms and sets of organisms coordinate with each other;
- The *geographical* dimension, encompassing the geographical areas occupied by the ecosystems, and the ways the ecosystems constitute themselves on those areas.

We use the four dimensions to define a series of concepts that leads to the concepts of *ecosystem*, *landscape*, *biome* and *biosphere*. We also characterize the concepts of *habitat* and *niche*[2].

- **Species**
    - **Populational Dimension:** Is a *group* of individuals characterized by the fact that individuals of a given species do not ordinarily breed with members of other species;
    - **Organizational and Functional Dimensions:** Species have no particular organizational or functional requirements;

    - **Geographical Dimension:** The members of a species are constrained to live in areas of a particular type, the possible niches of the species, each niche occupied by a particular population of the species.
- **Population**
    - **Populational Dimension:** The set of all the individuals of a given species that live in a particular habitat at a particular time;
    - **Organizational and Functional Dimensions:** Populations have no particular organizational or functional requirements;
    - **Geographical Dimension:** Populations are geographically characterized by the geographical area their individuals occupy, at a particular time;
- **Community**
    - **Populational Dimension:** The set of populations (each of a different species) living in a particular habitat at a particular time;
    - **Functional Dimension:** Biological communities support the performance of a variety of *functions* among its various populations, for instance:
        - Acquisition and use of food, space or other environmental resources;
        - Constitution of food webs, for nutrient cycling through all the members of the community;
        - Mutual regulation of population sizes;
    - **Organizational Dimension:** The main organizational features of biological communities is their biological diversity, the diversity of species of populations that they encompass; biological diversity implies a number of organizational properties:
        - The greater the diversity of its population, the greater the *complexity*, and greater the *stability* (*resilience*) of the community in face of disturbances in its set of populations;
        - The greater the diversity of the its population, the greater the *interconnectivity* of the *food web* operating among its populations;
    - **Geographical Dimension:** The geographical characterization of a biological community is given essentially by the geographical characterization of the area it occupies (we do not explore in detail, in this work, the issue of the geographical characterization of biological communities);
- **Habitat:** The combination of geographical, organizational and functional features that supports the existence of the populations of a community in a given geographical area; the geographical, organizational and functional features that supports the existence of a specific population is the niche of that population;
- **Niche**
    - **Populational Dimension:** A niche concerns a single population; it is the combination of organizational, functional and geographical features of the habitat within which the population exists;
    - **Organizational and Functional Dimensions:** There are no particular organizational or functional requirements for niches; however, a niche typically encompasses a reference to the whole biological community; that is, not only to the particular species to which it concerns, but also to all other species that should co-occur with that species, to support its permanence within the community;

    - **Geographical Dimension:** The geographical characterization of a niche is the geographical characterization of its corresponding habitat; a given geographical area can become the habitat of a community, by supporting the niches of its populations;
- **Ecosystem**
    - **Populational Dimension:** An ecosystem is a system composed of a biological community and an abiotic environment;
    - **Organizational Dimension:** Ecosystems can be endowed with a variety of organizational structures, characterized by the way the populations of its biological community relate to each other and to the components of the abiotic environment (we examine some of these possible organizational structures in Section 4);
    - **Functional Dimension:** The functional characterization of an ecosystem is directly related to its organizational characterization and to the characterization of its abiotic environment; among the relevant factors are:
        - Organizational factors = density of the set of populations;
        - Abiotic environmental factors = nutrient availability, temperature, light intensity;
    - **Geographical Dimension:** The geographical dimension of an ecosystem is given by its abiotic environment.
- **Landscape**
    - **Populational Dimension:** "Landscapes are spatially heterogeneous geographic areas characterized by diverse interacting patches or ecosystems." (Wu, 2012); as such, their population is the union of all populations of all ecosystems that compose it;
    - **Organizational Dimension:** The organizational features of a landscape are a combination of the organizational features of each of its ecosystems with the organizational features that arise from the interactions of those ecosystems;
    - **Functional Dimension:** The functional features of a landscape are a combination of the functional features of each of its ecosystems with the functional features that arise from the interactions of those ecosystems;
    - **Geographical Dimension:** The geographical dimension of a landscape encompasses all the geographical features of the geographical dimensions of the ecosystems that compose it;
- **Biome**
    - **Populational Dimension:** A biome is a *formation of species*, characterized typological in terms of the species that constitute it, and having a particular type of vegetation as a common factor among them; a biome occurs only in regions with certain definite characteristic climatic features;
    - **Organizational and Functional Dimensions:** No particular organizational or functional requirements are imposed on a biome, besides the requirement of a characteristic type of vegetation; but biomes admit that different species of the same family perform the same functions in different occurrences of the biome, in different regions;
    - **Geographical Dimension:** No particular geographical requirement is imposed on a biome, except for the fact that all its occurrences happen in regions with similar climates.

- **Biosphere**
  - **Populational Dimension:** The biosphere is the ecosystem whose abiotic environment is the whole abiotic environment of the Earth; it is the ecosystem of all living beings;
  - **Organizational and Functional Dimensions:** No particular organizational or functional requirements are imposed on the biosphere, besides that of being limited by the size of the whole abiotic environment of the Earth;
  - **Geographical Dimension:** The whole abiotic environment of the Earth determines the geographical dimension of the Biosphere.
- **Hierarchy of Ecosystem Organizational Levels**
  - The term "hierarchy" (or "pyramid") of "ecosystem levels" is often used to denote the hierarchically organized series of organisms and sets of organisms, from the individuals to the biosphere, each considered to be a *level* of the hierarchy:
    L1) **Individuals:** The individuals living the biosphere;
    L2) **Populations:** Sets of individuals of the same species, living in niches, inside habitats;
    L3) **Communities:** Organized sets of populations, living in habitats;
    L4) **Ecosystems:** Communities and their respective abiotic environments;
    L5) **Biomes:** Formations of types of species, each biome based on a particular type of vegetation and occupying only regions that have a certain characteristic climate;
    L6) **Biosphere:** The ecological system of the whole set of individuals existent on Earth.

The Organization and Functioning of an Ecosystem

Figure 1 illustrates the main elements of the *organization* and *functioning* of an ecosystem:

- The *sun* is the main source of energy of the ecosystem;
- The *atmosphere* provides O_2, CO_2 and water;
- The *soil* provides water and nutrients, besides a structural support and growing medium;
- The organisms are either *producers*, *consumers* or *decomposers*:
  - *producers* are plants, which are capable of producing their own food, through photosynthesis;
  - *consumers* are either *herbivores* or *carnivores*:
  - *Herbivores* feed on producers (plants);
  - *Carnivores* feed on herbivores and on carnivores;
  - *Decomposers* feed on dead producers or consumers, or on shed parts or tissues of them, and return O_2, CO_2 and water to the atmosphere, and water and nutrients to the soil;
- The *continuous lines* indicate flows of *continuous quantities* of O_2, CO_2, water and nutrients;
- The *dashed lines* indicate flows of *discrete quantities* of individuals, taken as food by others.

Figure 2 illustrates the organization of an ecosystem, seen as a *food web*, that is, a web of flows of individuals, taken as food, between the species of the ecosystem. The food is constituted by the individuals of the species. The species that are located at the starting points of the arrows operate as producers of individuals for the species that are located in the ending points of the arrows.

Figure 3 illustrates the *input-output functioning* of an ecosystem.

Essentially, the input-ouput functioning of an ecosystem amounts to:

Figure 1. The structure and functioning of an ecosystem
Source: PIDWIRNY, 2009

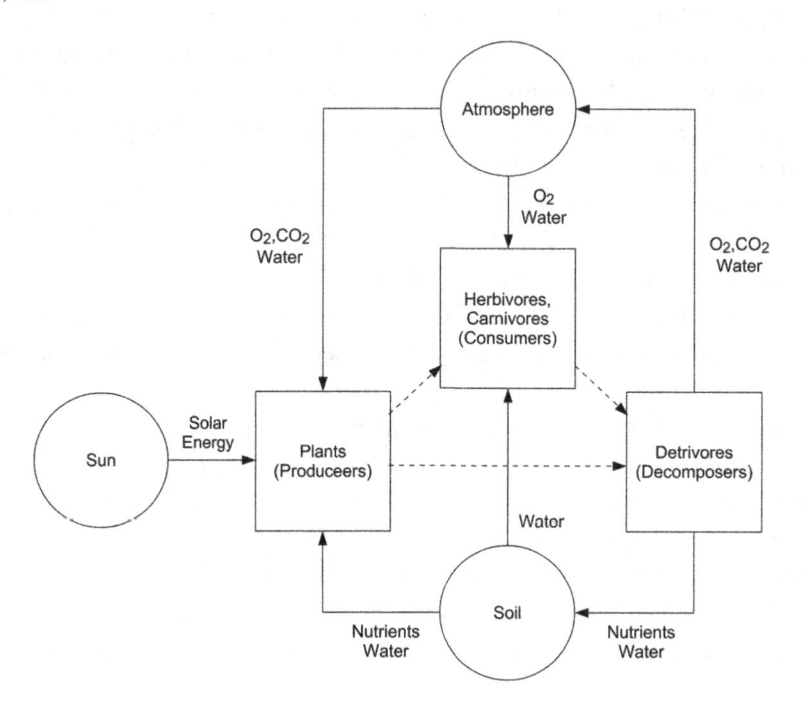

Figure 2. The food web of a sample ecosystem
Source: PIDWIRNY, 2009

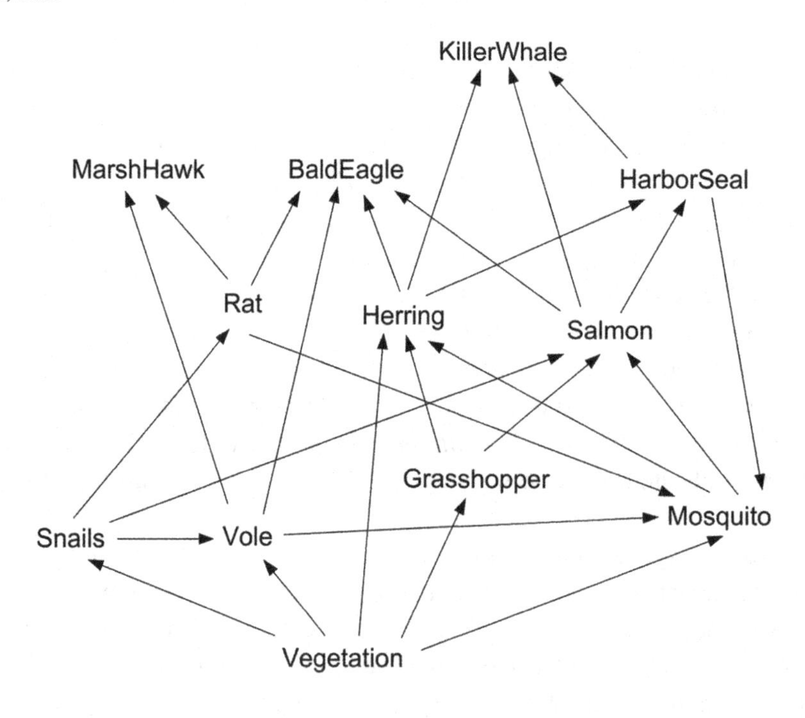

Figure 3. The input-output functioning of an ecosystem
Source: PIDWIRNY, 2009

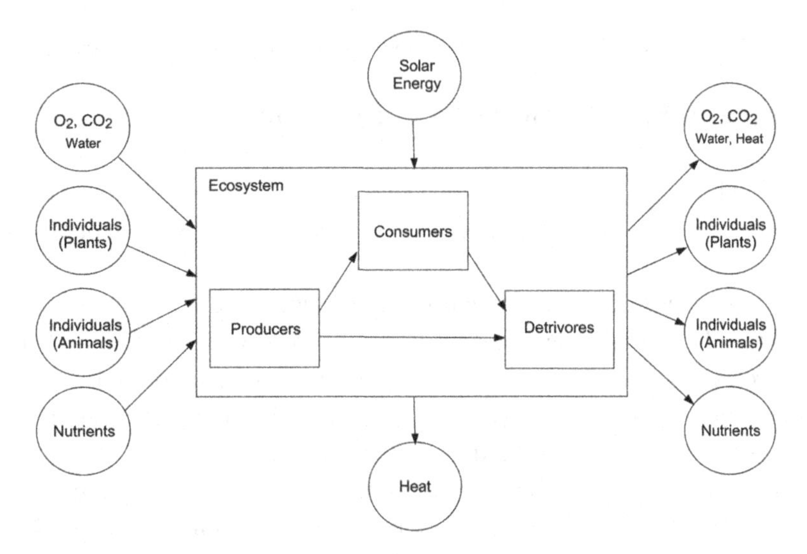

- *Inputting* from its exterior *individuals* (both animals and plants), *substances* (O_2, CO_2, water, nutrients), and *solar energy* from the sun;
- *Outputting* to its exterior *individuals* (both animals and plants), *substances* (O_2, CO_2, water, nutrients), and *heat* to the atmosphere and the outer space.

Figure 4 presents a schematic view of the input-output functioning of an ecosystem. Again, *dashed arrows* represent the flow of *discrete quantities*, and *continuous arrows* represent the flow of *continuous quantities* (we include heat and solar energy under the label "substances").

The *exchange* of individuals and substances serves as support for the connection of any two ecosystems, one operating as a *producer* of individuals and substances, the other as a *consumer* of individuals and substances.

Figure 4. The schematic input-output functioning of an ecosystem

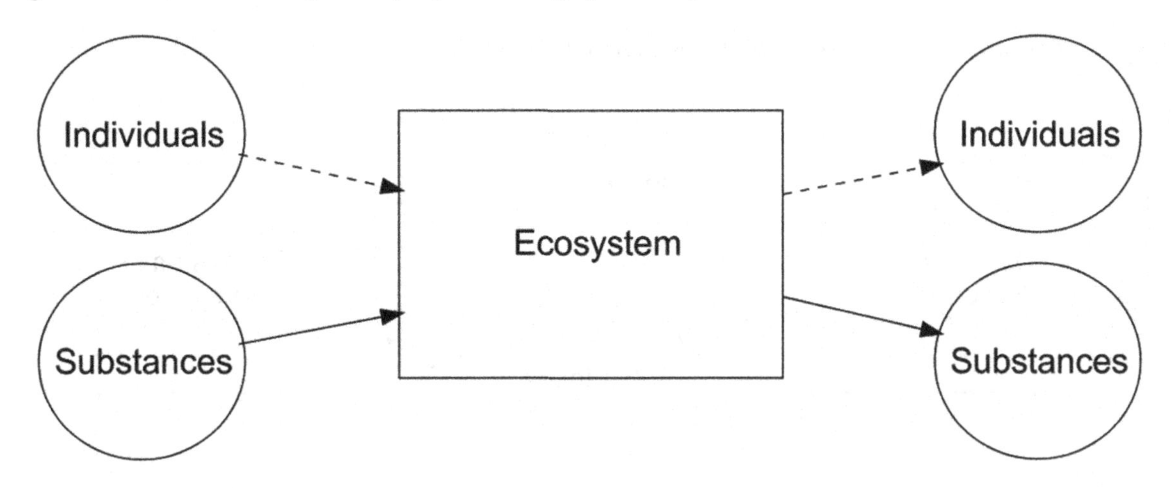

Figure 5 illustrates a *producer-consumer connection* between two ecosystems: ecosystem 1 operates as the producer, ecosystem 2 as the consumer. Although not indicated in the figure, simpler producer connections, involving only exchanges of individuals or substances, are also possible.

The Organization and Functioning of a Landscape

From the organizational and functional point of view, a *landscape* is a *network of ecosystems*, connected to each other through *producer-consumer connections*, each ecosystem operating as a producer, a consumer, or both.

Figure 6 illustrates the organization and functioning of a landscape. It is composed of six ecosystems (ES_1 to ES_6) and their producer-consumer connections (summarized here by simple dashed or continuous arrows, where the circles representing individuals and substances were omitted).

Notice how production-consumption cycles can be formed among ecosystems. Notice also how producer-consumer connections can be established with ecosystems situated outside the landscape.

Finally, notice the topological nature of this network of ecosystems. When mapped to the geographical area where the ecosystems are situated, the network of ecosystems induces a *network of ecotopes*, each ecotope with a uniform appearance, as illustrated by the pattern of fine lines in Figure 7. The network of ecotopes and the network of ecosystems of a landscape constitute, together, the *pattern* of the landscape.

Agent Societies and Multi-Societal Agent Systems

We use the term *agent society* to denote a particular type of multi-agent system, namely, as a multi-agent system that is:

- *Open*, meaning that agents can freely enter or leave it;
- *Organized*, meaning that systems processes can be identified which are essentially performed either by individual agents (individual processes) or by sets of agents (social processes);
- *Persistent*, meaning that the organization of the systems processes persists in time, independently (up to some point) of which agents enter or leave the system;

Figure 5. The producer-consumer connection between ecosystems

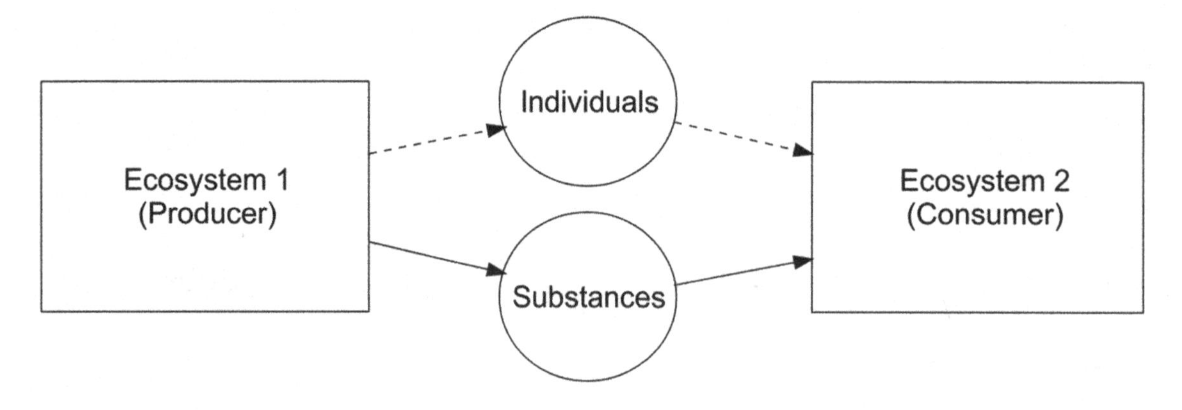

Figure 6. The network of ecosystems of a landscape, showing the producer-consumer connections among them

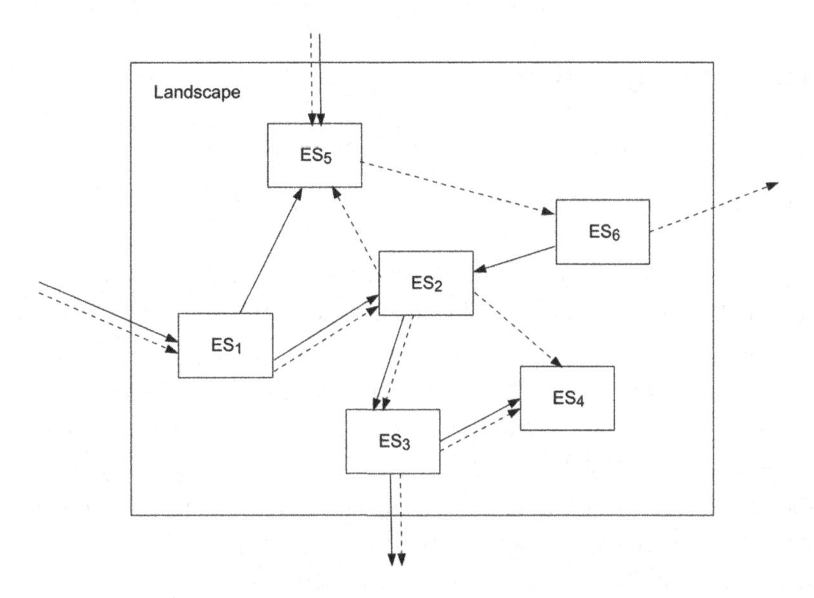

Figure 7. The pattern of a landscape, given by its network of ecotopes and its network of ecosystems

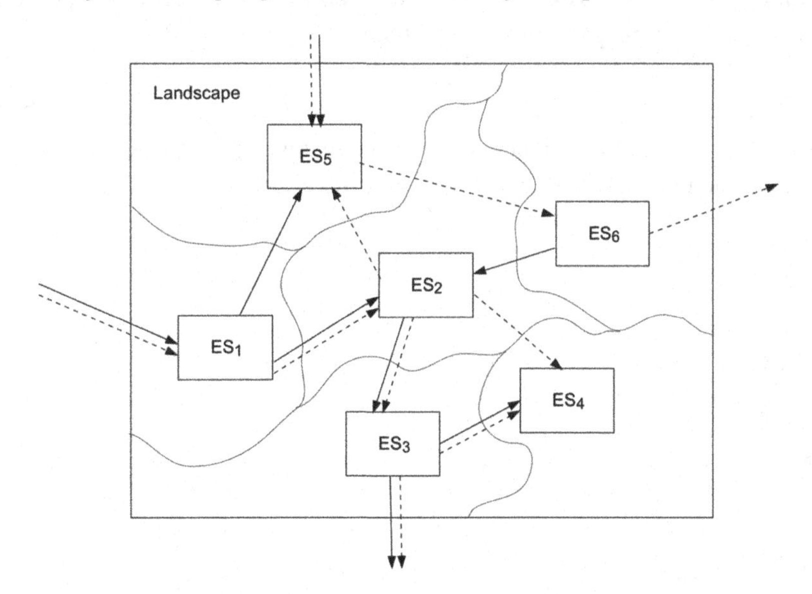

- *Situated*, meaning that the agents that constitute the system operate in determinate (physical or simulated) environment, whose objects the agents may make use of in their individual and social processes.

To model landscapes, we introduce here, the term *multi-societal agent system* to denote a *system of agent societies*, that is, a system whose components are agent societies. We require that the set of agent societies that constitute a multi-societal agent system explicitly define, at a minimum:

- For each agent society, the exchanges that it can perform with each of the other agent societies;
- That the graph of the resulting network of exchange processes is a connected graph.

Agent Societies, Formally Defined

We formally define the concept of agent society in the following way:

An *agent society* is a time-indexed structure $AgSoc = \left(Pop^t, Org^t, MEnv^t, SEnv^t\right)$ where:

- Pop^t is the *populational structure* of the society, at time t, composed of *individuals* and *sets of individuals* (the latter possibly structured in a hierarchically recursive way, that is, with sets of sets of individuals, etc.);
- Org^t is the *organizational structure* of the society, at time t, composed of the conducts performed in the society (individual behaviors, performed by the individuals; and social exchanges, performed by sets of individuals);
- $MEnv^t$ is the *material environment* of the society, at time t, that is, the *set of material structures* that are potentially accessible to all individuals and sets of individuals of the population and which they may make use of, in the performances of their conducts;
- $SEnv^t$ is the *symbolic environment* of the society, at time t, that is, the *set of symbolic structures* that are potentially accessible to all individuals and sets of individuals of the population and which they may make use of, in the performances of their conducts.

If we take Ag to be the universe of possible individuals of the society, `Cnd` to be the universe of possible conducts, `MStrc` to be the universe of possible material structures, and `SStrc` the universe of possible symbolic structures, then we have[3], for any time t and $n = 0,1,2,\ldots$:

$$Pop^t \subseteq \bigcup_n \wp^n\left(\mathtt{Ag}\right)$$

$$Org^t \subseteq \bigcup_n \wp^n\left(\mathtt{Cnd}\right)$$

$$MEnv^t \in \wp\left(\mathtt{MStrc}\right)$$

$$SEnv^t \in \wp\left(\mathtt{SStrc}\right)$$

In addition, we divide the organizational structure Org into three main *organizational levels*:

- Org^t_ω, the *micro-organizational level*, with the conducts that are performed by the *individuals* of the society at the time t;
- Org^t_μ, the *meso-organizational level*, with the conducts that are performed by (possibly recursively structured) *non-maximal* sets of individuals of the population at the time t;

- Org^t_Ω, the *macro-organizational level*, with the conducts that are performed by (possibly recursively structured) *maximal* sets of individuals of the population at the time t.

For any time t, we call:

- **Organizational Role:** The set of conducts of Org^t_ω, which individual agents perform;
- **Organizational Unit:** Non-Maximal sets of agents and the conducts of Org^t_μ that they perform collectively;
- **Social Subsystem:** Any maximal organizational unit, that is, any organizational unit that is not part of another organizational unit; organizational units perform conducts of Org^t_Ω .

Figure 8 (a) pictures the general architecture of an agent society and Figure 8(b) the internal hierarchy of the organizational levels of the its organizational structure.

Import-Export Agent Societies, Formally Defined

We call *import-export agent society* (or, *ie-agent society*, for short) any agent society *AgSoc* that is endowed with *import and export channels*. We assume that ie-agent societies can be connected to each other by the interconnection of their import-export channels. We denote by *ieAgSoc* any agent society *AgSoc* that is considered from the point of view of the *import-export connections* that it can have with other ie-agent societies.

Figure 9 illustrates that ie-agent societies can be endowed both with *discrete* and with *continuous* import and export channels, that is, import and export channels for *discrete and continuous flows* of

Figure 8. The general architecture of an agent society and the organizational levels of its organizational structure

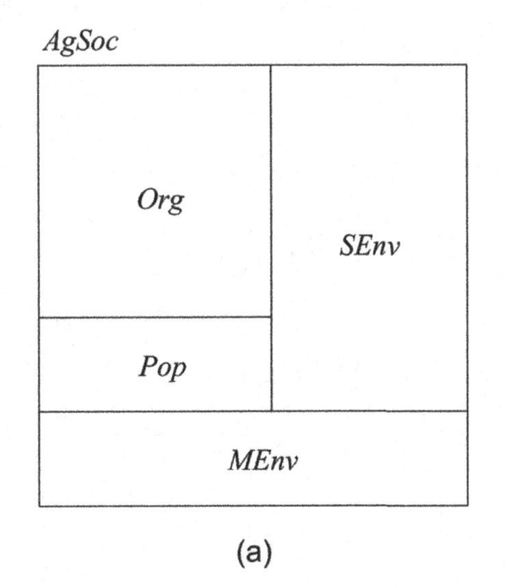

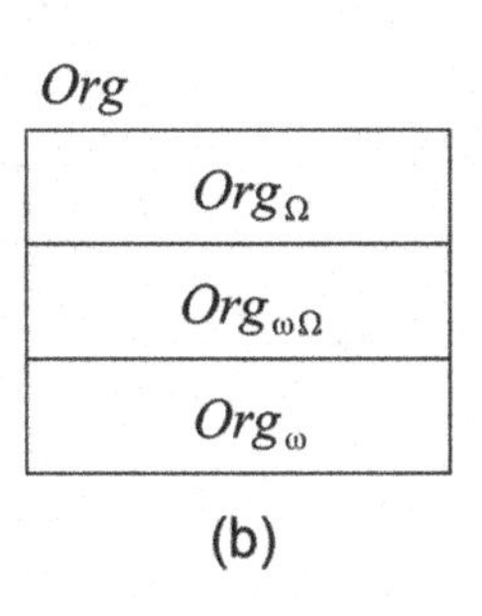

Figure 9. The import-export view of an agent society

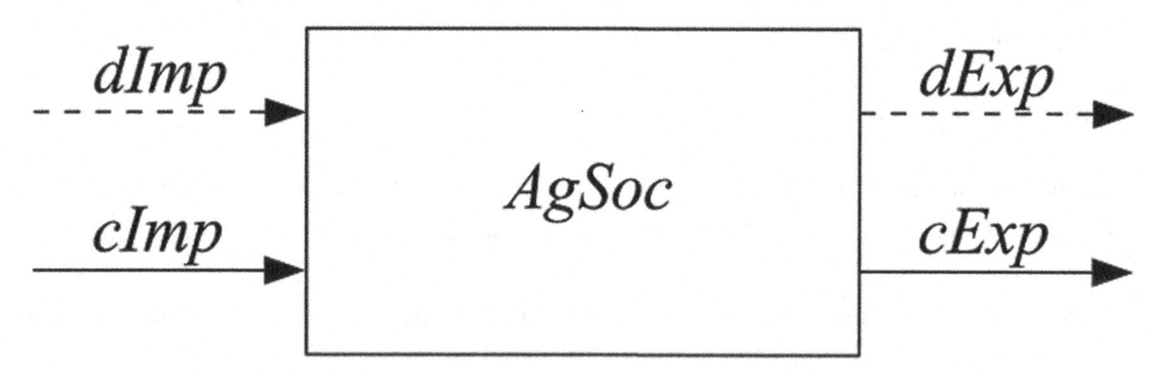

elements. Notice that not all agent society have to be endowed with all the four types of discrete and continuous import and export channels.

Agent societies can be connected to each other through import and export channels, as illustrated in Figure 10. Again, it is not necessary that both discrete and continuous connections be present.

Formally, we define:

An *import-export agent society* is a time-indexed structure $ieAgSoc^t = \left(AgSoc^t, ieChnls^t\right)$ where:

- $AgSoc^t$ is an agent society;
- $ieChnl^t = \left(dExp^t, dImp^t, cExp^t, cImp^t\right)$ is the *import-export channel structure* where:
    - $dExp^t$ is the set of *discrete export channels*;
    - $dImp^t$ is the set of *discrete import channels*;
    - $cExp^t$ is the set of *continuous export channels*;
    - $cImp^t$ is the set of *continuous import channels*.

Notice that agent societies, being open, are required to allow that individuals migrate between them. In the case of ie-agent societies, this is done through the discrete channels of their import-export connections.

We denote by `IE-AgSoc` the universe of ie-agent societies.

Figure 10. The import-export connection between two agent societies

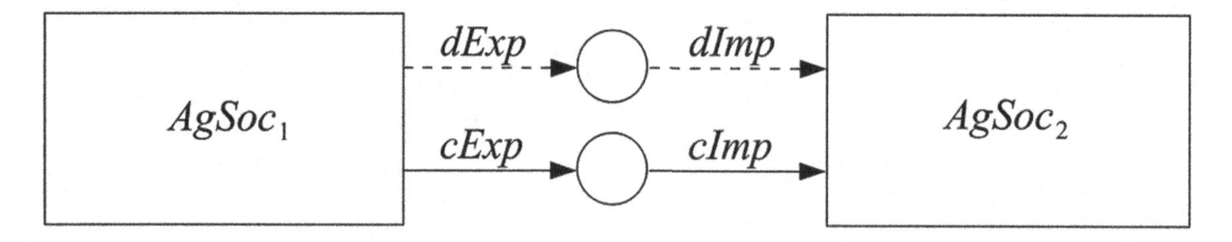

Multi-Societal Agent Systems, Formally Defined

Figure 11 illustrates a network of ie-agent societies, connected through their import-export channels. We call such type of network a *multi-societal agent system.*

Also, we define:

An *import-export connection* between two agent societies, $ieAgSoc_1^t$ and $ieAgSoc_2^t$, is any pair of import-export channels, denoted by $ie\text{-}conn\left(ieAgSoc_1^t, ieAgSoc_2^t\right)$, such that exactly one of the following holds:

$$ie\text{-}conn\left(ieAgSoc_1^t, ieAgSoc_2^t\right) \in dExp_1^t \times dImp_2^t;$$

$$ie\text{-}conn\left(ieAgSoc_1^t, ieAgSoc_2^t\right) \in dImp_1^t \times dExp_2^t;$$

$$ie\text{-}conn\left(ieAgSoc_1^t, ieAgSoc_2^t\right) \in cExp_1^t \times cImp_2^t;$$

$$ie\text{-}conn\left(ieAgSoc_1^t, ieAgSoc_2^t\right) \in cImp_1^t \times cExp_2^t.$$

We denote by $IE\text{ - }Conn\left(ieAgSoc_1^t, ieAgSoc_2^t\right)$ the set of import-export connections existent between the ie-agent societies $ieAgSoc_1^t$ and $ieAgSoc_2^t$, and by `IE-Conn` the universe of import-export connections between ie-agent societies.

Then, we define:

Figure 11. Sketch of a multi-societal agent system

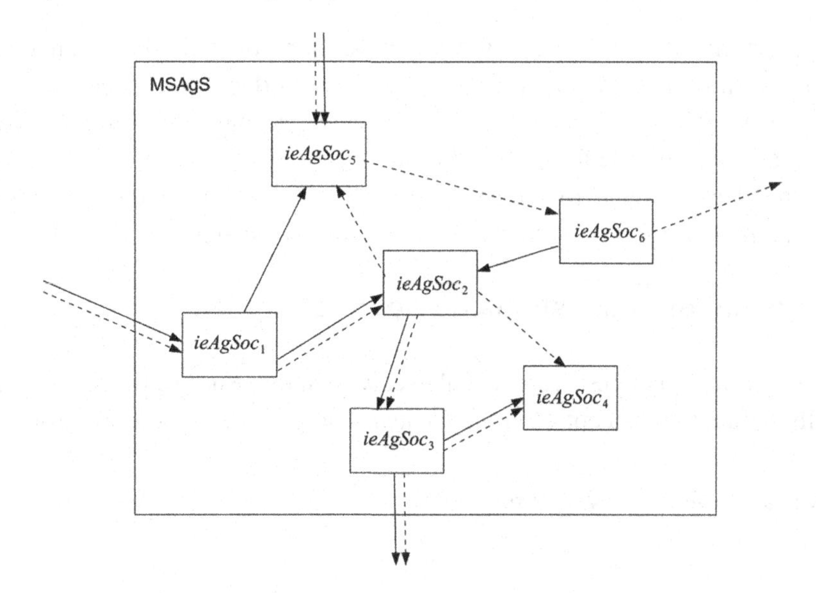

A *multi-societal agent system* is a time-indexed structure $MSocAgSys^t = \left(IE\text{-}AgSoc^t, IE\text{-}Conn^t\right)$ where, at each time t:

- $IE\text{-}AgSoc^t \in \wp\left(\texttt{IE-AgSoc}\right)$ is a non-empty set of import-export agent societies;
- $IE\text{-}Conn^t \in \wp(\texttt{IE-Conn})$, is the network of ie-connections among ie-agent societies, that is, $IE\text{-}Conn^t = \left\{ie\text{-}Conn(ieAgSoc_1^t, ieAgSoc_2^t) \mid ieAgSoc_1^t, ieAgSoc_2^t \in IE\text{-}AgSoc^t\right\}$.

We denote by $\texttt{MSocAgSys}$ the universe of multi-societal agent systems.

ECOSYSTEMS AS AGENT SOCIETIES

The Informal Mapping

Ecosystems can be seen as *import-export agent societies* through the following informal mapping of between *ecosystems components* and *components of agent societies*:

- At the *populational* level of agent societies, *populations* are seen as the *agents* of ecosystems;
- At the *micro-organizational* level of agent societies, *populations* are seen as implementing the *organizational roles* of the societies;
- At the *meso-organizational* level of agent societies, *communities* are seen as *organizational units*, with the species that compose a community seen as implementing the organizational roles of the organizational unit that corresponds to that community;
- At the *macro-organizational* level of agent societies, *ecosystems* are seen as *import-export agent societies*;
- At the *environmental* level of agent societies, the *habitats* and *niches* of *communities* and *species* are seen as the *part of the material environment* occupied by the species and communities.

The symbolic environment of an agent society will have an import to an ecosystem (and, for that matter) to a landscape, if *anthropic actions* are considered to have effects on that ecosystem or landscape. Then, the *symbolic environment* will capture, in the formal agent-based model, the *cultural aspects* of the human group that is responsible for those actions.

An ecosystem that does not relate to human groups, is seen as having no components that need to be mapped to the symbolic environment of the corresponding agent society.

The Formal Account of Ecosystems as Agent Societies

On the basis of the informal mapping presented above, we can make use of the formal concept of *agent society* to formally define the concept of an *ecosystem* that is not subject to anthropic action[4], in the following way:

An *ecosystem* is a time-indexed structure $EcoSys^t = \left(Pop^t, Org^t, Geo^t\right)$ where:

- Pop^t is the set of all *populations* that constitute the ecosystem, at the time t, each population of a given species;
- Org^t is the set of all *conducts* that are performed by the populations of the ecosystem, at the time t;
- Geo^t is the set of all geographical features of the *material environment* of the ecosystem, at the time t.

If we take $\mathtt{Pop}$ to be the universe of *populations* that can appear in any ecosystem, $\mathtt{Cnd}$ to be the universe of individual and social *conducts* that individuals and sets of individuals can perform, and $\mathtt{Geo}$ to be the universe of *geographical features* that the abiotic environment of ecosystems can present, then we have:

- $Pop^t \in \wp(\mathtt{Pop})$, a non-empty set of populations;
- $Org^t = \left(Org^t_\omega, Org^t_\mu, Org^t_\Omega\right)$ where:
  - $Org^t_\omega \in \wp(\mathtt{Cnd})$ is the set of conducts of the populations of the ecosystem;
  - $Org^t_\mu \in \wp(\mathtt{Cnd})$ is the set of conducts of sets of populations that operate as non-maximal organizational units of the ecosystem;
  - $Org^t_\Omega \in \wp(\mathtt{Cnd})$ is the set of conducts of sets of populations that operate as organizational sub-systems (maximal organizational units) of the ecosystem;
- $Geo^t \in \wp(\mathtt{Geo})$ is a non-empty set.

We require that the conducts Org^t_ω performed by each population be typical of the *species* of that population, and that the conducts Org^t_μ and Org^t_Ω performed by the organizational units constituted by sets of populations be typical of the way those populations interact with each other.

As mentioned above, we define the *habitat* of a population as the set of organizational and functional features present in the geographical area where it exists, together with the geographical (i.e., abiotic) features of that geographical area. The *niche* of a population is, on the other hand, the organizational and functional part of its habitat.

That is, given the population $pop \in Pop^t$ of the ecosystem $EcoSys^t$, we have that:

- $Hab^t_{pop} = \left(Nich^t_{pop}, Abio^t_{pop}\right)$ is the *habitat* of pop at the time t, where:
  - $Nich^t_{pop} = \left(Org^t_{\omega/pop}, Org^t_{\mu/pop}, Org^t_{\Omega/pop}\right)$ is pop's *niche*, at the time t, with non-empty $Org^t_{\omega/pop} \subseteq Org^t_\omega$, $Org^t_{\mu/pop} \subseteq Org^t_\mu$ and $Org^t_{\Omega/pop} \subseteq Org^t_\Omega$;
  - $Abio^t_{pop} \subset Geo^t$, non-empty, is the *abiotic environment* of pop, at the time t.

We denote by $\mathtt{EcoSys}$ the universe of ecosystems.

LANDSCAPES AS MULTI-SOCIETAL AGENT SYSTEMS

The Informal Mapping

Landscapes can be sees as *multi-societal agent systems*. *Biomes* and the *biosphere* can also be seen as multi-societal agent systems, the biosphere encompassing the whole of the network of ecosystems and biomes existing on Earth.

However, as indicated in the *Introduction*, we deal in this chapter only with the application of the concept of multi-societal agent systems to the formal account of landscapes, living the application to the formal account of biomes and the biosphere for further work.

Ecosystems as Import-Export Agent Societies

The ecosystems of a landscape interact with each other. We model such interaction by means of the concept of *import-export channels*, so that the ecosystems that constitute a landscape can be seen as *import-export agent societies*.

The Formal Account of Landscapes as Multi-Societal Agent Systems

On the basis of the informal mapping just presented, we can make use of the formal concept of *multi-societal agent system* to formally define the concept of *landscape*, in the following way:

A *landscape* is a structure $LndScp^t = \left(EcoSys^t, IE\text{-}Conn^t, Geo^t, ptrn^t\right)$ where:

- $EcoSys^t \in \wp\left(\texttt{EcoSys}\right)$ is the non-empty set of *ecosystems* that constitute the landscape, at the time t;
- $IE\text{-}Conn^t \in \wp\left(\texttt{IE-Conn}\right)$ is the set of *import-export connections* that tie together the network of ecosystems of the landscape, at the time t, each import-export connection relating a pair of ecosystems;
- $Geo^t \in \wp\left(\texttt{Geo}\right)$ is the set of all geographical features of the material environment of the landscape, at the time t;
- $ptrn^t : EcoSys^t \rightarrow \wp\left(Geo^t\right)$ is the mapping that determines the *pattern* of the landscape by giving, for each ecosystem $ecosys \in EcoSys^t$, the set $ptrn^t(ecosys)$ of geographical features to which it is associated, at the time t.

Patterns of landscapes have the general appearance illustrated in Fig. 7. We denote by ***LndScp*** the universe of landscapes.

Modeling Anthropic Action on Landscapes with Multi-Societal Agent Systems

One the main points of interest in *landscape ecology* is the study of the *effects of human action* on the landscapes (PIDWIRNY, 2009). We capture such aspect of landscapes by introducing one or more *agent societies* in the formal model of a landscape, in order to model the *human societies* that act on that landscape.

Formally, we say that a *landscape subject to anthropic action* is a time-indexed structure $AnthLndScp^t = \left(LndScp^t, HumSoc^t, humact\right)$ where:

- $LndScp^t$ is the *landscape* in question, as it stands at the time t;
- $HumSoc^t \in \wp\left(\mathtt{HumSoc}\right)$ is the non-empty set of *human societies* acting on $LndScp^t$, with the landscape $LndScp^t$ being a part of the material environment $MEnv^t$ of each of the human societies participating in $HumSoc^t$;
- $humact : \mathtt{HumSoc} \rightarrow \left(\mathtt{LndScp} \rightarrow \mathtt{LndScp}\right)$ is the mapping that determines the *effects of the actions* of each of those human societies on the landscape.

Figure 12 illustrates the network of ecosystems of a landscape, at a given time t, together with three human societies operating on it. The diamond headed arrows model the function $humact$, indicating which ecosystem is acted on by which human society[5].

We remark that a more complete account of the working of the mapping $humact$ than the one considered here can be taken in consideration, encompassing not only its effects on the landscape $LndScp^t$, but also its effects on the other components of the *material environments* and on the *symbolic environments* of the human societies, following the proposals in, e.g., (Costa, 2016).

This latter specification should allow, then, for the modeling of the *cultural effects* of the actions of the system of human societies present in $HumSoc^t$ on the landscape $LndScp^t$.

Figure 12. A picture of the action of human societies HS_1, HS_2 and HS_3 on a landscape

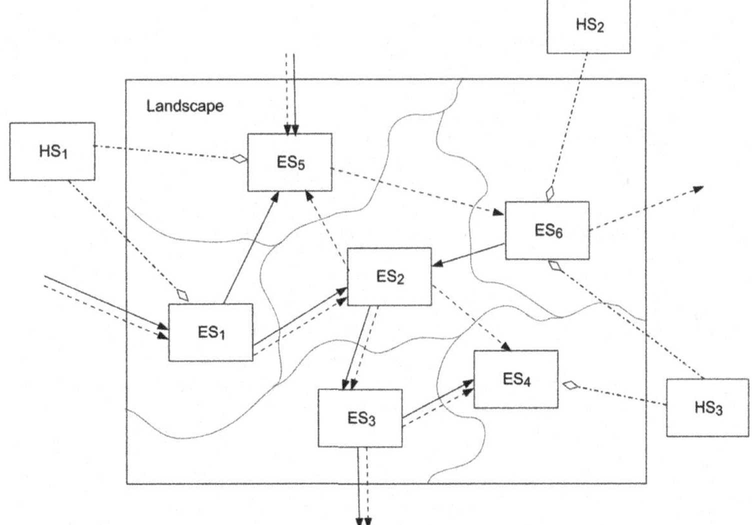

CONCLUSION

In this chapter, we have shown that the concepts of *import-export agent society* and *multi-societal agent system* are adequate to formally capture the concepts of *ecosystem* and *landscape*.

We submit that on the basis of this formalization, a sound agent-based approach can be developed to leverage *computational tools for landscape ecology*, in what concerns formal techniques for the agent-based modeling, analysis and simulation of ecosystems and landscapes.

Further work should provide the details, not tackled in the present, of methodological issues that the use of those formal models may raise, concerning the tasks of modeling and analysis of anthropic action on landscapes and the performance of services, for human societies, by the landscapes' ecosystems.

Also, further work should tackle the problem of the dynamics and the temporal evolution of landscapes and their ecosystems, either subject to anthropic action or not. The conceptualization of ecosystems and landscapes in terms of time-indexed structures should facilitate such task.

REFERENCES

Costa, A. C. R. (2009). DIMURO, Graçaliz Pereira. A Minimal Dynamical MAS Organization Model. In *Multiagent Systems - Semantics and Dynamics of Organization Models* (pp. 419–445). Hershey, PA: IGI Global. doi:10.4018/978-1-60566-256-5.ch017

Costa, A. C. R. (2016). Situated Ideological Systems: A Formal Concept, a Computational Notation, some Applications. *Axiomathes*.

Costa, A. C. R. (2014). *On the Bases of an Architectural Style for Agent Societies: Concept and Core Operational Structure*. Available at: www.ResearchGate.net

Odum, E. P. (1964). The New Ecology. *Bioscience, 14*(7), 14–16. doi:10.2307/1293228

Odum, H. T. (1994). *Ecological and General Systems - An Introduction to Systems Ecology*. Univ. Press of Colorado.

Odum, H. T., & Odum, B. (2003). Concepts and Methods of Ecological Engineering. *Ecological Engineering, 20*(5), 339–361. doi:10.1016/j.ecoleng.2003.08.008

Pidwirny, M. (2009). *Fundamentals of Physical Geography* (2nd ed.). Available at: http://www.physicalgeography.net

Waterman, T. (2009). *The Fundamentals of Landscape Architecture*. Lausanne: AVA Publishing.

Wu, J. (2012). Landscape Ecology. In A. Hastings & L. J. Gross (Eds.), *Encyclopedia of Theoretical Ecology* (pp. 392–396). Berkeley, CA: California Press.

ENDNOTES

1. In this work, we do not consider in detail the dynamical features of the evolutional dimension, which captures the way ecosystems evolve in time.
2. We adopt, for any given population, the distinction between a population's habitat (the whole of the geographical area where that population lives, and its niche in that geographical area) and the niche of a population (the particular organizational and functional conditions existing in its habitat, that allow that population to live inside habitat). We apply the notion of habitat also for communities.
3. For any set X, we denote by $\wp X$ the power-set of X.
4. Thus, with no counterpart for the symbolic environment $SEnv^t$ of the agent society.
5. The way to formally capture human societies in terms of agent societies should be clear from the definition of the latter concept, as it is given in Sect. 5.2. Additional information is available in our previous works, e.g., (Costa, 2014) and the references there in.

Chapter 3
Morphozoic, Cellular Automata with Nested Neighborhoods as a Metamorphic Representation of Morphogenesis

Thomas Portegys
Dialectek, USA

Gabriel Pascualy
University of Michigan, USA

Richard Gordon
Gulf Specimen Marine Laboratory and Wayne State University, USA

Stephen P McGrew
New Light Industries, USA

Bradly J. Alicea
OpenWorm, USA

ABSTRACT

A cellular automaton model, Morphozoic, is presented. Morphozoic may be used to investigate the computational power of morphogenetic fields to foster the development of structures and cell differentiation. The term morphogenetic field is used here to describe a generalized abstraction: a cell signals information about its state to its environment and is able to sense and act on signals from nested neighborhood of cells that can represent local to global morphogenetic effects. Neighborhood signals are compacted into aggregated quantities, capping the amount of information exchanged: signals from smaller, more local neighborhoods are thus more finely discriminated, while those from larger, more global neighborhoods are less so. An assembly of cells can thus cooperate to generate spatial and temporal structure. Morphozoic was found to be robust and noise tolerant. Applications of Morphozoic presented here include: 1) Conway's Game of Life, 2) Cell regeneration, 3) Evolution of a gastrulation-like sequence, 4) Neuron pathfinding, and 5) Turing's reaction-diffusion morphogenesis.

DOI: 10.4018/978-1-5225-1756-6.ch003

INTRODUCTION

Morphogenesis is a biological process by which cells move and differentiate into organs and tissues through genetic expression and collaborative, often physical mechanisms. They become different kinds of cells, perhaps as many as 7000 kinds in our bodies (Gordon, 1999). One of the most persistent concepts of morphogenesis is the morphogenetic field (Beloussov, Opitz, & Gilbert, 1997; Alberts et al., 2002; Levin, 2011, 2012; Morozova & Shubin, 2012; Vecchi & Hernández, 2014; Beloussov, 2015) with clinical significance for human birth defects (Opitz & Neri, 2013). A morphogenetic field is a region of an embryo that has the potential to develop into a specific structure. How this happens has been subject to much investigation and debate. Some mechanisms are better understood than others. As discussed in (Tyler, 2014), which reviews the full panoply of models of morphogenetic fields, we have added the idea that a morphogenetic field is the trajectory of a two dimensional differentiation wave that triggers a step of differentiation in each cell it traverses (Gordon, 1999; Gordon & Gordon, 2016b, a).

While the mechanisms behind the transformation of an egg into an embryo (embryogenesis) and then to an adult organism have traditionally been of great interest to biologists, as a pattern formation problem it is equally intriguing to computer scientists. Part of the allure involves the spontaneous attainment of order of great complexity from geometrically simple beginnings. Even though one of the founding fathers of computer science (Leavitt, 2006; Hodges, 2014; Tyldum, 2014), Alan Turing proposed a plausible model for understanding one level of the self-organizing aspect of morphogenesis (Turing, 1952; Gordon, 2015), the process of phenotype-building has a "ghost in the machine" (Koestler, 1967), a cybernetic aspect (Gordon & Gordon, 2016a; Gordon & Stone, 2016) that has gone underappreciated.

Recent chemical experiments have revealed that while Turing's original reaction-diffusion equations portray certain aspects of morphogenesis, they do not account for heterogeneity (Tompkins et al., 2014) or the multistep hierarchical differentiation of cells into different types (Gordon, 2015). In this study, we propose that given the right representation, simulated morphogenesis can yield solutions that are biologically plausible. Our approach, *Morphozoic*, models a hierarchical structure of cellular communities. Computationally, these communities are nested versions of Moore-like neighborhoods. A Moore neighborhood is the set of cells that are immediate neighbors to a cell, so in a two-dimensional square array the Moore neighborhoods contain eight cells (Weisstein, 2016b). In Morphozoic, a single higher-level cell houses an entire lower-level Moore neighborhood, down to single cells, and a set of lower-level neighborhoods compose a higher-level neighborhood (Gordon & Rangayyan, 1984a, b). While this serves as a constraint on cell-cell communication, it also serves as top-down information. This top-down information, when coupled with local, bottom-up information at different spatial scales, provides us with a mechanism for strongly emergent phenomena (Holland, 1992).

We originally called such nested neighborhoods "adaptive neighborhoods" in the context of image processing, because around each pixel a Moore neighborhood size was chosen that best fit the local features of the picture (Gordon & Rangayyan, 1984a) (Gordon & Rangayyan, 1984b). The field of adaptive neighborhood image processing now has over 400 publications, and an independent discovery of the idea has extended the idea in many additional directions (Katkovnik, Egiazarian, & Astola, 2006). In the field of cellular automata, adaptive neighborhoods have been used in the sense of changing neighborhood type rather than size (Mofrad et al., 2015). Our approach of nested neighborhoods has been combined with cellular automata for edge detection (Liu et al., 2012) and rule identification (Adamatzky, 1997; Sun, Rosin, & Martin, 2011; Zhao, Wei, & Billings, 2012). Irregular (Batty, 2003) and "extended" (Guan & Clarke, 2010) Moore neighborhoods have also been used for geographic cellular automata, though not with the nesting idea in mind. Irregular, grown adaptive neighborhoods have been called coalitions in cellular automata (Burguillo, 2013). Morphozoic appears to be unique in using nested neighborhoods, not to find an optimal neighborhood of a cell, but to provide information at many scales to that cell. It thus permits the study of local/global interactions.

Computer modeling of biological systems is widespread (Wyczalkowski et al., 2012; Tanaka, 2015). The Morphozoic approach is based on the Cellular Automaton (CA) architecture which exhibits computational universality (Dobrescu & Purcarea, 2011) (Wolfram, 2001) that is not well understood in the context of biological development. The aim of the Morphozoic project is to build an abstract model of morphogenetic fields to explore its computational capabilities. While the model may lead to some insights into biology, this is not a central goal of the project. Simulations presented here suggest that the model can be used to produce general self-organizing structures. In particular, many aspects of modern human life involve local/global interactions, so Morphozoic may contribute to the social sciences (Batten, 2001). We also show how Morphozoic may be used to reverse engineer a sequence of state changes of a system and derive an approximation to the rules governing that system. Morphozoic may therefore be used for reverse engineering (Gordon & Melvin, 2003; Deutsch, Maini, & Dormann, 2007; Elmenreich & Fehervari, 2011; Lobo & Levin, 2015).

Because of its local/global construction, Morphozoic may be a step towards meeting the challenge posed by Russ Abbott:

...when a glider appears in the Game of Life, it has no effect on the how the system behaves. The agents don't see a glider coming and duck. More significantly we don't know how to build systems so that agents will be able to notice gliders and duck. It would be an extraordinary achievement in artificial intelligence to build a modeling system that could notice emergent phenomena and see how they could be exploited. Yet we as human beings do this all the time (Abbott, 2006).

The Morphozoic platform can model a wide range of natural and artificial phenomena, but the question remains whether or not the software can exhibit biological realism. Certainly in terms of approximating morphogenesis, it is not clear whether patterns formed and identified by Morphozoic are produced by biologically-realistic mechanisms. However, in the realm of biological realism, Morphozoic is consistent with similar approximations of biological complexity. Particularly at the level of interacting cells, morphogenesis and Morphozoic alike exhibit what Abbott (Abbott, 2006) calls 'epiphenomena'. These epiphenomena, or emergent outcomes of collective interactions between agents, are as biologically realistic at the macro-scale as gene action is at the micro-scale. However, as morphogenesis is non-reductive (Abbott, 2006; Gordon & Gordon, 2016a), it becomes difficult to make exact predictions of behavior in existing biological systems. What makes for an excellent naturalistic pattern replication mechanism (Wolfram, 1984) might make for a poor descriptor of unfolding processes in the frog embryo.

Biological realism in modeling involves the degree to which selectively adding in features of the system you are attempting to model produces a useful representation. At a macro-level of description, Morphozoic exhibits *a priori* biological realism (Bourgine & Lesne, 2010) that captures the higher-order dynamics of a biological system rather than the lower-level causal mechanisms of complexity (Ruxton & Saravia, 1998). *A priori* realism incorporates known features in a minimal and abstract fashion (Bourgine & Lesne, 2010). In cases where the underlying system has many moving parts and layers of complexity, a high degree of abstraction is required to avoid transcomputational limits (Bremermann, 1967; Ashby, 1968; Gordon, 1970; Bower, 2005). Transcomputational limits are of particular concern in biological systems, where the myriad sources of variation can produce a very high dimensional problem space. While abstraction in the face of transcomputational limits is a supposed requirement for biologically-inspired simulations, nevertheless many models of collective behavior (Resnick, 1994) and evolution (Adami, 1998; Alicea & Gordon, 2014) also utilize highly abstract representation while producing realistic biological system dynamics. This is one reason why Morphozoic can reproduce patterns seen in morphogenesis without invoking mechanisms of gene expression.

Another reason why Morphozoic can exhibit "lifelike" patterns is due to the nature of Cellular Automata. In many cases what drives changes in the dynamics of the model are not lower-level control mechanisms, but the order in which key spatial and temporal events play out. While this is important for real biological systems as well, changes to spatial and temporal order of how events are executed lead to synergistic effects in Cellular Automata dynamics (Ruxton, 1996). Furthermore, the timing of events in a Cellular Automata model impact predictive ability, as Huberman and Glance (Huberman & Glance, 1993) have shown by implementing asynchronous behaviors in the Prisoner's Dilemma game. In Morphozoic, we keep the standard features of cellular automata: discrete time steps, each cell changes state simultaneously with all the others, and no cell moves from its initial position. These three aspects are clearly not biological. Lifting these constraints is a topic for future research. Morphozoic provides a rich starting point.

There are also parallels between biological and sociotechnical systems that demonstrate how adaptive behavior might be as much a consequence of temporal evolution than formal biological mechanisms. Sometimes this temporal evolution is intertwined with structural features of the system, such as when scientists can "unboil" an egg and return proteins to their original conformational state (Bijelic et al., 2015). The origin of evolutionary novelties also relies upon how temporal sequences interact with existing variation to produce innovations. This, evolutionary novelties are generated from what is already available through recombination and repurposing (Jacob, 1977). Assembling what already exists into new combinations is called combinatorial innovation, part of something Wagner and Rosen (Wagner & Rosen, 2014) refer to as innovability. Acting as a mechanism for new phenotypic possibilities, characterizing the innovability of emergent systems like Morphozoic might provide a very broad window into the systems-level mechanisms of developmental processes.

Innovability is a key driver in the functional diversity seen in Morphozoic, and can be analogous to the types of developmental plasticity observed in biological systems (West-Eberhard, 2002). In phenomena ranging from axonogenesis to generalized stress response (Bateson et al., 2004; Gluckman, Hanson & Low, 2011; Low, Gluckman & Hanson, 2012), organisms can exhibit a morphogenetic adaptive response to both local and global signals. In this sense, Morphozoic provides a means to explore how the timing and phenotypic consequences of these processes might unfold (Moczek et al., 2011). While Morphozoic does not possess the same mechanisms that drive biological plasticity, phenotypic changes can be approximated using discrete computation. While Morphozoic may be able to approximate the pattern-formation aspects of developmental plasticity, a concept known as the adjacent possible may be used to bridge the gap between biology and simulation. The adjacent possible was originally proposed by Stuart Kauffman as a way to describe possible new states for a system given historical contingency (Johnson, 2010). The adjacent possible is ever-expanding, and this non-random expansion into possibility space is what drives subsequent innovations (Tria et al., 2014).

Fortunately, the cellular automata representation is well-suited to producing novelties that result from interactivity. Identifying the specific mechanisms for the generation of novelties is a relatively unexplored question (Kier, Bonchev, & Buck, 2005). How can we discover these pathways and make parallels with developmental and other biological systems? On the one hand, the adjacent possible can be made salient using a method called fitness optimization. This allows us to understand patterns of innovation that conform to the adjacent possible idea as a series of moves towards a fitness peak (or functional optimum). On the other hand, many biological and technological systems also include significant constraints on their evolution such that the states making up the so-called adjacent possible are limited to relatively small portions of the whole system (Solé et al., 2013). One way to investigate these non-innovable portions of the system is to construct a neutral network (van Nimwegen, Crutchfield & Huynen, 1999) that defines all possible configurations of the system. To approximate the full set of possible states (which may or may not be transcomputational), we may rely upon computational information such as the neutral network, cellular automata rules, and knowledge about the constraints a given system might pose. Applying the concept of the adjacent possible then allows us to limit the range of plausible emergent mechanisms in biological and non-biological settings.

Coupling Morphozoic with other computational and data-driven models of embryogenesis may lead us to a middle-out approach (Noble, 2002). The middle-out approach can be defined as a combination of top-down and bottom-up approaches where specification begins at the level where the data are sufficient. Using inferential and other methods, we can then move towards other levels of analysis. This is ideal for systems that are too interactive for systems or reductionist approaches to be effective on their own (Noble, 2002). The middle-out approach is particularly useful to data-driven research in that it strikes a balance between reductionism and integration. As the reductionists have mechanisms, the systems people have overarching descriptions (Kohl & Noble, 2009). Morphozoic occupies a middle-ground between reductionism and integration in the sense that the states of single cells are influenced by local neighborhoods, but that each local neighborhood is interlinked to form a potentially large (global) problem space. This interlinking occurs at many levels, due to the use of nested neighborhoods. By occupying this middle ground, defined by neighborhoods, we can approach multiscalar biological processes in novel ways (Walker & Southgate, 2009). Multiscalarity can reach 8 orders of magnitude, as in diatoms, for instance (Ghobara et al., 2016).

One of the inspirations for Morphozoic is the *Morphone* model (Portegys, 2002), a programmable, modular signaling system for morphing complex patterns. One of the ways Morphozoic differs from Morphone is in its leveraging of local vs. global signaling fields as computing mechanisms. Another feature of Morphozoic, typical of computing systems but distinct from biological systems, is the use of cell states as signals to other cells instead of supporting separate signaling objects.

DESCRIPTION

Cellular Automaton

Morphozoic is built upon a two-dimensional (2D) cellular automaton (CA) architecture. CAs generally have these properties:

- A cell has a state value (e.g. on or off). However, the value can be a more complex entity, such as a real value or vector.
- A cell senses the states of adjacent cells that makes up its *neighborhood*. One example is a 3x3 Moore neighborhood.

- Action rules determine how a cell state changes based on the states of the cells in its neighborhood.

CAs date back to the 1940s and the work of Stanislaw Ulam, who, requiring a model to study the growth of crystals, developed a simple lattice network (Ulam, 1962). At the same time, John von Neumann, Ulam's colleague, was working on the problem of self-replicating systems (von Neumann & Burks, 1966). Von Neumann's initial design was to have robots build new robots out of a "sea of parts". This proved problematic, and a more abstract and discrete model was later developed that became a foundation for the CA approach. It was Ulam who suggested using a discrete system for creating a reductionist model of self-replication. Von Neumann's work in self-replication system is similar to what is probably the most famous cellular automaton: the "Game of Life," (Gardner, 1970) which is presented as an application of Morphozoic in a later section. We introduced long-range interactions in cellular automata in a simple model of two dimensional "snail" morphogenesis, showing they led to more robust pattern formation (Gordon, 1966). Long-range effects are also incorporated into human evacuation CA models (Kaji & Inohara, 2014).

In 2001, Stephen Wolfram's book *A New Kind of Science* was published (Wolfram, 2001). The book discusses how CAs are relevant to the study of biology, chemistry, physics, and all branches of science. In addition, CAs are relatively easy to create in software and are straightforward in performance evaluation. For these reasons the CA architecture was chosen as a platform for Morphozoic. As demonstrated in Wolfram's work, CAs can serve as a model of parallel distributed processing that fit many natural systems well. Some of these properties of the physical world:

- Contain a large number of simple parts, called cells, each of which is an automaton. This means it acts on its own (autonomously).
- Parallel operation.
- Global, emergent effects from local interactions.

Each cell (automaton) can respond to external signals in "deciding" what to do. Those decisions are restricted to certain choices, either finite in number or represented by continuous values. For example, CAs can simulate fluid or gas dynamics by storing individual molecules in the cells and implementing particle interactions by the local rules.

One of the main variations of CAs are in the number of dimensions they contain. A one-dimensional (1D) CA is simply a row of cells. A two-dimensional (2D) CA is a grid of cells. Three-dimensional (3D) CAs operate within a volume of discrete cells. The concept of a *neighborhood* is central to the description of a CA, as the states of a cell's neighborhood define the input to the local state transition rules. In a one-dimensional CA, a cell's neighborhood is usually its adjacent cells. In a 2D CA, typical neighborhood configurations are shown in Figure 1. It is also common, as is done for Morphozoic, to include the center cell in the definition of its neighborhood.

CA cells contain a *state* that can take on values defined by the automaton. For example, a binary state can be considered to be in a 0 (off)/1 (on) state, as shown in Figure 2.

State values change based on transition *rules*. A good way to understand how state transition rules work is to consider a one-dimensional CA with adjacent cell neighborhood, as shown in Figure 3. Using Wolfram's terminology, the eight possible neighborhood state configurations and center cell state transitions can be enumerated. So this rule set is number 30.

Figure 1. Typical two-dimensional CA neighborhoods
Source: Adapted from Figure 2, Espinola et.al, 2010.

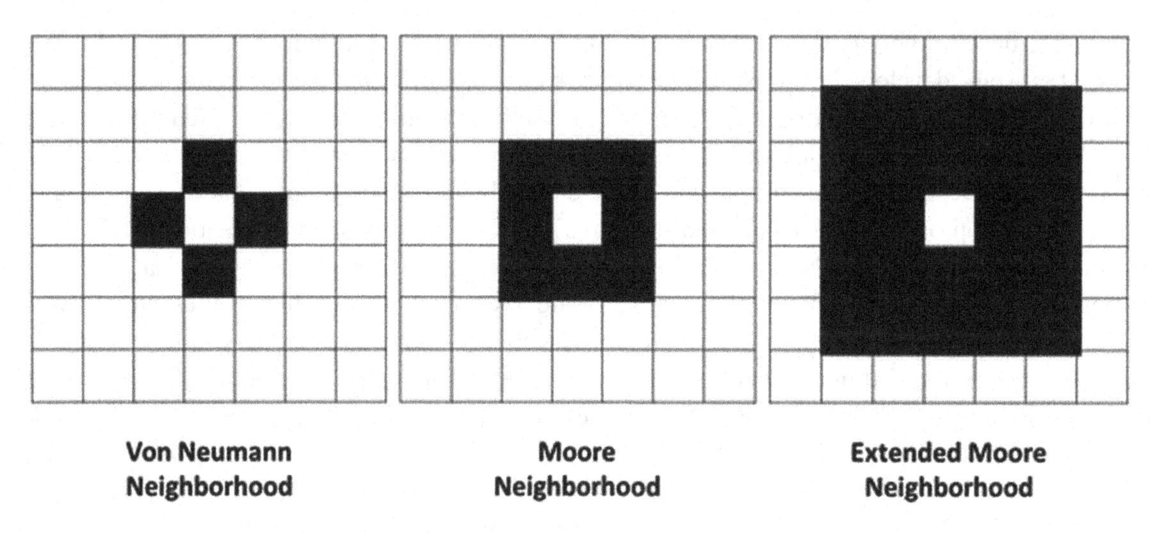

Figure 2. A 2-dimensional grid of cells, each in a discrete state of "0" or "1". Moore neighborhood of cells shown inside red bounding box. Automaton for Moore neighborhood is denoted with a red circle

0	0	1	0	1	1
1	0	0	0	1	1
1	0	1	1	1	0
0	0	1	0	1	1
1	1	0	0	1	0
1	1	1	0	0	1
1	0	0	1	1	1
0	0	1	0	1	0

Figure 3. State transition rule set 30, where 30 base 10 = 00011110 base 2
Source: Weisstein, 2016a with permission per Eric Wolfram's Notice of Copyright http://wolfram.org/copyright.html.

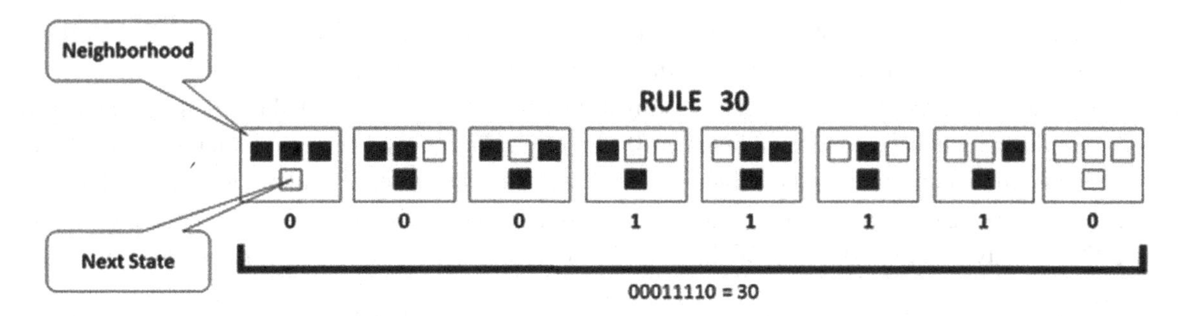

Figure 4 demonstrates how a CA develops over time from a single "on" cell using rule 30. Time advances from top to bottom.

There are many other possible rules to compute a cell's state from a group of cells. Consider blurring an image. A pixel's new state (i.e., its color) is the average of all of its neighbors' colors. Most image processing algorithms can be formulated as CAs. The rules define the functionality of the CA.

Variations of CAs

The following are some variations of the CA model:

Figure 4. Application of rule 30. A single cell is turned on (top row) and in the next time step (second row) all of the cells are subjected to rule 30. Two of them change state as a result, leaving three cells turned on in the second row. This process is iterated to generate the third row, etc. Patterns like these are actually generated by the pigment depositing line of cells in marine cone snails
*Source: Waddington & Cowe, 1969 The discrete version of such one dimensionally generated patterns became known as Lindenmeyer patterns from de Koster & Lindenmayer, 1987. From Weisstein, 2016a with permission per Eric Wolfram's Notice of Copyright http://wolfram.org/copyright.html_*

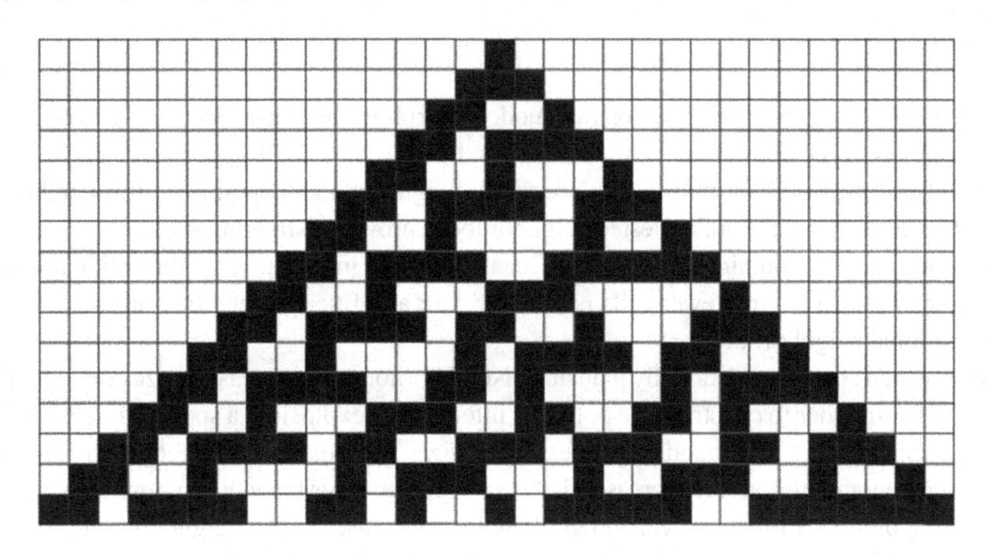

1. **Non-Rectangular Grids:** There is no essential reason why a CA must be confined to a rectangular grid or space.
2. **Probabilistic:** The rules of a CA need not necessarily work in a deterministic fashion (Gordon, 1966, 1980). An example is the Stochastic Game of Life (Monetti & Albano, 1997).
3. **Continuous:** The state of a cell can be something other than a discrete value, such as 0 or 1. For example, the values could range between 0 and 1. Of course the rules must then reflect how to calculate these continuous state values. Examples in materials science, traffic flow and earthquake analysis abound (Olami, Feder, & Christensen, 1992; Bubak & Czerwiński, 1999; Kitakawa, 2004, 2005; Gosálvez et al., 2009; Ferrando et al., 2011; Ferrando, Gosálvez, & Colóm, 2012; Li et al., 2015).
4. **Historical:** In the Game of Life CA, the current state configuration determines the next configuration. However, taking into account historical states is also possible. For example, Portegys and Wiles (2004) describe a CA that self-repairs in the presence of noise by use of historical cell states.
5. **Moving Cells:** In these examples, cells have a fixed position on a grid, but can move to other grid points (or, equivalently, their states can be transferred to other cells or switched with them) (Gordon et al., 1972, 1975; Chopard, 1990; Fukui & Ishibashi, 1996; Hochberger, Hoffmann, & Waldschmidt, 1999; Halbach & Hoffmann, 2005; Moussa, 2005).
6. **Nesting:** Another feature of complex systems is that they can be nested into hierarchies. For example, a city is a complex system of people, a person is a complex system of organs, an organ is a complex system of cells, etc. Cell values that reflect this hierarchical arrangement are possible (Weimar, 2001; Dunn & Majer, 2007; Kiester & Sahr, 2008; Dunn, 2010).

From a more speculative viewpoint (Ilachinski, 2001), researchers have raised the question of whether the universe is a cellular automaton. For example, mathematical models have shown the emergence of "particles" within CAs, such as the gliders in the Game of Life. This leads to conjectures that the natural world, which is well described by physics with particle-like objects, could actually be a CA. This hypothesis has led scholars to a perspective of nature existing within a discrete framework. Edward Fredkin (Fredkin, 1992), a strong proponent, has proposed the "finite nature hypothesis", i.e., the idea that "ultimately every quantity of physics, including space and time, will turn out to be discrete and finite."

Objectives

A main objective is to devise an abstraction that models morphogenesis in a CA using a nested neighborhood approach. As a type of multilayered scheme (Bandini & Mauri, 1999; Dascalu et al., 2011), nested neighborhoods, defined more formally below, are simply neighborhoods contained within neighborhoods, much like Russian matryoshka dolls. Nested neighborhoods provide a straightforward representation of a morphogenetic field that contains a hierarchy of local vs. global information. Information about a more local cell neighborhood having fewer cells is more precise and finer grained than information about a larger, more global neighborhood.

The scheme must be computationally plausible. Neighborhoods of increasing size contain increasing numbers of cells. In order to constrain the potential information explosion, a specific number of bits are used to represent each neighborhood, regardless of its size. The smallest neighborhood is represented precisely; larger neighborhoods are increasingly "fuzzy" because they cannot be completely represented by the available bits. This forms a precision gradient that decreases as the neighborhood grows in size.

An intended result of this plan is that more distant cells are sensed in aggregation, as described in the next section.

Additional objectives:

- Compact state change rules.
- Noise tolerant.
- Generalizable from exemplars.
- Evolvable.

MORPHOGENETIC FIELD SPECIFICATION

The morphogenetic field is specified in a CA by equipping cells with these properties:

- A cell state is its *type*.
- A cell emits, senses, and reacts to signals.
- Signals carry information about the types of neighborhood cells.
- A field is the confluence of signals sensed by each cell.

Rules are embodied in *metamorphs*, which encapsulate pattern-matching *morphogens* and cell state change actions.

Morphogen

A "morphogen" abstracts many types of morphogenesis mechanisms: chemical (the classical definition), physical, energy, etc. It also summarizes a morphogenetic field as a set of a cell's nested neighborhoods and their contents. This is shown in Figure 5. A neighborhood consists of an *NxN* set of *sectors* surrounding a lower level neighborhood:

Figure 5. Morphogen nested neighborhoods

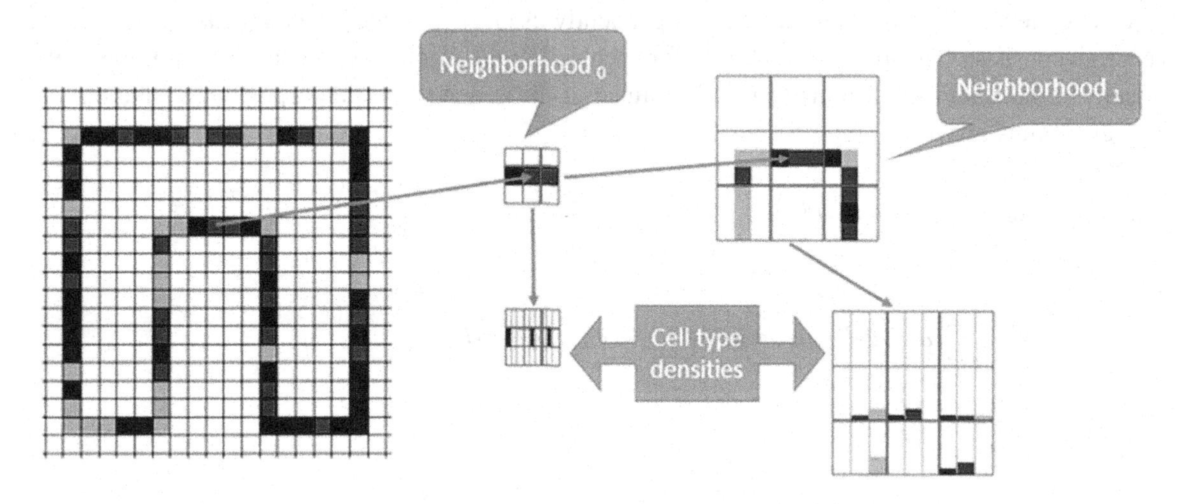

$neighborhood_i = NxN(neighborhood_{i-1})$

where *N* is a fixed odd positive number chosen by the user of Morphozoic, and $neighborhood_0$ is composed of *NxN* elementary cell sectors.

Hence the number of cells in $neighborhood_i = N^i x N^i = N^{2i}$.

A morphogen is composed of a set of nested neighborhoods:

$morphogen(cell) = \{ neighborhood_0(cell), neighborhood_1(neighborhood_0), \ldots neighborhood_n(neighborhood_{n-1}) \}$

The *value* of a sector is a vector representing a histogram of the cell type densities contained within it:

$value(sector) = [density(cell\text{-}type_0), density(cell\text{-}type_1), \ldots density(cell\text{-}type_n)]$

The number of cells contributing to the density histogram of a sector of $neighborhood_i = N^{i-1} x\, N^{i-1}$

Metamorph

A *metamorph* represents a cellular automaton morphogen→action agent, defined as a mapping from a morphogen to a cell type.

Generation

A set of metamorphs describing a pattern of cell activity can be generated from manual input or a programmed sequence of cellular automaton transitions. For example, the Game of Life application uses the programmed Game of Life rules to process the cell states. As the CA changes, the neighborhoods for each cell are used to construct morphogens, and the cell type transitions associated with the morphogens are actions. The morphogens and actions are composed into metamorphs.

Execution

Once generated, the metamorphs can be independently used to "execute" the application. Metamorph execution consists of creating a morphogen for each cell in the grid and comparing each of these morphogens to the stored set of morphogens contained in the generated metamorphs, where the distance between them is given by:

$$distance\left(metamorph_i, metamorph_j\right) =$$

$$\sum_{x}^{neighborhoods} \sum_{y}^{sectors} \sum_{z}^{cell\ types} abs\left(cell\ type\ density_{i,x,y,z} - cell\ type\ density_{j,x,y,z}\right)$$

The metamorph having the least morphogen distance is chosen as the cell action.

A unique feature afforded by the use of a distance metric to match morphogens is a noise-tolerant, self-healing capability. Metamorphs act on cells according to neighborhood similarity, which steers cell states toward patterns stored in the metamorphs. This feature is a hallmark of biological systems, and is quite distinct from typical rigid CA rule formulations, such as the Game of Life (see applications), where a minor introduction of noise often results in global disruptions.

Artificial Neural Network Implementation

A compact, fast, and noise tolerant representation of metamorphs can be implemented by an artificial neural network (ANN), a biologically-based learning machine that is particularly adept at classifying input patterns into output categories (Haykin, 2011).

ANN Background

ANNs are loosely modeled after the neuronal structure of biological nervous systems but on smaller scales. A large ANN might have thousands of processor units and interconnections, whereas a human brain, for example, can have 86 billion neurons. An African elephant brain contains 267 billion neurons (Wikipedia, 2016b), and human cerebral cortex contains 150 trillion synaptic interconnections (Drachman, 2005). ANN architectures often functionally diverge from their biological counterparts in important ways, including how learning is implemented.

An ANN is a subset of a general computing model known as connectionism. A connectionistic model features an interconnected network of simple units that produce emergent properties that are beyond the capabilities of the individual units (Medler, 1998). In the emergent respect, that the proverbial whole is greater than the sum of its parts, an ANN is similar to a cellular automaton.

A prevalent type of ANN is the multilayer perceptron (MLP). MLPs are organized in layers, shown in Figure 6. Layers are made up of a number of interconnected neurons. Patterns are presented to the network via the input layer, which connects to one or more hidden layers where the actual processing is done via a system of weights associated with the connections. The hidden layers then connect to an output layer where the output is represented by the activation of one or more neurons.

A *weight* value is associated with each connection in the network. The weights are multiplied by the outputs of the source neurons, the sum of which is input to an *activation function*, which computes a neuron's output.

Activation functions are typically sigmoid shaped, such as the *logistic* function shown in Figure 7. Here the output switches smoothly from 0 (off) to 1 (on) over an interval controlled by the β parameter. An important property of an activation function is that is it differentiable, which allows a network to be trained through learning.

Learning is the process of modifying the connection weights to produce outputs that differ least from "correct" outputs, i.e. minimized error. When the correct outputs are known, this type of learning is called *supervised learning*. Learning typically entails *backpropagating* the output error from the output to input neurons to modify the weights of the connections such that the error is reduced in subsequent computations. The most common modification algorithm for this is the *delta rule*.

Figure 6. Multilayer perceptron
Source: Cazala, 2015 with permission of Juan Cazala

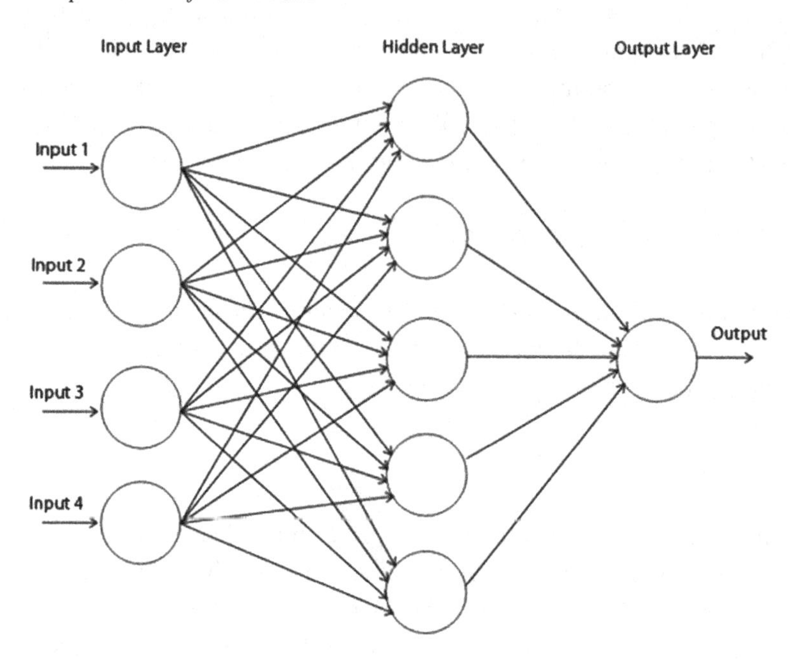

Figure 7. Logistic activation function
Source: Sayed, 2016 with kind permission of Saed Sayed

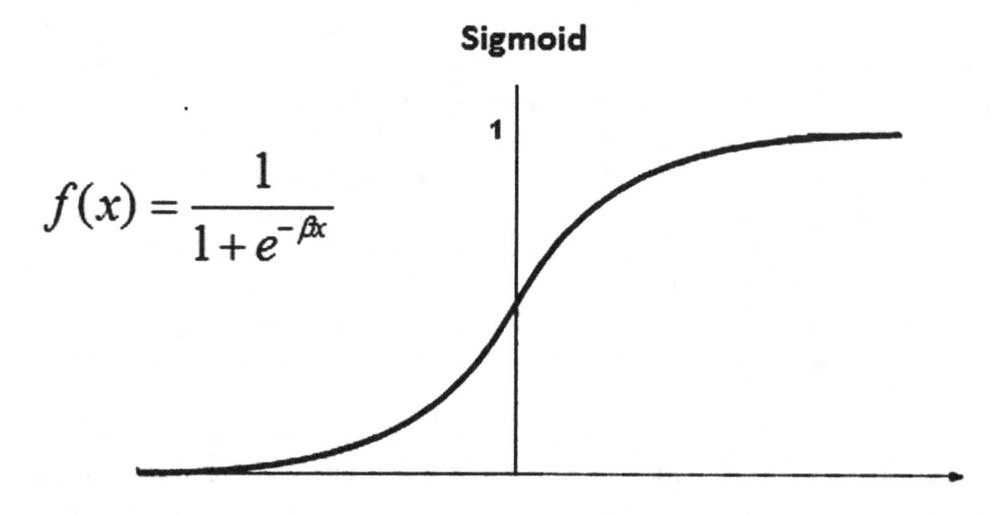

Training is a process that involves repetitive runs of input-output patterns through the network with an application of the learning rule performed for each pattern. A run through an entire training set is called an *epoch*. ANNs, like their biological counterparts, are known for their ability to produce correct outputs given noisy or similar inputs. After training, a separate test set of patterns with input variations can be evaluated to assess the effectiveness of training.

There are a number of ANN variations. For example, the activation function can be a Gaussian function instead of a sigmoid one. ANNs with many layers are also possible. In general these are known as *deep learning* networks. In a *convolutional* network, an input layer neuron feeds only a subset of neurons in the next layer. This resembles the architecture of the human retina. An important ANN variation, capable of learning input-output sequences such as those found in speech patterns, are called *recurrent* ANNs.

Implementation

By squashing the morphogen sector values into an input vector and considering the cell types as a set of outputs, an ANN can be trained to learn a set of metamorphs such that a morphogen derived from a cell in a CA will map to the metamorph closest matching it. This is shown in Figure 8.

The advantages of using an ANN are threefold:

1. **Speed:** Instead of searching a set of metamorphs during execution for the closest morphogen, an ANN performs a cascading set of arithmetic calculations to arrive at an output.
2. **Compact Representation:** An ANN is capable of retaining a large number of input-output mappings in the form of interconnection weights.
3. **Generalization Capability:** An ANN is capable of classifying inputs that are similar to training inputs. This capability will be exploited in subsequent applications.

Figure 8. Neural network implementation

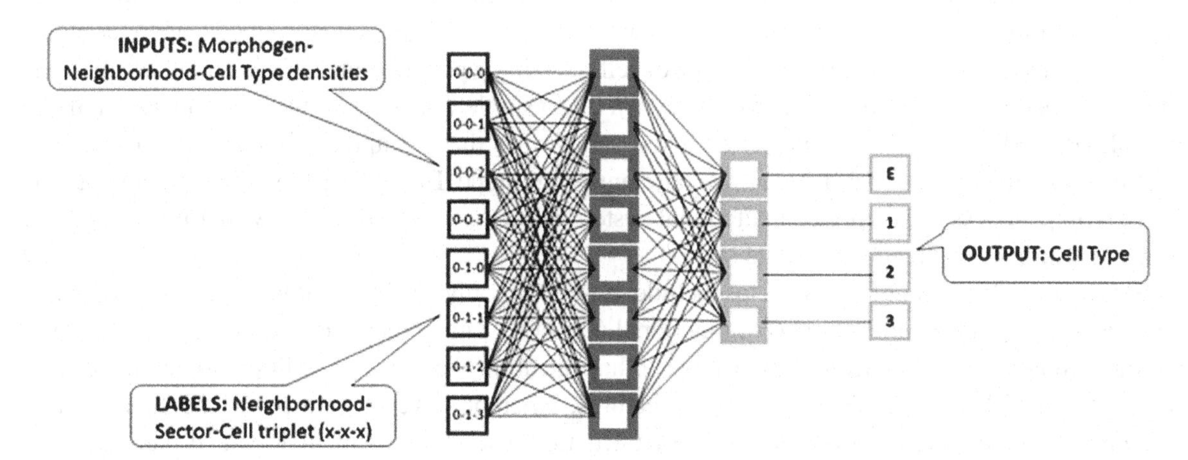

RESULTS

Various features of Morphozoic are illustrated by the following applications.

Conway's Game of Life

This well-known CA (Gardner, 1970) was chosen as a baseline capability test for Morphozoic. In the Game of Life (GoL), cells are in a rectangular array on a "game board". Each cell is either in an "alive" or "dead" state. The state change rules are as follows:

1. Any live cell with fewer than two live neighbors dies, as if caused by under-population.
2. Any live cell with two or three live neighbors lives on to the next generation.
3. Any live cell with more than three live neighbors dies, as if by overcrowding.
4. Any dead cell with exactly three live neighbors becomes a live cell, as if by reproduction.

A GoL "game" starts with an initial configuration of alive and dead cells. The rules are then applied to the cells at each time step. Usually, if the initial configuration is random, the pattern of live and dead cells appears to change chaotically for a while, then resolves itself into isolated clusters of cells that cycle through a repeating series of states.

A sample GoL configuration is shown in Figure 9. Despite its simple rules, the GoL produces many dynamic self-sustaining patterns (and is capable of producing an unlimited number of such patterns). One class of pattern, called "gliders", move across the game board as they go through their state changes and interact in various ways with other clusters when collisions occur (Rendell, 2002). More complex configurations called "guns" are a special type of self-sustaining pattern that produce gliders at a specific rate. Using gadgets such as these it has been shown that a single-tape Turing machine can be simulated by GoL (Rendell, 2002). In complexity theory, a set of rules that manipulates the state of a system is described as Turing complete if it can simulate any single-tape Turing Machine. To demonstrate the Turing completeness of another computational system all one needs to do is show that it is capable of simulating an existing Turing complete system.

Examples of Turing complete systems include most commonly used programming languages, which span paradigms such as procedural (C), functional (Haskell), and object-oriented (Java). A language's Turing completeness is determined by whether it has the ability to branch conditionally and load/store an arbitrary number of variables. These two features, inherent to general purpose languages, can be implemented in other paradigms using analogous structures such as using recursion in Haskell to implement repetition. Turing completeness research has resulted in the discovery of more and more simple systems that hold these properties. In fact, GoL is not the only cellular automata that has been proven to be Turing complete nor is it the simplest.

The elementary cellular automaton Rule 110, whose state transition rules are stated in Table 1, has also been shown to be Turing complete (Cook, 2004). Unlike GoL, Rule 110 functions in only one dimension, each cell's next state results only from the current states of its two neighbors.

Figure 9. A game of life configuration

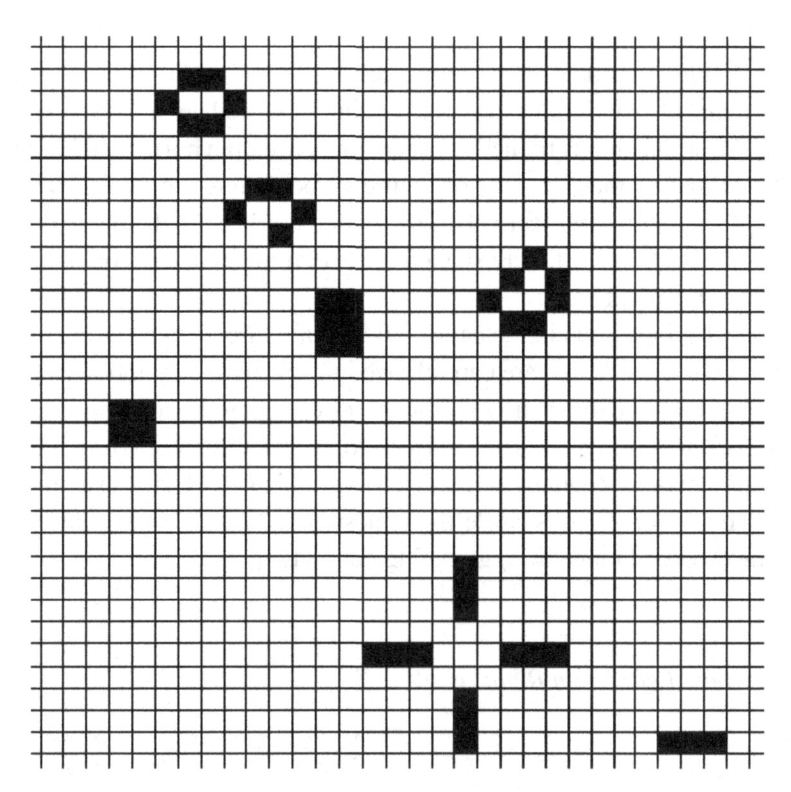

While proving that simple cellular automatons such as GoL and Rule 110 are Turing complete does not allow us to compute in novel or more efficient ways, it does allow us to demonstrate the Turing completeness of other systems such as Morphozoic. Proving that Morphozoic is Turing complete is significant because "Turing completeness" means that it is a universal computer. Given enough memory, Morphozoic would be able to perform any computation that any other Turing complete system is capable of. This says nothing about how efficiently the computation will be done; it only says that the computation *can* be done.

For Morphozoic, we "reverse engineer" a complete set of GoL rules, consisting of 512 3x3 Moore neighborhood configurations. Generating a metamorph for each rule, the configurations were correctly processed, both in the lookup and ANN implementations. Therefore, it follows that Morphozoic is also Turing complete. From the perspective of modeling complex systems like embryos, this means that Morphozoic is capable of modeling anything that can be computationally modeled.

Table 1. State transition from B_1 to B_1' given all possible starting states

$B_0 B_1 B_2$	111	110	101	100	011	010	001	000
B_1'	0	1	1	0	1	1	1	0

Cell Regeneration

The Morphozoic algorithm is capable of modeling cell regeneration. In order to demonstrate this, an apparatus was devised that also highlights the functionality of nested neighborhoods. Figures 10 and 11 show the apparatus. The automaton is trained to regenerate either the horizontal bar on the right side of Figure 10 or the vertical bar on the right side of Figure 11 beginning with the central block configurations on the left side of the figures, respectively. The shaded borders are the distinguishing features that determine regeneration direction: vertically shaded borders train for a horizontal bar and horizontally shaded borders train for a vertical bar.

To spotlight the results, only cells in the central 9x9 area in are allowed to create metamorphs during training, and thus cells only in this area are allowed to regenerate by modifying their type values. Two settings of the neighborhood dimension and number of neighborhoods were defined. The *unnested* setting was a single 9x9 neighborhood. The *nested* setting was three nested 3x3 neighborhoods. The unnested and nested automata thus equally contained 81 variable values. However, the nested automaton neighborhoods span an area of 27 cells while the unnested automaton spans only 9 cells. The nested automaton affords the extended range by aggregating cell values.

Figure 10. Horizontal bar: Left: begin Right: goal

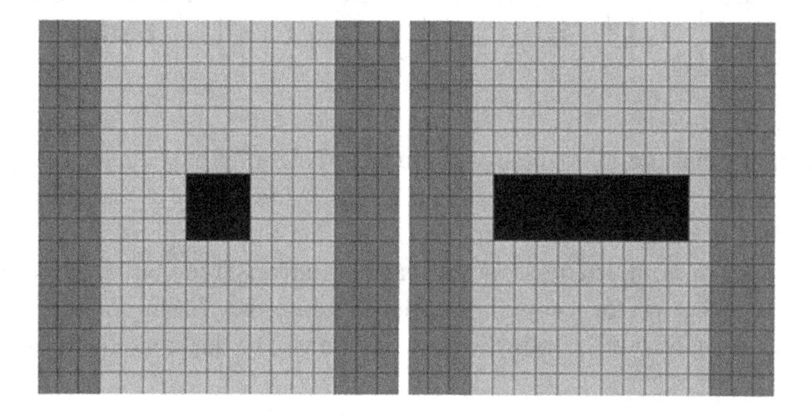

Figure 11. Vertical bar: Left: begin Right: goal

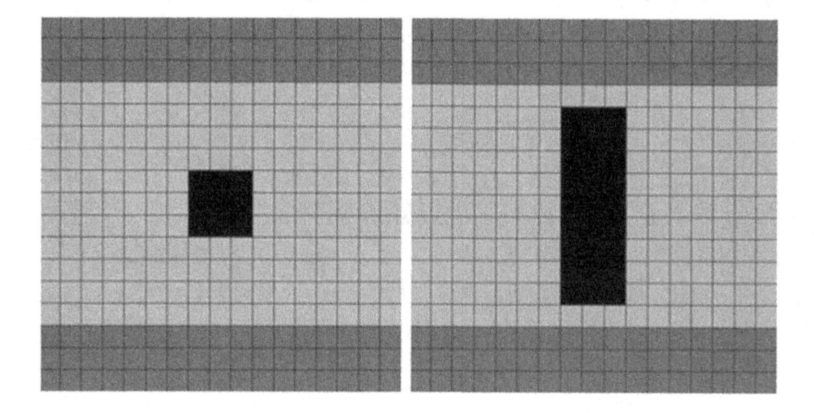

The automata were trained and tested for a variety of border lengths, which are manipulated by changing the dimensions of the grid: the larger the grid, the longer its borders. And as the border lengths increase, the distances from the central block of regenerating cells to the borders increase, eventually passing out of range of the morphogenetic fields associated with the metamorphs. This can be seen in Figure 12, which from left to right shows the morphogenetic field sizes for the 3x3, 9x9, and 27x27 nested neighborhoods. It can be seen that only the 27x27 neighborhood intersects with the left-right border, allowing morphogens associated with the central cells to generate the correctly aligned bar. Morphogens associated with the 3x3 and 9x9 neighborhoods do not intersect the border cells, and thus cannot determine the correct orientation.

For each border length setting, the automata were sequentially exposed to the correctly oriented bars for the vertical and horizontal borders, causing the creation of metamorphs that generate appropriate cell types. A test is correspondingly in two parts: the central block with vertical borders followed by the central block with horizontal borders. A successful test is defined as the regeneration of the correct bars for both parts.

The results are shown in Table 2 for border length increments of four cells. Both the unnested and nested automata correctly process the first two border lengths. However, at the third length, the unnested automaton cannot detect the borders and thus fails to regenerate both bars correctly. Eventually the border recedes beyond the range of the nested automaton as well, indicated in the last row of the table.

The cell regeneration presented here can also be understood as a type of tropism, wherein a directional movement or growth takes place in response to an environmental stimulus. Phototropism, for example, is a response to light (Goyal, Szarzynska, & Fankhauser, 2013). Another biological counterpart involving cell (re)generation stimulated by fields of chemical signals is paracrine signaling for wound healing (Hocking & Gibran, 2010; Dittmer & Leyh, 2014). Stem cells migrating to wound sites release bioactive factors that orchestrate wound healing.

When tissues are wounded, they are healed via a regenerative process called epithelialization (Hocking & Gibran, 2010). During this process, it has been found that stem cells, particularly mesenchymal stem cells (MSCs) aid in the speed of wound healing by releasing growth factors and other extracellular factors that facilitate regeneration (Hocking & Gibran, 2010). In general, stem cells are functionally plastic, and can differentiate into a variety of functional cell lineages depending on where they are recruited and the functional context. During organ regeneration, bone marrow stem cells migrate to sites that are damaged to provide various roles in the regeneration process (Morigi et al., 2004). It is this functional context where the parallels with Morphozoic's generative process can be found. For example, MSCs

Figure 12. Nested morphogen fields: left=3x3, center=9x9, right=27x27

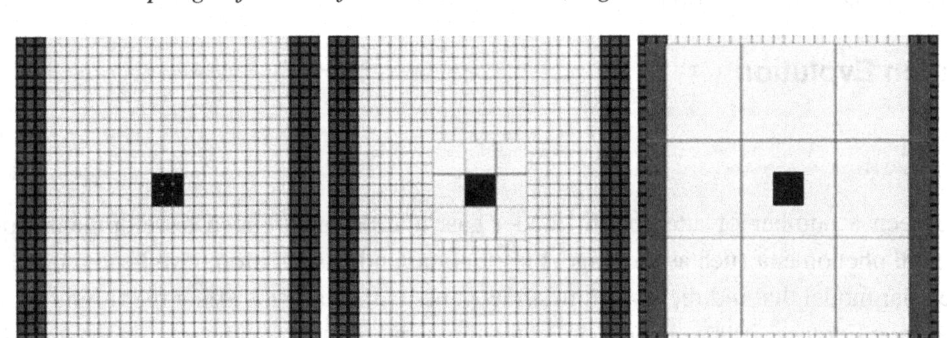

Table 2. Cell regeneration results

Border Length	Unnested Neighborhoods	Nested Neighborhoods
17	✓	✓
21	✓	✓
25	·	✓
29	·	✓
33	·	✓
37	·	✓
41	·	·

often act through paracrine signaling rather than direct replacement of the cells making up damaged tissue (Bruno et al., 2013) .

Stem cells also have the capacity to self-renew, which allow them to proliferate and thus persist for many more divisions than differentiated cells. Perhaps more importantly, stem cells can manipulate their local environments to favor regeneration (Dittmer & Leyh, 2014). This is done by acquiring the gene expression patterns of their new defined fates, as well as producing a host of secretion factors. It is this secretome (Salgado & Gimble, 2013) that most strongly influence the extracellular environment, and can act to coordinate the behavior of cells and tissues at multiple scales. Influence of the local environment is also accomplished via paracrine signaling (Baraniak & McDevitt, 2010), which involves short-range chemical signaling of growth factors between cells. In a very simple manner, Morphozoic can mimic these signaling mechanisms, and the architecture of Morphozoic could be used to implement these mechanisms at multiple spatial scales.

Cell regeneration is concerned with the restoral of missing information using nearby available information. A related process, digital image inpainting (DII) is a computer algorithm that restores missing information of images such as those of old oil paintings. A biological counterpart occurs in human visual systems as well in the form of blind spots (Satoh, 2012). As an illustration of the Morphozoic cell regeneration capability applied to image restoral, consider the problem of restoring an image from its edges, as shown in Figure 13.

This is done by casting the original image and edge-detected image into a CA grid. The edge cells, shown in the upper left part of Figure 14 (in reverse color for contrast), form an outline of the source pattern. The original image, shown in the bottom right of Figure 14, forms the target pattern. Only the dark edge cells are allowed to morph into target cells. With this restriction, five morphing steps are necessary to complete the transition.

Gastrulation Evolution

Background

There have been a number of attempts to model gastrulation and related developmental processes. Developmental phenomena such as cleavage, blastulation, and gastrulation have been modeled using a physical cellular model that indirectly assumes how genetic mechanisms effect the timing of these processes (Drasdo & Forgacs, 2000). A more direct technique would involve the use of Genetic Algorithms

Figure 13. Lenna image with edges
Source: Wikipedia, 2016a

Figure 14. Step-by-step image restoral from edges

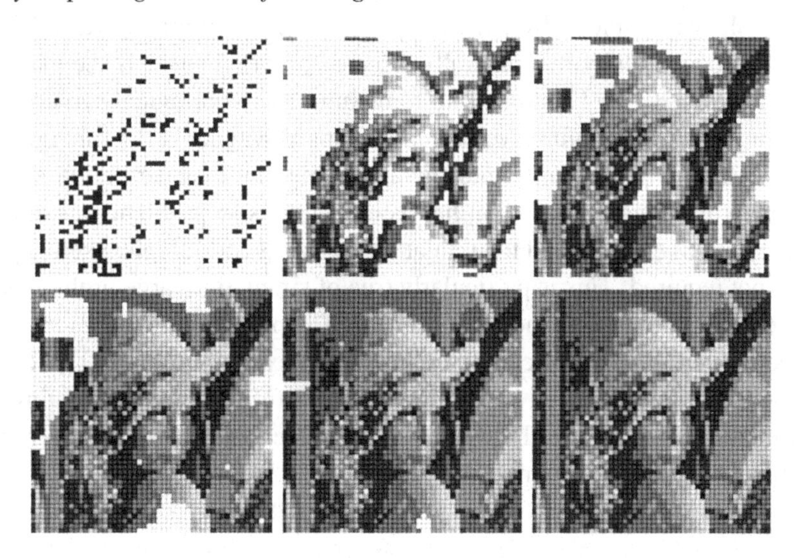

(GAs). GAs are a type of evolutionary computation that focuses on the construction and optimization of programs using natural selection. In this case, natural selection act to select programs that meet or exceed the criteria to reproduce and/or maintained in a population of programs. GAs have been used in conjunction with Young's cellular automata to approximate the reaction-diffusion driven biological pattern formation and replicate morphogenetic fields (Gravan & Lohoz-Beltra, 2004). Because they are an instance of adaptive computing, there is the potential for broader application of GAs to problems of biological development As an optimization procedure, GAs provide an adaptive method that can find solutions despite a rugged search landscape (Bornholdt, 1998). This is analogous to how biological populations evolve solutions to adaptive problems in a *de novo* fashion. We can successfully approximate complex, emergent phenomena in development specifically and biology more generally by more closely examining the relationship between computational representation and biological complexity.

A GA consists of three parts: a genetic representation, a set of operators for mutation and/or crossover operator, and a selection criterion (Holland, 1992). The algorithmic representation (genome) often consists of one or more chromosomes, while each chromosome consists of several genes. A genetic representation is a computational abstraction of a genome, each gene representing a compact encoding of some aspect or feature in a given system. Each gene consists of serial bits and acts as a bit register, which can be used to encode a wide variety of problems. This allows us to capture the power of population dynamics and heredity rather than expression of genes, although specialized genetic algorithms (Ferreira, 2001) can incorporate gene expression. In general, using genetic algorithms as a form of theoretical inquiry allows us to replicate the dynamics of a given biological system, which in turn allows us to better understand the complexity of the underlying system (Steventon & Arias, 2016).

The essential component of a genetic algorithm is the mutation and/or crossover operations. While mutation consists of bit flipping, recombination consists of exchanging entire blocks of bits either within or between programs. In both cases, an operator introduces variation in the function of a program. This variation is then selected upon based on fitness criterion, but is also subject to historical constrains of the problem space. This allows for the self-organization of patterns to emerge (Nizam & Shanmugham, 2013) without breaking the code or destroying the structure of the initial program. One way in which GAs avoid code fragility is to start with a population of programs that exhibit differential reproduction. In this way, a genetic algorithm selects from a variety of potential solutions. A related issue is the maintenance of structure and complexity in a naturally-selected program space is the existence of building blocks (Forrest & Mitchell, 2014). This is similar to modularity in biological evolution, in which parts of the program are protected from future evolutionary change (Wagner & Altenberg, 1996).

All of these issues figure prominently into the approximation of gastrulation as a developmental process. GAs are particularly good at finding heuristic solutions to problems residing in a large possibility space, so they are likewise suited to simulating instances of biological self-organization where an exact solution is not required. This is particularly true of simulating large problem spaces for where biological experimentation would yield no clear solution. The problem of emergence in biological self-organization can be viewed in terms of computational complexity (Grover, 2011). In gastrulation, we can see the usefulness of both properties, in addition to the usefulness of incorporating gene expression into a multi-scale model (Kaandorp et.al, 2012). Biological self-organization also relies on rules and constraints rather than explicit instructions.

Implementation

Gastrulation is a process by which the cells of an embryo form an invagination in an early spheroid shape in the course of tissue differentiation. A simple Morphozoic model of this is shown in Figure 15. This configuration was formed through a progression of cell "divisions" starting from a single central cell. Morphozoic generated metamorphs from a programmatically produced sequence. Two nested 3x3 neighborhoods were required; one neighborhood was found to be insufficient to reproduce the desired output. Once generated, the metamorphs executed the sequence correctly.

A model that purports to explain the workings of morphogenesis at any level of abstraction should be evolvable. In order to demonstrate the mutability and evolvability of Morphozoic, a genetic algorithm (GA) was used to evolve a population of "organisms" to gastrulate successfully. The fitness function was a count of the number of matching cell types summed over the sequence of cell type transitions.

Figure 15. Gastrulation configuration

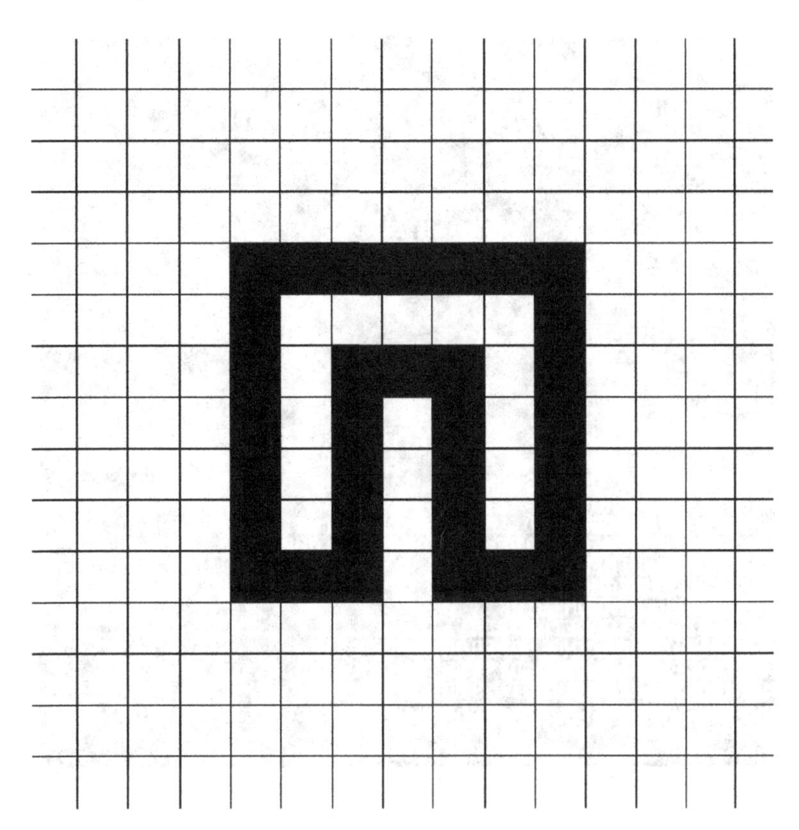

Each population member contained metamorphs that were initially constructed by randomly pairing morphogens generated from a programmed gastrulation sequence with actions from the sequence. The number of metamorphs in each member was determined by the number required to execute the sequence.

After a subset of the population was selected as fit members, the population was replenished with mutants and offspring of fit members. Offspring were created by mating randomly chosen parents members and performing a crossover operation on their metamorph sets. Crossover consisted of supplying the child with a set of metamorphs randomly selected from its parents. Mutants were created from individual fit members by discarding all metamorphs that were not executed and replacing them with random ones.

Additional GA parameters:

- Population size = 50
- Fit population size = 10
- Number of generations = 50
- Number of offspring from matings = 20

As Figure 16 indicates, the GA produced the required gastrulation morph sequence. The computation was done in approximately 12 hours on standard desktop computer.

Figure 16. Gastrulation evolution results

```
Fittest:
Member  Fitness
0       2475.0
_______________
_______________
_______________
_______________
____0000000____
____0_____0____
____0_000_0____
____0_0_0_0____
____0_0_0_0____
____0_0_0_0____
____000_000____
_______________
_______________
_______________
_______________
Parameters:
ORGANISM_DIMENSIONS = width:15 height:15
NUM_CELL_TYPES = 1
NEIGHBORHOOD_DIMENSION = 3
NUM_NEIGHBORHOODS = 2
NESTED_NEIGHBORHOOD_IMPORTANCE_WEIGHTS = null
MAX_MORPHOGEN_COMPARE_DISTANCE = 1000.0
METAMORPH_DIMENSION = 3
MAX_CELL_METAMORPHS = 1
METAMORPH_RANDOM_BIAS = 0.0
PROBABILISTIC_METAMORPH = true
INHIBIT_COMPETING_MORPHOGENS = false
MORPHOGENETIC_CELL_DISPERSION_MODULO = 1
METAMORPH_EXEC_TYPE = SEARCH_TREE
DEFAULT_ORGANISM = morphozoic.applications.Gastrulation
RANDOM_SEED = 48
Organism saved to morph_fittest.dat
```

Neuron Pathfinding

Also called axon guidance, neuron pathfinding is a process by which axons are guided by chemical signals to target neurons, a process essential for the formation of organized neural networks (Tessier-Lavigne & Goodman, 1996). While cellular automata have been used to simulate similar "branching" phenomena (Markus, Böhm, & Schmick, 1999), the generalization of reverse engineered rules is a novel application.

One of the benchmarks presented here is that of neuron pathfinding. Morphozoic is able to accomplish this gradually by exploiting the organizational feedback of cellular growth and spatial gradients embodied in the metamorphs. In biology, neuron pathfinding is accomplished a bit differently, instead using chemotropic mechanisms to guide axons to their target. Tropic cues are directional cues initiated by various stimuli. In the case of chemotropic cues, chemical signals and gradients (e.g. the axonal growth cone) guide regenerating tissues in the direction of its target (Huber et al., 2003). The chemotropic mode of action has not only been shown to exist in developmental morphogenesis (Tessier-Lavigne et al., 1988), but during tissue regeneration as well (Alto et al., 2009).

To simulate this, three types of cells are used: source neuron, target neuron, and axon. These can be seen in Figure 17. In this application, source neurons are allowed to have multiple axons. Metamorphs were generated from programmatic sequences of axon growth. For execution, however, the metamorphs were used to train an ANN (see Figure 8), which classified actions based on pattern-matching morphogens.

The training set was created by growing axons from source to target neurons. The source and target neurons were randomly placed along the left and right sides of the grid, respectively. Metamorphs were then generated from multiple random configurations.

Figure 17. Neuron pathfinding example

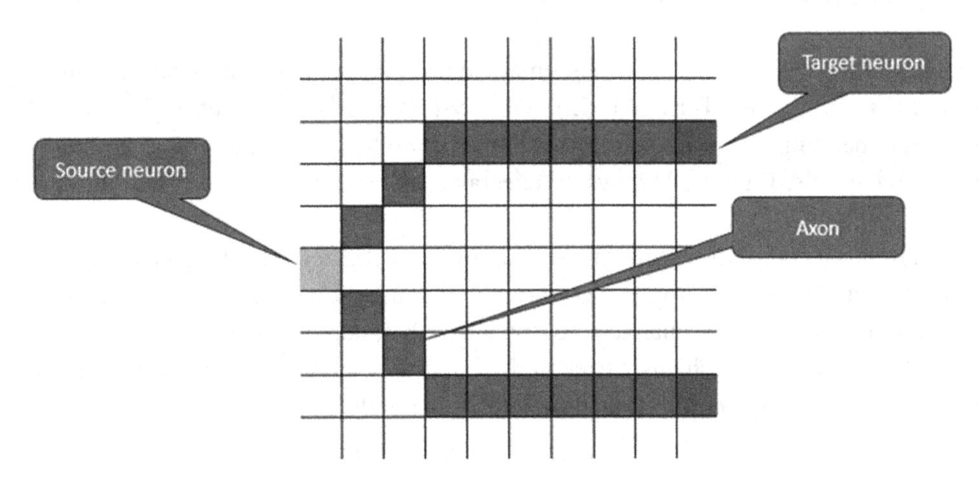

For testing, source and target neurons were randomly placed, morphogens generated from cell neighborhoods, and axons grown by pattern-matching the learned patterns. The challenge here is that axons must be grown from source to targets in different positions than those for which training was done, so training must generalize to handle novel neuron positions. The employment of an ANN as a classification mechanism was useful to accomplish this.

This procedure was done for a single source neuron, one and two target neurons, and with one, five, and ten training set exemplars. A successful trial is defined as axons connecting to the target neurons. Ten trials were run for each parameter configuration.

The results are shown in Figure 18. As might be expected, performance for the single target neuron exceeded that for two targets, the latter having a greater number of possible configurations. The number of exemplars also had a significant beneficial effect: one exemplar was insufficient to produce any successes for the two target neuron case.

Figure 18. Neuron pathfinding results

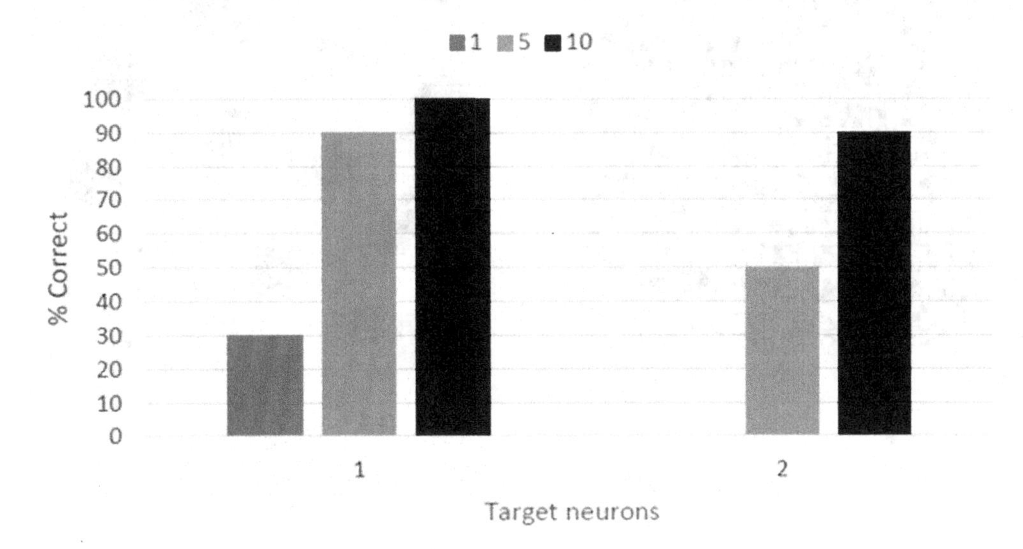

Turing's Reaction-Diffusion Morphogenesis

Alan Turing is credited with pioneering a mathematical formulation of morphogenesis generally known as a reaction-diffusion system (Turing, 1952). This system consists of a set of dynamically coupled substances that, depending on parameters, are capable of producing various complex patterns, such as stripes, spots, and spirals. Figure 19 shows a "cheetah coat" pattern generated from a random initial pattern of three cell types.

This application is a comparison with this well-known specialized morphogenesis algorithm. Metamorphs were generated from the Turing reaction-diffusion morphogenesis. These were used to train an ANN to learn appropriate cell type changes. To test, a freshly randomized pattern was generated and the metamorph pattern-matching mechanism executed. A typical result is shown in Figure 20. While not as smooth appearing as the original, the spotted pattern is distinctly visible.

CONCLUSION

Biology presents us with numerous cases of morphogenetic fields as a morphogenesis mechanism. The particular mechanisms and the depth of knowledge of these cases varies widely. Morphozoic is an attempt to implement an abstraction of the functional commonality of morphogenetic fields as a mechanism for self-organizing computation.

Morphozoic is a novel embodiment of a number of capabilities that are useful for morphogenesis: flexibility, compact and economical computability, evolvability, and generalizability, especially in association with the artificial neural network implementation. The cited applications were chosen to demonstrate these properties.

Figure 19. Reaction-diffusion morphogenesis of a "cheetah coat" pattern from random values

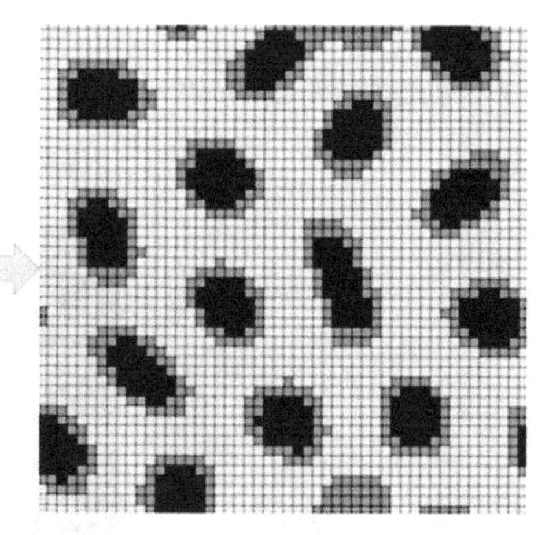

Figure 20. Simulation of reaction-diffusion morphogenesis

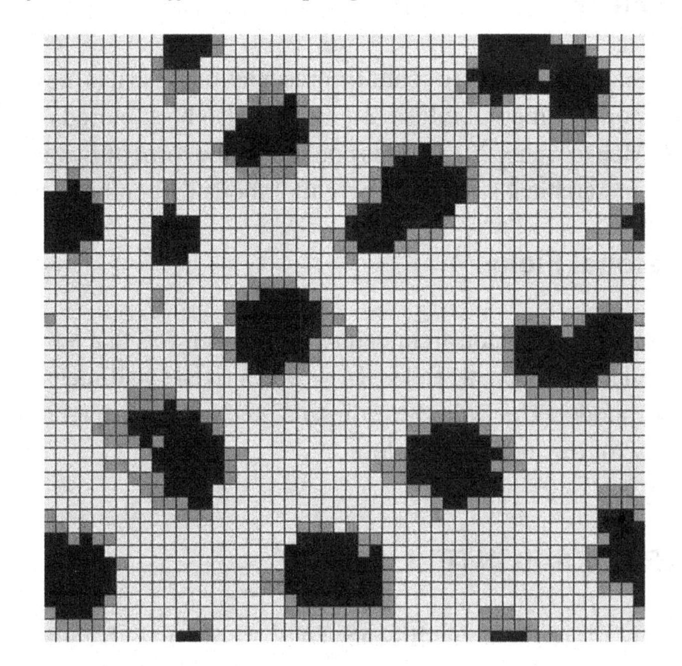

FUTURE DIRECTIONS

Neighborhood Sector Value

The assignment of a value to a sector could be any function of the types of its cellular components: average, mode, winner take all, etc. An alternative is to look at the change of cell type as an image processing operation, such as taking a Laplacian, Sobel and other edge enhancements, starbyte transformations (Sivaramakrishna & Gordon, 1997; Sivaramakrishna, 1998), contrast enhancement (Gordon & Rangayyan, 1984b, a), etc.

Field Signal Strength and Cell Type Variability

If cell types were continuous quantities instead of discrete, a straightforward proportional mapping of morphogen similarity to cell type value would be possible. On a related, more subtle possibility, morphogenetic fields are signaling constructs. Signals can vary in some ways, such as amplitude, while retaining invariant signatures, such as frequency spectra for electromagnetic waves. This suggests that variable action potentials proportional to field strength could be a fruitful topic for research. For example, might scaled, fractal-like structures be morphed?

Dynamic/Temporal Fields

Limiting morphogenetic fields exclusively to spatial representations restricts the model to a state machine. Dynamic fields that incorporate temporal information could also be explored (Portegys & Wiles, 2004; Martínez, Adamatzky & Alonso-Sanz, 2013). This could take the form of metamorphs that contain past neighborhood patterns or hierarchical streams of neighborhoods that embody context.

Fluid Three-Dimensional Fields

A major reason that the cellular automaton architecture was chosen was that it provides a straightforward mapping of local and global morphogenetic fields. However, a fixed two dimensional grid is not a crucial feature of the Morphozoic model. The fluid three dimensional medium that biological systems operate in points to an opportunity for the model to explore.

RELEVANCE TO ARTIFICIAL INTELLIGENCE

Morphozoic is also relevant for artificial intelligence (AI). AI research to a great degree focuses on the brain and behaviors that the brain generates. But the brain, an extremely complex structure resulting from millions of years of evolution, can be viewed as a solution to problems posed by the environment. There is a common and somewhat ironic tendency to describe AI inputs and outputs in human cognitive terms, i.e. post-processed brain output, such as symbolic variables (Hoffman, 2009).

An alternative approach, suggested by morphogenesis, is to view the environment as set of local/global, spatial/temporal signal fields, and that the processing of fields is what spurs brain development. Organisms are capable of performing amazing feats, such as navigation and nest-building, by the sensing of unique environmental signals, such as polarized light, magnetism, and chemical scent trails. Morphogenesis makes plain that signal fields can have powerful computing capabilities. Perhaps it is worth exploring artificial intelligence as a solution to environments composed of these fields. As Edmund Sinnott, author of *Plant Morphogenesis* (Sinnott, 1960), suggested, morphogenesis may be the key to understanding intelligence (Sinnott, 1961; Sinnott, 1962b; Sinnott, 1962a; Sinnott, 1966), starting with plant intelligence (Mancuso & Viola 2015).

In keeping with nature's penchant for extending rather than replacing, the sponge-like shape of the mammalian neocortex can be seen as symbolically apropos. For its purpose might be to soak up signals from far reaches of time and space and render them, as though yet near and present, to the old brain whose instinctual role has little changed over eons. The environmental gradients that clearly drive the behavior of simpler creatures then became internalized in the nervous systems of more neurologically complex ones.

The Morphozoic Java code is available here: https://github.com/pascualy/Morphozoic.

Morphozoic was developed in conjunction with these community projects that model and simulate the *C. elegans* nematode worm:

- DevoWorm (Alicea et al., 2014; Alicea, 2016) http://devoworm.weebly.com.
- OpenWorm (Szigeti et al., 2014)http://openworm.org/.

REFERENCES

Abbott, R. (2006). Emergence explained: Abstractions - Getting epiphenomena to do real work. *Complexity*, *12*(1), 13–26. doi:10.1002/cplx.20146

Adamatzky, A. (1997). Automatic programming of cellular automata: Identification approach. *Kybernetes*, *26*(2-3), 126–135. doi:10.1108/03684929710163074

Adami, C. (1998). *Introduction to Artificial Life*. Berlin: Springer. doi:10.1007/978-1-4612-1650-6

Alberts, B., Bray, D., Lewis, J., Raff, M., Roberts, K., & Watson, J. D. (2002). Universal mechanisms of animal development. In *Molecular Biology of the Cell*. New York: Garland Publishing. Retrieved from http://www.ncbi.nlm.nih.gov/books/NBK26825/

Alicea, B. (2016). *DevoWorm Project*. Retrieved from http://devoworm.weebly.com

Alicea, B. & Gordon. (2014). Toy models for macroevolutionary patterns and trends. *BioSystems, 122*, 25-37.

Alicea, B., McGrew, S., Gordon, R., Larson, S., Warrington, T., & Watts, M. (2014). *DevoWorm: differentiation waves and computation in* C. elegans *embryogenesis*. Retrieved from http://www.biorxiv.org/content/early/2014/10/03/009993

Alto, L. T., Havton, L. A., Conner, J. M., Hollis, E. R. II, Blesch, A., & Tuszynski, M. H. (2009). Chemotropic guidance facilitates axonal regeneration and synapse formation after spinal cord injury. *Nature Neuroscience*, *12*(9), 1106–U1108. doi:10.1038/nn.2365 PMID:19648914

Ashby, W. R. (1968). Some consequences of Bremermann's limit for information-processing systems. In H. L. Oestreicher & D. L. Moore (Eds.), *Problems in Bionics* (pp. 69–76). New York: Gordon and Breach Science Publishers, Inc.

Bandini, S., & Mauri, G. (1999). Multilayered cellular automata. *Theoretical Computer Science*, *217*(1), 99–113. doi:10.1016/S0304-3975(98)00152-2

Baraniak, P. R., & McDevitt, T. C. (2010). Stem cell paracrine actions and tissue regeneration. *Regenerative Medicine*, *5*(1), 121–143. doi:10.2217/rme.09.74 PMID:20017699

Bateson, P., Barker, D., Clutton-Brock, T., Deb, D., DUdine, B., Foley, R. A., & Sultan, S. E. et al. (2004). Developmental plasticity and human health. *Nature*, *430*(6998), 419–421. doi:10.1038/nature02725 PMID:15269759

Batten, D. F. (2001). Complex landscapes of spatial interaction. *The Annals of Regional Science*, *35*(1), 81–111. doi:10.1007/s001680000032

Batty, M. (2003). Geocomputation using cellular automata. In R. J. Abrahart, S. Openshaw, L. M. See, & C. R. C. Press (Eds.), Geocomputation (pp. 95–126). Academic Press.

Beloussov, L. V. (2015). Morphogenetic fields: History and relations to other concepts. In D. Fels, M. Cifra, & F. Scholkmann (Eds.), *Fields of the Cell* (pp. 271–282). Kerala, India: Research Signpost.

Beloussov, L. V., Opitz, J. M., & Gilbert, S. F. (1997). Life of Alexander G. Gurwitsch and his relevant contribution to the theory of morphogenetic fields. *The International Journal of Developmental Biology*, *41*(6), 771–779. PMID:9449452

Bijelic, A., Molitor, C., Mauracher, S. G., Al-Oweini, R., Kortz, U., & Rompel, A. (2015). Hen egg-white lysozyme crystallisation: Protein stacking and structure stability enhanced by a tellurium(VI)-centred polyoxotungstate. *ChemBioChem*, *16*(2), 233–241. doi:10.1002/cbic.201402597 PMID:25521080

Bornholdt, S. (1998). Genetic algorithm dynamics on a rugged landscape. *Physical Review E: Statistical Physics, Plasmas, Fluids, and Related Interdisciplinary Topics*, *57*(4), 3853–3860. doi:10.1103/PhysRevE.57.3853

Bourgine, P., & Lesne, A. (Eds.). (2010). *Morphogenesis: Origins of Patterns and Shapes*. Berlin: Springer Science & Business Media.

Bower, J.M. (2005). Looking for Newton: realistic modeling in modern biology. *Brains, Minds and Media, 1*(2).

Bremermann, H. J. (1967). Quantum noise and information. In *Proceedings of the Berkeley Symposium on Mathematical Statistics and Probability*. University of California Press.

Bruno, S., Collino, F., Tetta, C., & Camussi, G. (2013). Dissecting paracrine effectors for mesenchymal stem cells. In Mesenchymal Stem Cells: Basics and Clinical Application I. Academic Press.

Bubak, M., & Czerwiński, P. (1999). Traffic simulation using cellular automata and continuous models. *Computer Physics Communications*, *121*, 395–398. doi:10.1016/S0010-4655(99)00363-X

Burguillo, J. C. (2013). Playing with complexity: From cellular evolutionary algorithms with coalitions to self-organizing maps. *Computers & Mathematics with Applications (Oxford, England)*, *66*(2), 201–212. doi:10.1016/j.camwa.2013.01.020

Cazala, J. (2015). *Perceptron*. Retrieved from https://github.com/cazala/synaptic/wiki/Architect

Chopard, B. (1990). A cellular automata model of large-scale moving objects. *Journal of Physics. A, Mathematical and General*, *23*(10), 1671–1687. doi:10.1088/0305-4470/23/10/010

Cook, M. (2004). Universality in elementary cellular automata. *Complex Systems*, *15*(1), 1–40.

Dascalu, M., Stefan, G., Zafiu, A., & Plavitu, A. (2011). Applications of multilevel cellular automata in epidemiology. In *Proceedings of the 13th WSEAS international conference on Automatic control, modelling & simulation*. World Scientific and Engineering Academy and Society (WSEAS).

de Koster, C. G., & Lindenmayer, A. (1987). Discrete and continuous models for heterocyst differentiation in growing filaments of blue-green bacteria. *Acta Biotheoretica*, *36*(4), 249–273. doi:10.1007/BF02329786

Deutsch, A., Maini, P., & Dormann, S. (Eds.). (2007). *Cellular Automaton Modeling of Biological Pattern Formation: Characterization, Applications, and Analysis*. Birkhäuser Boston.

Dittmer, J., & Leyh, B. (2014). Paracrine effects of stem cells in wound healing and cancer progression[Review]. *International Journal of Oncology*, *44*(6), 1789–1798. PMID:24728412

Dobrescu, R. & V.I. Purcarea (2011). Emergence, self-organization and morphogenesis in biological structures. *Journal of Medicine and Life, 4*(1), 82-90.

Drachman, D. A. (2005). Do we have brain to spare? *Neurology*, *64*(12), 2004–2005. doi:10.1212/01.WNL.0000166914.38327.BB PMID:15985565

Drasdo, D., & Forgacs, G. (2000). Modeling the Interplay of Generic and Genetic Mechanisms in Cleavage, Blastulation, and Gastrulation. *Developmental Dynamics*, *219*(2), 182–191. doi:10.1002/1097-0177(200010)219:2<182::AID-DVDY1040>3.3.CO;2-1 PMID:11002338

Dunn, A. (2010). Hierarchical cellular automata methods. In A. G. Hoekstra, J. Kroc, & P. M. A. Sloot (Eds.), *Simulating Complex Systems by Cellular Automata* (pp. 59–80). doi:10.1007/978-3-642-12203-3_4

Dunn, A. G., & Majer, J. D. (2007). Simulating weed propagation via hierarchical, patch-based cellular automata. *Lecture Notes in Computer Science*, *4487*, 762–769. doi:10.1007/978-3-540-72584-8_101

Elmenreich, W., & Fehervari, I. (2011). Evolving self-organizing cellular automata based on neural network genotypes. *Lecture Notes in Computer Science*, *6557*, 16–25. doi:10.1007/978-3-642-19167-1_2

Ferrando, N., Gosálvez, M. A., Cerdá, J., Gadea, R., & Sato, K. (2011). Octree-based, GPU implementation of a continuous cellular automaton for the simulation of complex, evolving surfaces. *Computer Physics Communications*, *182*(3), 628–640. doi:10.1016/j.cpc.2010.11.004

Ferrando, N., Gosálvez, M. A., & Colóm, R. J. (2012). Evolutionary continuous cellular automaton for the simulation of wet etching of quartz. *Journal of Micromechanics and Microengineering*, *22*(2), 025021. doi:10.1088/0960-1317/22/2/025021

Ferreira, C. (2001). Gene expression programming: A new adaptive algorithm for solving problems. *Complex Systems*, *13*(2), 87–129.

Forrest, S., & Mitchell, M. (2014). Relative Building-Block Fitness and the Building Block Hypothesis. In *Foundations of Genetic Algorithms II* (pp. 109–126). San Mateo, CA: Morgan Kaufmann.

Fredkin, E. (1992). Finite nature. *Progress in Atomic Physics Neutrinos and Gravitation*, *72*, 345–354.

Fukui, M., & Ishibashi, Y. (1996). Traffic flow in 1D cellular automaton model including cars moving with high speed. *Journal of the Physical Society of Japan*, *65*(6), 1868–1870. doi:10.1143/JPSJ.65.1868

Gardner, M. (1970). Mathematical Games: The fantastic combinations of John Conway's new solitaire game "life". *Scientific American, 223*(4), 120-123.

Ghobara, Smith, Schoefs, Vinayak, Gebeshuber, & Gordon. (2016). On light and diatoms: A photonics and photobiology review. *Advances in Optics and Photonics*. Unpublished.

Gluckman, P. D., Hanson, M. A., & Low, F. M. (2011). The role of developmental plasticity and epigenetics in human health. *Birth Defects Research Part C-Embryo Today-Reviews*, *93*(1), 12–18. doi:10.1002/bdrc.20198 PMID:21425438

Gordon, N. K., & Gordon, R. (2016a). *Embryogenesis Explained*. Singapore: World Scientific Publishing. doi:10.1142/8152

Gordon, N. K., & Gordon, R. (2016b). The organelle of differentiation in embryos: the cell state splitter [invited review]. Theoretical Biology and Medical Modelling, 13.

Gordon, R. (1966). On stochastic growth and form. *Proceedings of the National Academy of Sciences of the United States of America*, *56*(5), 1497–1504. doi:10.1073/pnas.56.5.1497 PMID:5230309

Gordon, R. (1970). On Monte Carlo algebra. *Journal of Applied Probability*, *7*(02), 373–387. doi:10.1017/S002190020003494X

Gordon, R. (1980). Monte Carlo methods for cooperative Ising models. In Cooperative Phenomena in Biology. New York: Pergamon Press. doi:10.1016/B978-0-08-023186-0.50010-X

Gordon, R. (1999). *The Hierarchical Genome and Differentiation Waves: Novel Unification of Development, Genetics and Evolution*. London: World Scientific & Imperial College Press. doi:10.1142/2755

Gordon, R. (2015). Walking the tightrope: the dilemmas of hierarchical instabilities in Turing's morphogenesis. In S. B. Cooper & A. Hodges (Eds.), *The Once and Future Turing: Computing the World* (pp. 150–164). Cambridge, UK: Cambridge University Press.

Gordon, R., Goel, N. S., Steinberg, M. S., & Wiseman, L. L. (1972). A rheological mechanism sufficient to explain the kinetics of cell sorting. *Journal of Theoretical Biology*, *37*(1), 43–73. doi:10.1016/0022-5193(72)90114-2 PMID:4652421

Gordon, R., Goel, N. S., Steinberg, M. S., & Wiseman, L. L. (1975). A rheological mechanism sufficient to explain the kinetics of cell sorting. In G. D. Mostow (Ed.), *Mathematical Models for Cell Rearrangement* (pp. 196–230). New Haven, CT: Yale University Press.

Gordon, R., & Melvin, C. A. (2003). Reverse engineering the embryo: A graduate course in developmental biology for engineering students at the University of Manitoba, Canada. *The International Journal of Developmental Biology*, *47*(2/3), 183–187. PMID:12705668

Gordon, R., & Rangayyan, R. M. (1984a). Feature enhancement of film mammograms using fixed and adaptive neighborhoods. *Applied Optics*, *23*(4), 560–564. doi:10.1364/AO.23.000560 PMID:18204600

Gordon, R., & Rangayyan, R. M. (1984b). Correction: Feature enhancement of film mammograms using fixed and adaptive neighborhoods. *Applied Optics*, *23*(13), 2055. doi:10.1364/AO.23.002055 PMID:20424727

Gordon, R., & Stone, R. (2016in press). Cybernetic embryo. In R. Gordon & J. Seckbach (Eds.), *Biocommunication*. London: World Scientific Publishing. doi:10.1142/q0013

Gosálvez, M. A., Xing, Y., Sato, K., & Nieminen, R. M. (2009). Discrete and continuous cellular automata for the simulation of propagating surfaces. *Sensors and Actuators. A, Physical*, *155*(1), 98–112. doi:10.1016/j.sna.2009.08.012

Goyal, A., Szarzynska, B., & Fankhauser, C. (2013). Phototropism: At the crossroads of light-signaling pathways. *Trends in Plant Science*, *18*(7), 393–401. doi:10.1016/j.tplants.2013.03.002 PMID:23562459

Gravan, C. P., & Lohoz-Beltra, R. (2004). Evolving morphogenetic fieds in the zebra skin pattern based on Turing's morphogen hypothesis. *International Journal of Applied Mathematics and Computer Science, 14*(3), 351–361.

Grover, M. (2011). Parallels between Gluconeogenesis and Synchronous Machines. *International Journal on Computer Science and Engineering, 3*(1), 185–191.

Guan, Q., & Clarke, K. C. (2010). A general-purpose parallel raster processing programming library test application using a geographic cellular automata model. *International Journal of Geographical Information Science, 24*(5), 695–722. doi:10.1080/13658810902984228

Halbach, M., & Hoffmann, R. (2005). Optimal behavior of a moving creature in the cellular automata model. *Lecture Notes in Computer Science, 3606*, 129–140. doi:10.1007/11535294_11

Haykin, S.O. (2011). *Neural Networks and Learning Machines*. Pearson Education.

Hochberger, C., Hoffmann, R., & Waldschmidt, S. (1999). CDL++ for the description of moving objects in cellular automata. *Lecture Notes in Computer Science, 1662*, 428–435. doi:10.1007/3-540-48387-X_44

Hocking, A. M., & Gibran, N. S. (2010). Mesenchymal stem cells: Paracrine signaling and differentiation during cutaneous wound repair. *Experimental Cell Research, 316*(14), 2213–2219. doi:10.1016/j.yexcr.2010.05.009 PMID:20471978

Hodges, A. (2014). Alan Turing: The Enigma. New York: Simon & Schuster. doi:10.1515/9781400865123

Hoffman, D. D. (2009). The interface theory of perception: Natural selection drives true perception to swift extinction. In S. J. Dickinson, A. Leonardis, B. Schiele, & M. J. Tarr (Eds.), *Object Categorization: Computer and Human Vision Perspectives* (pp. 148–165). Cambridge, UK: Cambridge University Press. doi:10.1017/CBO9780511635465.009

Holland, J. H. (1992). Adaptation in Natural and Artificial Systems: An Introductory Analysis with Applications to Biology, Control, and Artificial Intelligence. Cambridge, MA: MIT Press.

Huber, A. B., Kolodkin, A. L., Ginty, D. D., & Cloutier, J. F. (2003). Signaling at the growth cone: Ligand-receptor complexes and the control of axon growth and guidance. *Annual Review of Neuroscience, 26*(1), 509–563. doi:10.1146/annurev.neuro.26.010302.081139 PMID:12677003

Huberman, B. A., & Glance, N. S. (1993). Evolutionary games and computer simulations. *Proceedings of the National Academy of Sciences of the United States of America, 90*(16), 7716–7718. doi:10.1073/pnas.90.16.7716 PMID:8356075

Ilachinski, A. (2001). *Cellular Automata: A Discrete Universe*. Retrieved from http://www.worldscientific.com/worldscibooks/10.1142/4702

Jacob, F. (1977). Evolution and tinkering. *Science, 196*(4295), 1161–1166. doi:10.1126/science.860134 PMID:860134

Johnson, S. (2010). *Where Good Ideas Come From*. Penguin Publishing Group.

Kaandorp, J. A., Botman, D., Tamulonis, C., & Dries, R. (2012). Multi-scale Modeling of Gene Regulation of Morphogenesis. In *How the World Computes: Turing Centenary conference and 8th conference on computability in Europe* (LNCS), (vol. 7318, pp. 355-362). Berlin: Springer. doi:10.1007/978-3-642-30870-3_36

Kaji, M., & Inohara, T. (2014). Numerical analysis focused on each agent's moving on refuge under congestion circumstances by using cellular automata. In *2014 IEEE International Conference on Systems, Man and Cybernetics*. doi:10.1109/SMC.2014.6974374

Katkovnik, V., Egiazarian, K., & Astola, J. (2006). *Local Approximation Techniques in Signal and Image Processing*. SPIE Press. doi:10.1117/3.660178

Kier, L. B., Bonchev, D., & Buck, G. A. (2005). Modeling biochemical networks: A cellular-automata approach. *Chemistry & Biodiversity*, *2*(2), 233–243. doi:10.1002/cbdv.200590006 PMID:17191976

Kiester, A. R., & Sahr, K. (2008). Planar and spherical hierarchical, multi-resolution cellular automata. *Computers, Environment and Urban Systems*, *32*(3), 204–213. doi:10.1016/j.compenvurbsys.2008.03.001

Kitakawa, A. (2004). On a segregation intensity parameter in continuous-velocity lattice gas cellular automata for immiscible binary fluid. *Chemical Engineering Science*, *59*(14), 3007–3012. doi:10.1016/j.ces.2004.04.032

Kitakawa, A. (2005). Simulation of break-up behavior of immiscible droplet under shear field by means of continuous-velocity lattice gas cellular automata. *Chemical Engineering Science*, *60*(20), 5612–5619. doi:10.1016/j.ces.2005.05.011

Koestler, A. (1967). *The Ghost in the Machine*. New York: Macmillan.

Kohl, P., & Noble, D. (2009). Systems biology and the virtual physiological human. *Molecular Systems Biology*, 5. PMID:19638973

Leavitt, D. (2006). *The Man Who Knew Too Much: Alan Turing and the Invention of the Computer*. New York: W. W. Norton.

Levin, M. (2011). The wisdom of the body: Future techniques and approaches to morphogenetic fields in regenerative medicine, developmental biology and cancer. *Regenerative Medicine*, *6*(6), 667–673. doi:10.2217/rme.11.69 PMID:22050517

Levin, M. (2012). Morphogenetic fields in embryogenesis, regeneration, and cancer: non-local control of complex patterning. *Biosystems, 109*(3), 243-261.

Li, Y., Gosálvez, M. A., Pal, P., Sato, K., & Xing, Y. (2015). Particle swarm optimization-based continuous cellular automaton for the simulation of deep reactive ion etching. *Journal of Micromechanics and Microengineering*, *25*(5), 055023. doi:10.1088/0960-1317/25/5/055023

Liu, Y., Cheng, H. D., Huang, J. H., Zhang, Y. T., & Tang, X. L. (2012). An effective approach of lesion segmentation within the breast ultrasound image based on the cellular automata principle. *Journal of Digital Imaging*, *25*(5), 580–590. doi:10.1007/s10278-011-9450-6 PMID:22237810

Lobo, D., & Levin, M. (2015). Inferring regulatory networks from experimental morphological phenotypes: A computational method reverse-engineers planarian regeneration. *PLoS Computational Biology, 11*(6), e1004295. doi:10.1371/journal.pcbi.1004295 PMID:26042810

Low, F. M., Gluckman, P. D., & Hanson, M. A. (2012). Developmental plasticity, epigenetics and human health. *Evolutionary Biology, 39*(4), 650–665. doi:10.1007/s11692-011-9157-0

Mancuso, S., & Viola, A. (2015). *Brilliant Green: The Surprising History and Science of Plant Intelligence*. Island Press.

Markus, M., Böhm, D., & Schmick, M. (1999). Simulation of vessel morphogenesis using cellular automata. *Mathematical Biosciences, 156*(1-2), 191–206. doi:10.1016/S0025-5564(98)10066-4 PMID:10204393

Martínez, G. J., Adamatzky, A., & Alonso-Sanz, R. (2013). Designing complex dynamics in cellular automata with memory. *International Journal of Bifurcation and Chaos in Applied Sciences and Engineering, 23*(10), 1330035. doi:10.1142/S0218127413300358

Medler, D. A. (1998). A brief history of connectionism. *Neural Computing Surveys, 1*, 18–72.

Moczek, A. P., Sultan, S., Foster, S., Ledon-Rettig, C., Dworkin, I., Nijhout, H. F., . . . Pfennig, D. W. (2011). The role of developmental plasticity in evolutionary innovation. *Proc. R. Soc. B-Biol. Sci., 278*(1719), 2705-2713. doi:10.1098/rspb.2011.0971

Mofrad, M. H., Sadeghi, S., Rezvanian, A., & Meybodi, M. R. (2015). Cellular edge detection: Combining cellular automata and cellular learning automata. *AEÜ. International Journal of Electronics and Communications, 69*(9), 1282–1290. doi:10.1016/j.aeue.2015.05.010

Monetti, R. A., & Albano, E. V. (1997). On the emergence of large-scale complex behavior in the dynamics of a society of living individuals: The Stochastic Game of Life. *Journal of Theoretical Biology, 187*(2), 183–194. doi:10.1006/jtbi.1997.0424 PMID:9405136

Morigi, M., Imberti, B., Zoja, C., Corna, D., Tomasoni, S., Abbate, M., & Remuzzi, G. et al. (2004). Mesenchymal stem cells are renotropic, helping to repair the kidney and improve function in acute renal failure. *Journal of the American Society of Nephrology, 15*(7), 1794–1804. doi:10.1097/01.ASN.0000128974.07460.34 PMID:15213267

Morozova, N., & Shubin, M. (2012). *The Geometry of Morphogenesis and the Morphogenetic Field Concept*. Retrieved from http://arxiv.org/abs/1205.1158v1

Moussa, N. (2005). A 2-dimensional cellular automaton for agents moving from origins to destinations. *International Journal of Modern Physics C, 16*(12), 1849–1860. doi:10.1142/S0129183105008370

Nizam, A. and Shanmugham, B. (2013). Self-organizing Genetic Algorithm: A survey. *International Journal of Computer Applications, 65*(18), 0975-8887.

Noble, D. (2002). The rise of computational biology. *Nature Reviews. Molecular Cell Biology, 3*(6), 459–463. doi:10.1038/nrm810 PMID:12042768

Olami, Z., Feder, H. J. S., & Christensen, K. (1992). Self-organized criticality in a continuous, non-conservative cellular automaton modeling earthquakes. *Physical Review Letters*, *68*(8), 1244–1247. doi:10.1103/PhysRevLett.68.1244 PMID:10046116

Opitz, J. M., & Neri, G. (2013). Historical perspective on developmental concepts and terminology. *American Journal of Medical Genetics. Part A*, *161*(11), 2711–2725. doi:10.1002/ajmg.a.36244 PMID:24123982

Portegys, T., & Wiles, J. (2004). A robust game of life. In *The International Conference on Complex Systems (ICCS2004)*. Retrieved from http://www.necsi.edu/events/iccs/openconf/author/abstractbook.php

Portegys, T. E. (2002). An abstraction of intercellular communication. In *Alife VIII Proceedings*.

Rendell, P. (2002). Turing universality of the game of life. In A. Adamatzky (Ed.), *Collision-Based Computing* (pp. 513–539). London: Springer London. doi:10.1007/978-1-4471-0129-1_18

Resnick, M. (1994). *Turtles, Termites, and Traffic Jams: Explorations in Massively Parallel Microworlds*. Cambridge, MA: MIT Press.

Ruxton, G. D. (1996). Effects of the spatial and temporal ordering of events on the behaviour of a simple cellular automaton. *Ecological Modelling*, *84*(1-3), 311–314. doi:10.1016/0304-3800(94)00145-6

Ruxton, G. D., & Saravia, L. A. (1998). The need for biological realism in the updating of cellular automata models. *Ecological Modelling*, *107*(2-3), 105–112. doi:10.1016/S0304-3800(97)00179-8

Salgado, A. J., & Gimble, J. M. (2013). Secretome of mesenchymal stem/stromal cells in regenerative medicine Foreword. *Biochimie*, *95*(12), 2195. doi:10.1016/j.biochi.2013.10.013 PMID:24210144

Satoh, S. (2012). Computational identity between digital image inpainting and filling-in process at the blind spot. *Neural Computing & Applications*, *21*(4), 613–621. doi:10.1007/s00521-011-0646-y

Sayad, S. (2016). *Artificial Neural Network*. Retrieved from http://www.saedsayad.com/artificial_neural_network.htm

Sinnott, E. W. (1960). *Plant Morphogenesis*. New York: McGraw-Hill Book Co. doi:10.5962/bhl.title.4649

Sinnott, E.W. (1961). *Cell and psych. The biology of purpose*. Academic Press.

Sinnott, E. W. (1962a). *The Biology of The Spirit*. New York: The Viking Press.

Sinnott, E. W. (1962b). *Matter, Mind and Man: The Biology of Human Nature*. New York: Atheneum.

Sinnott, E. W. (1966). *The Bridge of Life, From Matter to Spirit*. New York: Simon and Schuster.

Sivaramakrishna, R. (1998). Breast image registration using a textural transformation. *Medical Physics*, *25*(11), 2249. doi:10.1118/1.598426

Sivaramakrishna, R., & Gordon, R. (1997). Mammographic image registration using the Starbyte transformation. In P. G. McLaren & W. Kinsner (Eds.), *WESCSANEX'97* (pp. 144–149). Winnipeg, Canada: IEEE. doi:10.1109/WESCAN.1997.627128

Solé, R. V., Valverde, S., Rosas Casals, M., Kauffman, S. A., Farmer, D., & Eldredge, N. (2013). The evolutionary ecology of technological innovations. *Complexity*, *18*(4), 15–27. doi:10.1002/cplx.21436

Steventon, B. & Arias, A.M. (2016). *Evo-engineering and the cellular and molecular origins of the vertebrate spinal cord*. doi:10.1101/068882

Sun, X., Rosin, P. L., & Martin, R. R. (2011). Fast rule identification and neighborhood selection for cellular automata. *IEEE Transactions on Systems, Man, and Cybernetics. Part B, Cybernetics*, *41*(3), 749–760. doi:10.1109/TSMCB.2010.2091271 PMID:21134817

Szigeti, B., Gleeson, P., Vella, M., Khayrulin, S., Palyanov, A., Hokanson, J., & Larson, S. et al. (2014). OpenWorm: An open-science approach to modelling *Caenorhabditis elegans*. *Frontiers in Computational Neuroscience*, *8*(137). PMID:25404913

Tanaka, S. (2015). Simulation frameworks for morphogenetic problems. *Computation*, *3*(2), 197–221. doi:10.3390/computation3020197

Tessier-Lavigne, M., & Goodman, C. S. (1996). The molecular biology of axon guidance. *Science*, *274*(5290), 1123–1133. doi:10.1126/science.274.5290.1123 PMID:8895455

Tessier-Lavigne, M., Placzek, M., Lumsden, A. G. S., Dodd, J., & Jessell, T. M. (1988). Chemotropic guidance of developing axons in the mammalian central nervous system. *Nature*, *336*(6201), 775–778. doi:10.1038/336775a0 PMID:3205306

Tompkins, N., Li, N., Girabawe, C., Heymann, M., Ermentrout, G. B., Epstein, I. R., & Fraden, S. (2014). Testing Turings theory of morphogenesis in chemical cells. *Proceedings of the National Academy of Sciences of the United States of America*, *111*(12), 4397–4402. doi:10.1073/pnas.1322005111 PMID:24616508

Tria, F., Loreto, V., Servedio, V. D. P., & Strogatz, S. H. (2014). The dynamics of correlated novelties. *Scientific Reports*, 4. PMID:25080941

Turing, A. M. (1952). The chemical basis of morphogenesis. *Philosophical Transactions of the Royal Society of London. Series B, Biological Sciences*, *237*(641), 37–72. doi:10.1098/rstb.1952.0012

Tyldum, M. (2014). *The Imitation Game* [movie]. Retrieved from http://theimitationgamemovie.com

Tyler, S. E. B. (2014). The work surfaces of morphogenesis: The role of the morphogenetic field. *Biological Theory*, *9*(2), 194–208. doi:10.1007/s13752-014-0177-8

Ulam, S. (1962). On some mathematical problems connected with patterns of growth of figures.*Symposium in Applied Mathematics,14*, 215-224. doi:10.1090/psapm/014/9947

van Nimwegen, E., Crutchfield, J. P., & Huynen, M. (1999). Neutral evolution of mutational robustness. *Proceedings of the National Academy of Sciences of the United States of America*, *96*(17), 9716–9720. doi:10.1073/pnas.96.17.9716 PMID:10449760

Vecchi, D., & Hernández, I. (2014). The epistemological resilience of the concept of morphogenetic field. In A. Minelli & T. Pradeu (Eds.), *Towards a Theory of Development*. doi:10.1093/acprof:oso/9780199671427.003.0005

von Neumann, J., & Burks, A. W. (1966). *Theory of Self-Reproducing Automata*. Urbana, IL: University of Illinois Press.

Waddington, C. H., & Cowe, R. J. (1969). Computer simulation of a mulluscan pigmentation pattern. *Journal of Theoretical Biology*, *25*(2), 219–225. doi:10.1016/S0022-5193(69)80060-3 PMID:5383503

Wagner, A., & Rosen, W. (2014). Spaces of the possible: Universal Darwinism and the wall between technological and biological innovation. *Journal of the Royal Society, Interface*, *11*(97), 20131190. doi:10.1098/rsif.2013.1190 PMID:24850903

Wagner, G. P., & Altenberg, L. (1996). Complex Adaptations and the Evolution of Evolvability. *Evolution; International Journal of Organic Evolution*, *50*(3), 967–976. doi:10.2307/2410639

Walker, D. C., & Southgate, J. (2009). The virtual cell-a candidate co-ordinator for middle-out modelling of biological systems. *Briefings in Bioinformatics*, *10*(4), 450–461. doi:10.1093/bib/bbp010 PMID:19293250

Weimar, J. R. (2001). Coupling microscopic and macroscopic cellular automata. *Parallel Computing*, *27*(5), 601–611. doi:10.1016/S0167-8191(00)00080-6

Weisstein, E. W. (2016a). *Cellular Automaton*. Retrieved from http://mathworld.wolfram.com/CellularAutomaton.html

Weisstein, E. W. (2016b). *Moore Neighborhood*. Retrieved from http://mathworld.wolfram.com/MooreNeighborhood.html

West-Eberhard, M. J. (2002). *Developmental Plasticity and Evolution*. Oxford, UK: Oxford University Press.

Wikipedia. (2016a). *Lenna*. Retrieved from https://en.wikipedia.org/wiki/Lenna

Wikipedia. (2016b). *List of animals by number of neurons*. Retrieved from https://en.wikipedia.org/wiki/List_of_animals_by_number_of_neurons

Wolfram, S. (1984). Cellular automata as models of complexity. *Nature*, *311*(5985), 419–424. doi:10.1038/311419a0

Wolfram, S. (2001). *A New Kind of Science*. Champaign, IL: Wolfram Media.

Wyczalkowski, M. A., Chen, Z., Filas, B. A., Varner, V. D., & Taber, L. A. (2012). Computational models for mechanics of morphogenesis. *Birth Defects Research Part C-Embryo Today-Reviews*, *96*(2), 132–152. doi:10.1002/bdrc.21013 PMID:22692887

Zhao, Y., Wei, H. L., & Billings, S. A. (2012). A new adaptive fast cellular automaton neighborhood detection and rule identification algorithm. *IEEE Transactions on Systems, Man, and Cybernetics. Part B, Cybernetics*, *42*(4), 1283–1287. doi:10.1109/TSMCB.2012.2185790 PMID:22695356

Chapter 4
A Scalable Multiagent Architecture for Monitoring Biodiversity Scenarios

Vladimir Rocha
University of São Paulo, Brazil

Anarosa Alves Franco Brandão
University of São Paulo, Brazil

ABSTRACT

Recently, there has been an explosive growth in the use of wireless devices, mainly due to the decrease in cost, size, and energy consumption. Researches into Internet of Things have focused on how to continuously monitor these devices in different scenarios, such as environmental and biodiversity tracking, considering both scalability and efficiency while searching and updating the devices information. For this, a combination of an efficient distributed structure and data aggregation method is used, allowing a device to manage a group of devices, minimizing the number of transmissions and saving energy. However, scalability is still a key challenge when the group is composed of a large number of devices. In this chapter, the authors propose a scalable architecture that distributes the data aggregation responsibility to the devices of the boundary of the group, and creates agents to manage groups and the interaction among them, such as merging and splitting. Experimental results showed the viability of adopting this architecture if compared with the most widely used approaches.

INTRODUCTION

In the last years, an explosive growth has been observed in researches and real-world applications to ecological monitoring and wildlife tracking, spread over several scenarios, such as biodiversity, environmental, agriculture, and social biology, among others (Huang et al., 2010; Hart & Martinez, 2015; Dlodlo & Kalezhi, 2015). This growth is sustained mainly due to the decrease in cost, size, and energy consumption of wireless sensor devices used for monitoring those scenarios (Galluccio et al., 2011).

DOI: 10.4018/978-1-5225-1756-6.ch004

For example, in the biodiversity scenarios, sensor devices allows to collect and to monitor the behavior exhibited by a group of birds flying in a certain area, known as flocking, in order to provide a sustainable ecological development. For technologically supporting the high demand of information collected, shared and transmitted by the sensors devices, a relative new paradigm has emerged, called Internet of Things, whose vision is to integrate these physical devices (or *things*) into the virtual environment (Kaukalias & Chatzimisios, 2016). A *thing* is a real world device which provides services to sense, to communicate, and to cooperate with other devices and with its environment. In this paradigm, researches have focused on how to continuously monitor a large number of devices, considering both scalability and efficiency while searching and updating information of a group of devices, because the large number of them imply an explosion in the traffic communication and energy consumption (Borgia, 2014).

Current alternatives use a combination of a widely recognized method, called data aggregation (Ren et al., 2013), and a widely adopted distributed structure, called Distributed Hash Table (DHT) (Stoica et al., 2001), in order to monitor these devices. The data aggregation method chooses one device for collecting and for managing the information of all devices that form a group, minimizing the number of transmission and saving energy. The distributed structure is used for efficiently retrieving the chosen device when another device needs to join a group. However, there are still some open issues faced by the method: single point of failure (also known as SPOF) and scalability.

On the one hand, single point of failure issues arise when the chosen device interrupts its aggregation data responsibility due to a malfunction (e.g., the total consumption of energy, the transmission signal is lost in some areas, among others) which must be considered in the device behavior. As a consequence, the group information is inaccessible until a new device is chosen. On the other hand, scalability issues arise when the group increases its members, either because new devices join the group or when various group of devices need to be merged in one. This increment let the chosen device overloaded when collecting information until a point at which the device is unable to continue managing the group. As a consequence, the group information is inaccessible even if a new device is chosen.

In order to know how the system will behave when its members are overloaded, it is necessary to deploy and to test such system in the real world, that could be sometimes not feasible due to the characteristics of the scenario. For example, it is complex to test how the different path of the clouds or the wind could affects the climate on some region (Duraccio et al., 2007). For overcoming the problems mentioned above, simulations come as an interesting alternative. The idea is to create computational models of the scenario that simulate the real world behavior (Ilhan et al., 2015). Different approaches have proved to be useful for simulating the real world behavior in different scenarios, such as meteorological conditions (Duraccio et al., 2007), how the agricultural affects the environment (Prada et al., 2015) or how to predict bee colonies behavior under adverse temperature conditions (Kridi et al., 2014). Nevertheless, these works do not focus on analyze the scalability of the system.

In this chapter, a scalable architecture has been proposed for discovering and updating the localization and movement of group of devices. The architecture is composed of two layers, a device and a multiagent layer. The device layer consists in groups of connected devices geographically closer. To support scalability in the updating information process, each group assigns the responsibility for aggregating and for sending its information -such as localization and movement- to the devices in its boundary. The multiagent layer consists in agents responsible for managing the information of groups and for verifying if there is any interaction with other groups -such as merging and splitting. To support scalability in the updating information process, agents can create new agents (agentifying devices) in order to share the responsibility for managing the group.

The novelties of this work are:

1. Scalability support on the data aggregation method;
2. Interaction support among groups, managing splitting and merging;
3. Scalability support in the distributed structure;
4. Architectural multiagent details for supporting different monitoring and tracking scenarios.

In the first novelty, the scalability of the data aggregation method is extended for discovering and updating information of a group of devices. Instead of one device, the responsibility is distributed to the boundary of the group, avoiding the single point of failure and turning the method more scalable than previous alternatives. For this, devices in the boundary of each group are responsible for collecting and for aggregating the information of their neighbors and sending it to an agent.

In the second novelty, agents communicate with other agents to effectively exchange inter-groups interactions (merging and splitting). For this, agents maintain relationships with other agents whose boundary are closer enough to create these interactions. When there is an interaction, agents collaborate among them to generate the new boundary and to choose the agent responsible for managing this group.

In the third novelty, the distributed structure is extended, allowing agents to agentify (converting into agent) some of the group devices in order to distribute the responsibility for managing the group information. This extension increments the system scalability when the number of updates of a certain group overcomes the requests an agent can process.

In the fourth novelty, the internal behavior of each component of the architecture (i.e., device and agent) is provided as well as the steps to build and to deploy the architecture in other scenarios. With that, researchers will understand how both layers interact (from the system initialization to the system lifecycle) and how to apply the architecture in their experiments simulations.

The chapter is organized as follows: first, existing solutions are presented for data aggregation and Distributed Hash Tables in the IoT context, comparing them with the proposed solution. Then, the broad picture of the architecture is explained. Next, the device and the multiagent layers is detailed, analyzing how the architecture is built during the system lifecycle. Then, the results of the experiments are presented, evaluating the architecture viability through simulations. Finally, the authors conclude this work and propose some future directions.

BACKGROUND

Since this proposal is applied to monitor biological and environmental scenarios, extending the two techniques mentioned in the introduction (i.e., data aggregation and distributed structure) for dealing with the problem of scalability in the monitoring context. In this section the authors first review works that use agents and IoT for monitoring these scenarios. Next, works that use data aggregation are reviewed for supporting the group updating information in an scalable way. Finally, works that use Distributed Hash Tables are reviewed for efficiently discovering devices responsible for a group. It is also discussed how this work differs from previous ones.

Monitoring Biological and Environmental Scenarios

Several researches are motivated to monitor biological and environmental scenarios using different approaches besides IoT, such as centralized servers, agents, among others (Naumowicz et al., 2008; Ilhan et al., 2015; Dyo et al., 2012). Centralized servers, also called base station or data sink, are used to receive and to store the data transmitted by the devices, either for detecting birds' activity (Naumowicz et al., 2008), for knowing how primates form social group (Ilhan et al., 2015) or for analyzing the badgers' mobility patterns (Dyo et al., 2012). On the other hand, agents are deployed on the devices, with the responsibility for monitoring and learning the user behavior in order to react when similar situations arise, and to interact with other agents that are near to it, either for detecting air pollution (Zenonos, Stein, & Jennings, 2015) or for maintaining the agents connected in disaster zones (Yadgar, 2007).

In the IoT context, Huang et al (Huang et al., 2010) propose Echonet, a system that collects GPS sensor nodes information, carried by wildlife (such as buffalos, sheep, and eagles) living in an observation area, in order to provide wildlife rehabilitation, disease control, and sustainable ecological development. Liu et al. (Liu et al., 2011) propose GreenOrb, responsible for collecting sensor information (such as temperature, humidity, illumination, and content of carbon dioxide) in order to analyze the biological activity in the forest, providing these information to different research on biodiversity. In the agricultural scenario, Jiang and Zhang (Jiang & Zhang, 2013) and Wu et al. (Wu, Xu, Li, Zha, & Li, 2014) also collect sensor information (such as temperature, humidity and illumination information) allowing control irrigation, fertilization and pest control when adverse conditions arise. Following this direction, Haider et al. (Haider, ur Rehman, & Durrani, 2013) show that IoT could optimize and aid farm management (e.g., productivity and cost-effectiveness) while preserving different aspects, such as environmental independence, scalability, and adaptability. In the environmental scenario, there are several systems, based on sensor monitoring, which provide hazard warnings (such as forest fires (L. Yu et al., 2005), tsunami alerts (Meinig et al., 2005), among others) in order to help population remain safe.

Despite the works explained in this section allow centralized servers, agents, and IoT permeate several biological and environmental scenarios, most of them do not focus on scalability issues, since they work with a small number of sensors in its simulated and real-life experiments.

Data Aggregation for Updating Information

Recent researches on IoT have focused on how to exploit data aggregation for supporting the discovery and for updating information of a group of devices, in an efficient and scalable way. Galluccio et al. (Galluccio et al., 2011) propose that a mobile device (called Group Master or GM) is responsible for collecting, aggregating and sending the information of all the devices (called slaves) lying in its radio coverage area. This approach was evaluated in (D'Oro et al., 2015), whereby a large number of simulations were performed in logistic and human scenarios. The advantages of this solution are: slave position displacements do not need to be updated if they remain in the GM coverage area (only the GM position needs to be updated); to join a group, a new device only needs to find the GM and to connect with it (how to find the GM will be covered in the next section). The disadvantages of this solution are: a group area is limited by the GM coverage area, and the devices outside this area could not be considered part of the group (even if they are); a GM could be overloaded if many devices join its group affecting the system scalability; the GM is as a single point of failure, affecting the system availability. To deal with the disadvantages, Tang et al. (2014) propose to group devices in rectangular sectors, using a distributed and scalable index tree structure (see Zhang et al., 2011) for a review of the structure), called ECH-tree, to efficiently find them. As in the previous alternative, each sector has a device (Head Node) with the same responsibility as the GM. The improvements over the previous alternative are: the Head Node is rotated among all the devices in the sector, avoiding overloading it; devices on the boundary extend the group area by connecting to devices of adjacent sectors. The disadvantages of this solution are: device displacements could make the sector information outdated, needing to re-index the tree from time to time; the Head Node is a single point of failure until the rotation process chooses another device.

The present work improves scalability, maintaining the advantages of current alternatives. First, a group is composed of all the devices within the polygon formed by the devices in the boundary, not depending on the GM radio coverage area (i.e., as ECH-tree does, the GM range is extended too, but as explained in the architecture section, was used both rectangular sectors and the boundary of the group to do that). Second, the data aggregation responsibility is rotated on devices near the boundary, avoiding overloading one device and preventing the single point of failure issue.

Distributed Hash Table for GM Choosing and Retrieval

The use of Distributed Hash Table for finding the GM has emerged as a promising alternative to avoid the scalability issues given by centralized servers solutions (Manzanares-Lopez et al., 2011; Zhao et al., 2011), which receive all devices position information. In this structure, which acts as a regular hash table storing pair *(key, value)*, the geographic space to be monitored is divided into sectors and each one (associated with a unique identifier) is assigned to a GM which will be responsible for managing (i.e., collecting and aggregating) the information of all devices lying in that sector, just as the ECH-tree does. When a GM needs to be located, the DHT structure provides efficient algorithms to find it, given a sector identifier (Stoica et al., 2001).

Recent researches on DHT for IoT have focused on how to choose the GM responsible for a sector (which will be part of the DHT) avoiding constantly changing it, since, as its energy is consumed in each action -e.g., collecting information- at a certain point, it will have to be replaced. For DHT, avoiding changing a member is critical because this behavior results in a structure reorganization, spending energy resources while performing it. The Chord for Sensor Networks (CSN) (Ali & Uzmi, 2004) proposes randomly choosing the GM among all the devices lying in the managed sector. Changing a member will be performed just when the device is close to consuming its energy. When this occurs, the system chooses another one, which will be the new GM (causing the GM rotation). The Tiered Chord System (T-Chord) (Ali & Langendoen, 2007), on the other hand, analyzes the properties of all devices in a certain sector (such as, energy capacity, communication stability, among others) and chooses the most useful for being the GM. However, the system does not uses the GM rotation, which could cause the sector unavailability if the GM consumes all its energy (thus, loosing the devices information lying in that sector). The Two-Tiered Chord System (C2WSN) (J. Yu, Liu, & Song, 2007) also analyzes the device properties (as T-Chord does), but, when the device is close to consuming its energy, the system chooses another GM with similar properties among the devices lying in the sector, providing the GM rotation.

Although the works explained in this section allow choosing and rotating a GM device without decreasing the search efficiency for a certain sector, scalability problems still remains because one sector is managed by a single GM. In this context, if the group of devices lying in that sector increments its members, the GM could be overloaded and be unable to manage them. This work improves scalability creating new agents responsible for the same sector, hence distributing the load applied on the GM.

PROPOSED ARCHITECTURE

The proposed architecture creates relationships among agents and devices, allowing a scalable information updating of a group of devices while maintaining the Distributed Hash Table efficiency in discovering the agent responsible for a region. In this architecture, the following concepts are defined:

- The location of a device *d* is the geographic location point of *d* according to a coordinate system such as IR2 or IR3. The device location could be obtained using the GPS (Global Positioning System) sensor installed in the device.
- The influence zone of a device *d* is the circle centered on the location of *d* and radius r_d. The r_d radius could be seen as the transmission range a sensor has and that allows it to communicate with other devices (for simplicity, we assume for this architecture that all devices have the same transmission range). A device that lies within the influence zone of *d* is called a *neighbor* and is somehow related to *d*.
- The connection between devices d_i and d_j is the communication channel (i.e., information transmission) established between devices d_i and d_j, such that the location of d_j lies in the influence zone of d_i. Concretely, the connection could be seen as a segment starting at the location of d_i and ending at the location of d_j.
- The sector sec is a subset of a rectangular partition of the geographic space to be monitored, defined a priori.
- The sector of agent *a* is the geographic area managed by agent *a*. By "managed" is meant that agent *a* will be responsible for collecting and for aggregating data from the swarms lying in its sector. Adjacent sectors are the ones which share boundaries. It is important to note that the same sector could be managed by one or more agents.
- A swarm S is the closed region composed by all devices d_i whose location lies inside S.
- A swarm *S* is managed by an agent *a*. Note that agent *a* manages both a sector and the swarm located in that sector.
- The hull of a swarm *S* is the set *H* of device connections that bound the swarm. In this context, each connection between every two consecutive devices which belongs to *H* is a segment of the hull.
- The boundary of a hull *H* is the set *B* of devices such as the Euclidean distance from a device d_b (from *B*) to a device d_h (from hull *H*) is lesser than radius r_d.
- A relationship is established between agents a_i and a_j. If a_i and a_j manages the same sector or if the sector managed by a_j is an adjacent sector of the sector managed by a_i, agents a_i and a_j have a sequential relationship. On the other hand, if a_i and a_j manages non-adjacent sectors, agents a_i and a_j have a jumping relationship.

An example of this architecture, composed of two layers, is shown in Figure 1. In the device layer, devices in the boundary establish connections among them to decide which ones are responsible for aggregating and for sending the group location and movement information to the agent that manages that group (devices d_1, d_2 and d_3 from swarm S_1). In the multiagent layer, each agent monitors its swarm behavior (agent a_1 manages swarm S_1). In this layer, agents establish relationships among them, exchanging information about their swarms, i.e., if they are merging (agent a_2 interacting with agent a_3 about S_2 and S_3 swarms merge) or splitting (agent a_4 interacting with agent a_2 and a_3 about S_4 swarm split).

Figure 1. Overall architecture

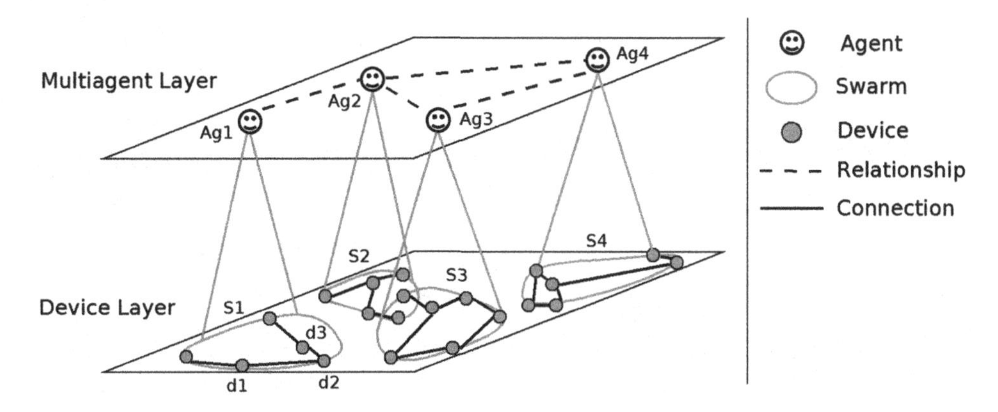

The device layer, using a IR2 coordinate system for simplicity, is detailed in Figure 2. In this figure there is a swarm composed of a large number of devices. The hull of this swarm is the set of devices dH_i, with i from 1 to 18, from where every consecutive pair of devices (dH_i, dH_{i+1}) has a connection stablished between them, including the connection between the first and the last device dH_1, dH_{18}. The boundary of this swarm is the set of devices dB_k, with k from 1 to 16, which are in the influence zone of some devices of the hull (e.g., the location of dB_3 lies in the influence zone of dH_2).

The multiagent layer is detailed in Figure 3, in which, differently from Figure 1, is presented a layer composed of devices that were agentified (i.e., converted into agents) and that belong to a certain sector. In this figure, agents a_1 and a_2 are managing their respective swarms S_1 and S_2 in the same sector $Sector_1$ (note that a sector area is represented in the device layer and could be greater than the area covered by a swarm). As those agents manage the same sector, they have a sequential relationship. The same kind of relationship exists between agents a_1 and a_3 and between agents a_2 and a_3, because the sector managed by

Figure 2. Device layer

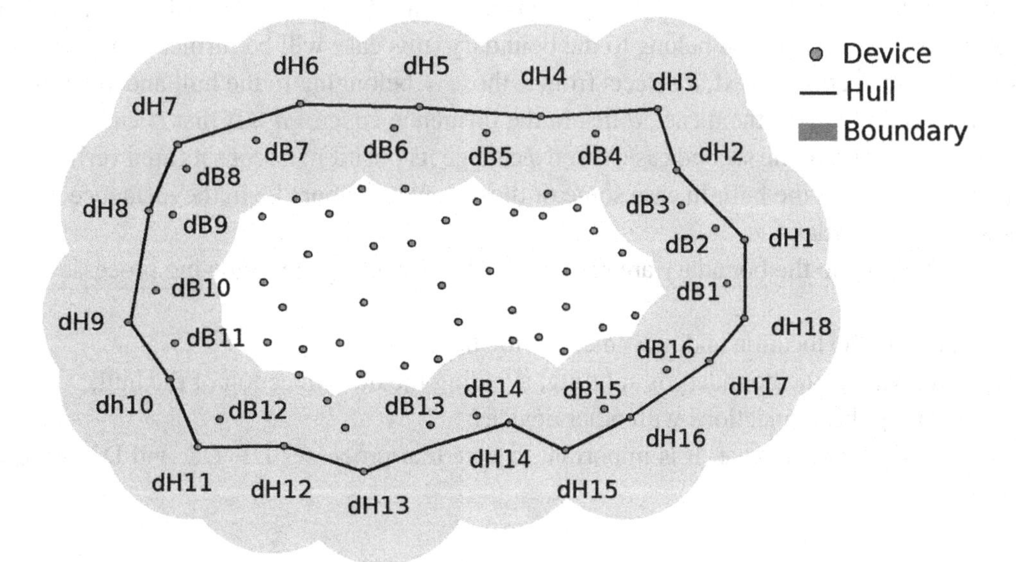

Figure 3. Agent layer

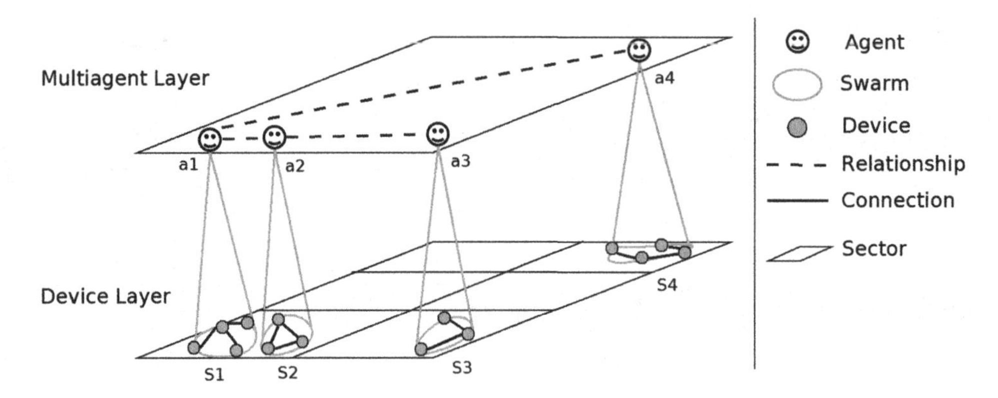

a_3 is adjacent to the sector managed by a_2 - which is the same sector managed by a_1. Agent a_4, on the other hand, has a jumping relationship only with agent a_1, because their respective sectors are non-adjacent.

DEVICE LAYER

The device layer consists of inter-connected devices in which each one has its own movement, but if seen as a whole, they move as swarms. In this layer, to support scalability, only devices belonging to the boundary will be responsible for executing certain processes related to the swarm behavior. In this context, first is analyzed how to characterize a device as a boundary device and second, what processes (this boundary device) should be executed.

Choosing a Boundary Device

Each device of a swarm must analyzes if it belongs to the boundary: when it joins the swarm, and during its lifecycle. In the first case, when a device *d* joins a swarm, the agent responsible for it will provide a list *L* of devices near *d* and that belong to the boundary (this case will be further explained in the D2 process of the next section). Next, *d* selects from *L* those d_i belonging to the hull and verifies if *d* is in the influence zone of any of them (i.e., *d* lies in the influence zone of d_i). If that occurs, device *d* will belong to the boundary. In the second case, when *d* change its position, it scans its area verifying if some new neighbors belong to the hull. In case some of those new neighbors lie in the influence zone of *d*, *d* will belong to the boundary.

Devices belonging to the boundary are responsible for executing the following processes:

- (D1) updating the location and movement of the boundary;
- (D2) perceiving if the boundary is splitting, avoiding the disconnection of the hull;
- (D3) updating the connections with other devices;
- (D4) agentifying the device. It is important to note that processes D1, D2, and D3 could be performed in parallel.

The algorithm presented below shows the updating boundary process (D1). First, the device verifies if it belongs to the boundary (Line 2), obtaining the positions of its neighbors, and sending this information to the agent (Lines 3-4). If the device does not belong to the boundary (Line 5), it verifies if any of its neighbors belong to the hull, changing its current state (Lines 6-8). Note that only devices in the boundary will send information to the agent, decreasing the information transferred.

1. **Algorithm UpdatingBoundary()**
2. **if** belongsToBoundary **is true**
3. positions ← neighbors.positions
4. agent.updateBoundary(positions)
5. **else**
6. **for each** neigbor **in** neighbors
7. **if** neighbor.belongsToHull **is true**
8. belongsToBoundary ← true

Among all these processes, the split device process (D2) is a critical one and it is sketched in Figures 4 and 5. In Figure 4, non-connected devices d_1 and d_2 generate an open hull. Figure 5 shows the connected path that close the hull, going through the devices belonging to the boundary between d_1 and d_2. In order to close the hull, device d_1 performs a recursive and distributed search, asking any of its neighbors d_n if it has d_2 as a neighbor, and creating a new alternative path between d_1, d_n, and d_2. If none of them has d_2 as a neighbor, d_1 forwards the search to the device d_n which is closer to the last location of d_2, repeating the process until the formed path close the hull or d_2 is found

Figure 4. Device-boundary process – open hull

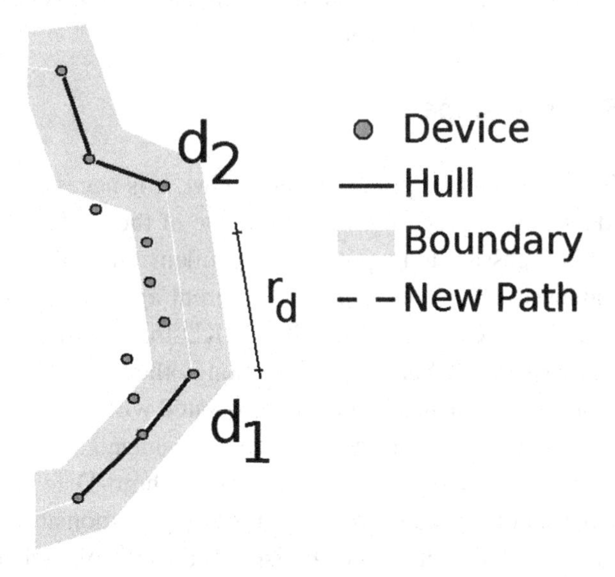

Figure 5. Device-boundary process – alternative path

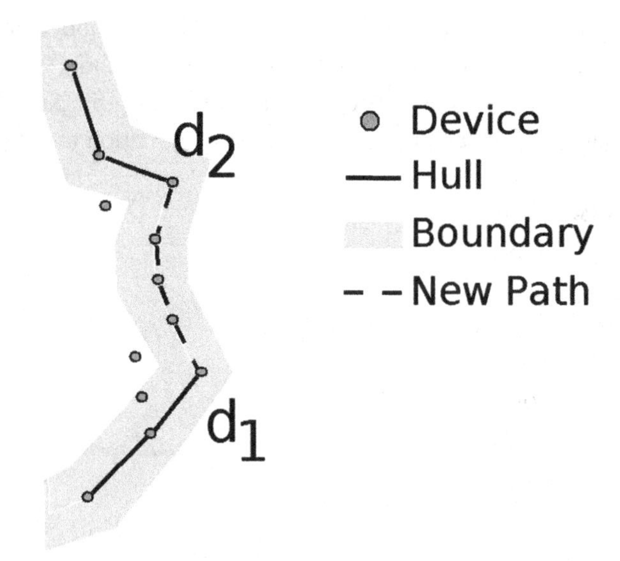

MULTIAGENT LAYER

The multiagent layer consists of agents responsible for managing swarms of devices. As mentioned, each agent has two kinds of relationship with other agents: sequential and jumping. Sequential relationships are established to support scalability, allowing the closer agents (those whose agents' swarms are about to merge) to exchange information among them. Jumping relationships are established to support efficiency, allowing agents to access swarms locations that are beyond the sequential relationships. In this context, first is analyzed how to create the jumping relationships and how to search a specific sector, and second, the processes the agent should execute.

Creating Jumping Relationships

As mentioned in the DHT background section, to locate a GM, it is necessary to use its corresponding sector identifier. As the identifier is, generally, the coordinate of the center of the sector (i.e., a two or three dimensional point), it is necessary to map it to a one-dimensional DHT space, since this structure uses a one-dimensional namespace for keys. For this, different approaches are used, such as: Hilbert Curve (Schmidt & Parashar, 2003; Ganesan et al., 2004), KD-tree (Zhang, Krishnamurthy, & Wang, 2004), QUAD-tree (Tanin, Harwood, & Samet, 2007), among others.

Having the identifiers for each sector, each agent is associated with a sector where its corresponding agentified device is located (in this case, one sector could be managed by more than one agent). Then, jumping relationships will be created using the same concept as Chord-DHT (Stoica et al., 2001). Figure 6 illustrates the initial formation of the agent relationships (agent responsible for a sector identified as *r*). In this architecture, jumping relationships can be visualized as a table, where agents are matched to sector identifiers that precede and succeed sector *r*. As defined by Chord-DHT, in the successor table of an agent, each registry *i* of this table (*r* and *i* are positive integers) corresponds to a successor sector

Figure 6. Initial jumping relationships

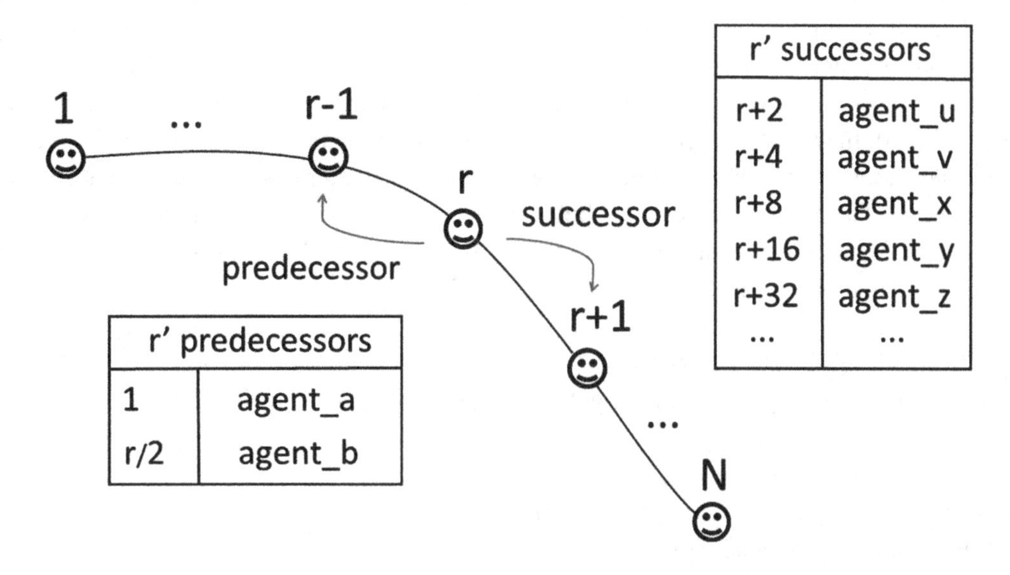

whose value is equal to or greater than $r + 2^{i-1}$. On the other hand, in the predecessor table there are only two registries. The first one points to the first region (identified as 1). The second registry points to the sector located in the middle between the first sector and *r* (i.e., pointing to the agent responsible for the sector identified as *r*/2).

When a device requests sector *s* to an agent a_r (responsible for sector *r*), the search for *s* is performed following the well known binary search process. This process is performed in *O(log N)* where *N* is the number of sectors that comprise the whole space. The steps executed by the process are:

1. In case agent a_r is already responsible for s (thus, r = s), the device is notified that a_r is responsible for s.
2. Otherwise, a_r needs to find, in its successor or predecessor table, the agent ac responsible for the closest sector to s (called sector c).
3. In case ac is responsible for s (thus, c = s), the device is notified that ac is responsible for s.
4. Otherwise, a_r needs to recursively search s, but using the closest agent ac as the new starting point. In this case, the closest agent ac will search s repeating the four steps mentioned above.

Each agent of this layer is responsible for executing the following processes:

- (A1) searching for a specific location requested by a device;
- (A2) monitoring the swarm behavior;
- (A3) receiving the location of its boundary;
- (A4) perceiving if its swarm is merging with others;
- (A5) perceiving if its swarm is splitting, preventing the hull from opening;
- (A6) perceiving if its swarm must be split. It is important to note that processes A2 and A4 to A6 could be performed in parallel.

Among all these processes, the merge agent-swarm process (A4) is a critical one and it is sketched in Figures 7 and 8. In Figure 7, swarms S_1, S_2 and S_3 have a boundary intersection perceived by their respective manager agents a_1, a_2 and a_3. In the process, every agent verifies, at predetermined times, if an intersection occurs with the boundaries managed by its sequential relationships. When the intersection occurs (Line 3), the agents merge their swarms, calculating the new hull generated by the merging (Lines 3-7), and choosing by consensus the agent which has the greater uptime among them, eliminating the others (Line 8). The algorithm presented below shown this process, assuming that a_i has the greater uptime. Note that the merging is not mandatory in case the amount of devices is greater than some value that could cause scalability problems to the agent (Line 5).

1. **Algorithm MergingSwarms(a_i, a_j)**
2. **if** existsIntersection(a_i.getBoundary, a_j.getBoundary)
3. $qtyDevices_i$ ← a_i.getBoundary.size
4. $qtyDevices_j$ ← a_j.getBoundary.size
5. **if** $qtyDevices_i$ + $qtyDevices_j$ < MAX_DEVICES
6. a_i.addBoundary(a_j.getBoundary)
7. a_i.recalculateNewBoundary

Figure 7. Swarms with boundary intersection

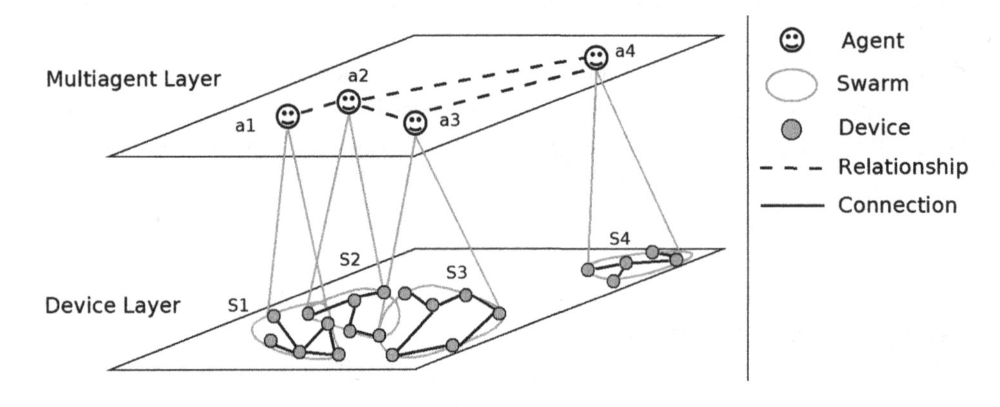

Figure 8. Swarms merged in one

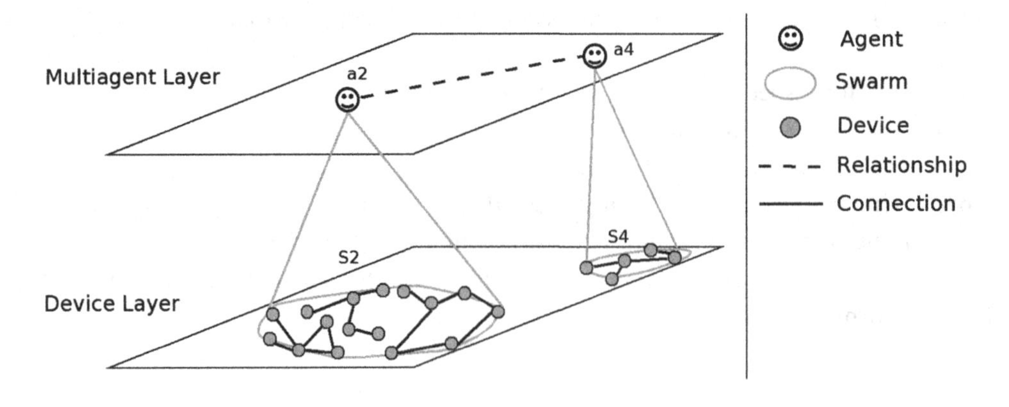

8. removeAgent(a_j)

Finally, Figure 8 shows the result of the merging process, which generates only one agent a_2 and one swarm S_2.

BUILDING THE DEVICE AND MAS LAYERS

After presenting the architecture and its internal components, in this section is described how its layers are built. As seen in the previous section, the device layer must perform some activities while others must be performed by the agent layer. To understand how both layers interact, the building process is divided into eight steps, the first being related to the system initialization and the remaining ones to the system lifecycle. The steps are:

1. The first device joins the system and the first agent is created;
2. New devices join the system in swarms and begin to collect its location;
3. A device finds its neighbors in the swarm;
4. Devices in the hull send the boundary location to the agent;
5. A device maintain the hull closed in case it has been opened;
6. Agents merge their swarms if they are overlapping;
7. An agent monitors and splits its swarm if this is affecting the system scalability;
8. An agent leaves the system and another one needs to replace it. Note that steps 2 to 8 are performed in parallel, until the last device leaves the system.

Although the steps could be performed in parallel, Figure 9 shows the building steps as if they are flowing sequentially.

The first step is the system set up, which initializes both layers. The device layer is initiated when the first device joins the layer, providing its location. In this case, the device is agentified (converted to an agent), using the convert device-agent process (D4), also initiating the multiagent layer, which creates and register an agent in a global repository. In order to implement this repository, several approaches could be adopted, such as: agents directory (Dimakopoulos & Pitoura, 2003), agents distributed structures (Kambayashi & Harada, 2007), among others. The proposed architecture is prepared to use any of them.

In the second step, new devices join the system and form swarms. Each new device *d* uses the global repository to find any agent a_i to request joining the agent's swarm. Then, agent a_i, based on the device location, uses the search agent-swarm process (A1) to find agent a_j responsible for the sector where *d* is located. Next, agent a_j may accept (or not) joining it to its swarm, based in the number of members, avoiding scalability issues that could be generated. If the agent accepts *d*, it creates a communication channel with the device, informing it with a list *L* of devices near the boundary which are closer to device *d* location. If the agent does not accepts *d*, it can separate the swarm in two to allow the device *d* to join one of those swarms. To separate the swarm, agent a_j uses the divide agent-swarm process (A6).

In the third step, each new device *d* chooses some devices (from the list *L* received by the agent in the previous step) to find those which will be its neighbors, using the update device-neighbors process (D3). This step is required because the devices received by the agent (which belong to the boundary) are not necessarily in the influence zone of *d*.

Figure 9. Building layers steps

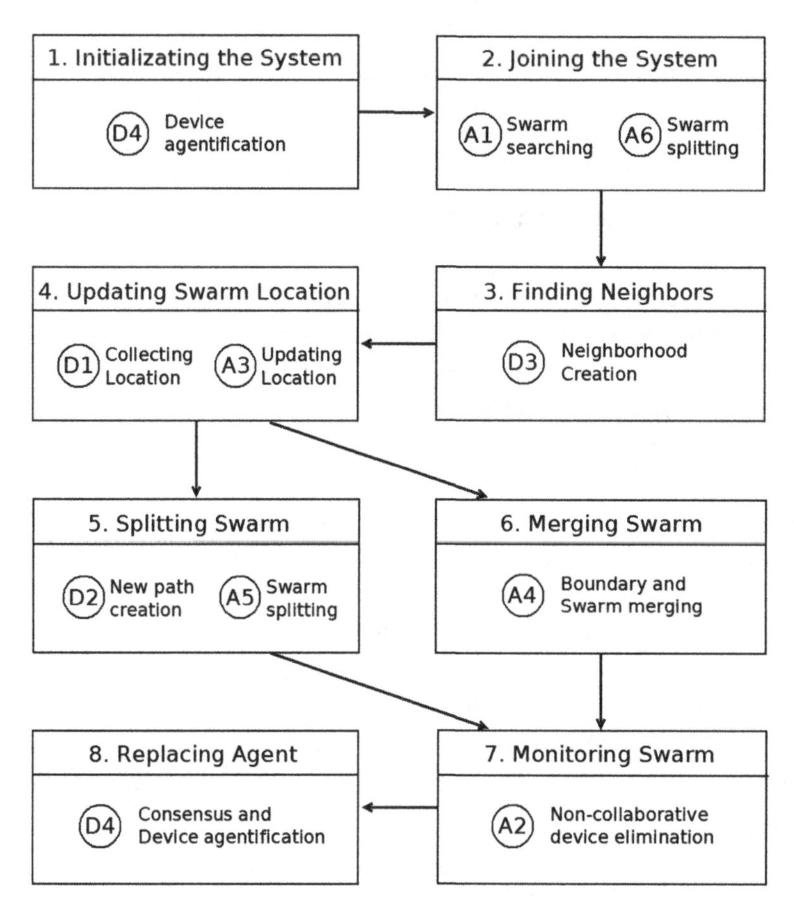

In the fourth step, devices belonging to the hull collect the location of their boundary neighbors, sending this information to the agent through the communication channel created in the second step. Once the device reaches a certain number of sent messages, it interchanges this responsibility with its neighbors belonging to the boundary, using the update device-boundary process (D1). The agent, with the information received, updates the boundary of the swarm using the update agent-swarm process (A3).

In the fifth step, a device d_i belonging to the hull perceives that its closer neighbor d_j (also belonging to the hull) is moving away or has left the system, which could open the hull. In this case, d_i tries to create an alternative path with d_j or d_{j+1} (i.e., the nearest device to d_j which also belongs to the hull) using the split device-boundary process (D2). If the path is not found, d_i sends this information to the agent (using the channel created in the second step), which analyzes if other devices near d_i also perceive this situation. In case the separation among devices does not affect the hull (either because device d_j is moving from the boundary to the inside of the swarm or because they are leaving the swarm), the agent closes the communication channel with device d_j. If the separation opens the hull, the agent splits the swarm into two using the split agent-swarm process (A5).

In the sixth step, sequential agents exchange information about their swarms. If the agents perceive that their swarms are intersecting (analyzing the current position of each boundary) or near to intersect (analyzing the last positions of each boundary) the agents could merge their swarms, using the merge agent-swarm process (A4), as long as this merge does not affect the system scalability.

In the seventh step, each agent monitors the behavior of the devices belonging to the boundary, purging those which are not performing their responsibilities, using the monitor agent-swarm process (A2).

In the eighth step, an agent leaves the system and is replaced by a boundary device of its swarm that is agentified. When an agent leaves the system, its associated device also does and each device *d* of the boundary is aware of this situation (because the channel with the agent is closed). In this case, each *d* analyzes if it could be considered a stable device (e.g., by its energy level properties) and sends this information to its neighbors. As various devices could be stable, a consensus is needed for choosing which will be agentified. In this case, the device with the highest energy level is selected. Finally, the device will be agentified with the convert device-agent process (D4).

SOLUTIONS AND RECOMMENDATIONS

In this section, the scalability of the proposed architecture is evaluated using simulations. The layers were implemented in Java, using PeerSim (http://peersim.sourceforge.net) (Montresor & Jelasity, 2009) as the simulation framework to evaluate the protocols developed for the Multiagent and Device layers. Each experiment was evaluated conducting 20 simulations.

As the devices and agents use the Internet to communicate among them, a network topology based on King (Gummadi et al., 2002) was used to simulate their connectivity in PeerSim. This topology, used in various scientific researches, represents a realistic situation of Internet hosts with their bandwidth and latency constraints. In this topology, the average end-to-end latency among all the nodes is approximately 200 milliseconds, with a peak of 300 milliseconds. Also, each node was set with 250 Kbps inbound and 250 Kbps outbound access link bandwidth, unless otherwise stated, as used in protocols for monitoring and control, such as Zigbee (Alliance, 2005) and IEEE 802.15.4 (IEEE 802.15, 2005) standard.

In the simulated experiments, device parameters were set as follows:

5. The size of the neighbor list is 50;
6. 30 seconds is the time interval between boundary device updating information related to the location collected to its agent;
7. 60 seconds is the time interval between each non-boundary device scans its influence zone to find hull neighbors which will transform the non-boundary device into a boundary device;
8. 30 seconds is the time interval between each hull device to verify if the connections established with its hull neighbors are lost, producing the openness of the hull;
9. Devices joining and leaving the network at arbitrary times, independent of each other;
10. As usual on the Internet, and as other approaches do (Paganelli & Parlanti, 2012; Stoica et al., 2001), the device exit was modeled without informing anybody (MAS, neighbors) about its departure.

In the agent context, the parameters were set as follows:

1. Sizes of the sequential relationship list and jumping relationship list are 5 and 7, respectively, unless otherwise stated;
2. 30 seconds is the time interval between boundary device's updating information related to the location collected to its agent;
3. Agents inform the state of their swarm to their sequential relationships every 30 seconds;

4. Agents monitor the state of the swarm every 30 seconds;
5. When a device requests to join the swarm, the agent returns 50 IP:port device addresses, totaling approximately 3000 bytes of information.

These parameters are set following the peer default configuration of Chord-DHT (http://tomp2p.net/doc/advanced) protocol implementation used as the baseline for the simulations.

In the scalability experiments, it was necessary to understand how the number of messages received by the agent (e.g., requests for joining a swarm or for updating the device location) decreases the performance of the system to answer these messages. In order to do that, first, the network bandwidth (other possibilities could be the RAM or the CPU) was defined as the computer resource which limits the system performance, as proposed in (Schroeder & Harchol-Balter, 2006), the performance being defined in terms of the time needed to answer these messages. Second, the behavior executed by an agent was defined when the messages are received. The behaviors defined were: the agent answers all the messages (exactly as the works using DHT do); the agent limits the simultaneous requests, agentifying devices to answer those requests not answered by the agent (used by the proposed architecture). Third, it was defined how the messages will be sent by the devices. For that, at predetermined moments, all the devices of the network send a message for joining the same region *r*, and the agent must returns a list of 50 devices near the boundary in a time close to 1 second, as proposed by (Jimenez, Osmani, & Knutsson, 2011).

With these definitions, the scalability had two aspects analyzed:

1. The messages sent by the swarm compared with the messages sent by the boundary;
2. The time consumed by the agent to answer the messages when it is overloaded.

In order to analyze aspect A, different swarm shapes were obtained from running the *2D* flocking model defined by Craig Reynolds (Reynolds, 1987). As presented in the algorithm UpdatingBoundary, Figure 10 shows that using the boundary decreases the number of messages received by the agent (which is managing the swarm) if compared with the strategy used by current alternatives in which all the devices of the swarm send the messages to the agent. For example, in swarms whereby the sum of its members totalize a network of 5,000 devices, the presented architecture outperforms other works, just sending 2,175 messages instead of 5,000 (corresponding to one message for each device), which represents a reduction of approximately 57%. On the other hand, the proposed approach uses the boundary just after the swarm shape is formed. In the meantime, the architecture uses the same strategy as the other alternatives, in which all devices send messages to the agent.

In order to analyze aspect B, it was first necessary to understand if the agent behavior (i.e., agentifying devices for answering requests) can answer all the messages received, just as performed by a DHT-node and second, the time consumed by each approach to answer those messages. Figure 11 shows the former, comparing the messages sent by the devices and the messages answered by the system (Figure 12 only shows the messages not answered by the agent behavior). For example, in a network of 200,000 devices, which represents 200,000 messages sent, a DHT node answers all of them. In turn, the agent behavior, for an agent with 250 kbps and 1250 kbps of outbound bandwidth, only answers 198,283 and 199,843 messages, respectively (using a sequential relationship list of 50 agents). This situation occurs because the agent forwards the messages it cannot handle (i.e., those that could compromise its scalability) to other agents which are managing the same region (or agentifying new agents). In this context, the number of messages an agent can handle (i.e. the bandwidth network used to return the list

Figure 10. Scalability test using the boundary to send messages

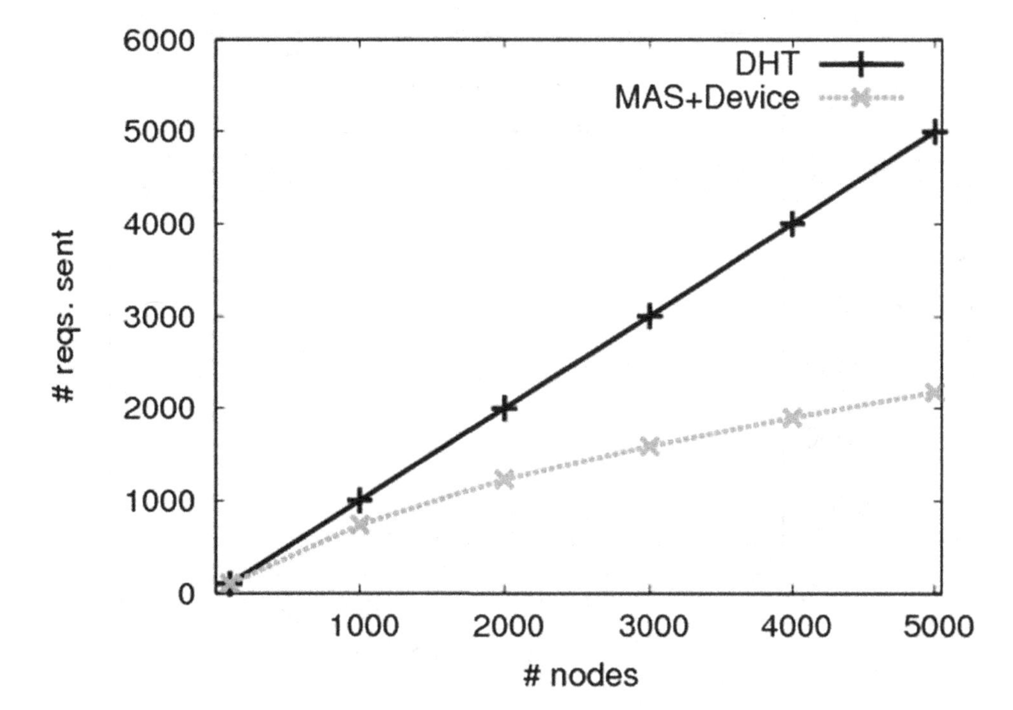

in a time close to 1 second) must also consider the bandwidth network consumed by the forwarding. As each agent could executes the receive-and-forward process to maintain the scalability, this proposal must drop some messages (Figure 12), needing to be sent again by the devices. This behavior produces a drawback if compared to the DHT alternative because there are 1717 and 157 messages lost for the same output bandwidth presented above. Even that these quantities are small if compared to the total messages sent, if all of them are lost by the devices located at the same swarm, it is probable that the swarm could become inaccessible until the new messages sent are processed.

Figure 13 shows the latter, comparing the time consumed for answering the messages (Figure 14 zooms in the first 50,000 messages received). This figure shows that, when 2,000 messages are sent, a DHT node takes 3,000 ms to return the list (three times the established 1 second limit). From that point, a DHT node cannot handle more messages in proper time. Conversely, the agents can handle more messages in proper time. For example, the 200,000 messages received by the agent were answered in 3,073 ms and 1,031 ms for an agent with 250 kbps and 1250 kbps of outbound bandwidth, respectively.

CONCLUSION

In the biodiversity scenarios, such as flocking, several researches are using sensor devices for collecting and monitoring the wildlife behavior. In order to support the high demand of information transmitted, a new paradigm (called Internet of Things or IoT) has emerged, but scalability is still an issue that can be solved when the amount of sensors carried by wildlife is large.

Figure 11. Scalability test on messages answered

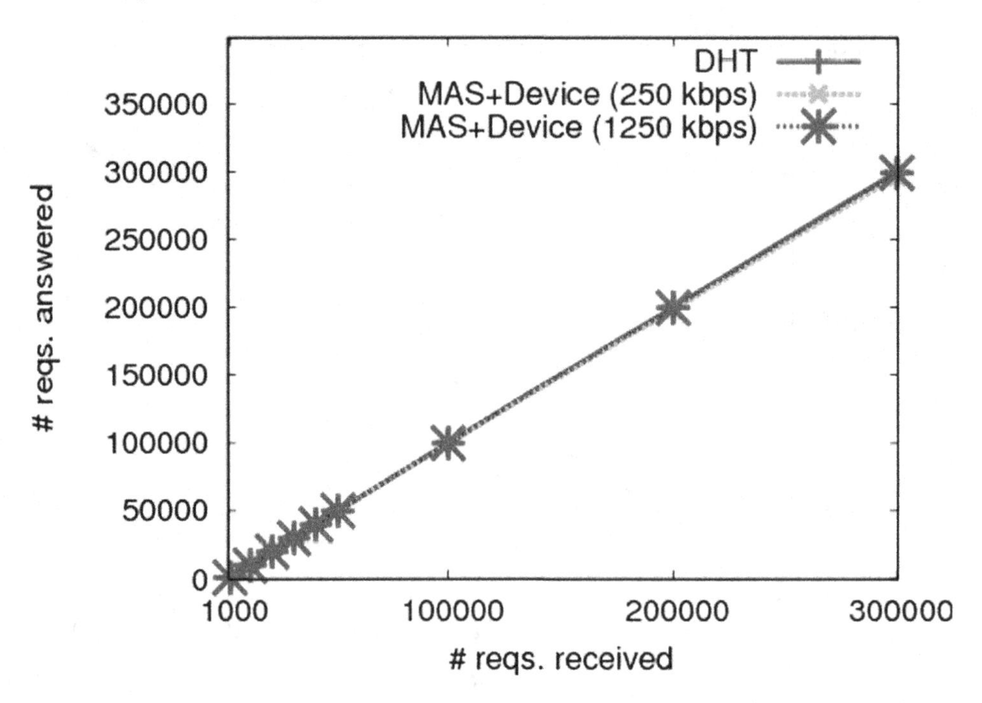

Figure 12. Scalability test on messages not answered

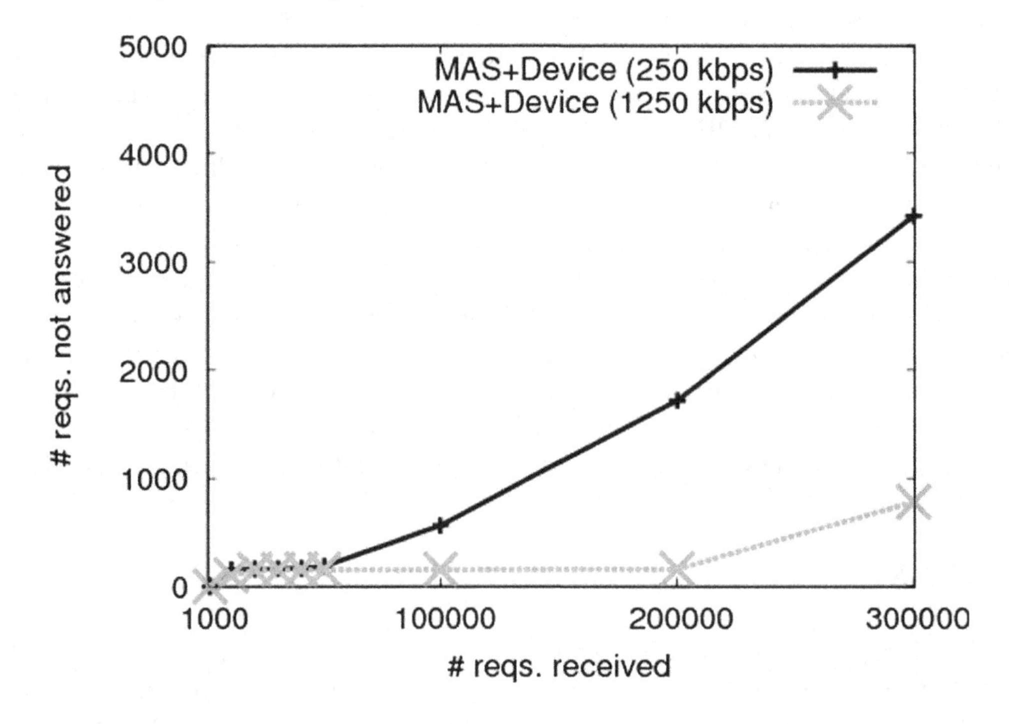

Figure 13. Time needed to answer the messages

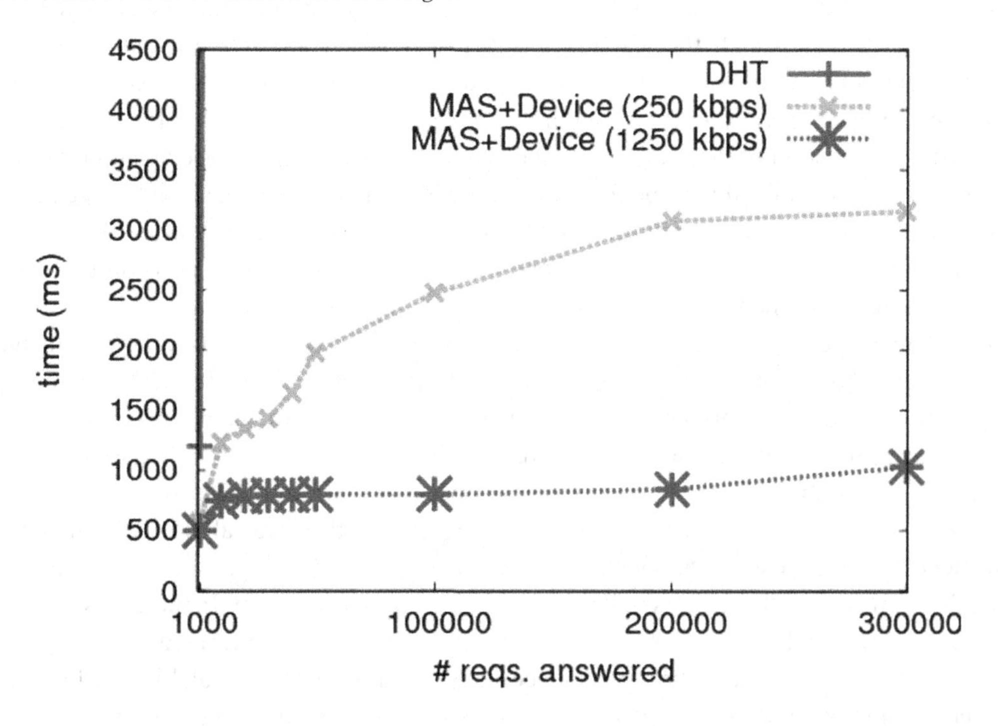

Figure 14. Zooms in the first 50.000 messages

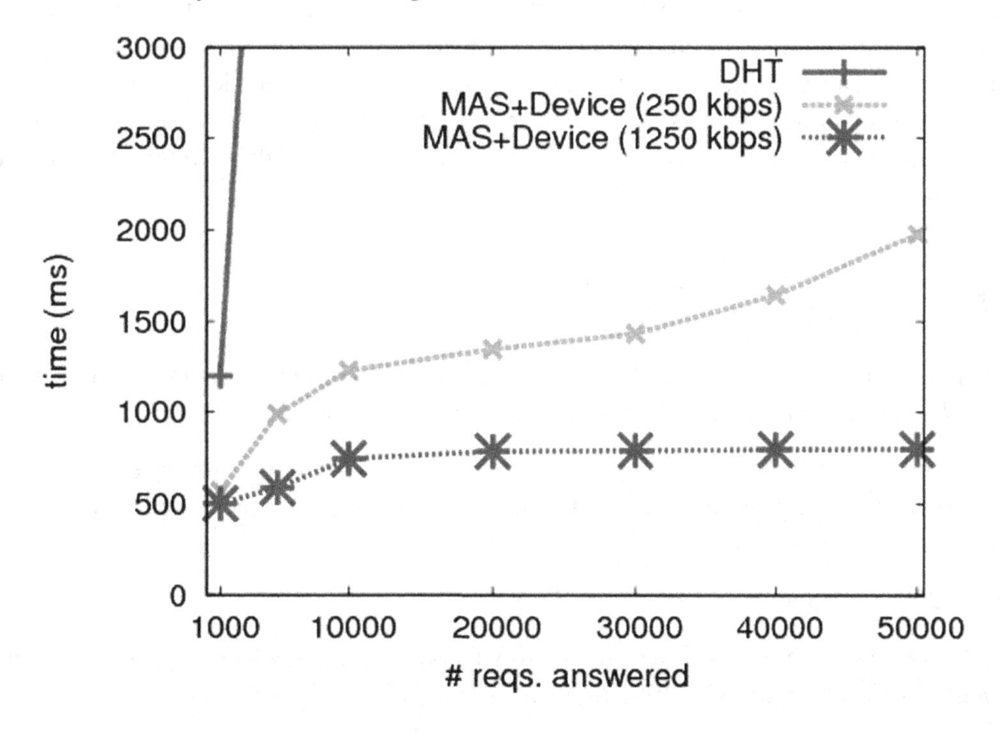

To increase the system scalability, the authors presented a layered architecture for IoT, using an innovative approach that extends the scalability of the data aggregation method used in combination with the Distributed Hash Table for discovering and updating information of a group of devices. For this, in the device layer, devices in the boundary of each group are responsible for collecting and for aggregating the information of their neighbors and sending it to an agent. In the agent layer, agents are responsible for managing the groups and for communicating with other agents to exchange inter-groups interactions, such as merging and splitting. When the scalability issue arise, agents are able to create new agents -agentifying devices of its swarm- in order to share the management responsibility among them.

According to the experimental results, in the device layer, using the boundary of the swarm decreases the number of messages sent to the agent, increasing the scalability of the agent which manages the swarm. In the agent layer, the addition of new responsibilities to the DHT node (which, in the proposed architecture, also acts as an agent) increases the scalability of the system, mainly because the agent distributes the load applied on it (i.e., the messages received) among its sequential neighbors. However, in the presented solution, the boundary can only be used when the swarm if formed, meaning that, in the meantime, it is used the same approach as the data aggregation method, i.e., all devices sent messages to the agent. Besides, there are a number of messages that are lost, i.e., messages the agent or its neighbors cannot handle. In this case, the devices will have to send the messages again. In biodiversity monitoring scenarios, this is acceptable considering the fact that this architecture avoids the data aggregator single point of failure, which could make the swarm inaccessible, and the gains in scalability, which is considerably higher than in other scenarios, such as video-on-demand (Rocha & Brandão, 2015).

In addition to the scalability advantage and the inter-groups interaction the architecture considers, an additional benefit of using the boundary is that it allows knowing the shape of the area covered by the group, which could be used for specific application scenarios, such as flocking and schooling tracking.

The authors are currently studying alternatives to the *2D* flocking models used, applied to other scenarios, such as ant or bee colonies. Also, they are improving the simulation experiments in order to analyze how the architecture behaves considering the network bandwidth as well as the CPU and the RAM as computational resources.

REFERENCES

Ali, M., & Langendoen, K. (2007). A Case for Peer-to-Peer Network Overlays in Sensor Networks. In *International Workshop on Wireless Sensor Network Architecture* (pp. 56–61).

Ali, M., & Uzmi, Z. (2004, May). CSN: a network protocol for serving dynamic queries in large-scale wireless sensor networks. In Proceedings of Communication Networks and Services. (pp. 165–174). doi:10.1109/DNSR.2004.1344725

Alliance, Z. (2005). *ZigBee Specification 1.1* (Tech. Rep.).

Borgia, E. (2014). The Internet of Things vision: Key features, applications and open issues. *Computer Communications*, *54*, 1–31. doi:10.1016/j.comcom.2014.09.008

Dimakopoulos, V. V., & Pitoura, E. (2003). A Peer-to-Peer Approach to Resource Discovery in Multi-agent Systems. In *Cooperative information agents* (Vol. 2782, pp. 62–77). Springer. doi:10.1007/978-3-540-45217-1_5

Dlodlo, N., & Kalezhi, J. (2015, May). The internet of things in agriculture for sustainable rural development. In *International Conference on Emerging Trends in Networks and Computer Communications (ETNCC)* (pp. 13–18). doi:10.1109/ETNCC.2015.7184801

DOro, S., Galluccio, L., Morabito, G., & Palazzo, S. (2015). Exploiting Object Group Localization in the Internet of Things: Performance Analysis. *IEEE Transactions on Vehicular Technology*, *64*(8), 3645–3656. doi:10.1109/TVT.2014.2356231

Duraccio, V., Falcone, D., Silvestri, A., & Di Bona, G. (2007). Use of Simulation for the Prevention of Environmental Problems. In *Proceedings of the 2007 summer computer simulation conference* (pp. 863–866). San Diego, CA: Society for Computer Simulation International.

Dyo, V., Ellwood, S. A., Macdonald, D. W., Markham, A., Trigoni, N., Wohlers, R., & Yousef, K. (2012, September). WILDSENSING: Design and Deployment of a Sustainable Sensor Network for Wildlife Monitoring. *ACM Trans. Sen. Netw., 8*(4), 29:1–29:33.

Galluccio, L. (2011, June). On the potentials of object group localization in the Internet of Things. In *IEEE International Symposium on a World of Wireless, Mobile and Multimedia Networks* (pp. 1–9). doi:10.1109/WoWMoM.2011.5986489

Ganesan, P., Bawa, M., & Garcia-Molina, H. (2004). Online Balancing of Range-partitioned Data with Applications to Peer-to-peer Systems. In *Proceedings of the Thirtieth International Conference on Very Large Data Bases* (Vol. 30, pp. 444–455). doi:10.1016/B978-012088469-8.50041-3

Gummadi, K. P., Saroiu, S., & Gribble, S. D. (2002). King: Estimating Latency Between Arbitrary Internet End Hosts. In *Proceedings of the Second ACM SIGCOMM Workshop on Internet measurment* (pp. 5–18). doi:10.1145/637201.637203

Haider, W., ur Rehman, A., & Durrani, N. M. (2013, Sept). Towards decision support model for ubiquitous agriculture. In *International Conference on Digital Information Management (ICDIM)* (pp. 308–313). doi:10.1109/ICDIM.2013.6693987

Hart, J. K., & Martinez, K. (2015). Toward an environmental Internet of Things. *Earth and Space Science*, *2*(5), 194–200. doi:10.1002/2014EA000044

Huang, J. H., Chen, Y. Y., Huang, Y. T., Lin, P. Y., Chen, Y. C., Lin, Y. F., & Chen, L. J. et al. (2010, June). Rapid Prototyping for Wildlife and Ecological Monitoring. *IEEE Systems Journal*, *4*(2), 198–209. doi:10.1109/JSYST.2010.2047294

IEEE 802.15. (2005). *IEEE Standard for Information Technology - Telecommunications and Information Exchange Between Systems - Local and Metropolitan Area Networks - Specific Requirements Part 15.4: Wireless Medium Access Control (MAC) and Physical Layer (PHY) Specifications for Low-Rate Wireless Personal Area Networks (LR-WPANs)* (Tech. Rep.).

Ilhan Akbas, M., Brust, M. R., Turgut, D., & Ribeiro, C. H. (2015). A preferential attachment model for primate social networks. *Computer Networks, 76*, 207–226. doi:10.1016/j.comnet.2014.11.009

Jiang, R., & Zhang, Y. (2013, September). Research of Agricultural Information Service Platform Based on Internet of Things. In *International Symposium on Distributed Computing and Applications to Business Engineering and Science*.

Jimenez, R., Osmani, F., & Knutsson, B. (2011, August). Sub-Second lookups on a Large-Scale Kademlia-Based overlay. In *11th IEEE Conference on Peer-to-Peer Computing*. doi:10.1109/P2P.2011.6038665

Kambayashi, Y., & Harada, Y. (2007). A Resource Discovery Method Based on Multi-agents in P2P Systems. In Agent and Multi-Agent Systems: Technologies and Applications (Vol. 4496, pp. 364–374). Springer. doi:10.1007/978-3-540-72830-6_38

Kaukalias, T., & Chatzimisios, P. (2016). Internet of Things (IoT). In Encyclopedia of Information Science and Technology, (3rd ed.; pp. 7623–7632).

Kridi, D., Carvalho, C., & Gomes, D. (2014). A Predictive Algorithm for Mitigate Swarming Bees Through Proactive Monitoring via Wireless Sensor Networks. In *Proceedings of the 11th ACM symposium on Performance evaluation of wireless ad hoc, sensor, & ubiquitous networks* (pp. 41–47). New York, NY: ACM. doi:10.1145/2653481.2653482

Liu, Y., He, Y., Li, M., Wang, J., Liu, K., Mo, L., & Li, X.-Y. (2011, April). Does wireless sensor network scale? A measurement study on GreenOrbs. In INFOCOM, 2011 Proceedings IEEE (pp. 873–881).

Manzanares-Lopez, P., Muñoz Gea, J. P., Malgosa-Sanahuja, J., & Sanchez-Aarnoutse, J. C. (2011, May). An Efficient Distributed Discovery Service for EPCglobal Network in Nested Package Scenarios. *Journal of Network and Computer Applications, 34*(3), 925–937. doi:10.1016/j.jnca.2010.04.018

Meinig, C., Stalin, S. E., Nakamura, A. I., Gonzalez, F., & Milburn, H. B. (2005, Sept). Technology developments in real-time tsunami measuring, monitoring and forecasting. In *OCEANS.Proceedings of MTS/IEEE* (Vol. 2, pp. 1673–1679). doi:10.1109/OCEANS.2005.1639996

Montresor, A., & Jelasity, M. (2009, September). PeerSim: A scalable P2P simulator. In *Proceedings of the 9th Int. Conference on Peer-to-Peer(P2P'09)* (pp. 99–100).

Naumowicz, T., Freeman, R., Heil, A., Calsyn, M., Hellmich, E., Brandle, A., & Schiller, J. (2008). Autonomous Monitoring of Vulnerable Habitats Using a Wireless Sensor Network. In *Proceedings of the workshop on real-world wireless sensor networks* (pp. 51–55). New York, NY: ACM. doi:10.1145/1435473.1435488

Paganelli, F., & Parlanti, D. (2012). A DHT-Based Discovery Service for the Internet of Things. *Journal of Computer Networks and Communications, 2012*, 107041:1-107041:11.

Prada, R., Prendinger, H., Yongyuth, P., Nakasoneb, A., & Kawtrakulc, A. (2015, February). AgriVillage: A Game to Foster Awareness of the Environmental Impact of Agriculture. *Comput. Entertain., 12*(2), 3:1–3:18.

Ren, F., Zhang, J., Wu, Y., He, T., Chen, C., & Lin, C. (2013, May). Attribute-Aware Data Aggregation Using Potential-Based Dynamic Routing in Wireless Sensor Networks. *IEEE Transactions on Parallel and Distributed Systems*, *24*(5), 881–892. doi:10.1109/TPDS.2012.209

Reynolds, C. W. (1987, August). Flocks, Herds and Schools: A Distributed Behavioral Model. *SIGGRAPH Comput. Graph.*, *21*(4), 25–34. doi:10.1145/37402.37406

Rocha, V., & Brandão, A. A. F. (2015). Towards conscientious peers: Combining agents and peers for efficient and scalable video segment retrieval for VoD services. *Engineering Applications of Artificial Intelligence*, *45*, 180–191. doi:10.1016/j.engappai.2015.07.001

Schmidt, C., & Parashar, M. (2003, June). Flexible information discovery in decentralized distributed systems. In *IEEE International Symposium on High Performance Distributed Computing* (pp. 226–235). doi:10.1109/HPDC.2003.1210032

Schroeder, B., & Harchol-Balter, M. (2006, February). Web Servers Under Overload: How Scheduling Can Help. *ACM Transactions on Internet Technology*, *6*(1), 20–52. doi:10.1145/1125274.1125276

Stoica, I., Morris, R., Karger, D., Kaashoek, M. F., & Balakrishnan, H. (2001). Chord: A Scalable Peer-to-peer Lookup Service for Internet Applications. In *Proceedings of the 2001 Conference on Applications, Technologies, Architectures, and Protocols for Computer Communications* (pp. 149–160). New York, NY: ACM. doi:10.1145/383059.383071

Tang, J., Zhou, Z. B., Niu, J., & Wang, Q. (2014). An energy efficient hierarchical clustering index tree for facilitating time-correlated region queries in the Internet of Things. *Journal of Network and Computer Applications*, *40*, 1–11. doi:10.1016/j.jnca.2013.07.009

Tanin, E., Harwood, A., & Samet, H. (2007, April). Using a Distributed Quadtree Index in Peer-to-peer Networks. *The VLDB Journal*, *16*(2), 165–178. doi:10.1007/s00778-005-0001-y

Wu, R., Xu, Y., Li, L., Zha, J., & Li, R. (2014). Advanced Technologies in Ad Hoc and Sensor Networks. In *Proceedings of the 7th China Conference on Wireless Sensor Networks*. Berlin: Springer Berlin Heidelberg.

Yadgar, O. (2007). Emergent Ad Hoc Sensor Network Connectivity in Large-scale Disaster Zones. In *Proceedings of the 6th International Joint Conference on Autonomous Agents and Multiagent Systems* (pp. 269:1–269:2). New York, NY: ACM. doi:10.1145/1329125.1329450

Yu, J., Liu, W., & Song, J. (2007, Dec). C2WSN: A Two-Tier Chord Overlay Serving for Efficient Queries in Large-Scale Wireless Sensor Networks. In *International Conference on Advanced Computing and Communications* (pp. 237–242). doi:10.1109/ADCOM.2007.25

Yu, L., Wang, N., & Meng, X. (2005, Sept). Real-time forest fire detection with wireless sensor networks. In *International Conference on Wireless Communications, Networking and Mobile Computing* (Vol. 2, pp. 1214–1217).

Zenonos, A., Stein, S., & Jennings, N. R. (2015). Coordinating Measurements for Air Pollution Monitoring in Participatory Sensing Settings. In *Proceedings of the 2015 International Conference on Autonomous Agents and Multiagent Systems* (pp. 493–501).

Zhang, C., Krishnamurthy, A., & Wang, R. Y. (2004, May). *SkipIndex: Towards a Scalable Peer-to-Peer Index Service for High Dimensional Data* (Vol. TR-703-04; Tech. Rep.). Department of Computer Science, Princeton University.

Zhang, C., Xiao, W., Tang, D., & Tang, J. (2011). P2P-based Multidimensional Indexing Methods: A Survey. *Journal of Systems and Software*, *84*(12), 2348–2362. doi:10.1016/j.jss.2011.07.027

Zhao, W., Liu, X. Y., Ma, S., Yuan, C. Y., & Wang, L. F. (2011, Aug). A Distributed RFID Discovery System: Architecture, Component and Application. In *IEEE International Conference on Computational Science and Engineering (CSE)*, (pp. 518–525). doi:10.1109/CSE.2011.93

ADDITIONAL READING

Abid, S. A., Othman, M., & Shah, N. (2014, May). 3D P2P Overlay over MANETs. *Computer Networks*, *64*, 89–111. doi:10.1016/j.comnet.2014.02.006

Gubbi, J., Buyya, R., Marusic, S., & Palaniswami, M. (2013, September). Internet of Things (IoT): A Vision, Architectural Elements, and Future Directions. *Future Generation Computer Systems*, *29*(7), 1645–1660. doi:10.1016/j.future.2013.01.010

Ma, Y. et al.. (2012). An Efficient Index for Massive IOT Data in Cloud Environment. In *Proceedings of the 21st acm international conference on information and knowledge management* (pp. 2129–2133). New York, NY, USA: ACM. doi:10.1145/2396761.2398587

Muller, J., Oberst, J., Wehrmeyer, S., Witt, J., Zeier, A., & Plattner, H. (2010, Jan). An Aggregating Discovery Service for the EPCglobal Network. In *System sciences (hicss), 2010 43rd hawaii international conference on* (p. 1-9). doi:10.1109/HICSS.2010.47

Perera, C., Zaslavsky, A., Christen, P., Compton, M., & Georgakopoulos, D. (2013). Context-Aware Sensor Search, Selection and Ranking Model for Internet of Things Middleware. In *Proceedings of the 2013 ieee 14th international conference on mobile data management*-volume 01 (pp. 314–322). Washington, DC, USA: IEEE Computer Society. doi:10.1109/MDM.2013.46

Sánchez López, T., Ranasinghe, D. C., Harrison, M., & Mcfarlane, D. (2012, March). Adding Sense to the Internet of Things. *Personal and Ubiquitous Computing*, *16*(3), 291–308. doi:10.1007/s00779-011-0399-8

Tsai, C.-W., Lai, C.-F., & Vasilakos, A. V. (2014, November). Future Internet of Things: Open Issues and Challenges. *Wireless Networks*, *20*(8), 2201–2217. doi:10.1007/s11276-014-0731-0

Wei, Q., & Jin, Z. (2012). Service Discovery for Internet of Things: A Context-awareness Perspective. In *Proceedings of the fourth asia-pacific symposium on internetware* (pp. 25:1–25:6). New York, NY, USA: ACM. doi:10.1145/2430475.2430500

Wu, Y., Sheng, Q. Z., Ranasinghe, D., & Yao, L. (2012). PeerTrack: A Platform for Tracking and Tracing Objects in Largescale Traceability Networks. In *Proceedings of the 15th international conference on extending database technology* (pp. 586–589). New York, NY, USA: ACM. doi:10.1145/2247596.2247672

Wu, Y., Sheng, Q. Z., Shen, H., & Zeadally, S. (2013, October). Modeling Object Flows from Distributed and Federated RFID Data Streams for Efficient Tracking and Tracing. *IEEE Transactions on Parallel and Distributed Systems*, *24*(10), 2036–2045. doi:10.1109/TPDS.2013.99

KEY TERMS AND DEFINITIONS

Agentification: A process that allows software to be converted into an agent in order to integrate its functionalities into a multiagent system.

Data Aggregation: A method that allows a device, in the Internet of Thing context, to collect and to summarize information of a group of devices.

DHT: The Distributed Hash Table (or DHT) is a structured Peer-to-Peer overlay which acts as a regular hash table, but storing (key, value) pair in a distributed fashion, and recovering the value in the $O(log\ n)$ number of messages.

Flocking: A group of agents, such as, birds or sheep, which seems to act as one, be it for traveling, living or feeding together.

IoT: The Internet of Things (or IoT) is a network of physical and virtual devices, connected to the Internet, which provides services to sense, to communicate, and to cooperate among them in order to achieve a goal.

Multiagent System: A group of autonomous software agents that interact among them to solve problems.

SPOF: The Single Point Of Failure (or SPOF) is a software or hardware component in which its malfunction or failure causes the entire system to cease operation.

System Scalability: A process that allows a system to continue to operate when the demand increases without affecting its performance.

Chapter 5
Mase:
A Multi-Agent-Based Environmental Simulator

Celia G. Ralha
University of Brasilia, Brazil

Carolina G. Abreu
University of Brasília, Brazil

ABSTRACT

This chapter presents research carried out under the MASE project, including the definition of a conceptual model to characterize the behavior of individuals that interact in the dynamics of land-use and cover change. A computational tool for analyzing environmental scenarios of land change was developed, called MASE - Multi-Agent System for Environmental Simulation. MASE enables agent-based simulation scenarios and integrates the influence of socio-economic and political dynamics through the interaction of agents with rules of land-use and planning policies and the environmental physical and spatial variables. MASE simulator was extended to implement the Belief-Desire-Intention (BDI) model, called MASE-BDI. MASE and MASE-BDI are discussed including the conceptual model complexity and statistical techniques of map comparison to land change models. Two real cases of the Brazilian Cerrado validate quantitative and qualitative aspects of MASE and MASE-BDI simulators. Finally, the authors present some auto-tuning aspects of adjusting simulation parameters of MASE-BDI.

INTRODUCTION

Models are crucial to represent complex and detailed reality since they can explicitly account for the state of knowledge, predict results, or act as objects of experiments. A model can represent real environmental patterns that embody essential and interesting aspects of the reality to be studied. Even simple conceptual models are critical components of biological disciplines, e.g., the earliest Lotka-Volterra model of species competition and predator-prey relationships (Lotka, 1925). Although this model is recognized

DOI: 10.4018/978-1-5225-1756-6.ch005

as not being very realistic, it plays a role in organizing themes such as the dynamics of a hierarchically organized system, competitive exclusion, and the cycling behavior in predator-prey interactions.

Since all models are by definition incomplete, the central issue of modeling process is whether the fundamental aspects of the phenomenon are represented. Considering ecological phenomena, what is interesting and significant is usually a set of individuals and their relationships - from the interaction of two individuals to the behavior of a population in its environment. But the limited human comprehension of complex biological systems arises a problem when attempting to dissect a phenomenon into more easily understood components. This challenge is compounded by human's current inability to understand relationships between the elements as they occur in reality. The real environment present multiple competing influences related to the broader context of time and space.

Models are used in land change science to better understand the dynamics of systems, to develop hypotheses to be tested empirically, and to make predictions or evaluate scenarios for use in assessment activities. The nature of the models, the selection of appropriate models and the nature of the abbreviations can be highlighted by the diversification of simulated models. In this sense, all modelers are using models whenever they use statistical tests or define a mathematical schema. The simulation of these models can help human comprehension of complex environmental systems. The land use and land cover change (LUCC) modeling strengthen the increasing need for tools able to capture the dynamics and outcomes of human actions through the use of individual-based and multi-agent models. Thus, LUCC dynamics are considered to remain a significant ecological challenge.

As cited by Brown et al. (2004), modeling is a major component of each of the three foci outlined in the science plan of the LUCC project (Turner II et al., 1995) of the International Geosphere- Biosphere Program and the International Human Dimensions Program.

- In Focus 1, on comparative land-use dynamics, models are used to help to improve our understanding of the dynamics of land-use that arise from human decision-making at all levels (households to nations). Surveys and decision maker interviews support these models.
- Focus 2 emphasizes the development of empirical diagnostic models based on aerial and satellite observations of spatial and temporal land cover dynamics.
- Finally, Focus 3 focuses specifically on the development of models of LUCC that can be used for prediction and scenario generation in the context of integrative assessments of global change.

This chapter presents the MASE (Multi-Agent System for Environmental simulation) project and focus on the importance of computational frameworks for modeling and simulating of environmental changes. The MASE project was developed in the Computer Science Department in association with the Biological Sciences Faculty of the University of Brasilia (UnB) during 2009-2015. The MASE project was defined considering Focus2 of LUCC project (Turner II et al., 1995) that emphasizes the development of empirical models based on satellite images that include observations of spatial and temporal land cover dynamics. The primary objective of MASE project is to define a conceptual model to characterize the behavior of individuals that interact in the dynamics of LUCC (Ralha et al., 2013).

The rest of the chapter includes a short discussion of basic modeling approaches for integrated environmental assessment, illustrated by some environmental multi-agent based simulators in the Background Section. In the Section MASE Simulator, the MASE conceptual model is presented focusing the evaluation of land-use change models, including dynamics of space, time and human choice with two different implementations: MASE and MASE-BDI. Next, MASE and MASE-BDI are illustrated with two real environmental cases using data from the Brazilian Cerrado Biome. In the Section Results, the MASE and MASE- BDI outcomes are validated using a spatially explicit model with statistical techniques of map comparison to land change models. Finally, some conclusions and future research directions are outlined.

BACKGROUND

In this section, some of the existing modeling approaches for integrated environmental assessment are presented. For a complete review of modeling approaches see Letcher et al. (2013). Each approach has its set of techniques to model and aggregate different systems processes, making possible for the user to have a grasp of the reality, analyzing the alternatives assessing the outcomes and communicating the results. All of these techniques may be used to develop models of complex systems, such as LUCC. To determine which technique is the best fit for a problem, it is needed to observe the varied purposes of modeling, the available data, and choose the appropriate approach given the requirements placed on the model or the system definition. Just to cite some well know approaches:

- System Dynamics (SD) modeling is a system formalism based on ordinary differential equations, which is formulated when the modeler convert the hypothesis into a 'stock and flow' representation. SD makes good learning tools and is generally user-friendly. The upside in LUCC investigation is that the treatment of space is extremely limited in the existing frameworks and building tools.
- Bayesian Networks (BN) use probabilistic relationships to describe the connections among system variables and are often used in decision-making and management applications, mainly when uncertainty is under investigation. BN definitions are useful to integrate different sources of information and derive the conditional probability among variables, reducing constraints imposed by lack of data. BN often require discretization of continuous variables, which may add substantial imprecision to variable relationships and model predictions, especially in LUCC. Couple component models involve combing models from different disciplines to come up with an integrated outcome, and therefore can incorporate deeper representations of systems. This characteristic may also result in compromising the breadth of the system to be represented, adding time and resources constraints.
- Individual-based Models (IBM) focus on the representation of the interactions between autonomous entities in a system portraying most often humans or biophysical entities. IBM are based on the multi-agent system paradigm that features a communal environment where a group of autonomous entities with its goals acts and communicate (Wooldridge, 2009). IBM biggest advantage is the discovery of emergent behavior, that is, large-scale outcomes that result from simple interactions and learning among individuals. Some of the IBM limitation is its highly complex structure, a large number of parameters, a high computational cost, and the not-so-easy reproducible results.
- Knowledge-based Models (KBM) are usually encoded into a knowledge base where an inference engine uses logic to infer conclusions. Although it can be very effective to use human experts to train a system, some problems may be too complex or uncertain to be formalized using KBM.

Models of land use tools to support the analysis of the causes and consequences of land use changes to better understand the functioning of the land use system and to support land use planning and policy. Models are useful for untangling the complex suite of socio-economic and biophysical forces that influence the rate and spatial pattern of land use change and for estimating the impacts of changes in land use. Furthermore, models can support the exploration of future land use changes under different scenario conditions (Verburg et al., 2004). Summarizing, land use models are useful and reproducible tools, supplementing our existing mental capabilities to analyze land use change and to make more informed decisions (Constanza & Ruth, 1998).

The distinctive characteristic of LUCC models is the difference between the projections for the quantity of change and the location where these changes will take place (Veldkamp & Lambin, 1996). Veldkamp (1996) summarizes the common functional structure presented by LUCC models comprising four steps:

1. They answer the *when?* question, and establish its temporal extent and resolution;
2. LUCC models have rules that manage the amount of change (the *how much?* question);
3. The models determine where the projected change will take place (the *where?* question);
4. The models apply the changes in an appropriate way, including external restrictions (the *how?* question).

The IBM are increasingly popular techniques for describing LUCC since it explicitly deals with the diversity of land-use decisions (Valbuena et al., 2008). This characteristic is particularly useful in LUCC scenarios where the combined interaction of all entities can create rich emergent behaviors (Bonabeau, 2002). The main disadvantage of IBM compared to standard mathematical modeling is in testing the robustness of the results and the theorems which they are trying to prove. IBM (and other modeling approaches) have been integrated with automated tools to facilitate assessment and deal with the interdisciplinary nature of complex environment management issues. Thus, some environmental multi-agent based simulators are briefly discussed here, e.g., *Cormas*, *NetLogo,* and *TerraME.*

Simulation Tools

This review will focus on agent-based modeling software that provides a framework for users to create, execute and visualize the model and simulation output. We choose to discuss simulation tools based on the multi-agent system (MAS) paradigm, highlighting its features and capabilities. A complete comparison analysis is beyond the scope of this chapter and can be found in Castle and Crooks (2006), Railsback et al. (2006), Nikolai & Madey (2009), Allan (2011) and Salamon (2011).

The Cormas (Common-Pool Resources and Multi-agent Systems) was developed by the French Agricultural Research Center for International Development (CIRAD) as a multi-agent simulation tool that focuses on natural resource management, namely studying the interaction of human societies with the Earth's ecosystem (Le Page et al., 2000). The Cormas simulation platform is based on the VisualWorks programming environment that allows the development of applications in the Smalltalk object-oriented language. Cormas pre-defined entities are represented as Smalltalk generic classes from which, by specialization and refinement, users can create distinct entities for their model. In Cormas there are three pre-defined programmable agents:

1. Space agents -- units of space (cell) characterized by its field of perception. Departure and arrival methods enable the user to program the agent's departure from a cell and its arrival at another cell;
2. Communicating agents -- entities that have internal state and the capacity of exchanging messages; and
3. Space -- communicating agents: a combination of the entities above.

Cormas also works with steps of predefined interval of times. At every step, the agents perform their actions in a sequential manner and update their status. There is no other rational feature in the agents, other than the behavior set by the agent rules. The software is now on its 2014b version. There are a various number of model applications available, in addition to some written material.

The NetLogo is currently the most popular agent-based simulation environmental tool. It is a programmable modeling environment for simulating natural and social phenomena (Wilensky,1999). NetLogo has been developed at the Center for Connected Learning and Computer-Based Modeling at the Northwestern University. Modelers can give instructions to hundreds or thousands of independent agents all operating concurrently, which makes it possible to explore the connection between the micro-level behavior of individuals and the macro-level patterns that emerge from the interaction of many individuals. It also includes a user interface builder and other tools such as an SD modeler. A NetLogo program has three parts:

1. A section that says what kinds of agents there will be and names the variables that will be available to all agents (the global variables);
2. A setup procedure that initializes the simulation; and
3. A 'go' procedure, which is repeatedly executed by the system to run the simulation.

The NetLogo platform offers support for reactive agent systems. Modeling Belief-Desire-Intention (BDI) agents (Bratman, 1987) and explicit symbolic message exchange are not supported. Nevertheless, a few non-official BDI and FIPA (Foundation for Intelligent Physical Agents) Agent Communication Language (ACL) library extensions were developed by the community. There are a few plugins that allow GIS integration as well. Its programming language includes many high-level structures and primitives that significantly reduce programming effort. The language is based on Logo, a dialect of Lisp, and contains many of the control and structuring capabilities of a standard programming language. NetLogo was designed to allow users with almost no training to rapidly prototype an IBM model. It comes with an extensive library of sample models and code examples that help beginning users get started authoring models. There are extensive documentation and a highly active community being in its 5.0.5 version covered by a GLP license.

The Terra Modeling Environment (TerraME) is an open source software developed by the Brazilian National Institute for Space Research (INPE) and distributed under the GNU LGPL license (Carneiro et al., 2013). It supports multi-paradigm and multi-scale modeling of coupled human-environmental systems and enables models that combine agent-based, cellular automata, SD, and discrete event simulation paradigms. TerraME's central design principle is flexibility. It does not enforce a unique modeling paradigm but provides the tools needed by the modeler.

TerraME has a GIS interface for managing real-world geospatial data and uses Lua, an expressive scripting language. It has building blocks for model development, allowing the user to specify the spatial, temporal and behavioral parts of a model independently. When defining behavior, agents are purely reactive. The software provides a genetic algorithm for model calibration. Lucc-ME is an extension of TerraME for spatially explicit LUCC modeling. Most of TerraME applications are regarding Nature-Society Interactions, and it is widely used in INPE's research projects. There is some written documentation and tutorials to guide new users.

Table 1 summarizes some general characteristics of the presented tools and shows MASE for comparison purposes. MASE features are going to be presented in Section 3. According to this brief review, most IBM software implement individuals as an abstraction of a computational object using classical object-oriented programming. The exchange of parameters simulates the interactions and autonomy among individuals. CORMAS implements a MAS using reactive agents. Other tools, such as Netlogo, has extensions that simulate MAS and BDI. A standard feature in all the individuals is the purely reactive behavior. A practical-reasoning approach is not usual among the IBM software and motivated the MASE project group to develop a framework capable of executing goal-oriented, practical reasoning and cognitive behavior for agents.

MASE SIMULATOR

The conceptual model of MASE is presented based on two different dimensions:

1. The systematic and structured empirical characterization of the model; and
2. The conceptual structure definition.

Regarding the implementations, two different versions of the simulation tool are going to be presented (MASE and MASE-BDI), highlighting their similarities and differences (Abreu et al., 2015). Also, the complexity of MASE conceptual model, focusing the assessment of land-use change models, including dynamics of space, time and human choice is addressed (Agarwal et al., 2002). Finally, some auto-tuning aspects are discussed, which are very relevant to adjust simulation parameters in land change frameworks.

Conceptual Model Dimensions

Related to the systematic and structured empirical characterization of the model, the LUCC simulation tools need to represent and simulate human decision-making processes, whether the multi-agent models are emphasized by LUCC modeling. By observing the dynamics of LUCC variables, it is possible to

Table 1. General characteristics of IBM simulation tools

Simulation Tool	Type of Model	Agent's Rationality	Spatial Complexity
Cormas	IBM with multi-agent simulation	Reactive	Spatially explicit Raster maps
NetLogo	IBM with a BDI plugin	Reactive	Raster maps with GIS extension
TerraME	Cellular automata and IBM	Reactive	Vector and Raster maps
MASE	IBM with multi-agent simulation	Objective oriented	Spatially explicit Raster maps and variables
MASE-BDI	IBM with multi-agent simulation and BDI	Objective oriented	Spatially explicit Raster maps and variables

explore causes, consequences and formulate useful scenarios during the decision-making process and planning. The implementation of human decision-making processes is the main strength of agent-based models. The behavioral response functions (that represent these processes) require knowledge support from qualitative and quantitative empirical sources. Thus, the MASE project have used a set of specific methods, commonly used in most models, based on the methodological parameterization process for human agents discussed in Smajgl et al., 2011.

Smajgl et al. (2011) built a comprehensive framework with the set of the most commonly used methods for the challenge of parameterizing behavioral responses of humans empirically. Some examples are the use of expert knowledge, participant observation, social surveys, interviews, census data, field or lab experiments, role-playing games, cluster analysis, dasymetric mapping and the Monte Carlo method. The parameterization process of empirical data requires the use of techniques to develop the behavioral categories and for scaling the whole population. To produce the MASE conceptual model we used: expert knowledge, census data and field or lab experiments. Sample-based data was gathered and upscaled to represent the entire populations.

The expert knowledge was captured by a variety of methods from experts from the Brazilian Institute of Environment and Renewable Natural Resources (IBAMA) and the Ecology Department of the Institute of Biology of the University of Brasília. The aggregated census data, including GIS data, was gathered from the Agricultural Census of 2006, provided by the Brazilian Institute of Geography and Statistics (IBGE). For the lab experiments, a series of simulations were made to calibrate the model, exploring how each model variable affects the simulation results, to adjust the conceptual model to the documented expected behavior. For a full description of the MASE agent's parameterization see Section 2.1 of Ralha et al. (2013). In this section, only a brief overview is provided just to clarify MASE components and operation.

MASE purpose is to explore human agents and physical environment behavior regarding the LUCC dynamics. The model is structured through transformation agents, mainly farmers and ranchers, representing humans and the socioeconomic changes. The biophysical aspects are mapped into the multi-agent environment, where the maps of the location are transformed into a grid to allow agents interaction (move and alter the land use and cover). The biophysical variables may be spatially explicit (maps and locations), such as slope, roads, watercourses or static values.

The political aspect is also drawn into the model through governmental policies in the space, such as maps of the spatial plan that determines the land use to each region, or general conservation policies, e.g. each farm must preserve 20% of the original natural coverage of its property. All of these parameters are defined by the user and can be altered to explore different scenarios.

The interaction of the IBM in the simulation is done with cycles of perception and action of the transformation agents in the environment. The perception may be of the physical aspects of the grid cell the agent are in, or may consider the neighborhood. It can contemplate a single moment of time or take into account the history of actions of a single agent or a group of agents. The transformation agents' actions can vary from a purely reactive behavior to objective-oriented. In the MASE-BDI model, transformation agents have individual knowledge bases and intentional behavior. The BDI implementation will be detailed in Section 3.2. The agents may interact, cooperating or competing for the land resources.

Each cycle is a simulation step. The emergence behavior arises from each time-step. At the end of a simulation (user defines its duration), a collection of goodness-of-fit measures are generated, as well as a set of maps showing the simulated land use and land cover changes. The thorough ODD documentation of MASE individual based model is presented in Sections 2.2.1 to 2.2.3 of Ralha et al. (2013). The

Overview, Design concepts and Details (ODD) protocol proposed by Grimm et al., (2006, 2010) is used to keep consistency in MASE documentation, according to standards to agent-based models. For the ODD protocol, the agent-based information follows a structured order of seven elements grouped into three blocks (called the ODD sequence):

1. **Overview:** the context and general information are provided first. The aim is to give a very quickly idea of the model's focus, resolution, and complexity, including three elements: purpose, state variables and scales, process overview and scheduling.
2. **Design Concepts:** more strategic considerations. The aim is to provide a common framework for developing and communicating agent-based models. The purpose is to link model design to general concepts identified in the field of Complex Adaptive Systems, including questions about behavior emergence, the type interactions among individuals, whether individuals consider predictions about future conditions, or why and how stochasticity is considered. We have used five items: emergence, sensing, fitness (goal-based), interaction and observation.
3. **Details:** more technical details include three elements: initialization, input, submodels, to present the details that were omitted in the overview. The submodels implementing the model processes are described in length, and the information required to completely re-implement the model and run the baseline simulations is provided in this block.

MASE and MASE-BDI Implementations

The MASE simulator is based on a MAS approach. The Tropos methodology was used for the design project (Bresciani et al., 2004). For the implementation, the Java Agent Development Framework (JADE), version 4.0 (release date 04/20/2010), was used (Bellifemine et al., 2001, 2008).

The MASE architecture is organized in three hierarchical layers: from the top, User Interface (UI), Pre-processing and Agent layer. The UI layer is responsible for the interactions with the user through the graphical interface. It is also the layer where the user defines the model characteristics and visualizes the ongoing simulation and its results (maps and metrics). The second layer is responsible for all the pre-processing needed to support the Agent's layer: the image processing of the simulation, the configuration of the user model and the auto-tunning module.

The Agent layer is the agent implementation of the multi-agent system. It is composed of classes of agents organized in a hierarchical order, which holds the logic and behavior of the transformation agents. MASE uses four different agent classes: grid manager (GRIDM), spatial manager (SM), transformation manager (TM), and transformation agent (TA). The GRIDM, SM, and TM are goal-based agents, while the TA are reactive agents (reflexive with an internal state).

The environment characterization of the multi-agent system was defined according to Wooldridge (2009) and Russel (2010). The simulation grid is partially observable by agents, stochastic, sequential, dynamic, continuous, multi-agent and uses a competitive approach to interact. The detailed MASE architectural aspects are presented in Section 2.3.1 of Ralha et al. (2013).

JADE framework was adopted in the MASE project since it is FIPA compliant. FIPA is an organization connected to the Institute of Electrical and Electronics Engineers (IEEE) that is responsible for specifying standards for the development of technologies based on intelligent agents. The ACL used to provide the exchange of messages among agents is the FIPA-ACL. Considering the agents' behavior aspects, MASE implements simple approaches (OneShotBehaviour and CyclicBehaviour) and composed approaches

(SequentialBehaviour and FiniteStateMachineBehaviour) from JADE middleware. Considering the interaction protocol, MASE implements a hybrid approach using cooperative and competitive relations.

The MASE-BDI is an extended version of MASE. Thus it followed the general definitions, functionalities and adapted the architecture to the mentalistic BDI model of agent reasoning (Bratman, 1987). The MASE-BDI deploy agents with individual knowledge bases and intentional behavior to represent the complex, dynamic and error-prone LUCC environment. MASE-BDI has refactored MASE code regarding the agents, including their internal logic, communication and interaction protocols, and the way agents perceive and act in the environment.

The MASE-BDI was developed using the JADEX framework (Pokahr et al., 2003). JADEX is an extension of JADE that provides both a middleware and a reasoning-oriented system. The reasoning oriented feature emphasizes on rationality and goal-directedness, with a reasoning engine based on the BDI model. To access functionality of the JADEX system, a Java API is provided for basic actions such as sending messages, manipulating beliefs or creating subgoals.

In MASE-BDI implementation, each agent is coupled with an individual Belief Database. From the outside perspective, an agent is perceived as an entity that receives and sends messages. From the BDI reason engine perspective, incoming messages, as well as internal events and new goals serve as input to the agent's internal reaction and deliberation mechanism.

Based on the results of the deliberation process these events are dispatched to already running plans, or to new plans instantiated from the plan library. Running plans may access and modify the belief base, send messages to other agents, create new top-level or subgoals and cause internal events. MASE-BDI uses the plan-library approach defined by JADEX to represent the plans for each agent. For each plan, a plan head defines the circumstances under which the plan may be selected and, a planning body specifies the actions to be executed.

The reaction and deliberation mechanisms are generally the same for all agents. The behavior of a specific agent is therefore determined solely by its concrete beliefs, goals and plans, set by the user in the conceptual model. The beliefs are represented by Java Objects contained in the belief base, as named facts that can be directly manipulated by setting, adding or removing facts. The beliefs are used as input for reasoning engine by specifying certain belief states, e.g., as preconditions for plans.

Goals are concrete, momentary desires of an agent. A deliberation mechanism is responsible for managing the state transitions of all goals. In MASE-BDI, a perform goal is directly related to the execution of actions. Therefore, the goal is considered to be complete, when the related actions have been executed, regardless of the outcome of these actions.

Complexity Model

The assessment of land-use change models, including dynamics of space, time and human choice is addressed by Agarwal et al. (2002) in a detailed study of environmental tools that analyzed tendencies and methodologies. According to this study, the ultimate goal of modeling the dynamics of man and the environment involves high complexity in the three dimensions. As a result, they define six increasing levels of complexity for human decision-making:

1. Without human decision-making (only the biophysical variables in the model);
2. Human decision-making determined by population size or population density;

3. Human decision-making is seen as a probability function depending on socioeconomic or biophysical variables beyond population variables, without feedback from the environment to the function that determines the decision;
4. Similar to (3) but with feedback;
5. A type of agent, whose decisions are modeled in relation to choices made and based on variables that affect other processes and outcomes; and
6. Similar to (5), but with multiple agents, where the model may also be able to make decisions on domains, as the time steps are processed, or about interactions between agents decision makers.

Considering the aforementioned classification, MASE achieved the level (6) of complexity for modeling the dynamics between man and the environment when considering the dimensions of dynamics of space, time and human choice. As presented, MASE has multiple agents that can make decisions of LUCC exploration considering different domains, including areas that are already occupied (roads), water courses or protected areas. MASE agents maintain their autonomy to move and explore using different exploitation rates in the simulation space keeping interactions among the various agent types (individuals) and groups of agents (condominium). By the other hand, the three pre-defined agents in Cormas do not achieve the level (6) of complexity since they have no autonomy or rational feature to decide about their space exploitation.

Auto-Tunning Aspects

Considering the parameter tuning methods, there are many algorithms in the literature, such as Active Harmony (Ţăpuş, 2002), Parallel Rank Order (Tabatabaee, et al., 2005) and Nelder-Mead Simplex (Nelder & Madey,1965). These algorithms were primarily designed and employed for high-performance applications and kernels with the goal of maximizing performance: Dakota (Adams, et al., 2014), OSKI (Vuduc et al., 2005), OpenTuner (Ansel et al., 2014). By the other hand, the land change modeling and simulator frameworks involve a large number of parameters related to individual interactions resulting in complex and significant processing demands. Thus, the use of parameter tuning to improve application analysis results of land change simulations is a novel use of auto-tuning techniques.

In MASE project the application of auto-tuning has presented initial, but notable results, with the reduction of a 6 million combination search space parameters to less than 0.00115% evaluations. With the quick tune of the simulation parameters associated with the application of parallel execution MASE-BDI reduced the simulation time compared to MASE, using the Brazilian Cerrado DF-Case (presented in Section 4) by at least 82x while maintaining the simulations quality.

Also, experimental results show that executing auto-tuning algorithms in parallel leads to speedups of approximately 11x compared to sequential execution with 16-CPU cores. The experiments design and results of the tunning use in the MASE Project are discussed in Coelho et al. (2016).

REAL ENVIRONMENTAL CASES

Two real environmental cases are going to be presented to illustrate the MASE and MASE-BDI simulators. An important area for LUCC studies is the process of deforestation on the Brazilian Amazon Rainforest. Nevertheless, LUCC research has pointed out an even more endangered biome in Brazil: the Brazilian

Cerrado (woodland savanna). The first case is related to the Federal District (Distrito Federal - DF) area (5,789 km2) using the Federal District Spatial Plan (Plano Diretor de Ordenamento Territorial - PDOT) that sets out the specific areas that can accommodate agriculture or urbanization for the agent's behavior (Abreu et al., 2014). The second case covers the Integrated Region Economic Development of the Federal District (Região Integrada de Desenvolvimento do Distrito Federal e Entorno - RIDE-DF), comprising 56,400 km2, covering the region of 35 Brazilian municipalities distributed in the states of Goiás, Minas Gerais, and DF.

The cases are related to the Cerrado, the second largest biome in South America, covering 22% of the Brazilian territory with 204,7 million hectares (Sano et al., 2008, 2010). The Cerrado harbors outstanding biodiversity, being included in the world's 25 principal hotspot areas, with considerable endemism and less than 30% remaining natural vegetation, requiring urgent action for conservation and land use policies. The Cerrado is also unique in the sense that they serve as corridors for species inhabiting neighboring biomes. Unfortunately, this biologically diverse ecosystem is poorly protected. Only 2.2% of Cerrado's original area has specific policies for environment protection, contrasting 80% of Amazon total area with specific environmental policies (Klink & Machado, 2005).

These changes come mainly from the replacement of native vegetation for agricultural activities since the Cerrado is considered one of the last agricultural production frontiers in the world. The main activities in the region focus on soybean production and large-scale grazing. The conversion of savannas for soy production is accelerating to meet the demand for feed rations for cattle and swine feedlots (McAlpine et al., 2009). Soybeans and soy products are amongst the largest of Brazil's export commodities, and the Cerrado support the largest cattle herd in the country (Klink & Moreira, 2002). The expansion of these activities is driven by a series of interconnected socio-economic factors, often encouraged by government policy. A great extent of LUCC studies tried to determine proximate causes and driving forces of Cerrado deforestation, in an attempt to understand the land use dynamics and its consequences (Geist & Lambin, 2001, 2002) (Chowdhury, 2006) (Morton et al., 2006) (Boucher et al., 2011).

Despite much research, there is currently no agreement as to the main causes of Brazilian biomes deforestation. This incertitude is partly due to the lack of an established theory on human-environment interaction. On the other hand, there is a clear sense of the LUCC scientific community that human activities play a central role in the land use system (Lambin & Geist, 2001). These studies reinforce the thesis that LUCC modeling efforts should attempt to represent the multiples drivers of human-environment interaction at different spatial and temporal scales (Lambin et al., 2003) (Rindfuss et al., 2004).

The Brazilian Federal District case study developed in MASE and its experiments are detailed in Ralha et al. (2013). The simulations were performed in steps, where each step corresponds to a week time in chronological time. The total area of study was divided into cells of one hectare. The physical environment is spatially represented by a set of layers of GIS data, also called proximal variables. The layer set determines the physical environment of any given point in the simulation grid.

Both the landscape system and the human system have their perceptions relating to the characterization of the physical environment. The physical state of the cells corresponds to the set of real spatial data including six proximal variables:

1. Water courses (rivers);
2. Water bodies (lakes);
3. Buildings;
4. Highways;

5. Streets; and
6. Protected areas.

In this first experiment using cognitive agents, the human factor over the land is represented by two types of transformation agents: farmers and ranchers.

The political aspects are also taken as a compelling force in the simulation, translating the PDOT onto an influence matrix for the transformation agents. Figure 1 illustrates the details of a rural Macrozone established in the PDOT. The two images show land use areas (green/gray colored areas) with a political and economic incentive for agricultural exploration (a- Zone of Diversified Land Use) or areas where there are law restrictions for agricultural exploration (b- Zone of Controlled Land Use). To simulate this political influence into the agent's perception, the PDOT maps are taken as parameters of the simulation and transformed in proximal variables in the simulation. The different pieces of information on the maps are used to create a probabilistic model that reflects the likelihood of exploration under this spatial policy. For example, a rancher agent perceives that it is located in a Zone of Diversified Land Use and that there is some economic incentive for agricultural exploration there. PDOT is set as a belief in the BDI model, and it will shape the agent behavior.

The second study area is related to the RIDE-DF (56,400 km2), covering the region of the Federal District, 18 municipalities of the state of Goiás (Abadiânia, Água Fria de Goiás, Águas Lindas de Goiás, Alexânia, Cabeceiras, Cidade Ocidental, Cocalzinho de Goiás, Corumbá de Goiás, Cristalina, Formosa, Luziânia, Mimoso de Goiás, Novo Gama, Padre Bernardo, Pirenópolis, Planaltina, Santo Antônio do Descoberto, Valparaiso de Goiás and Vila Boa) and three municipalities of the state of Minas Gerais (Buritis, Cabeceira Grande e Unaí). The RIDE-DF is an area entirely covered by the Cerrado (the more pressured biome from the intensification of land use in Brazil). Figure 2 shows the perspective of Brazil, the Cerrado Biome (light gray), and the RIDE-DF (dark gray).

The RIDE-DF has 26 areas with some level of environmental protection, 11 Strict Nature Reserve (Protected Area) with sustainable use of natural resources, representing approximately 12.5% of its territory or nearly 866,144 hectares. Even with a significant number of protected areas, the region has suffered several environmental conflicts, a fact perceived by the poor state of preservation of many Parks, Reserves, and even the Environmental Protection Areas. There are many cases of protected areas invaded by illegal constructions, natural water sources contaminated by construction garbage and sewage, besides

Figure 1. Details of Rural Macrozone - PDOT DF
Source: GDF, 2009

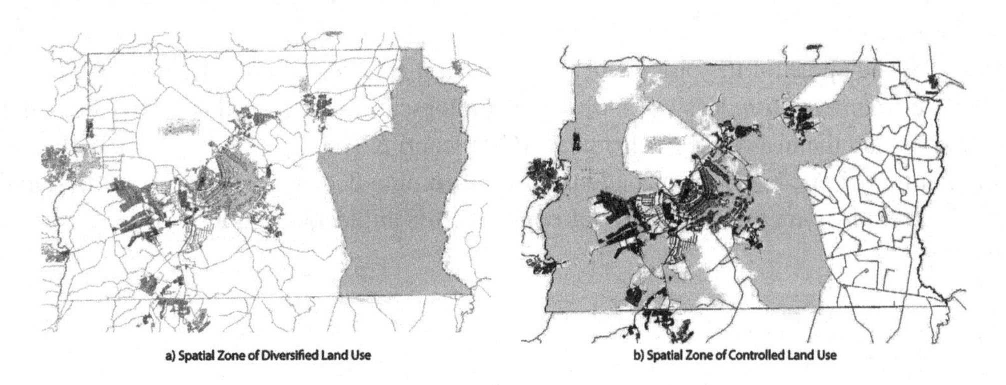

Figure 2. Overview of the Brazilian map with the Cerrado biome and RIDE-DF

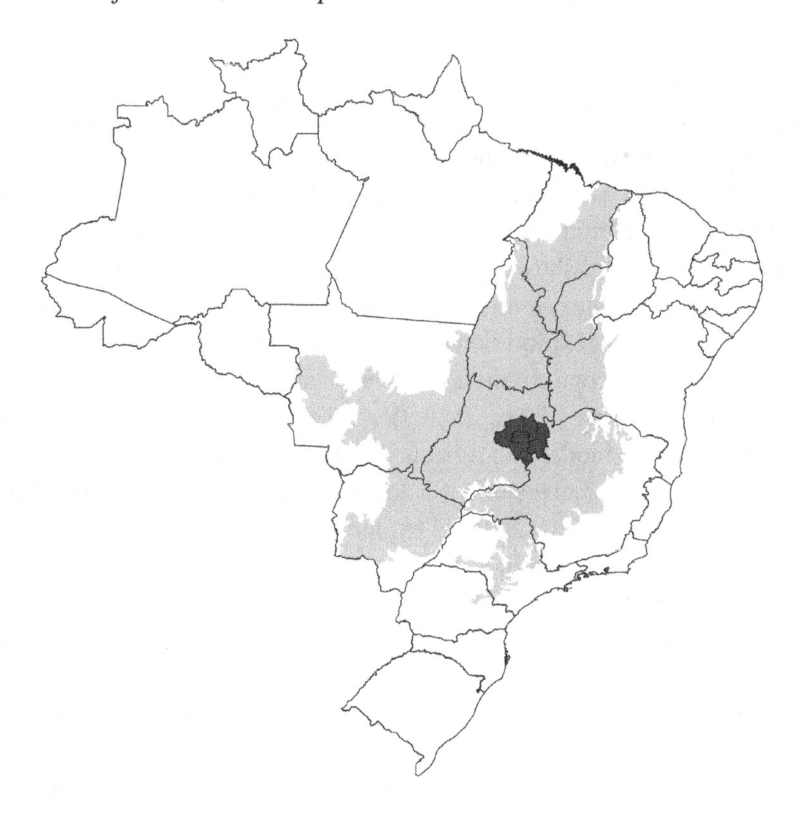

the removal of the land cover. The process of land occupation in the Federal District and surrounding areas for the past ten years has generated many conflicts derived from the illegal appropriation of public lands and disobedience to law threaten protected areas. The disordered occupation has worsened the quality of life of many RIDE-DF residents.

The RIDE-DF was created by law in 1998, but it was only formally recognized by the government in the National Policy for Regional Development (*Política Nacional de Desenvolvimento Regional* - PNDR) in 2011. The primary objective of the government was to develop regional policies that could strengthen strategic capacity, monitoring and evaluation practices in integrated regional plans and programs. PNDR text informs the population that specific regional policies on land use could "reduce regional inequalities and unlock the potential of the country's regions." Unfortunately, "unlock the potential" comes with high social and environmental costs, but no studies are assessing the real outcomes of regional land use policies within RIDE-DF.

The intensity and speed of the process of destruction of the Cerrado require urgent action to slow down landscape fragmentation, loss of biodiversity, biological invasion, soil erosion, water pollution, land degradation, and the heavy use of chemicals (Cavalcanti & Joly, 2002). To reconcile land use and conservation of this biome is the main challenge for the coming decades. Therefore, understanding the dynamics of these phenomena is an essential condition for an adequate policy of sustainable use of resources.

The RIDE-DF data has two milestones involving the period 2002-2008. Thus, we used these two time instants with satellite images of LANDSAT ETM, classified by the PROBIO initiative (IBAMA/Environment Ministry (MMA) or Remote Sensing Center (CSR)-IBAMA/MMA), as presented in Figure 3. Remaining natural vegetation is presented in green and anthropic activity area in yellow.

Different from the six proximal variables used in the Brazilian Federal District case study, in the RIDE-DF case we used eleven variables:

1. Railways;
2. Highways;
3. Water courses;
4. Body of water;
5. Strict nature reserve (IUCN – Category Ia)- integral protection conservation units;
6. Protected area with sustainable use of natural resources (IUCN - Category VI);
7. Terrain - land relief;
8. Soil;
9. Urban area;
10. Reforestation areas; and
11. Degraded areas.

RESULTS

In this section, MASE and MASE-BDI outcomes are validated by evaluating their behavior, using a spatially explicit model with statistical techniques of map comparison to land change models (Pontius et al., 44). MASE and MASE-BDI obtained good results when compared to similar applications. For this comparison, we used two statistics: Null Model (NM) resolution and the Figure of Merit (FoM).

Figure 3. The RIDE-DF land cover LANDSAT images classified by the PROBIO program

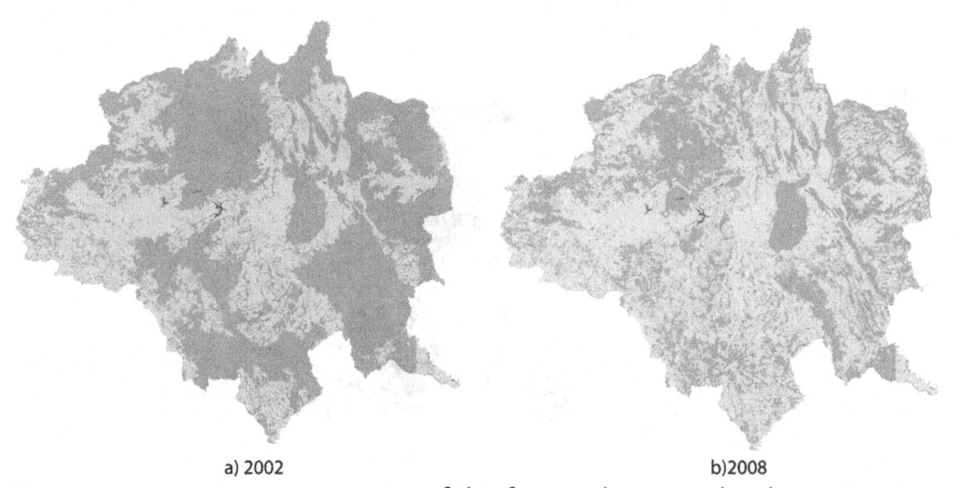

**For a more accurate representation of this figure, please see the electronic version.*

Some assessment of simulation models, concerning quantity and allocation, are also discussed (Olmedo et al., 2015).

At the end of each simulation, MASE and MASE-BDI presents a set of maps and metrics of the land use changes in a very straightforward and clean user interface. Figure 4 illustrates the user interface and one of the maps generated as a result. In Figure 4, the Graphic Results show the predicted changes in the land cover, rendered after the actions of the land transformation agents in the initial map. It is possible to render a final map highlighting the different land use types, according to the research question. This characteristic is why MASE-BDI provides a spatially explicit simulation model.

The simulation framework calculates the two paramount components of any LUCC simulation, such as quantity disagreement (i.e., net change) and location disagreement (i.e., swap change), which sum to the total disagreement. While the quantity disagreement derives from differences between the maps with regards to the number of pixels for each land use category, the location disagreement is the disagreement that could be resolved by shifting the pixels spatially within one map. The maps are generated assigning different color codes for each type of land use. With any standard GIS framework, it is possible to classify those raw maps produced at the end of each simulation, highlighting various characteristics of the results, as illustrated in Figure 5. The example shows the classification of crops and pastures following a simulation on the RIDE-DF.

For each run of the simulation, the software was set with a variable number of transformation agents. In order to compare quantitative and qualitative aspects of the multi-agent model proposed we applied two scientifically rigorous statistical techniques of map comparison to land change models developed by Pontius et al. (2008). These techniques allow different applications to be summarized and compared: the NM resolution and the FoM. The definition of the NM is a prediction of complete persistence, i.e. no change, between the initial and the subsequent time. Therefore, the accuracy of the NM is 100% minus

Figure 4. MASE-BDI user interface and graphical results

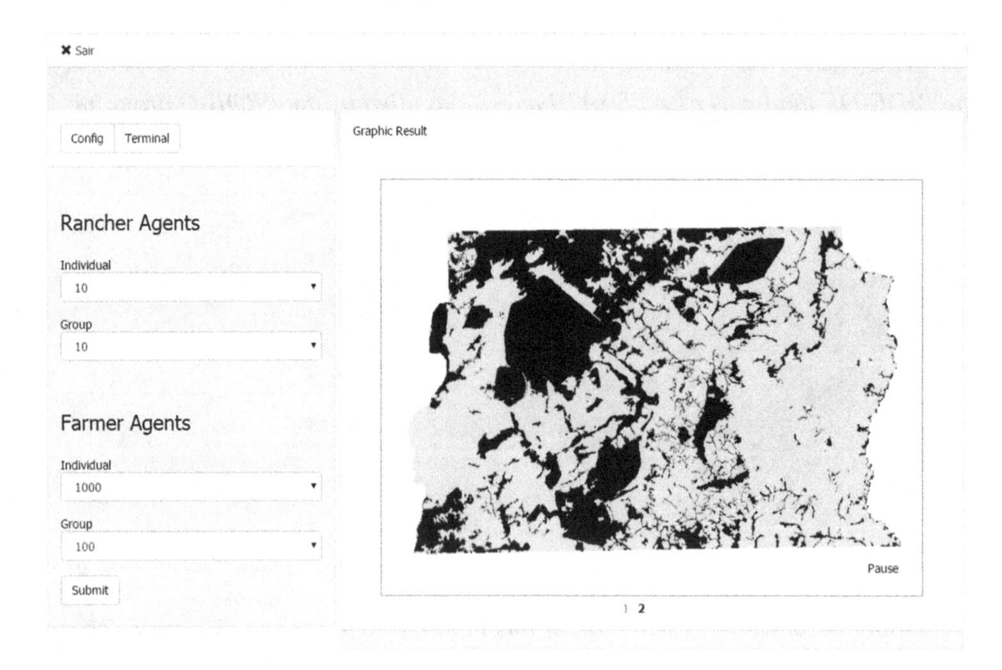

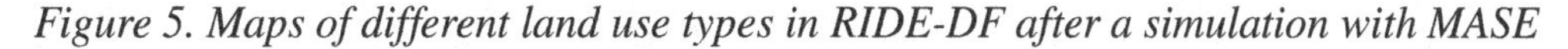

Figure 5. Maps of different land use types in RIDE-DF after a simulation with MASE

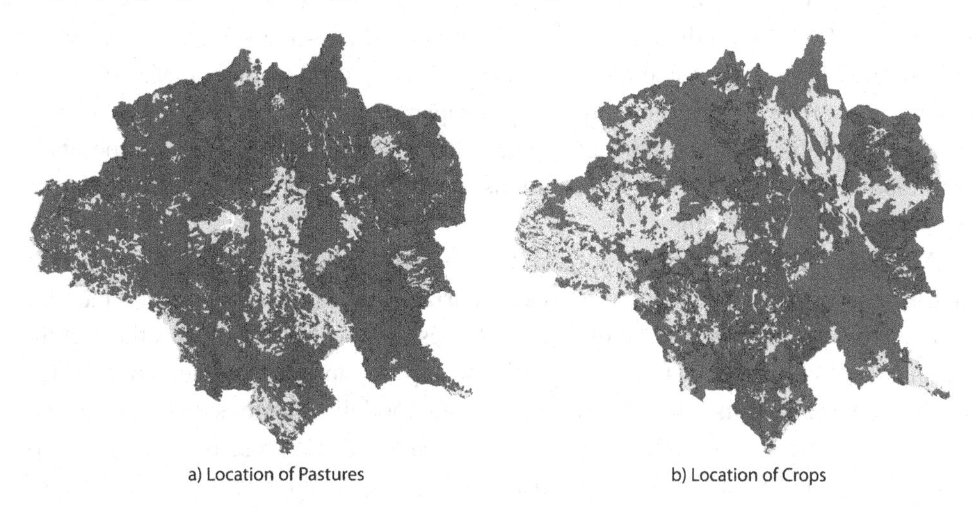

the amount of observed change. MASE results were consistent and statistically better than other similar frameworks, reviewed and compared by Pontius et al. (2008), where the review shows that 50% of the simulation frameworks are worse than the NM.

The second metric, the FoM is a statistical measurement that derives from the information of the different colors and its location in Figure 6. FoM is the ratio of the intersection of the observed change and predicted change to the union of the observed change and predicted change, which can range from 0% (no overlap between observed and predicted change) to 100% (perfect overlap between observed and predicted change - an absolutely accurate prediction). Considering the FoM, the most accurate applications are the ones where the amount of observed net change in the reference maps is larger. MASE was able to use the correct or nearly correct net quantities for the categories in the prediction map. A complete analysis of MASE results is published in Ralha et al. (2013).

Figure 6. MASE-BDI model predictions for DF showing the predicted change 2002-2008

Some simulation results are illustrated in Figure 6 and Figure 7. Figure 6 presents the DF area considering the 2008 MASE-BDI predicted map. Figure 7 shows a three-map comparison using: a reference map of initial time (2002); a reference map of subsequent time (2008); and an MASE-BDI prediction map of the subsequent time (2008). The three-map comparison specifies the amount of the predictions accuracy that is attributable to land persistence versus land change. It is possible to pinpoint the locations where the simulation was accurate and the locations where the predicted land cover changes different from the observed land cover change.

Every MASE and MASE-BDI shows a final map to sum up all land changes that occurred during the simulated period of time. The statistic also summarizes the final results of the land change of the simulation. The next step to improve the quality of the results would be to assess the simulated transition among land categories through time. A significant step to improve the results of LUCC simulators would be to compare the outputs from models and how the modeling options influence the quantity and allocation of simulated transitions. Although MASE and MASE-BDI present good results compared to similar simulators, it is not yet possible to compare output maps from pairs of model runs with respect to a reference map of transitions during the validation interval. According to Olmedo et al. (2015), the separation of quantity and allocation of the transitions is a helpful approach to demonstrate how models work and to describe pattern validation. They recommend that the first step is to assess the quantity of each transition and to determine the cause of the variation in quantity among model runs. The second step is to assess the allocation of transitions and to determine the cause of the variation in allocation among model runs. This latter validation process is intended for future investigation.

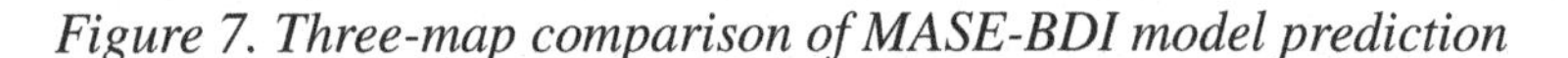

Figure 7. Three-map comparison of MASE-BDI model prediction

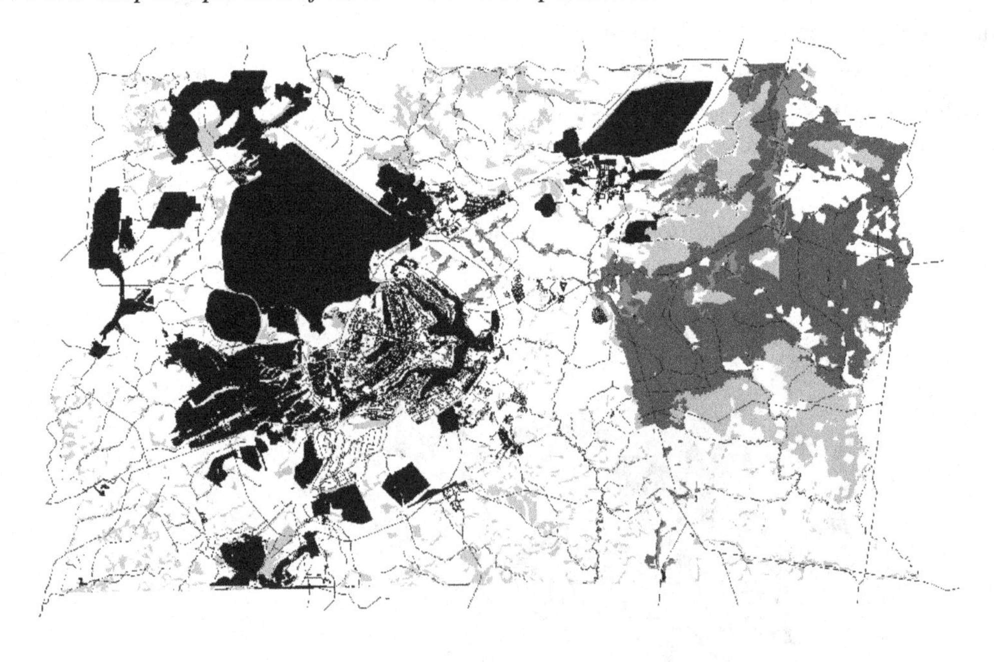

CONCLUSION

This chapter gives a concise outline of the research carried out in the MASE Project. The experience demonstrates that the definition and implementation of a LUCC framework is a challenging task to integrate computer and biology researchers. As presented, the focus of MASE and MASE-BDI agent-based simulators can represent the complexity of individual interactions through the general rule for ecological applications, a feature that is necessary for adequately support decision-making processes as described in Section 3.

Although, there are many popular and peer-reviewed land change frameworks in the literature, e.g., IDRISI, ArcGIS, DinamicaEGO, CORMAS, Netlogo, RePast, none implement the practical reasoning engine with goal-based agents. Thus, the MASE-BDI land change simulator uses an internal architecture based on the practical reasoning model being the only agent-based simulator that implements rational agents with the BDI mentalistic approach. Another important aspect of MASE-BDI simulator is the use of the auto-tuning module to reach efficient and parallel simulation executions, what is a novel use of auto-tuning techniques to improve application analysis results of land change simulations.

The Brazilian Cerrado cases presented by the MASE and MASE-BDI simulators illustrate the correctly predicted land change according to well-established statistical evaluation methods using the NM resolution and the FoM metrics (Section 5). Compared to popular land change frameworks, MASE-BDI is the only multi-agent simulator that implements the BDI mentalistic model with an auto-tuning module. In this sense, researchers and practitioners can benefit from MASE Project overview when focusing the problem to model and simulate environmental phenomena.

It is generally agreed by the ecology community that better decisions are implemented with less conflict and more success when driven by stakeholders. Although we had the participation of end users in model development, there is still the need to involve stakeholders into validation and elaboration of simulation scenarios. Stakeholders familiar with real land use decision-making could provide insight about the important relationships between the key entities in the model. As a future work, we aim to gather end user knowledge and the already implemented techniques (tunning and statistics) to improve the MASE framework, to create better awareness for the real problem and to identify future research priorities.

REFERENCES

Abreu, C. G., Coelho, C. G. C., & Ralha, C. G. (2015). MASE-BDI: Agents with Practical Reasoning for Land Use and Cover Change Simulation. In *Proceedings of the 6th Workshop of Applied Computing for the Management of the Environment and Natural Resources*. XXXV Congress of the Brazilian Computer Society.

Abreu, C. G., Coelho, C. G. C., Ralha, C. G., & Macchiavello, B. (2014). A Model and Simulation Framework for Exploring Potential Impacts of Land Use Policies: The Brazilian Cerrado Case. In *Proceedings of the 47th Hawaii International Conference on System Sciences (HICSS)*, (pp. 847-856). doi:10.1109/HICSS.2014.113

Adams, B. M., Bauman, L. E., Bohnhoff, W. J., Dalbey, K. R., Ebeida, M. S., Eddy, J. P., . . . Wildey, T. M. (2014). *Dakota, A Multilevel Parallel Object-Oriented Framework for Design Optimization, Parameter Estimation, Uncertainty Quantification, and Sensitivity Analysis: Version 6.0 User's Manual.* Sandia Technical Report SAND2014-4633.

Agarwal, C., Green, G. M., Grove, J. M., Evans, T. P., & Schweik, C. M. (2002). *A Review and Assessment of Land-use Change Models: Dynamics of Space, Time, and Human Choice*. Gen. Tech. Rep. NE-297. Newton Square, PA: U.S. Department of Agriculture, Forest Service, Northeastern Research Station.

Allan, R. (2011). *Survey of Agent Based Modelling and Simulation Tools. Version 1.1.* Available at http://www.grids.ac.uk/Complex/ABMS/

Ansel, J., Kamil, S., Veeramachaneni, K., Ragan-Kelley, J., Bosboom, J., O'Reilly, U. M., & Amarasinghe, S. (2014). OpenTuner: An Extensible Framework for Program Autotuning. In *Proceedings of the 23rd International Conference on Parallel Architectures and Compilation*, (pp. 303–316). New York, NY: ACM. doi:10.1145/2628071.2628092

Bellifemine, F., Caire, G., Poggi, A., & Rimassa, G. (2008). JADE: A Software Framework for Developing Multi-agent Applications. Lessons Learned. *Information and Software Technology*, *50*(1-2), 10–21. doi:10.1016/j.infsof.2007.10.008

Bellifemine, F., Poggi, A., & Rimassa, G. (2001). Developing multi-agent systems with JADE. *Software, Practice & Experience*, *31*(2), 103–128. doi:10.1002/1097-024X(200102)31:2<103::AID-SPE358>3.0.CO;2-O

Bonabeau, E. (2002). Agent-based modeling: Methods and techniques for simulating human systems. *Proceedings of the National Academy of Sciences of the United States of America*, *99*(Suppl 3), 7280–7287. doi:10.1073/pnas.082080899 PMID:12011407

Boucher, D., Elias, P., Lininger, K., May-Tobin, C., Roquemore, S., & Saxon, E. (2011). *The root of the problem. What's driving tropical deforestation today? Technical Report. Tropical Forest and Climate Initiative*. Union of Concerned Scientists.

Bratman, M. (1987). *Intentions, Plans and Practical Reason*. Harvard University Press.

Bresciani, P., Perini, A., Giorgini, P., Giunchiglia, F., & Mylopoulos, J. (2004). Tropos: An Agent-Oriented Software Development Methodology. *Autonomous Agents and Multi-Agent Systems*, *8*(3), 203–236. doi:10.1023/B:AGNT.0000018806.20944.ef

Brown, D. G., Walker, R., Manson, S., & Seto, K. (2004). Modeling Land Use and Land Cover Change. In Land Change Science: Observing, Monitoring and Understanding Trajectories of Change on the Earth's Surface, (pp. 395–409). Dordrecht: Springer Netherlands.

Carneiro, T. G. S., de Andrade, P. R., Câmara, G., Monteiro, A. M. V., & Pereira, R. R. (2013). TerraME: An extensible toolbox for modeling nature society interactions. *Environmental Modelling & Software*, *46*, 104–117. doi:10.1016/j.envsoft.2013.03.002

Castle, C. J. E., & Crooks, A. T. (2006). *Principles and Concepts of Agent-Based Modelling for Developing Geospatial Simulations. Technical Report, UCL Centre For Advanced Spatial Analysis*. London: UCL.

Cavalcanti, R. B., & Joly, C. A. (2002). Biodiversity and Conservation Priorities in the Cerrado Region Biodiversity and Conservation Priorities in the Cerrado Region. In The Cerrados of Brazil: Ecology and Natural History of a Neotropical Savana, (pp. 351–367). New York: Columbia University Press.

Chowdhury, R. R. (2006). Driving forces of tropical deforestation: The role of remote sensing and spatial models. *Singapore Journal of Tropical Geography*, *27*(1), 82–101. doi:10.1111/j.1467-9493.2006.00241.x

Coelho, C. G. C., Abreu, C. G., & Ramos, R. M. (2016). MASE-BDI: Agent-based simulator for environmental land change with efficient and parallel auto-tuning. *Applied Intelligence*, 1–19. doi:10.1007/s10489-016-0797-8

Costanza, R., & Ruth, M. (1998). Using Dynamic Modeling to Scope Environmental Problems and Build Consensus. *Environmental Management*, *22*(2), 183–195. doi:10.1007/s002679900095 PMID:9465128

Geist, H. J., & Lambin, E. F. (2001). What Drives Tropical Deforestation? A meta-analysis of proximate and underlying causes of deforestation based on subnational case study evidence. LUCC International Project Office.

Geist, H.J., & Lambin, E.F. (2002). Proximate Causes and Underlying Driving Forces of Tropical Deforestation. *BioScience, 52*(2),143-150.

Governo do Distrito Federal, G. D. F. (2009). *Plano Diretor de Ordenamento Territorial do Distrito Federal: documento técnico. Technical Report*. Brasília, DF, Brazil: Secretaria de Habitação, Regularização e Desenvolvimento Urbano - SEDHAB.

Grimm, V., Berger, U., Bastiansen, F., Eliassen, S., Ginot, V., Giske, J., & Huse, G. (2006). A standard protocol for describing individual-based and agent-based models. *Ecological Modelling*, *198*(1-2), 115–126. doi:10.1016/j.ecolmodel.2006.04.023

Grimm, V., Berger, U., DeAngelis, D. L., Polhill, J. G., Giske, J., & Railsback, S. F. (2010). The ODD protocol: A review and first update. *Ecological Modelling*, *221*(23), 2760–2768. doi:10.1016/j.ecolmodel.2010.08.019

Klink, C. A., & Machado, R. B. (2005). A conservação do Cerrado brasileiro. *Megadiversidade*, *1*(1), 147–155.

Klink, C. A., & Moreira, A. G. (2002). Past and Current Human Occupation, and Land Use. In R. J. Marquis & P. S. Oliveira (Eds.), *The Cerrados of Brazil: Ecology and Natural History of a Neotropical Savana*. New York: Columbia University Press. doi:10.7312/oliv12042-004

Lambin, E. F., & Geist, H. J. (2001). Global land-use and land-cover change: What have we learned so far? *Global Change Newsletter*, *29*(46), 27–30.

Lambin, E. F., Geist, H. J., & Lepers, E. (2003). Dynamics of Land-Use and Land-Cover Change in Tropical. *Annual Review of Environment and Resources*, *28*(1), 205–241. doi:10.1146/annurev.energy.28.050302.105459

Le Page, C., Bousquet, F., Bakam, I., Bah, A., & Baron, C. (2000). CORMAS: A multiagent simulation toolkit to model natural and social dynamics at multiple scales. In Workshop The ecology of scales, Wageningen, The Netherlands.

Letcher, R. A. K., Jakeman, A. J., Barreteau, O., Borsuk, M. E., ElSawah, S., Hamilton, S. H., & Voinov, A. A. et al. (2013). Selecting among five common modelling approaches for integrated environmental assessment and management. *Environmental Modelling & Software*, *47*, 159–181. doi:10.1016/j.envsoft.2013.05.005

Lotka, A. J. (1925). *Elements of Physical Biology*. Williams & Wilkins Company.

McAlpine, C. A., Etter, A., Fearnside, P. M., Seabrook, L., & Laurance, W. F. (2009). Increasing world consumption of beef as a driver of regional and global change: A call for policy action based on evidence from Queensland (Australia), Colombia and Brazil. *Global Environmental Change*, *19*(1), 21–33. doi:10.1016/j.gloenvcha.2008.10.008

Morton, D. C., Defries, R. S., Shimabukuro, Y. E., Anderson, L. O., Arai, E., Espirito-santo, F. B., & Espirito-santo, B. (2006). Cropland expansion changes deforestation dynamics in the southern Brazilian Amazon. *Proceedings of the National Academy of Sciences of the United States of America*, *103*(39), 14637–14641. doi:10.1073/pnas.0606377103 PMID:16973742

Nelder, J. A., & Mead, R. (1965). A Simplex Method for Function Minimization. *The Computer Journal*, *7*(4), 308–313. doi:10.1093/comjnl/7.4.308

Nikolai, C., & Madey, G. (2009). Tools of the Trade: A Survey of Various Agent Based Modeling. *Journal of Artificial Societies and Social Simulation*, *12*(2).

Olmedo, M. T. C., Pontius, R. G., Paegelow, M., & Mas, J.-F. (2015). Comparison of simulation models in terms of quantity and allocation of land. *Environmental Modelling & Software*, *69*, 214–221. doi:10.1016/j.envsoft.2015.03.003

Parker, D. C., Berger, T., & Manson, S. M. (2001). *Agent-Based Models of Land- Use and Land- Cover Change*. Technical Report No.6. Report and Review of an International Workshop. L. R. No.6, Irvine, CA.

Pokahr, A., Braubach, L., & Lamersdorf, W. (2003). Jadex: Implementing a BDI Infrastructure for JADE Agents. *EXP*, *3*(3), 76–85.

Pontius, R. G., Boersma, W., Castella, J. C., Clarke, K., de Nijs, T., Dietzel, C., & Verburg, P. H. (2008). Comparing the input, output, and validation maps for several models of land change. *The Annals of Regional Science*, *42*(1), 11–37. doi:10.1007/s00168-007-0138-2

Railsback, S. F., Lytinen, S. L., & Jackson, S. K. (2006). Agent-based Simulation Platforms: Review and Development Recommendations. *Simulation*, *82*(9), 609–623. doi:10.1177/0037549706073695

Ralha, C. G., Abreu, C. G., Coelho, C. G., Zaghetto, A., Macchiavello, B., & Machado, R. B. (2013). A multi-agent model system for land-use change simulation. *Environmental Modelling & Software*, *42*, 30–46. doi:10.1016/j.envsoft.2012.12.003

Rindfuss, R. R., Walsh, S. J., Turner, B. L., Fox, J., & Mishra, V. (2004). Developing a science of land change: Challenges and methodological issues. *Proceedings of the National Academy of Sciences of the United States of America*, *101*(39), 13976–13981. doi:10.1073/pnas.0401545101 PMID:15383671

Russell, S. J., & Norvig, P. (2010). *Artificial Intelligence: A Modern Approach* (3rd ed.). Upper Saddle River, NJ: Pearson Education.

Salamon, T. (2011). *Design of Agent-Based Models: Developing Computer Simulations for a Better Understanding of Social Processes. Academic Series*. Repin, Czech Republic: Bruckner Publishing.

Sano, E. E., Rosa, R., Brito, J. L. S., & Ferreira, L. G. (2008). Mapeamento Semi Detalhado do Uso da Terra do Bioma. *Pesquisa Agropecuaria Brasileira, 43*(1), 153–156. doi:10.1590/S0100-204X2008000100020

Sano, E. E., Rosa, R., Brito, J. L. S., & Ferreira, L. G. (2010). *Mapeamento do Uso do Solo e Cobertura Vegetal–Bioma Cerrado: ano base 2002*. Report MMA/SBF.

Smajgl, A., Brown, D. G., Valbuena, D., & Huigen, M. G. A. (2011). Empirical characterisation of agent behaviours in socio-ecological systems. *Environmental Modelling & Software, 26*(7), 837–844. doi:10.1016/j.envsoft.2011.02.011

Tabatabaee, V., Tiwari, A., & Hollingsworth, J. K. (2005). Parallel Parameter Tuning for Applications with Performance Variability. In *Proceedings of the 2005 ACM/IEEE Conference on Supercomputing*. Washington, DC: IEEE Computer Society. doi:10.1109/SC.2005.52

Ţăpuş, C., Chung, I. H., & Hollingsworth, J. K. (2002). Active Harmony: Towards Automated Performance Tuning. In *Proceedings of the 2002 ACM/IEEE Conference on Supercomputing, SC '02*, (pp. 1-11). Los Alamitos, CA: IEEE Computer Society Press.

Turner, B. L., II, Skole, D., Sanderson, S., Fischer, G., Fresco, L., & Leemans, R. (1995). *Land-use and land-cover change, science/research plan. Iiasa policy report*. International Geosphere-Biosphere Programme (IGBP) Report No. 35/HDP Report No. 7.

Valbuena, D., Verburg, P. H., & Bregt, A. K. (2008). A method to define a typology for agent-based analysis in regional land-use research. *Agriculture, Ecosystems & Environment, 1281*(1-2), 27–36. doi:10.1016/j.agee.2008.04.015

Veldkamp, A., & Fresco, L. O. (1996). CLUE: A conceptual model to study the Conversion of Land Use and its Effects. *Ecological Modelling, 85*(2-3), 253–170. doi:10.1016/0304-3800(94)00151-0

Veldkamp, A., & Lambin, E. F. (2001). Predicting land-use change. *Agriculture, Ecosystems & Environment, 85*(1-3), 1–6. doi:10.1016/S0167-8809(01)00199-2

Verburg, P. H., Schot, P. P., Dijst, M. J., & Veldkamp, A. (2004). Land use change modelling: Current practice & research priorities. *GeoJournal, 61*(4), 309–324. doi:10.1007/s10708-004-4946-y

Vuduc, R., Demmel, J. W., & Yelick, K. A. (2005). OSKI: A library of automatically tuned sparse matrix kernels. *Journal of Physics: Conference Series, 16*(1), 521–530. doi:10.1088/1742-6596/16/1/071

Wilensky, U. (1999). *NetLogo*. Retrieved from http://ccl.northwestern.edu/netlogo/

Wooldridge, M. (2009). *An Introduction to MultiAgent Systems* (2nd ed.). Wiley Publishing.

Chapter 6
Modelling and Simulating Complex Systems in Biology:
Introducing NetBioDyn – A Pedagogical and Intuitive Agent-Based Software

Pascal Ballet
LaTIM, INSERM, UMR 1101, Université de Bretagne Occidentale, France

Jérémy Rivière
Lab-STICC, CNRS, UMR 6285, Université de Bretagne Occidentale, France

Alain Pothet
Académie de Créteil, France

Michaël Theron
ORPHY, EA 4324, Université de Bretagne Occidentale, France

Karine Pichavant
ORPHY, EA 4324, Université de Bretagne Occidentale, France

Frank Abautret
Collège Max Jacob, France

Alexandra Fronville
Université de Bretagne Occidentale, France

Vincent Rodin
Lab-STICC, CNRS, UMR 6285, Université de Bretagne Occidentale, France

ABSTRACT

Modelling and teaching complex biological systems is a difficult process. Multi-Agent Based Simulations (MABS) have proved to be an appropriate approach both in research and education when dealing with such systems including emergent, self-organizing phenomena. This chapter presents NetBioDyn, an original software aimed at biologists (students, teachers, researchers) to easily build and simulate complex biological mechanisms observed in multicellular and molecular systems. Thanks to its specific graphical user interface guided by the multi-agent paradigm, this software does not need any prerequisite in computer programming. It thus allows users to create in a simple way bottom-up models where unexpected behaviours can emerge from many interacting entities. This multi-platform software has been used in middle schools, high schools and universities since 2010. A qualitative survey is also presented, showing its ability to adapt to a wide and heterogeneous audience. The Java executable and the source code are available online at http://virtulab.univ-brest.fr.

DOI: 10.4018/978-1-5225-1756-6.ch006

INTRODUCTION

The theory of complex systems has become more and more relevant over the last thirty years to understand concepts such as emergence and self-organization and their role in biological, physical, chemical and social systems (Jacobson & Wilensky, 2006). Modelling and simulating these concepts behind the theory of complex systems is now very important, especially for biologists (students, teachers, researchers). For example, recent works have highlighted the complementarity of modelling both bottom-up and top-down (aggregated) approaches (Stroup & Wilensky, 2014). It has been shown though that teaching and learning complex systems are challenging tasks, and raise a lot of pedagogical issues. The two main issues are:

1. A widely taught top-down approach that have led students to think in that unique way; and
2. System dynamics that are difficult to understand and can be non-linear and counter-intuitive (Jacobson & Wilensky, 2006).

Among different strategies, the use of Multi-Agent Based Simulations (MABS) in research, teaching and learning has proved to be an efficient way of answering (some of) these issues (Epstein, 1999; Jacobson & Wilensky, 2006; Ginovart, 2014). Indeed, the individual-based approach, in opposition to a population-based approach (such as ordinary differential equations), focuses on the entities of the system, their behaviour and their local interactions to explain the global system's behaviour. As such, this concept is easier to understand, more intuitive and do not require advanced mathematical skills (Ginovart, 2014). The use of computer tools to implement these models in virtual environments and simulate them allows researchers and students to generate hypothesis, test them and, especially for students, build their knowledge more efficiently (in a process of *abduction* (Houser, 1992)), see Section 2.2). NetLogo (Wilensky, 1999) is a good example of such agent-based software, and has been used in a lot of works, both in education (Gammack, 2015) and research (Banitz, Gras, & Ginovart, 2015).

This chapter focuses on how to teach complex biological systems to, for example, university students in Biology, middle school and high school students; students who often do not have any experience in computer programming, and/or are reluctant to implement any code. The problem is that most of agent-based software require knowledge and skills in programming languages such as Java (*e.g.* the Repast software (North, Collier, & Vos, 2006)) or the NetLogo environment. This is an issue that occurs more and more as the use of agent-based software, both in research and teaching, grows in social and life sciences. In (Ginovart, 2014) and (Gkiolmas, Karamanos, Chalkidis, Skordoulis, & Papaconstantinou, 2013), a model of a predator-prey system is used with NetLogo to help respectively first-year and high-school students to understand individual-based approaches and eco-systemic mechanisms. Despite its interest, this model is a ready-made one, coded in NetLogo, where the students can change some parameters of the simulation, but are not implied in the process of creating their own simulation from scratch. Indeed, implementing this model is considered as "a hard task" by the students (Ginovart, 2014) and could be counterproductive: looking for errors in the code instead of thinking at the rightness of the model. The authors argue that using pre-existent models could have a part in preventing students for building their own system's *representation* (this point will be developed in Section 2.2). The gap between most of agent-based software prerequisites and the actual programming skills of students (and sometimes researchers) in these fields has to be dramatically reduced and the processes of implementing a model and simulate it should be more intuitive.

This chapter essentially looks at the modelling of biological systems at and around the cellular scale. However, it is interesting to note that a lot of works have been focusing on modelling and simulating systems at the molecular scale, resulting in several software developed for students. For example, it is possible to find molecular visualisation systems like PyMol (Schrödinger, 2015) or software that help the docking of molecules like AutoDock Vina (Trott & Olson, 2010). There is also an automatic plotting of protein-ligand interactions with Lig-Plot (Wallace, Laskowski, & Thornton, 1996). It is possible to visualise molecular interaction networks and biological pathways thanks to software like CytoScape (Shannon et al., 2003). When focusing on DNA, a complete suite in Python can be used for the sequence alignment (BioPython - Cock et al., 2009). All these software are sources of inspiration when dealing at the cellular level.

In this chapter, the authors propose a software called NetBioDyn working like a game construction kit, where entities are located in a 3D grid environment and follow simple local transformation rules. In the name NetBioDyn, Bio stands for Biology, Dyn for Dynamic and Net comes from Internet because initially NetBioDyn was a Java Applet. Though this software is dedicated to cellular systems where several cells are in interaction, it can also model and simulate molecular mechanisms in a simple and symbolic way. The process of building models and simulations is facilitated by a graphical user interface similar to a drawing software like Microsoft Paint. Researchers or students can build their own models, continuously visualize a clear representation of these models, and see them evolving in time and space. The results are displayed like a chess board with moving and changing coloured entities and can be analysed thanks to recorded population curves. From the point of view of the researcher (who usually is a teacher too), the software can help to explain non-intuitive and complex biological systems in a visual way, easily understandable by his peers or students. The authors present in this chapter the key principles of the software in rigorous but comprehensive language and develop three examples coming from real biological systems. The first example is a predator-prey system involving two bacteria. The second one simulates the mechanism of oxygen exchange in the zebra fish gills between water and blood. The third one simulates the blood coagulation mechanisms in a small section of a vein. A discussion is then engaged to show the interests of such software as well as the limits and drawbacks. A qualitative survey is also presented to show how students in biology and students in computer sciences have received the software. After the conclusion, different perspectives are presented in order to improve both the software and its use.

BACKGROUND

The use of computers in teaching biology can be divided into three common practices:

1. The development of models and simulations using a programming language like C++, Python or Java. This practice is powerful but takes time and requires advanced programming skills.
2. The use of customizable software, where students work on pre-existent models and can only change some parameters in simulations. This approach is very easy to use with students, but is not as efficient as building their own models.
3. The use of modelling software to help students develop their own models without any programming skills.

This third practice requires though the development of complex software that simplifies the process of modelling, by proposing for example, a graphical representation and construction of virtual agents, their behaviours and their environment.

Naturally, these three practices can also be found in the use of Multi-Agent Based Simulations and agent-based software in teaching biology. A quick overview of:

1. Agent-based modelling software that allow the development of models and simulations thanks to a programming language,
2. Agent-based modelling software that propose pre-made simulations, or
3. Computer aided-design modelling software is presented here.

A more complete review of agent-based environments has been made by Nikolai and Madey (2009), where more than 50 platforms are classified and discussed. Readers who wish to deepen the potential of the multi-agent approach for the simulation of complex systems can refer to the book edited by Weiss (2000), especially the chapter "Search Algorithms for Agents".

Agent-Based Software in Biology

Programming Languages

NetLogo (Wilensky, 1999) is one of the most well-known software allowing the development of agent-based programs, including simulations. It has a simplified programming language and simple mechanisms to create interfaces. This software is widely used in the community of biologists and many biological systems have been developed thanks to it. It has also been specifically used to help students understand the mechanisms behind complex systems (Stroup, 2014, Jacobson & Wilensky 2006). Nevertheless, pupils or students have to learn a programming language if they want to create their own models (Ginovart, 2014, Gkiolmas et al., 2013). One can also mention AgentSheets (Repenning, & Sumner, 1995) and AgentCubes (Repenning, Smith, Owen, & Repenning, 2012), which are simple programming environments allowing middle school pupils to learn programming in an appealing manner. These software are designed for multi-purpose applications and mostly games. They are based on a multi-agent system where behaviours are created by drag and drop condition-action algorithms. Although the programming is easy, some code is still required. Among generic multi-agent platforms, one can cite Breve (Klein, 2014; Klein & Spector, 2009) and Repast (North et al., 2006). Despite the fact that they have integrated development environments, the creation of models and simulations needs the knowledge of a programming language like Python, Groovy or Java. More recently, Biocellion (Kang, Kahan, McDermott, Flann, & Shmulevich, 2014) allows accelerating discrete agent-based simulation of biological systems with millions to trillions of cells. It uses cluster computers by partitioning the work to run on multiple compute nodes. It requires though mathematical modelling backgrounds and C++ programming skills.

Pre-Maid Simulations

Different simulations of multi-agent systems have been developed with NetLogo. The community models website of NetLogo proposes, for example, a customizable simulation of a predator - prey - poison system with 11 parameters: initial number rabbits and coyotes, rabbit reproducing probability, grass regrowth time, poison quantity, etc. (Jenkins, 2015). Concerning the simulation of multicellular systems, one can cite VirtualLeaf (Merks, Guravage, Inzé, & Beemstaer, 2001) which allows the simulation of tree-like structures based on agent representing cells and regulation network. It includes 12 different models and about 60 parameters that can be adjusted (e.g. cell mechanic or mathematical integration parameters).

Computer Aided-Design Software

In addition to a simplified programming language, NetLogo provides a "System Dynamic Modeler" that can be used to represent the dynamic of the system, entities inter-relations, feedback loops, flows and stocks. However, this tool requires a global understanding of the system (that is often missing with an agent-based approach) and is quite difficult to use as an intuitive modelling tool. Ioda (Interaction-Oriented Design of Agent simulations) and Jedi (an implementation of the Ioda concepts), allow the creation of multi-agent simulations in a very simple manner (without any coding skill), where predefined interactions can be added between agents (Kubera, Mathieu, & Picault, 2008). As Jedi is a demonstration software, it cannot be used to create complex simulations (4 agents maximum can be manipulated at the same time and there is no graphical user interface to precisely build the initial state). AnyLogic (XJ Technologies Company Ltd., 2012) is a multipurpose simulation software with a visual development environment. It has pre-built object libraries allowing users to create models by a drag-and-drop of elements. Its generic architecture makes it a powerful software but it could be relatively complex to understand for some students, especially the secondary ones. Another very interesting human-computer interaction is developed in the SimSketch software (Bollen, 2013). In this approach, the user draws the agents using an integrated paint tool then adds a name, a type and different pre-defined behaviours. This software focuses on primary and early secondary school students whereas this work focuses on the end of the secondary school to the university. Note that, this tool is not yet totally usable to create multicellular systems because there is no if...then...else structure. The software called CompuCell3D (Izaguirre et al., 2004) allows the simulation of multicellular systems with a user-friendly interface. The Python programming language can be used to create models but an XML file can also be edited to create simulations without coding. SimBioDyn (Ballet, Zemirline, & Marcé, 2004) is a software based on a hybrid multi-agent and mass-spring system designed to graphically create multicellular systems. Although rich in features (cell deformation, migration, adhesion and differentiation, molecular production and consumption and configurable through scripts in Python), it remains hard to use by non-specialists without a course of several hours (Ballet et al., 2004).

Among computer aided-design software, there are also some interesting works that have focused on modelling biological systems following other bottom-up or top-down approaches. For example, one can cite Virtual Cell (Shin, Liu, Loew, & Schaff, 1998), a software allowing to model intra-cellular mechanisms thanks to differential equations. It has a rich web-based Java interface to specify compartmental topology and geometry, molecular data and interaction parameters. However, the purpose of this software is to simulate a single whole cell or sub-parts of a cell but not multicellular systems, and it is not fitted for non-specialist users as advanced skills in mathematics are needed.

In this chapter, the authors focus on teaching pupils or students with no programming knowledge complex biological systems, such as multicellular systems.

In this context, the interests of using computer aided-design software are multiple:

- The student do not focus on how to program but on what to model (they remain in the field of biology).
- They cannot make syntactic or lexical errors: their models are always functional despite the fact that they can make semantic errors.
- They are no longer in charge of just some simulation parameters, but become the main actors of the model building, thus favouring the building of their knowledge.
- The teacher visualizes quickly where a student has insufficient knowledge.

So what is needed is a modelling software designed to be used by students with no programming knowledge and dedicated to agent-based multicellular systems modelling and simulation. However, this survey shows that many computer aided-design software exist, but to our knowledge none of them are intuitive enough to be used in this context. In order to answer these issues and give students the ability to easily create their own models, the authors have developed the software called NetBioDyn.

Reasoning About Complexity

Presenting essential concepts on system dynamics has become very important for students in biology. Teaching and reasoning about complex systems still raise a lot of pedagogical issues as shown in particular by Jacobson and Wilensky (2006).

Lynch and Ferguson (2014) argue that grasping the complexity requires a representational framework able both to describe executable computer models *and* to support reasoning as a simply readable "external representation", as opposed to internal representations that are the user's mental models (Zhang and Patel, 2006). According to them, a good external representation affords the classic scientific approach, called *abduction*, emphasizing the observation, the build of hypothesis and the tests to verify the relevance of assumptions (Houser, 1992): it promotes the exploration and understanding of a phenomena thanks to a practical approach. It should be too constructed by the students themselves for two principal reasons: the first one is that it improves their learning (better memorization, quasi-physical experimentation of the model and appropriation of fundamental mechanisms) and the second one concerns the amelioration of their self-commitment (Prain, & Waldrip, 2006). Lynch and Ferguson define an ideal representation as "some written, external representation to complement user's internal representations [...], producing an external cognitive artefact which facilitates exploration and discovery [...] while also offering the ability to be executed".

Lynch and Ferguson summarize the 9 properties of a suitable external representation that were previously defined by Zhang and Patel (2006):

- Short-term or long-term memory aids.
- Directly perceived information.
- Knowledge and skills that are unavailable internally.
- Support for easy recognition and inference.
- Unconscious support for cognitive behaviour.
- More efficient action sequences generated.
- Facilities to stop time and make information visible and sustainable.
- Need for abstraction reduced.
- Aids to decision making through high accuracy and low effort.

Following these properties, the authors claim that NetBioDyn, thanks to its graphic interface and its simulation tools proposes a suitable external representation. It gives to students and teacher a simple way to create and experiment and test models of biological systems, while providing at any time a simplified and complete view of the system's state.

More generally, readers who want to improve their knowledge on "learning by doing" in virtual environments (VE) can read the book of Aldrich (2009).

NETBIODYN, A PEDAGOGICAL AND INTUITIVE AGENT-BASED SOFTWARE

NetBioDyn is a fully open source Integrated Development Environment (IDE) dedicated to teaching activities in biological complex systems, mainly at the cellular level and secondarily at the molecular scale. It gives the possibility to teachers and students to create simulations of multicellular and multimolecular systems in a simple and visual manner. NetBioDyn does not require any skill in computing and lays on the agent-based approach. Besides being simple to understand for students, this approach allowed the authors to design an intuitive graphical user interface to easily manipulate the agent-based concepts: entity, behaviour and interaction.

In this section the graphical user interface of the software is first presented. Then, the section goes inside the software architecture with the description of the entities populating the simulated environment and how they can evolve thanks to algorithmic behaviours (movement, differentiation, division, apoptosis, *etc.*). In the next section, three examples of easily reproducible simulations are detailed. The code and the executable (for UNIX, Mac and Windows Operating Systems) are available for download on the website http://virtulab.univ-brest.fr.

Graphical User Interface

For educational purpose, the design of a graphical user interface is a crucial point. For the student or the teacher who creates a simulation it must be:

- Intuitive, to prevent the user from being lost in multiple buttons or lists.
- Logic, to build a structured simulation.
- Clear, to show important data and hide less important ones.
- Accessible, to concentrate on "what to simulate" instead of "how to simulate".
- Complete, to have a powerful tool that can simulate advanced biological behaviours.
- Attractive, to stimulate student's imagination and motivation.

For the teacher who uses the software in a course, it must:

- Be robust, avoiding the teacher to help students in case of inappropriate action (for example, no source code is required, avoiding student to make syntax mistakes - only semantic errors can be made, which are the most relevant ones to test the student comprehension of the modelled biological system).
- Show relevant information, to quickly see where each student is and if he is mistaken.
- Allow a short learning curve, not to spend more than one hour to explain how the software works.

To try to achieve these points, the authors decided to use Multi-Agent Based Simulations (MABS) to guide the development of the graphical user interface. MABS are naturally close to how biological objects like cells are described, reducing the abstraction needed to make simulations. In this chapter, the authors use the term "entity" to represent biological objects (or group of objects) and the term "behaviour" to explain how the entities evolve in time. The word "environment" is used to represent the space where simulations occur. A good analogy can be done with board games like chess: an entity is a piece, a behaviour is a game rule, the environment is the board and the initial state of the simulation corresponds to the initial positions of the pieces.

Figure 1 shows the different parts of the interface. The first part of the graphical user interface is made of a list of entities, followed by a list of behaviours and then a button to change the environment's properties (such as size and textual description). The position order of the two lists is logic: entities must be defined before their behaviours. Nevertheless, note that a new entity or a new behaviour can be added, changed, renamed or removed at any time by the user to support the creation process which is rarely linear. The environment button is located below the two lists because the default properties do not change frequently.

The second part of the graphical user interface is the placement and the display of entities in their environment. A central panel serves as a paint area where entities can be individually or collectively located or removed thanks to classic drawing tools like pen, spray, line and eraser, plus a random initializer tool. Entities are located in several layers that can be displayed or hidden using the Z slider. The simulator is controlled by three buttons:

Figure 1. NetBioDyn graphical user interface

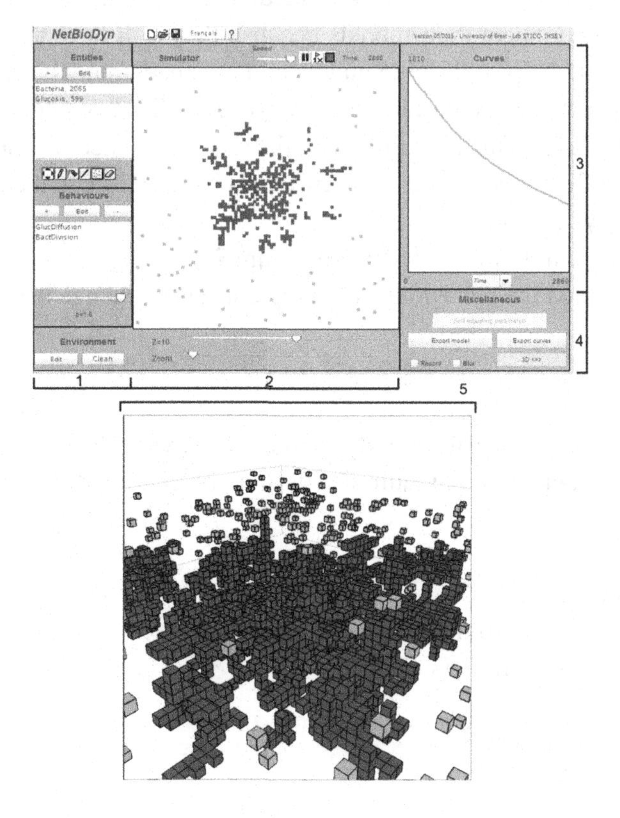

- Play (a green right arrow),
- Play one step (an orange right arrow), and
- Stop (a red square).

The simulator can also be paused at any time to observe a specific state or make a screen capture, and its speed can be reduced or increased.

The third part displays the population curves of entities evolving in real time. Several curves can be displayed at the same time to compare their values according to the simulation steps. The user can also display several population curves according to a specific entity (for an example, see next section Examples).

The fourth part enables the user to change miscellaneous properties and export simulation data in the well-used formats in Biology CSV (spreadsheet) and R programming language. A tool to automatically adjust the simulation's parameters according to *in-vitro* results is a work in progress and will be available soon.

The fifth and last part display in a specific frame the environment in 3D in real-time.

In the next subsection, the NetBioDyn entities are described in details.

Entities

An entity in NetBioDyn is a simple cube of size 1 x 1 x 1, located as a grid element of the environment (which is described in details below), having a unique name and a unique colour. Other forms (triangle, disc, star) or images can also replace the coloured cube.

An entity also has a half-life parameter indicating its speed of degradation or destruction.

For each entity, a short description can also be added. This is made for two main reasons. The first is to encourage students to comment their models because they will have to maintain and share their work. The second reason is to help students to understand pre-made models that the teacher created.

Finally a cleanable property can be unselected to prevent entities from being removed when the environment is cleaned by the user. This is useful when the teacher created a simulation where the initial state contains a minimum of entities located at specific positions: the student can clean his own entities but the teacher's ones always remain.

The entity configuration window can be seen on Figure 2.

In the following sub-section the behaviours allowing entities to change their states are explained.

Behaviours

In NetBioDyn, a behaviour aims to change entity's states. For example an entity can be put in motion, differentiated or destroyed. Four parts are required to define a behaviour, three conditions and one action:

- Conditions
  - Which entities are involved?
  - Where are they located from each other?
  - Is the probability validated?
- Action
  - What will happen to them?

Figure 2. The name, the half-life and the appearance of an entity is made in the entity configuration window. Note also the description field on the right and the cleanable check-box at the bottom-left

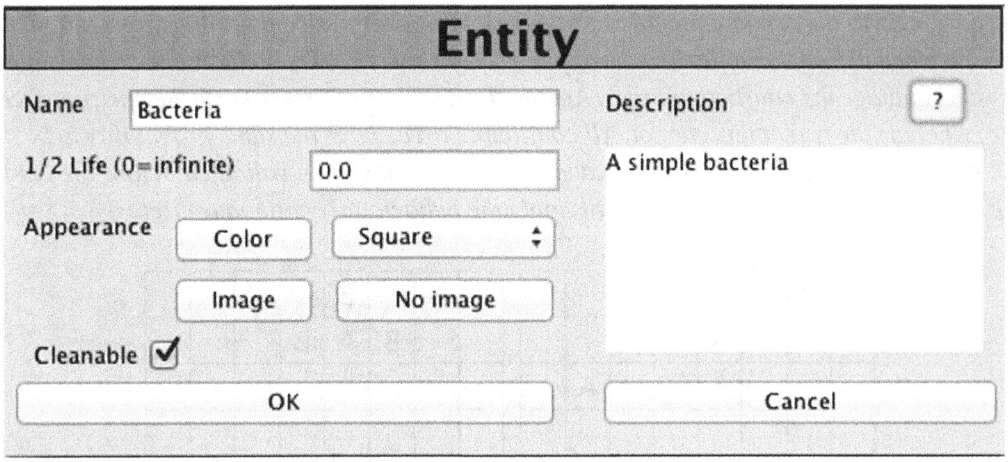

The first condition identifies all entities that are involved in the simulation. Only those that are in a good position from each other are kept (condition two). Thanks to the third condition, a behaviour is probabilistically performed: a random number between 0.0 and 1.0 is generated and must be inferior to the probability of the behaviour.

If the conditions are validated at different areas in the environment, a behaviour applies several times during one simulation step. Conflicting areas are detected where a same entity can be used two or more times by behaviours. In this case, only one area is randomly chosen among the conflicting areas (see Figure 3).

In order to ease the design of behaviours, the authors created a graphical representation consisting of two main parts (see Figure 4). The first part containing three lines indicates the behaviour's name, its description and its probability to occur. The second part is divided in three columns: the left one contains the entities involved in the behaviour (reactive), the middle one indicates where the entities are located one from each other and the right column indicates how entities are changed.

In the software NetBioDyn, the graphical user interface is very close in terms of displayed data. The entities are selected thanks to a combo-box containing all the created entities. The positions are selected by clicking on the light or dark squares. The cross at the left allows the user to remove a full line of reactive and products (see Figure 5).

The next subsection describes the environment, where entities are placed and behaviours occur.

Environment

The environment represents a 2D or 3D space where entities evolve and behaviours take place. It is a simple grid where a grid cell contains none or a maximum of one entity. An entity occupies a whole grid cell. It follows Von Neumann neighbourhood: it has four neighbours in 2D and six in 3D. The size of the environment is by default 100 x 100 x 2 but can be changed by the user. The view is only 2D when the user put, move or remove entities but a 3D view can be activated for a better and more intuitive rendering.

Figure 3. In (1), a behaviour can be applied several times during one simulation step. In this example the system detects that the entities A and B are in contact. The probability to perform the behaviour, if the entities are found in contact, is 0.8. So, of the six areas where the behaviour can be performed, four are kept (continuous circles) whereas two are not (dotted circle). Note that this area configuration can change accordingly to the random numbers generated. Moreover notice that three areas are in conflict and must be treated before the action can be applied (continuous circles including the same entity A). In (2), the system manage the conflicting areas. Among them, only one can be kept (the horizontal continuous circle) whereas the two others cannot. All conflicting areas have the same probability to be kept (or not kept). Note that the probabilities of behaviours have already been validated before the conflicting areas are treated. Finally, only two areas will apply the behaviour (continuous circles)

(1) Probabilistic behaviour selection (2) Conflicting areas resolution

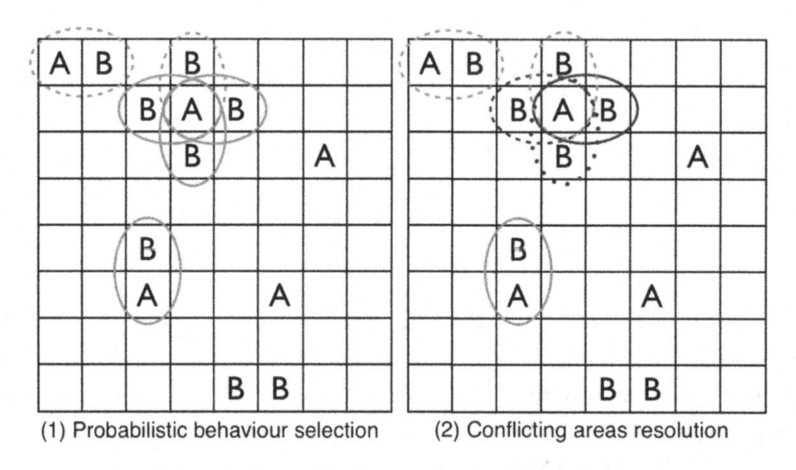

Figure 4. For this example, the behaviour is read like this: "the entity A, which is the centre of the behaviour (+) must be in contact with one entity B and one entity C. B can be anywhere (up, down, left, right, front, behind) around A and C can be at the left, the front or behind A. Moreover, if the probability is validated and the possible conflicts resolved, A will become A', B will become B' and C will be changed in C' ". The first reactive is always at the centre of the reaction. A reactive or a product can be 0 which means no entity. The diagonals are not taken into account in the positions

Name	behaviour	
Description	Short text of explanation	
Probability	Decimal number in [0, 1]	
Reactives	Positions	Products
A	+	A'
B	⊙ ↑ ← A → ↓ ⊗	B'
C	↑ ← A ↓	C'

*Figure 5. The graphical user interface of behaviours in NetBioDyn has the same properties as seen previously. In this example, an entity A is removed from its current position (first line of the lower part), to be placed at a free location (second line). The third line contains the symbol * as reactive (which means anything) and the symbol - as product (which means do nothing). This behaviour applies a random walk to all A entities in the simulation*

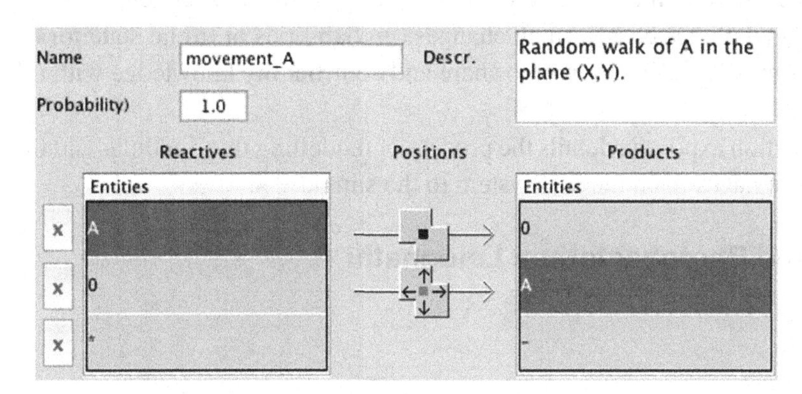

From the principles presented, the next section develops three examples that can be realised by pupils or students.

EXAMPLES

The examples detailed in this section can be performed using the software NetBioDyn freely available at http://virtulab.univ-brest.fr.

It can be made by pupils from age 14 to university students. The first example deals with a prey-predator system involving two species of bacteria: Bdellovibrio and Photobacterium leiognathi. The second example models the dioxygen exchanges between water and the blood of zebra fishes. Finally, the third one aims to reproduce and understand the blood coagulation mechanisms.

These examples show to the students how to create a model in four stages:

- **The Real System**
  - What is the current knowledge?
  - What part of the system would I model?
  - What are the limits of my study (dimensions, time, energy, *etc.*)?
  - Am I able to distinguish, among the current knowledge, what are hypothesis and what are facts?
- **The Simplifications**
  - What have I to keep absolutely from the real system?
  - What can be neglected?
- **The Model**
  - How to represent real entities using virtual ones?
  - How to model biological mechanisms thanks to algorithms or equations?
  - What are the dimensions and the topology of the environment and how to represent it on a computer?

- **The Simulation**
  - Can I trust the results?
  - What are the limits concerning explanation and prediction?
  - Are the results qualitative or quantitative?
  - Does the simulation inspire me in order to prepare real experiments?
  - Is the simulation robust to small changes (probabilities or initial state for example)?
  - Is the simulation clear enough to share and confront my knowledge with others?

The next sub-section expose in details the process of modelling three cellular and molecular systems, following these four stages from the real system to the simulation.

Bdellovibrio and Photobacterium Leiognathi

Real System

Bdellovibrio and Photobacterium leiognathi are marine bacteria. The interactions between these two bacteria represent a good example of a predator-prey system. Bdellovibrio is a predator of Photobacterium leiognathi. It penetrates inside its prey, proliferates, kills the host and finally releases its progeny. This cycle time is about 4 hours. Bdellovibrio is a bacterium of 1 µm in size and thanks to its flagellum, moves very fast (up to 160 µm/s). The size of Photobacterium leiognathi is about 2 µm. They usually are in symbiosis with fishes and, in this case, have not shown to be motile.

Biological systems involve too many parameters, objects and behaviours to be fully modelled and simulated. A simplification of the system described here is needed in order to focus on the predator-prey principles.

Simplification

In order to reproduce the Bdellovibrio and Photobacterium leiognathi relationship, a simplified model which takes into account five objects and five behaviours is proposed:

- Biological entities:
  - Bdellovibrio (Bdel).
  - Photobacterium leiognathi (Leio).
  - Photobacterium leiognathi infected (LeioInfected).
  - Fish tissue where Photobacterium leiognathi lives (FishT).
  - Nutriment for Photobacterium leiognathi coming from the fish (Nutri).
- Biological behaviours:
  - Bdellovibrio movement (BdelMvt).
  - Infection of Photobacterium leiognathi by Bdellovibrio (LeioInfection).
  - Photobacterium leiognathi lyse and Bdellovibrio release (LeioLyse).
  - Photobacterium leiognathi division (LeioDiv).
  - Nutriment production for Photobacterium leiognathi (NutriProd).

This model has to be also adapted to our simulator. The next section explains how to put the model in the software NetBioDyn, which has its own possibilities (*i.e.* multiple entities and behaviours) and constraints (*i.e.* the environment is a 3D grid and a grid element has a maximum of six neighbours).

Model

The entities chosen for the simulation are Bdel, Leio and LeioInfected. The fish tissue and the nutriments are implicit in our simulation. This is a choice which can be changed if the fish tissue or the nutriments must be quantitatively taken into account or if their spatial distribution has to be studied. The graphical representation used for entities are simple coloured cubes of volume 1 μm³. In this example Bdel is red, Leio is blue and LeioInfected is purple. Bdel and Leio have a half-life of 100 simulation steps (which means that after 100 simulation steps a Bdel or a Leio entity has 50% of chance to die, or said differently, after 100 simulation steps for a non-reproductive population of N individuals, only N/2 individuals will remain).

Four behaviours have been kept to simulate the bacterial system (see Figure 6):

1. The movement of Bdel entities which is a simple 3D random walk. The probability of this behaviour is 1.0, the maximum possible, because no other behaviour is as fast as this one. The bibliography gives the speed of 160 μm/s for Bdellovibrio. This is a maximum and an instantaneous speed. Due to its frequent directional changes, its average speed is closer to few μm/s. A grid cell can contain one bacterial entity at the same time, so a grid cell represents a cube of 1 x 1 x 1 μm³.

Figure 6. The four behaviours described using the previously seen representation

Name	BdelMvt	
Description	Random walk of Bdellovibrio bacteria.	
Probability	1.0	
Reactives	Positions	Products
Bdel	+	0
0	⊙ ↑ ← Bdel → ↓ ⊗	Bdel
.		-

Name	LeioInfection	
Description	Infection of photobacterium Leiognathi by Bdellovibrio.	
Probability	0.1	
Reactives	Positions	Products
Leio	+	LeioInfected
Bdel	⊙ ↑ ← Leio → ↓ ⊗	0
.		-

Name	LeioLyse	
Description	Lyse of photobacterium Leiognathi and release of two Bdellovibrio.	
Probability	0.1	
Reactives	Positions	Products
LeioInfected	+	Bdel
0	⊙ ↑ ← LeioInfected → ↓ ⊗	Bdel
.		-

Name	LeioDiv	
Description	Division of photobacterium Leiognathi following the nutriment intake.	
Probability	0.01	
Reactives	Positions	Products
Leio	+	Leio
0	↑ ← Leio → ↓	Leio
.		-

2. The infection of Leio by Bdel. The probability of infection is 0.1, indicating that one contact over ten can produce an infection.
3. The lysis of Leio and the release of two Bdel. The probability is chosen to reproduce the 4 hours of incubation for Bdel. A time step is equivalent to one second, so 4 hours are simulated by 14400 steps. At this time, this is supposed that 99% of the LeioInfected have been killed.
4. The division of Leio following the nutriment intake.

The environment is a 100 x 100 x 2 grid. All the Leio entities are at z=0 in order to imitate the fact that they are on the fish tissue.

Simulation and Results

The initial state of the simulation is given by Figure 7. The initial state is made of 1000 Leio entities and 100 Bdel entities randomly located in the environment. Leio entities have been placed at z=0 while Bdel entities are located at z=1. The number of preys is ten times inferior to the number of predators in order to initiate the system without having the preys eaten by too numerous predators at the beginning. They all are located at random in the same 3D environment.

Figure 8 shows two snapshots of the simulation at time steps 1490 and 12840. At time step 1490 (upper screen capture), the first predator-prey cycle is shown, where the Leio population (preys) grows initially, just followed by the growth of the Bdel population (predators). Then, the Leio population decreases because they are eaten by the Bdel, and then the Bdel population decreases too by lack of food, resulting in an increase of Leio due to the fact there are less Bdel entities. The two peaks of the curves are shifted in time: the peak of the preys (Leio) occurs before the peak of predators (Bdel) because the predators need the preys to grow. Small spots of Leio entities appear because of their bacterial division. Five other cycles are shown at the lower part of the figure (time step 12840). The stochastic aspect of

Figure 7. This screen capture of NetBioDyn shows the initial state of the simulation where 1000 Leio (triangles) and 100 Bdel (squares) have been placed at random in the environment using the random initializer tool. Only entities placed at z=0 are here visible

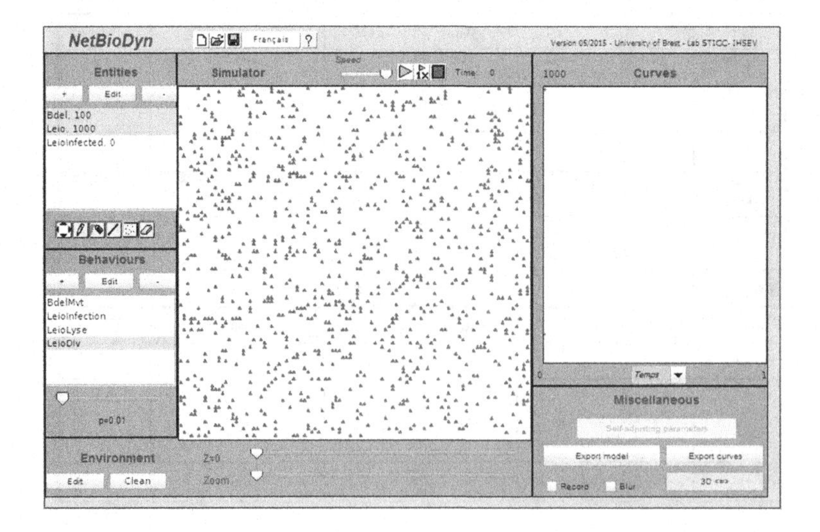

Figure 8. Screen captures of the simulation at time steps 1490 and 12840. From left to right: the environment states, the population curves of entities Bdel (dark grey) and Leio (light grey) according to time, and the population curve of Leio according to Bdel (after change of the abscissa)

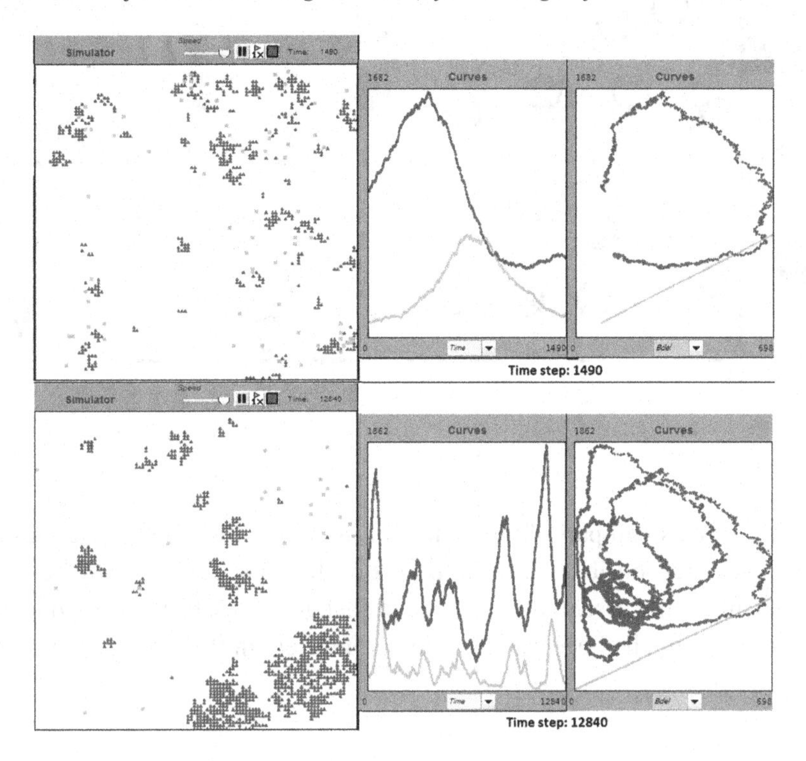

the simulation is well shown in the irregularities of curves and entity positions. The predators are close to the preys not because they are attracted by them but simply because they divide inside them. Smaller cycles inside the main cycles can also be seen. Those emerging global behaviours can also be observed in real predator-prey systems (see the Canadian lynx and snowshoe hare pelt-trading records of the Hudson Bay Company over almost a century (Odum, 1953)).

The simulation expresses classical cycles of predator-prey systems, indicating that the results are qualitatively correct.

Water-Blood Dioxygen Exchanges in Zebra Fish

Real System

Zebra fishes have an efficient respiratory organ able to extract dioxygen from the water to their blood. Their gills use a counter-current system to improve the dioxygen exchange. Figure 9 represents how the system works. The water contains dioxygen molecules which circulate from the left to the right. A permeable membrane, that only gazes can cross (dotted line), separates the blood from the water. The dioxygen can cross the membrane from the water to the blood or from the blood to the water with exactly the same probability. In the real system, the blood circulates in the opposite direction. However, by using the simulation, the benefits of such mechanism will be studied as compared with the co-current one. The simulation will aim to evaluate the impact of the current direction (counter-current co-current) on the blood oxygenation.

Figure 9. Mechanisms of dioxygen exchange inside the zebra fish gills

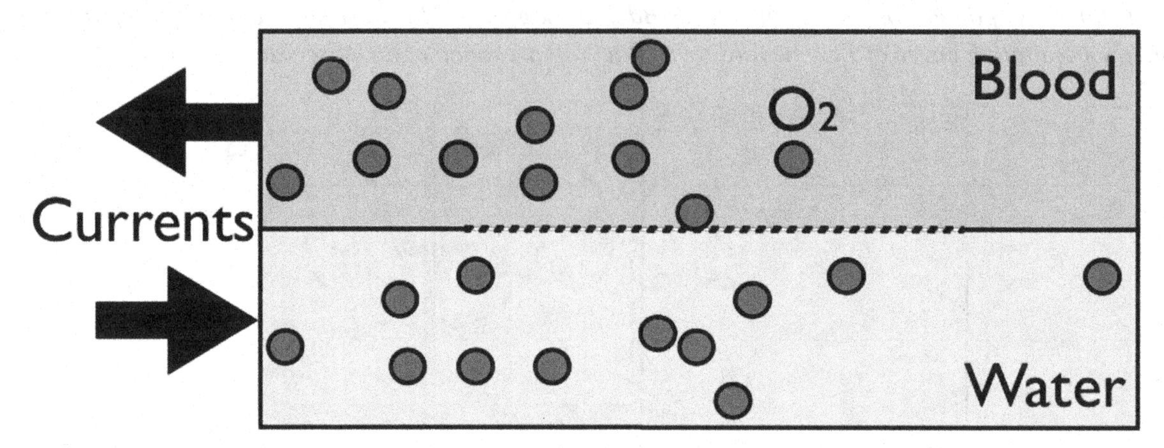

Simplification

The simulation focuses on a small part of the system of the Zebra Fish, where the dioxygen exchanges occur. The red blood cells responsible for the transport of the dioxygen in the blood are not modelled, because this mechanism occurs essentially after the exchange. Other blood components, like immune cells for example, are not modelled because they do not interfere in the dioxygen exchange.

Model

The model is made to test the efficiency of the zebra fish counter-current respiratory system where the O_2 is central. The chosen model has three types of biological entities, the O_2 in various states, and two membranes:

- The O_2 is divided into three types of entity in order to give them different behaviours and count them according to their locations:
  - The O_2 in water (O2Water).
  - The O_2 in blood (O2Blood).
  - The O_2 in water which can no longer go to the blood (O2WaterLost) because it has been evacuated outside the respiratory system.
- The permeable membrane that can be crossed by the O_2 (Membrane).
- The impermeable membrane (Wall).

To complete the model, four dedicated types of entity have been used:

1. Entities O2WaterIn, which perform the addition of O2Water thanks to the behaviour called O2WaterGetIn.
2. Entities O2WaterOut, which remove O2Water thanks to the behaviour O2WaterGetOut.
3. Entities O2BloodOut, which remove O2Blood thanks to the behaviour O2BloodGetOut.

4. Entities O2WaterLostBorder, which detect O2Water entities that can no more go into the blood thanks to the behaviour O2WaterIsLost.

Figure 10 shows how entities are located in their grid environment at t=0.

Only two biological entities are placed: the impermeable membrane (Wall) and the permeable membrane (Membrane) which creates the upper row of blood circulation and the lower row of water circulation. There is also the four ad-hoc entities (non-biological entities). Note that the O_2 is not present at t=0 but will be added at the right to the ad-hoc entities O2WaterIn. When O_2 entities in the water diffuse to the right, they can cross the Membrane to go in the blood. The O_2 in the blood can also go back to the water by crossing again the Membrane. The O_2 entities in water reaching the O2WaterLostBorder can no longer cross the Membrane, and they become lost for the blood (O2WaterLost). The region between O2WaterLostBorder and O2WaterOut only contains O2WaterLost entities that will be counted to test the efficiency of the current and counter-current mechanisms. The O2BloodOut are located at both left and right sides of the upper row in order to evacuate the O_2 in the blood whatever the direction the current follows.

Nine behaviours have been made to add movement, membrane crossing, transformation, creation and removal of entities. O_2 entities move up, down or right at random in water at each simulation step (probability set to 1). The same mechanism is used to define the movements of O_2 in blood and O2WaterLost entities. Note that this behaviour can be modified by changing the left direction by the right, giving the possibility to test the impact of the O_2 transfer efficiency from the water to the blood according to the relative direction of both flows.

Figure 11 shows how the O_2 crosses the membrane. The behaviour indicating how the O_2entities in water become lost for the blood is described in Figure 12.

Finally, Figure 13 shows the only one behaviour that adds entities in the environment. It creates O2Water entities just on the right of O2WaterIn entities with a probability of 0.1. It means that one time step over 10 (on average), one O2WaterIn adds one O2Water (if there is an empty space at its right). The same mechanisms are used to define two more behaviours: the behaviour removing the O_2 entities in the water that are too far from the membrane to be useful for the blood (O2WaterLost), and the behaviour removing the O_2 entities in blood (probabilities at 1). Defining these behaviours led to create dedicated entities O2WaterIn, O2WaterOut and O2BloodOut that do not exist in the real system. They only are

Figure 10. The gill model at its initial state (t=0)

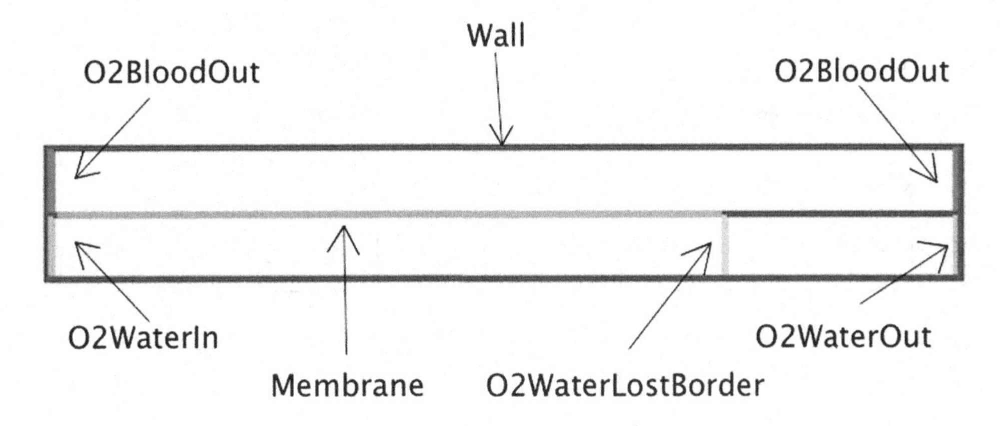

Figure 11. The two behaviours of this figure indicate how O_2 entities cross the membrane separating the blood and the water. The left one concerns the crossing of O_2 in water (O2Water) to the blood (O2Blood). The right one describes the crossing of O_2 located in blood (O2Blood) toward the water. Both of these behaviours have the same probability: 0.3

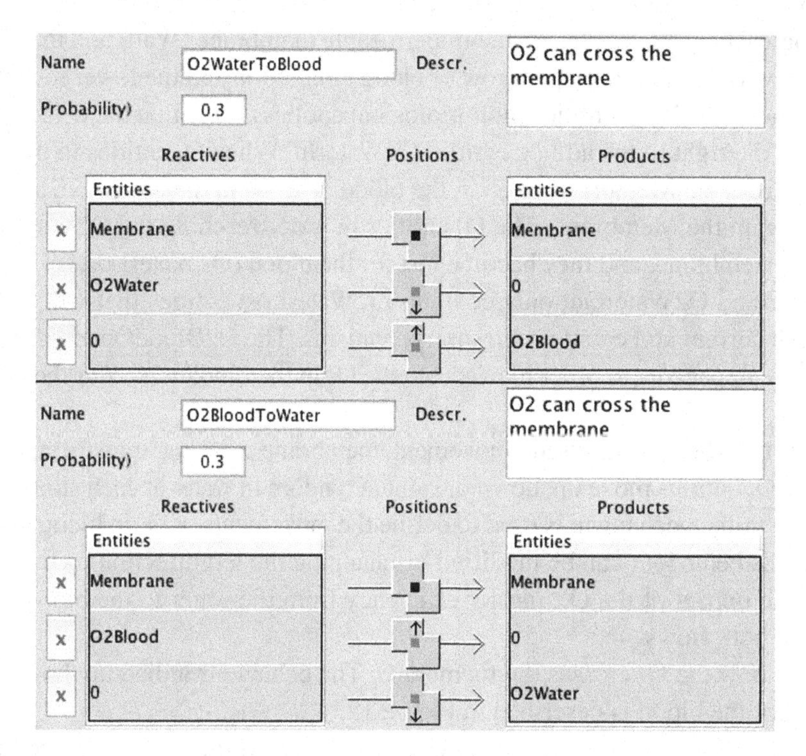

Figure 12. This behaviour changes O2Water entities touching the O2WaterLostBorder into O2Water-Lost entities. Thus, O_2 that cannot be used by the blood can be counted to evaluate the efficiency of the counter-current mechanism as compared with the current one. The less O2WaterLost entities are present, the more the gills are efficient

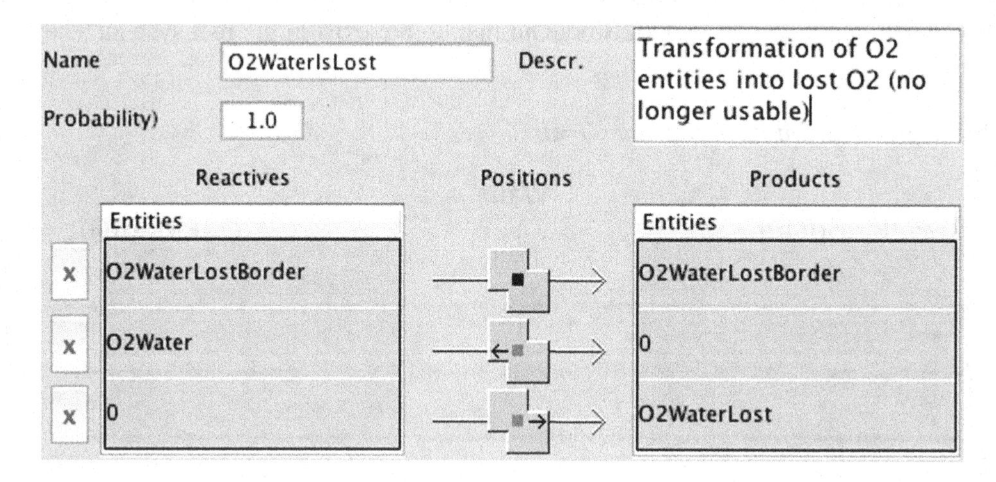

Figure 13. The behaviour adding O2Water entities in the environment

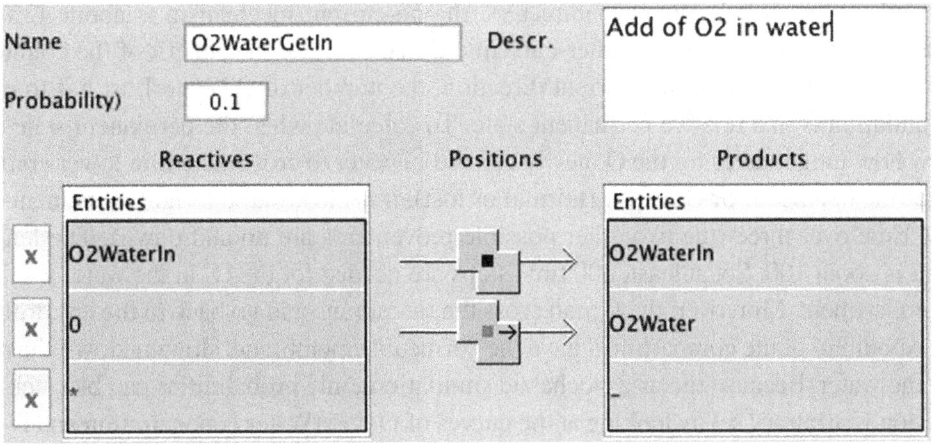

Name O2WaterGetIn Descr. Add of O2 in water

Probability) 0.1

Reactives	Positions	Products
Entities		Entities
O2WaterIn		O2WaterIn
0		O2Water
*		-

here to simulate the continuous flow of O_2 in the water coming from the left and disappearing on the right, when they go out of the simulation area.

Simulation and Results

The simulation using the counter-current behaviour is shown as an example of the dynamics of the gills (see figure 14).

Figure 14. Three screen captures show how the system evolves when the currents are in opposition (water comes from the left and circulates to the right; blood circulates from the right to the left). At t=150, the O2Water created by the O2WaterGetIn behaviour starts to diffuse to the right of the water compartment (the lower one). The O2Blood entities that reach the O2BloodOut are removed from the simulation thanks to the O2BloodGetOut behaviour. Few O2Blood, coming from the O2Water having crossed the membrane, move in the opposite direction. At t=350, O2Water entities reach the O2WaterLostBorder. At t=2000, the system is in a relatively stable state and O2WaterLost entities can be seen

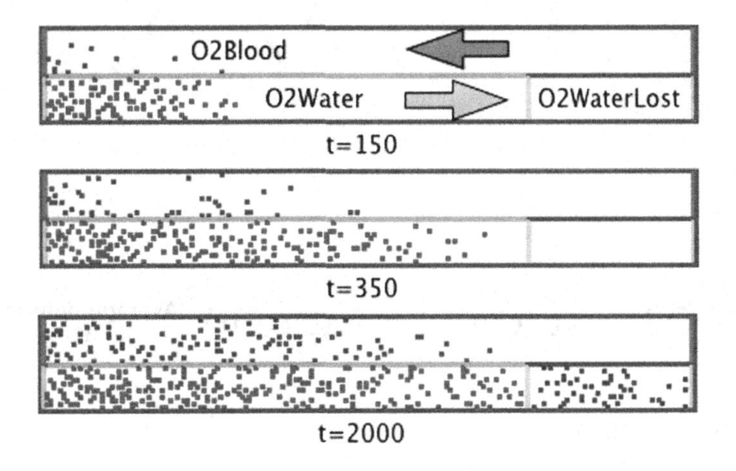

The counting of O2WaterLost gives the efficiency of the system and shows that, for the given initial state and the given behaviours' parameters, the co-current mechanism is about 42% worse in O2WaterLost as compared with the counter-current one. To evaluate the benefit of the counter-current direction as compared with the same current direction, the number of O2WaterLost has to be counted when the simulation is in a relative permanent state. To calculate when the permanent state arises, we need to know how long it takes for the O_2 newly created in water to cross the entire lower compartment. By looking at the behaviour of O2Water (normal or lost), it is clear that the right movement occurs, in average, one time over three (the two other possible movements are up and down). The length of the compartment is about 100. So, at least, 300 time steps are needed for the O_2 in the water to run through the lower compartment. Moreover, the O_2 can cross the membrane and go back to the left, following the blood flow. About 3/4 of the compartment have the permeable membrane, slowing down the movement of the O_2 in the water. Because this is a stochastic simulation, only probabilities can be calculated. The chosen criterion is arbitrary set by looking at the curves of O2LostWater concentration and by choosing a time from which the state looks permanent enough. The average number of O2LostWater is calculated between 1000 and 4000, which seem consistent (see Figure 15).

Blood Coagulation

Real System

The real system simulated in this example is a small section of a vein, in which a lesion is present. The aim of this model is to reproduce some of the molecular and cellular mechanisms during the blood

Figure 15. The average number of O2WaterLost under the co-current condition is 63.4 and 36.7 under the counter-current condition. That makes a loss of O_2 which is 42% superior in the case of co-current. Note that this percentage depends on the size of the compartments and the size of the membrane: this can be easily seen on the screen captures where the density of O_2 in water decreases according to the distance from the O_2 creation, at the left

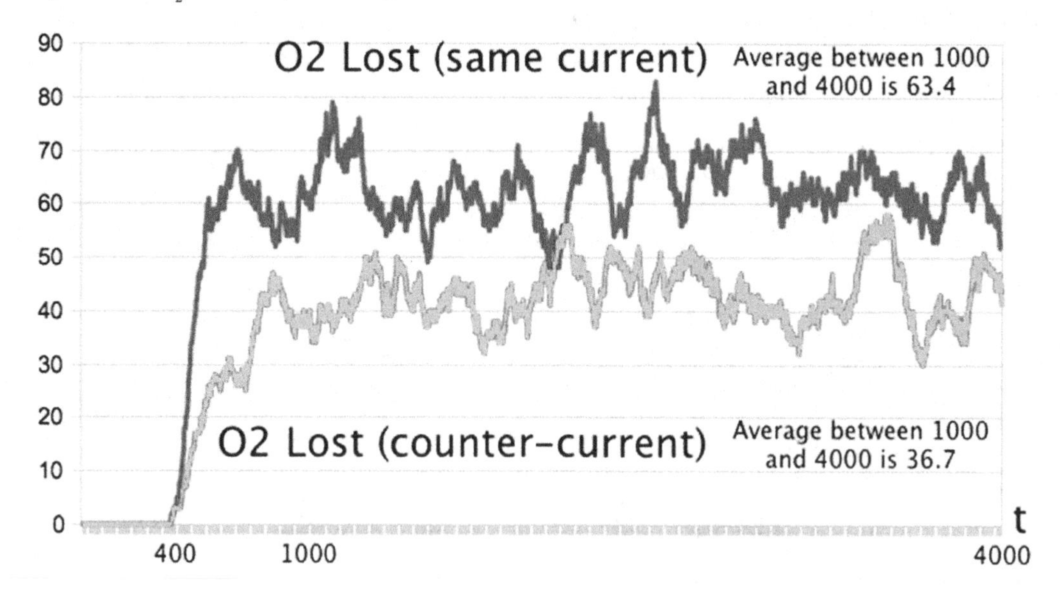

coagulation leading to the formation of a blood clot. Three stages are usually described. The primary hemostasis starts with the vein lesion and finishes with the formation of a plug covering the lesion. The plug is made of platelets and soluble rod-like shaped proteins called fibrinogen. The secondary hemostasis is the transformation of the platelets plug into a blood clot thanks to the conversion of the fibrinogen into insoluble strands called fibrins. The third and final stage is the fibrinolysis, when the clot is broken down.

Model

The model presented here includes the primary hemostasis but not the secondary one nor the fibrinolysis. Moreover, not all the known mechanisms are simulated to improve the model clarity. The model is described using all the involved entities (cells, molecules and others), their initial state (stored into a 2D matrix) and their behaviours (interactions, transformations, creations and blood flow).

Two types of entities can be distinguished. The first ones are those which directly represent real biological entities like molecules or cells:

1. Fibro (for the fibroblast cells).
2. Endo (for the endothelial cells).
3. PlaC (for the circulating platelets).
4. PlaF (for the fixated platelets).
5. Fibri (for the fibrinogen molecule).

The second ones are those which do not exist as such, but are needed to simulate abstract biological mechanisms, like the blood flow in this case:

6. FlowS (for start of the blood flow).
7. FlowE (for end of the blood flow).

The initial state looks like a rectangle in which the upper edge consists of one layer of endothelial cells. The lower edge contains one layer of endothelial cells (with a discontinuity representing the lesion in the vein) plus one layer of fibroblast cells, below the endothelial layer. The left edge is made of entities regulating the blood flow entrance and the right side consists of entities regulating the blood flow exit (see Figure 16).

The behaviours developed in this simulation aim to model the blood flow and the core mechanisms of primary hemostasis. Firstly, five behaviours are needed to model the blood flow (see Figure 17):

- One behaviour to model the apparition of the platelets coming from the left of the vein section.
- Two behaviours to model the blood flow circulating from the left to the right which are applied to the platelets (PlaC entities) and the fibrinogens (Fibri entities).
- Two behaviours to model the disappearance of the circulating entities (PlaC and Fibri) when they reach the right of the vein section.

Secondly, three behaviours are needed to model the coagulation (see Figure 18):

Figure 16. The initial state, and more generally all the simulation, takes place into a 100x28 matrix where entities are distributed according to a simple rectangle corresponding to a vein section. The vein is made-up of three layers, two of endothelial cells (Endo entities) and one layer of fibroblast cells (Fibro entities). Note that the bottom layer of endothelial cells has a hole in order to simulate a lesion, resulting in a direct contact between the fibroblast cells and the circulating blood. The two vertical edges do not exist as real biological entities but are necessary to represent, at the left side of the vein, the incoming of circulating entities (FlowS entities) like the platelets and, at the right side of the vein, the disappearing entities (FlowE entities). The flow of circulating entities moves from the left (Start) to the right (End)

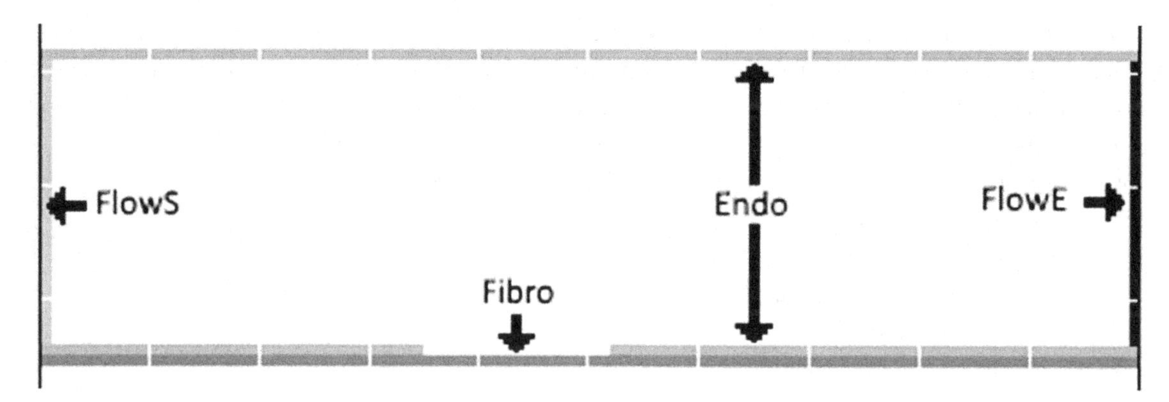

- The first behaviour models the fixation of circulating platelets on the lesion (on the fibroblasts).
- The second one models the fixation of circulating platelets on the plug of platelets in presence of fibrinogens.
- The third behaviour models the indirect production of fibrinogens by fibroblasts. Note that this last behaviour is a simplification of a more complex molecular pathway.

The end of the growth of the platelet plug is not modeled thanks to a behaviour but emerges from the simulation. More precisely, when all the fibroblast cells are covered with the anchored platelets, they simply cannot produce fibrinogen, stopping the growth of the plug.

Simulation and Results

In the simulation, 1 step corresponds to 1 second. The size of the vein is about 3 millimeter long and 1 millimeter large. As outputs, the shape of the plug is observed and the quantity of circulating fibrinogen is measured at each time step (see Figure 19).

The simulation shows that the plug of platelets starts after 139 steps (2 minutes and 19 seconds) and ends at step 254 (4 minutes and 14 seconds). After this step, no more fibrinogen entity is produced because the fibroblasts are no more in contact with the blood flow. The remaining fibrinogen disappears at the right of the vein section. The simulation is stopped at 360 steps, when all fibrinogen entities have disappeared.

With such a model, it could be possible to use the results of different simulations to build a synthetic and formalised mathematical model of the primary hemostasis. The virtual experiments can be performed quickly and at a very little price. Another interesting feature of such a model is that during a

Figure 17. The blood flow is simulated thanks to five behaviours. The two firsts (upper-left) are the movements of circulating platelets and fibrinogens entities from the left to the right. Note that they can also move vertically to reproduce the non-linearity of the blood flow. The third behaviour (upper-right) is the addition of platelets from FlowS (start of flow at the left of the vein section). The probability is 1% and can be changed to simulate different concentrations of platelets (according to the patient for example). The fourth and fifth ones (bottom of the figure) simply remove circulating entities that reach the end of the simulated vein section

Name	PlatCMvt / FibriMvt	
Description	Movement from the left to the right of the circulating platelets / fibrinogens	
Probability	1.0	
Reactives	Positions	Products
PlaC / Fibri	+	0
0	↑ PlaC / Fibri → ↓	PlaC / Fibri
•		-

Name	AddPlaC	
Description	"Creation" of circulating platelets from the start of the flow	
Probability	0.01	
Reactives	Positions	Products
FlowS	+	FlowS
0	FlowS →	PlaC
•		-

Name	DelPlaC	
Description	Removal of circulating platelets arriving at the right of the vein section	
Probability	1.0	
Reactives	Positions	Products
FlowE	+	FlowE
PlaC	← FlowE	0
•		-

Name	DelFibri	
Description	Removal of circulating fibrinogens arriving at the right of the vein section	
Probability	1.0	
Reactives	Positions	Products
FlowE	+	FlowE
Fibri	← FlowE	0
•		-

simulation it is possible to create new lesions, change the size or the shape of lesions, modify the platelet concentration, etc. The user can experiment many different scenario and hypothesis and see the validity but also the limits of the model.

DISCUSSION

Here, the authors discuss the interests and the drawbacks of NetBioDyn.

NetBioDyn Interests

The use of software in biology and education is interesting, both for teachers and students. It enables teachers to explain complex systems in an intuitive and pseudo-experimental way. It also helps students to create their own computational biological systems and see the important stages by creating a virtual system that rebuilds the real system. They can learn efficiently how the real system works.

Figure 18. The core mechanism governing the apparition of a plug of platelets is only made with three behaviours. The first behaviour (upper-left) is the creation of fibrinogen entities from fibroblast ones. In the real system, it does not occur like this, but it is a convenient simplification for this simple example. The second one is the fixation of circulating platelets that enter in contact with the fibroblast cells of the lesion. The third one, involving three reactive entities, simulates the fixation of circulating platelets onto the plug of anchored platelets, allowing it to grow

Name	AddFibri	
Description	The fibroblasts *indirectly* create fibrinogens	
Probability	1.0	
Reactives	Positions	Products
Fibro	+	Fibro
0	↑ Fibro	Fibri
*		-

Name	PlugBeginning	
Description	Fixation of circulating platelets onto the fibroblasts in the lesion	
Probability	1.0	
Reactives	Positions	Products
PlaC	+	PlaF
Fibro	↑ ← PlaC → ↓	Fibro
*		-

Name	PlugGrowth	
Description	Fixation of circulating platelets on the platelets plug in presence of fibrinogens	
Probability	1.0	
Reactives	Positions	Products
PlaC	+	PlaF
PlaF	↑ ← PlaC → ↓	PlaF
Fibri	↑ ← PlaC → ↓	0

From the teacher point of view, the modelling and simulation using such a software enable to:

- Create explicative and predictive simulations which can be easily handled by students
- Teach the students how to "program" complex systems with very little knowledge in computer sciences.
- Improve the learning curve of students.
- Stimulate students' creativity.
- Quickly find out which students have assimilated the course and which have not.
- See where, in the system, a student is right or wrong.

For the students, NetBioDyn gives the possibility to:

- Build or explore biological systems in an intuitive way with entities, behaviours and environments.

Figure 19. After 360 steps of simulation, corresponding to 6 minutes, the plug is formed and the circulating platelets continue to flow from the left to the right (upper diagram of the figure). The quantity of circulating fibrinogen is drawn on the lower part of the figure, showing its growth up to the value of 96 which could be multiplied by a given factor to match real experiments

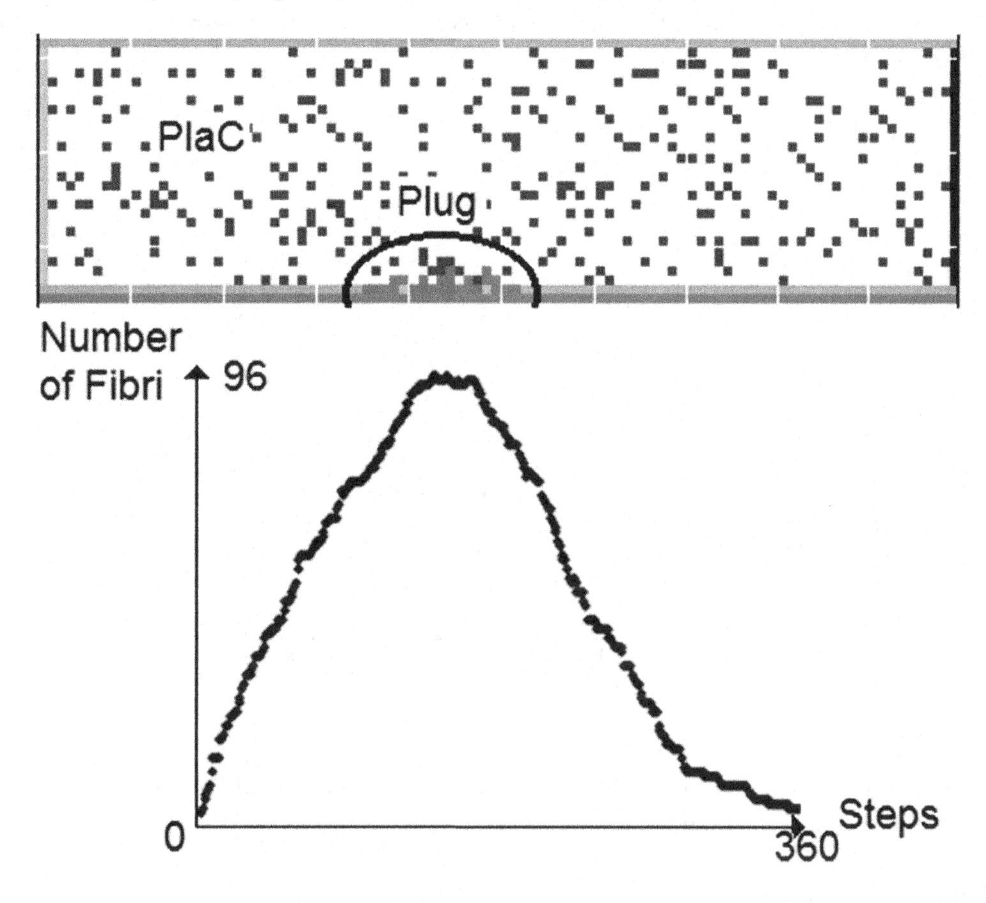

- Focus on the main mechanisms of a system and acquire a synthetic view.
- Pay attention to important mechanisms even if they look like details.
- Learn partially by themselves what is important and what is negligible.
- Organise their data and knowledge.
- Play at a motivating construction-like game.
- Explore the impact of parameters or behaviour variations in a simple way.
- Show the teacher what is their comprehension or misunderstanding of the real system.
- Share their knowledge with other students in a formal way.

A survey of 30 students (between 22 and 25 years old) has been made to assess the interest of the use of NetBioDyn in class. The results are presented in Figure 20. What is noteworthy is that biologist as well as computer scientist students are mostly interested by such an approach. The question of ease of use for NetBioDyn obtains the most "Very Good" answers for the two master students. Note that this survey is anonymous and that no teacher was present.

Figure 20. Survey of 24 students in Master 1 Biology and 6 students in Master 2 Computer Science at the Université de Bretagne Occidentale. 10 questions have to be answered on the course they had on computational biology and on the use of the software NetBioDyn. Answers show that both the course and the software are considered important for their training. Moreover, the software is adapted to their skills (no code is required). Finally, the course on modelling and the software interest them and are useful for their studies

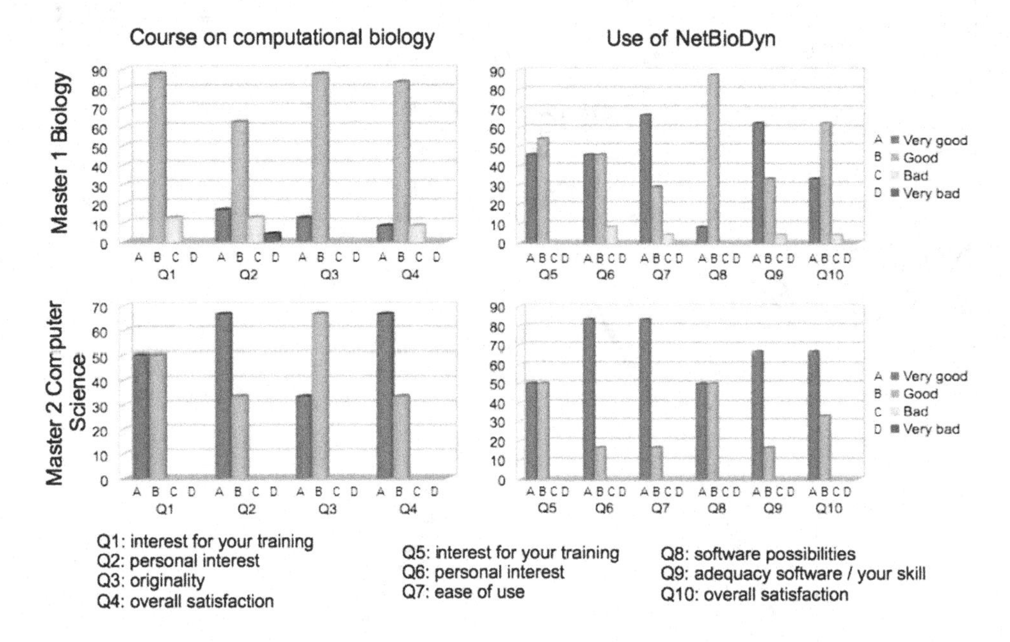

This software has also been successfully used by pupils (between 14 and 16 years old) of the middle school Max Jacob in Quimper (France) in May 2011.

NetBioDyn Drawbacks

The limitations of NetBioDyn can be divided into two kinds. The first concerns the multi-agent approach itself and the second specifically concerns the software.

Multi-agent systems involve many interacting agents. This usually requires powerful computers, in term of computation and memory, to perform all the entities and their interactions. NetBioDyn reduces this problem by only implementing algorithms of complexity N+B (where N is the number of entities and B the number of behaviours). This is possible because the simulator only uses a small local neighbourhood on a grid. Another problem with multi-agent systems is the number of parameters that can be very important. The simulations made with NetBioDyn could become difficult to control and understand if too many types of entities and behaviours are added. Usually the authors recommend to their students not to add more than 20 types of entity or more than 20 different behaviours. Note that a type of entity is not an entity: a type of entity is a prototype that can be instantiated many times into entities without any problem of parameters.

Regarding the software limitations, the most important one is the entity shape that can only be a cube, implying that the simulations cannot be multi-scale. This last point can be partially circumvented as seen in all the presented examples where molecules, cells and tissues are mingled. For instance, an entity can represent far more than just one molecule, but a group of molecules. An entity can represent a single cell, but also a part of a tissue. What is important here is that an intrinsically multi-scale problem can generally be flattened.

CONCLUSION

The simulation of complex multi-cellular systems using computers represents a great challenge, both scientifically and technically. Scientifically, it serves the comprehension of fundamental mechanisms governing cells, from single-cell behaviours to whole systems involving numerous interacting cells.

Technically, the creation of computational models imitating multicellular systems requires advanced computational skills.

This is why the authors have developed and presented in this chapter a software making it accessible for non-computer specialists by simplifying the process of modelling and simulation.

This simulator is made to create and manipulate models whereas most of educational software are already-made models that researchers or students can test by only changing some parameters or initial state (see for example the simulator presented in (Meir, Perry, Stal, Maruca, & Klopfer, 2005)).

Three simulations have been presented to show the diversity of models that can be done and to explain how they can be made.

To support this statement, this software has been evaluated by pupils of middle school and higher education students, especially those in the field of biology.

The interest of this software has been discussed to bring out both positive contributions like focusing in what to model instead of how to model but also negative ones like the absence of multi-scale entities.

The perspectives are multiple and can be divided into two parts. Firstly the software itself can be improved. One of the most important features could be the possibility to factorise similar behaviours into one. This could be performed for the entities by adding generic behaviours like Random Walk. Another one is to increase the number of graphical tools to quickly draw shapes of entities like squares, circles or filler. The authors are also working on a self-adjusting system to find the proper values for all the parameters involved in a simulation (behaviours' probabilities, entities half-lives etc.), in order to stay close to real results obtained for example *in vitro*.

Secondly, in terms of communication between NetBioDyn users, the main point to address will be the development of an internet social network where researchers, students and teachers will share their works, models and experiences (a starting professional network can be found at http://lefil.univ-brest.fr then NetBioDyn group).

REFERENCES

Aldrich, C. (2009). The Complete Guide to Simulations and Serious Games. John Wiley & Sons.

Ballet, P., Zemirline, A., & Marcé, L. (2004). The Biodyn language and simulators, application to an immune response and E. coli and phage interaction. In *Proceeding of the spring school on Modelling and simulation of biological processes in the context of genomics*.

Banitz, T., Gras, A., & Ginovart, M. (2015). Individual-based modeling of soil organic matter in NetLogo: Transparent, user-friendly, and open. *Environmental Modelling & Software*, *71*, 39–45. doi:10.1016/j.envsoft.2015.05.007

Bollen, L., & van Joolingen, W. R. (2013). Simsketch: Multiagent simulations based on learner-created sketches for early science education. *IEEE Transactions on Learning Technologies*, *6*(3), 208–216. doi:10.1109/TLT.2013.9

Cock, P. J. A., Antao, T., Chang, J. T., Chapman, B. A., Cox, C. J., Dalke, A., & de Hoon, M. J. L. et al. (2009). Biopython: Freely available Python tools for computational molecular biology and bioinformatics. *Bioinformatics (Oxford, England)*, *25*(11), 1422–1423. doi:10.1093/bioinformatics/btp163 PMID:19304878

Epstein, J. M. (1999). Agent-based computational models and generative social science. *Complexity*, *4*(5), 41–60. doi:10.1002/(SICI)1099-0526(199905/06)4:5<41::AID-CPLX9>3.0.CO;2-F

Gammack, D. (2015). Using NetLogo as a tool to encourage scientific thinking across disciplines. *Journal of Teaching and Learning with Technology*, *4*(1), 22–39. doi:10.14434/jotlt.v4n1.12946

Ginovart, M. (2014). Discovering the power of individual-based modelling in teaching and learning: The study of a predator-prey system. *Journal of Science Education and Technology*, *23*(4), 496–513. doi:10.1007/s10956-013-9480-6

Gkiolmas, A., Karamanos, K., Chalkidis, A., Skordoulis, C., & Papaconstantinou, M. D. S. (2013). Using simulation of Netlogo as a tool for introducing Greek high-school students to eco-systemic thinking. *Advances in Systems Science and Application*, *13*(3), 275–297.

Houser, N. (1992). *The Essential Peirce: Selected Philosophical Writings (1867-1893)* (Vol. 1). Indiana University Press.

Izaguirre, J. A., Chaturvedi, R., Huang, C., Cickovski, T., Coffland, J., Thomas, G., & Glazier, J. A. (2004). Compucell, a multi-model framework for simulation of morphogenesis. *Bioinformatics (Oxford, England)*, *20*(7), 1129–1137. doi:10.1093/bioinformatics/bth050 PMID:14764549

Jacobson, M. J., & Wilensky, U. (2006). Complex systems in education: Scientific and educational importance and implications for the learning science. *Journal of the Learning Sciences*, *15*(1), 11–34. doi:10.1207/s15327809jls1501_4

Jenkins, S. (2015). *Tools for critical thinking in biology*. New York, NY: Oxford University Press.

Kang, S., Kahan, S., McDermott, J., Flann, N., & Shmulevich, I. (2014). Biocellion: Accelerating computer simulation of multicellular biological system models. *Bioinformatics (Oxford, England)*, *30*(21), 3101–3108. doi:10.1093/bioinformatics/btu498 PMID:25064572

Kang, S., Kahan, S., & Momeni, B. (2014). Simulating microbial community patterning using biocellion. In Engineering and Analyzing Multicellular Systems (pp. 233-253). Springer. doi:10.1007/978-1-4939-0554-6_16

Klein, J. (2014). *Breve* [computer software]. Available from http://www.spiderland.org/breve/

Klein, J., & Spector, L. (2009). 3D Multi-Agent Simulations in the Breve Simulation Environment. In M. Komosinski & A. Adamatzky (Eds.), *Artificial Life Models in Software* (pp. 79–106). Springer. doi:10.1007/978-1-84882-285-6_4

Kubera, Y., Mathieu, P., & Picault, S. (2008). Interaction-oriented agent simulations: From theory to implementation. In *Proceedings of the 18th European Conference on Artificial Intelligence* (pp. 383-387). IOS Press.

Lynch, S. C., & Ferguson, J. (2014). Reasoning about complexity - software models as external representations. In *Proceedings of the 25th Workshop of The Psychology of Programming Interest Group*.

Meir, E., Perry, J., Stal, D., Maruca, S., & Klopfer, E. (2005). How effective are simulated molecular-level experiments for teaching diffusion and osmosis? *Cell Biology Education*, *4*(3), 235–248. doi:10.1187/cbe.04-09-0049 PMID:16220144

Merks, R. M. H., Guravage, M., Inzé, D., & Beemstaer, G. T. S. (2001). VirtualLeaf: An open-source framework for cell-based modelling of plant tissue growth and development. *Plant Physiology*, *155*(2), 656–666. doi:10.1104/pp.110.167619 PMID:21148415

Nikolai, C., & Madey, G. (2009). Tools of the trade: A survey of various agent based modelling platforms. *Journal of Artificial Societies and Social Simulation*, *12*(2), 2.

North, M. J., Collier, N. T., & Vos, J. R. (2006). Experiences Creating Three Implementations of the Repast Agent Modeling Toolkit. *ACM Transactions on Modeling and Computer Simulation*, *16*(1), 1–25. doi:10.1145/1122012.1122013

Odum, E. P. (1953). Fundamentals of Ecology. Cengage Learning.

Prain, V., & Waldrip, B. (2006). An exploratory study of teachers and students use of multimodal representations of concepts in primary science. *International Journal of Science Education*, *28*(15), 1843–1866. doi:10.1080/09500690600718294

Repenning, A., Smith, C., Owen, B., & Repenning, N. (2012). Agentcubes: Enabling 3d creativity by addressing cognitive and affective programming challenges. In *Proceedings of World Conference on Educational Multimedia, Hypermedia and Telecommunications* (pp. 2762-2771). Chesapeake, VA: AACE.

Repenning, A., & Sumner, T. (1995). Agentsheets: A medium for creating domain-oriented visual languages. *IEEE Computer*, *28*(3), 17–25. doi:10.1109/2.366152

Schrödinger, L. L. C. (2015). *The PyMOL Molecular Graphics System, Version 1.8*. Available at http://pymol.org

Shannon, P., Markiel, A., Ozier, O., Baliga, N. S., Wang, J. T., Ramage, D., & Ideker, T. et al. (2003). Cytoscape: A software environment for integrated models of biomolecular interaction networks. *Genome Research*, *13*(11), 2498–2504. doi:10.1101/gr.1239303 PMID:14597658

Shin, D. G., Liu, L., Loew, M., & Schaff, J. (1998). Virtual cell: A general framework for simulating and visualizing cellular physiology. *Visual Database Systems*, *4*, 214–220.

Stroup, W. M., & Wilensky, U. (2014). On the embedded complementary of agent-based and aggregate reasoning in students developing understanding of dynamic systems. *Technology. Knowledge and Learning*, *19*(1-2), 19–52. doi:10.1007/s10758-014-9218-4

Trott, O., & Olson, A. J. (2010). AutoDock Vina: Improving the speed and accuracy of docking with a new scoring function, efficient optimization and multithreading. *Journal of Computational Chemistry*, *31*, 455–461. PMID:19499576

Wallace, A. C., Laskowski, R. A., & Thornton, J. M. (1996). LIGPLOT: A program to generate schematic diagrams of protein-ligand interactions. *Protein Engineering*, *8*(2), 127–134. doi:10.1093/protein/8.2.127 PMID:7630882

Weiss, G. (Ed.). (1999). *Multiagent Systems: A Modern Approach to Distributed Artificial Intelligence*. Cambridge, MA: MIT Press.

Wilensky, U. (1999). *Netlogo*. Center for Connected Learning and Computer-Based Modeling. Available at http://ccl.northwestern.edu/netlogo/

XJ Technologies Company Ltd. (2012). *AnyLogic* [computer software]. Available from www.xjtek.com

Zhang, J., & Patel, V. L. (2006). Distributed cognition, representation, and affordance. *Pragmatics & Cognition*, *14*(2), 333–341. doi:10.1075/pc.14.2.12zha

KEY TERMS AND DEFINITIONS

Behaviour: A rule that transform or move entities in the simulated environment.

Bdellovibrio: A motile, vibrio shaped, monoflagellated, gram negative bacteria. It is a predator of Photobacterium leiognathi.

Entity: A virtual entity representing a cell or a molecule.

Integrated Development Environment (IDE): A software with an advanced graphical user interface (GUI) aiding the user to create other software applications.

Lysis: The death of an entity representing a cell.

Multi-Agent System: A set of interacting agents (also called entities). In our system, each agent has behaviours to perform simple reactive tasks.

Photobacterium Leiognathi: A bacteria which is bioluminescent and a prey for Bdellovibrio.

Chapter 7
Agent-Based Modelling in Multicellular Systems Biology

Sara Montagna
Università di Bologna, Italy

Andrea Omicini
Università di Bologna, Italy

ABSTRACT

This chapter aims at discussing the content of multi-agent based simulation (MABS) applied to computational biology i.e., to modelling and simulating biological systems by means of computational models, methodologies, and frameworks. In particular, the adoption of agent-based modelling (ABM) in the field of multicellular systems biology is explored, focussing on the challenging scenarios of developmental biology. After motivating why agent-based abstractions are critical in representing multicellular systems behaviour, MABS is discussed as the source of the most natural and appropriate mechanism for analysing the self-organising behaviour of systems of cells. As a case study, an application of MABS to the development of Drosophila Melanogaster is finally presented, which exploits the ALCHEMIST platform for agent-based simulation.

INTRODUCTION

The chapter reviews the role of *agent-based modelling* (ABM) in the simulation of biological systems, focussing on the simulation of multicellular systems. Modelling multicellular systems requires tools that can support multi-scale models, where different cells form large-scale, dynamic networked systems – as, e.g., in tissues of cells, organs, and even full embryos – and where both the biochemical reactions that occur inside each cell and the molecules diffusion along the tissue (mediating the interaction among nuclei/cells) can be reproduced (Dada & Mendes, 2011; Deisboeck, Wang, Macklin, & Cristini, 2011).

Despite its recognised value for modelling complex systems (Bonabeau, 2002), and also for modelling biological systems (Holcombe et al., 2012), ABM is still not completely accepted as a suitable modelling approach in the literature on multicellular systems—whereas, dually, multicellular systems are not usu-

DOI: 10.4018/978-1-5225-1756-6.ch007

ally considered as a valid application domain in the ABM literature. This provides the first motivation behind this chapter: that is, pointing out the potential of ABM in the context of multicellular systems.

ABM grounds on top of the three fundamental abstractions for multi-agent systems (MAS) (Zambonelli & Omicini, 2004):

- **Agent:** The autonomous source of activity in the system,
- **Society:** A group of interacting agents, and
- **Environment:** The context where agents and societies live and interact. As such, agent-based models are perfectly suited for studying the systemic and emergent properties that characterise a multicellular system, meant to be reproduced *in virtuo*.

ABM looks a promising approach in this context given the direct mapping between agent-based abstractions and multicellular systems: agent-cell (with internal and interacting behaviour), agent society-multicellular system, environment-extracellular matrix.

As an example, *developmental biology* is used as a reference scenario to demonstrate the power of *multi-agent based simulation* (MABS). Developmental biology is a challenging branch of life science that studies the process by which organisms develop, focussing on the genetic control of cell growth, differentiation and movement: as such, it analyses multicellular systems during their formation and development. In particular, the chapter illustrates the application of MABS to the development of *Drosophila Melanogaster* based on the ALCHEMIST agent-based simulation framework (Pianini, Montagna & Viroli, 2013; Montagna, Omicini & Pianini, 2016). In the overall, the contribution of the chapter can be articulated as follows:

1. Reviewing the actual state-of-the-art in MABS applied to multicellular systems biology, as well as making an argument for the value of agent-based approaches in the field;
2. Reporting on a case-study of agent modelling and simulation;
3. Discussing the specific issue of developmental biology – that is, one of the hottest and most challenging field in biology –, and verifying which specific achievements were obtained in the last years by adopting the agent metaphor within biological studies;
4. Providing an example of MABS platform for biological systems; and
5. Supplying biologists with a tool possibly helping them finding insights into the biological phenomena, experimenting working hypotheses, performing the so called *what-if* analysis, and making prediction on system behaviour.

BACKGROUND

The field of research devoted to understanding biological cell functionalities by means of modelling and simulation techniques is called *Systems Biology*. The field originally emerged around year 2000, in particular the pioneering work of Kitano (2002). Systems biology is grounded on the idea that the high complexity of phenomena and mechanisms regulating biological system dynamics cannot be understood by *in-vivo* experiments alone—mostly, due to the limitation of experimental techniques in investigating the coordinated activities among system components. In order to understand biological systems, other concepts and tools are required that could enable the execution of *in-silico* experiments, *i.e.*, modelling and simulation techniques to understand how the integration and interaction of components can yield emergent behaviours.

Systems biology mainly focuses on the analysis of mechanisms at the intracellular scale—such as pathways, genetic, and metabolic regulation. Recently, a massive trend of research moved towards the analysis of the entire multicellular system (Hoehme & Drasdo, 2010), and a new field of research emerged: multicellular systems biology. Accordingly, the remainder of this section provides a description of the main biological facts involving multicellular systems, and points out the critical issues that make such systems particularly interesting from the research viewpoint.

Multicellular Systems Biology

Multicellular systems biology studies multicellular systems. *Multicellular systems are living organisms that are composed of numerous interacting cells.*[1] Humans and most of the animals are multicellular organisms. Research focuses in analysing specific tissues of the entire organism, such as neural system, immune system, adult stem cells etc., that are leading specific biological functions. Moreover, many research initiatives are devoted to analysing systems during embryogenesis and population of cancer cells.

All of the aforementioned systems share some features and properties, which make their understanding particularly challenging. The main property is that they are inherently *multi-scale*—see Figure 1. As such, the global behaviour emerges by mechanisms working across multiple space and time scales. Each scale integrates information from strata above and below, according to the principle of *upward and downward causation* (Uhrmacher, Degenring, & Zeigler, 2005). The building block for the vast majority of mechanisms at each level is the *interactions* among components.

Figure 1. A biological system is articulated over at least four levels with different spatial and temporal scales. Each level influences the dynamic at the level above (upward causation) and below (downward causation)

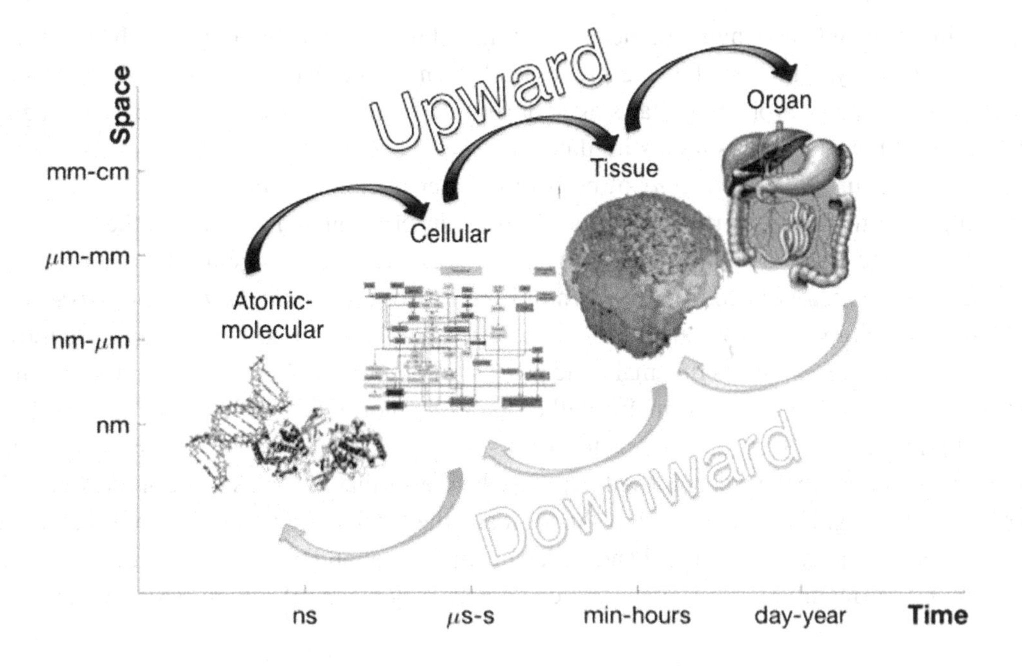

It is possible to identify three main hierarchical levels for multicellular systems (Setty, 2012): molecular, cellular, and tissue.

Intracellular regulatory network controls molecular mechanisms, such as gene expression, receptor activity and protein degradation. The dynamic at intracellular level brings each cell into a state that affects the architecture and function on the cell level. However, the intracellular regulatory processes is not enough for understanding the emergent processes: an individual cell decides on its next developmental step, proliferation, fate determination, and motility, according to intracellular state as well as to constrains deriving from the level above. Cell population acts in concert to develop its anatomy and function.

Computational Model Requirements

In the literature, a *middle-out modelling approach* is suggested as a possible technique to reproduce the behaviour of these systems. It starts with an intermediate scale (the cell, the basic unit of life), and it gradually expands to include both smaller and larger scales for reproducing the intra- and inter-scale interactions and integration. It should then be multi-scale in nature, so as to capture several spatial and temporal scales.

Also, it should feature abstractions for modelling diffusion of molecules across cells and in extracellular matrix, for studying the effects of short and long range signals, and for modelling the compartment membrane. Moreover, it should be stochastic, in order to capture the aleatory behaviour characteristic of those systems involving few entities; also, the cell topology should be dynamic, for modelling cell division and movement. Finally, it should be able to capture cell heterogeneity, so as to model individual structures and behaviours of different entities of the biological system.

From the validation perspective, the field is particularly critical since it requires multiple data: molecular data such as gene expression profiles and image data such as spatial-temporal growth pattern.

Developmental Biology

Among the different and challenging branches of multicellular systems biology, the chapter focuses on developmental biology, which studies the processes by which multicellular organisms develop. Multicellular organisms pass through similar stages during development. The life of animals begins with the fertilisation of the egg; *fertilisation* stimulates the egg to begin development. A series of extremely rapid and synchronous *mitotic divisions* of cells or cellular nuclei characterises the cleavage stage of development. After that, cells start the process of gene transcription and protein synthesis: cells *differentiate* in that each type expresses a diverse set of genes. Cell diversity is organised into a precise spatial distribution (*embryo regionalisation*), as shown in Figure 2 for the *Drosophila Melanogaster* embryo.

Developmental biology focuses on the genetic control of cell growth, differentiation, and movement, for understanding the mechanisms that make the process of embryo regionalisation so robust, making it possible for an organism to evolve from one cell to the same morphology every time (Hohm & Zitzler, 2009). Answering the big questions of development – such as "How do local interactions among cells and inside cells give rise to the emergent self-organised patterns that are observable at the system level?" – has a double meaning. On the one hand, it should make the fascinating process of multicellular organisms morphogenesis clear. On the other hand, since the molecular and biochemical processes regulating cell proliferation, differentiation, survival, and death also play a central role in tissue homeostasis and

Figure 2. The pair-rule gene even-skipped (red) together with hb (blue) and Kr (green) in Drosophila embryo at the cleavage cycle 14A temporal class 8. Reconstructed image from (Pisarev et al., 2009). Embryo name: ba3

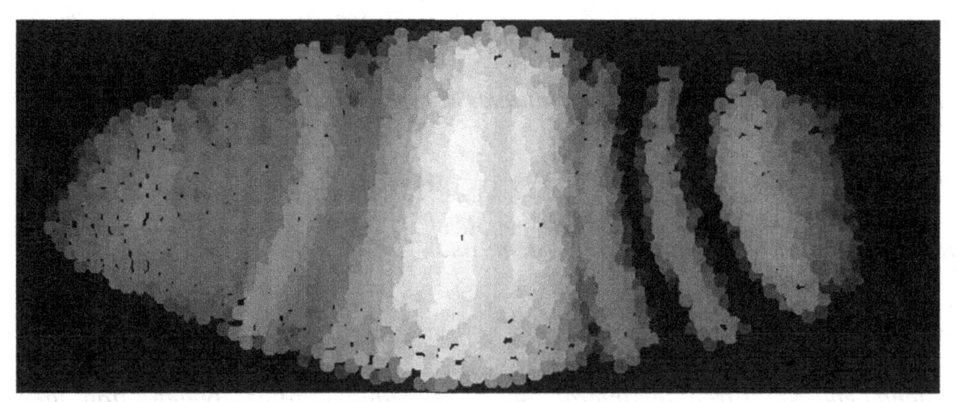

*For a more accurate representation of this figure, please see the electronic version.

cancer cells, it could reveal the key mechanisms that drive cancer progression, paving the way towards the identification of new therapies that can control tumour growth.

ABM IN MULTICELLULAR SYSTEMS BIOLOGY

The interdependent nature of multicellular processes often makes it difficult to apply standard mathematical techniques to separate out the scales, uncouple the physical processes or average over contributions from discrete components. (Cooper & Osborne, 2013)

Individual-based models (IBM) are often presented in literature as one of the main alternative to mathematical models based on differential equations. Both cellular automata (CA) and agent-based modelling (ABM) are representative of the class of IBM. Altogether, their computational models are similar, since they are organised around an abstraction of *individual* entities – *cells* in CA, agents in ABM – and the concept of *space*, defining the notion of neighbourhood for each entity. On the other hand, cells in CA are simpler then agents in ABM, since their behaviour is defined with update rules that change their states, switching between boolean values in the basic case, based on the state of the cell itself and of its neighbours. Also, the grid of cells is usually static, so that movement or cell replication is allowed. However, CA cannot model with sufficient accuracy cell proliferation and movement, which are crucial mechanisms especially in tumourogenesis.

This is where the theory of agents and multi-agent systems (MAS) steps in, promoting IBM where individual entities can express a more articulated autonomous behaviour, living, moving, and replicating in a possibly dynamic and complex environment. In the literature, agent-based systems, and MAS in particular, are considered as an effective paradigm for modelling, understanding, and engineering *complex systems* (Omicini & Zambonelli, 2004), and biological systems in particular (Cannata, Corradini, Merelli, Omicini, & Ricci, 2005), since they provide a basic set of high-level abstractions that make it possible to directly capture and represent the main aspects of complex systems, such as interaction, multiplicity and decentralisation of control, openness, and dynamism (Uhrmacher & Weyns, 2009; Merelli et al., 2007; Macal & North, 2010; An, Mi, Dutta-Moscato & Vodovotz, 2009; Gelfand, 2013).

As from the pioneer work of Bonabeau (2002), an ABM describes the system *from the perspective of its constituent units*. Moreover he states that (Bonabeau, 2002:

The benefits of ABM over other modeling techniques can be captured in three statements:

1. ABM captures emergent phenomena;

2. ABM provides a natural description of a system; and

3. ABM is flexible.

Emergent phenomena result from the interactions of individual entities. By definition, they cannot be reduced to the systems parts: the whole is more than the sum of its parts because of the interactions between the parts. An emergent phenomenon can have properties that are decoupled from the properties of the part. [...] ABM is, by its very nature, the canonical approach to modeling emergent phenomena: in ABM, one models and simulates the behavior of the systems constituent units (the agents) and their interactions, capturing emergence from the bottom up when the simulation is run.

Given that the above definition shares its crucial concepts with the one of systems biology, ABM looks particularly adequate for modelling multicellular systems.

By adopting ABM, biological systems can be modelled as a set of interacting autonomous components, i.e., a set of agents, and their chemical environment can be modelled by suitable environment abstractions, enabling and mediating agent interactions (Weyns, Omicini, & Odell, 2007). In particular, ABM provides a straightforward way to model:

1. The diverse individual structures and behaviours of different entities of the biological system as different agents (heterogeneity);
2. The heterogeneous – in space and time – environment structure and its dynamics; and
3. The local interactions between biological entities/agents (locality) and their environment.

Executing an agent-based model makes it possible to study its evolution through time, in particular:

1. Observing individual and environment evolution;
2. Observing global system properties as emergent properties from agent-environment and inter-agent local interaction; and
3. Performing in-silico experiments.

In ABM each cell is typically modelled as an agent, accounting for independent cell behaviour as well as for cell interactions. In the context of biological system, agent-based models can therefore account for individual cell biochemical mechanisms – gene regulatory network, protein synthesis, secretion and absorption, mitosis and so on – as well as the extracellular matrix dynamic – diffusion of morphogens, degradation and so on – and their dynamic influences on cell behaviour. On top of the cells/agents system, then, there is the whole population, whose behaviour can be observed through simulation, thus showing the relationship between microscopic behaviours of the single agents and macroscopic behaviours of cell population. A further benefit of ABM in multicellular systems biology is that each agent can be associated to a different autonomous behaviour, thus capturing cell heterogeneity that plays a crucial role in a lot of multicellular phenomena.

Different ABM techniques can be found in the literature, such as lattice-based, lattice-free, cellular Potts, lattice-gas, and subcellular element modelling methods (Anderson, Chaplain, & Rejniak, 2007). Typically, they mostly differ for how they model the agent distribution over the environment: for instance, lattice-based models, such as cellular Potts models, organise agents into a lattice structure.

In the Literature

The idea presented in this chapter is supported by related literature that reports about platforms specifically developed for building agent-based models of multicellular systems and running simulation. Generally, a significant number of applications of ABM to biological systems can be found in the literature—even though some of them mainly specifically refer to intracellular networks only (Montagna, Riccim & Omicini, 2008; González et al., 2003; González Pérez, Omicini, & Sbaraglia, 2013; Pogson, Smallwood, Qwarnstrom, & Holcombe, 2006). However, for their very nature, ABM fits multicellular systems, becoming a common method in multicellular system biology (Gorochowski et al., 2012). This is why the remainder of this section will only mention the works related to multicellular systems. In particular, most of the works couple ABM with reaction-diffusion systems that model intracellular or microenvironmental dynamics of substrates.

Platforms

In the following the main ABM platforms are mentioned and shortly described. Since providing details on how to implement a model within each platform is out of the scope of this chapter, interested readers can refer to the original platform documentation.

- **COMPUCELL3D:** (Cickovski et al., 2005) is probably the best example of a hybrid platform that combines a lattice-based ABM (to model cell interactions) and continuous models based on reaction-diffusion equation (to model chemical diffusion). Each compartment – which can be a cell, or even a subcellular compartment – is associated to a set of attributes (like cell type, volume, area), is characterised by a state, and has a behaviour defined by a set of formula changing the value of a variable called *effective energy*, which implements most cell properties, and also governs internal and interacting dynamics. For instance, it accounts for cell adhesion, or elastic constraints that consider volume and surface constraints. In the overall, COMPUCELL3D looks like a very promising framework, whose main limitation is the lack of a realistic model for cell internal behaviour.
- **Morpheus:** (Starruß, de Back, Brusch & Deutsch, 2014) is also is a hybrid framework embodying both ABM and mathematical models such as ordinary differential equation and reaction diffusion systems. Moreover, an integration of the two modelling approaches is possible: each compartment is an agent that, according to the cellular Potts model, defines compartment shape and motility, and whose internal behaviour is usually linked to models of intracellular biological networks

specified in the form of ordinary or stochastic differential equations. Morpheus is enhanced with a full-featured graphical user interface that supports the user during model definition and editing as well as during simulation, providing tools for interacting with the simulation and visualising simulation output and results in various ways.

- **ALCHEMIST:** (Pianini, Montagna & Viroli, 2013; Montagna, Omicini & Pianini 2016) is a novel platform based on the concept of nodes/agents, whose behaviour is specified by chemical-like rules. It is particularly suited for modelling biological systems, being its main abstractions very close to the main components of such systems. Details on that will be given in the next section.
- **Other Platforms:**. Beyond the aforementioned platforms, which are the best examples of ABM platforms specifically developed for multicellular systems, other examples are worth to be mentioned shortly in the following. CellSys (Hoehme & Drasdo, 2010) implements a class of lattice-free agent-based models permitting realistic simulations of tissue growth. Currently, the platform seems to be not maintained anymore.

Other platforms are built for developing ABM-MABS in diverse domains, thus without specific abstractions devoted to modelling biochemical mechanisms. For instance, FLAME (Holcombe et al., 2012) implements a pure agent-based model. The behaviour is here modelled in terms of state machines, i.e., a set of states with transition rules or functions that determine how to make the agent evolve. In biology, it has been applied to the modelling of stem cells. Repast (North et al., 2013), MASON (Luke, Cioffi-Revilla, Panait, Sullivan & Balan, 2005), NetLogo (Wilensky, 1999) are the most popular MABS platforms. As such, they provide tools for implementing the typical ABM abstractions. Agent behaviours are defined in different ways: the most general-purpose one is by programming languages (Java for MASON and Repast; a Logo-inspired language for NetLogo) by which any possible behaviour can be given to the agent and to the environment that contains the agents.

Successful Applications

Mostly because it makes it possible to model cell diversity, ABM has been frequently adopted in order to simulate cancer growth, also accounting for the stochasticity of genetic mutations. Wang et al. (2015) provide a review of the most recent ABM approaches developed for studying different aspects of cancer growth. Accordingly, in the following the most representative works showing how ABM can capture the main mechanisms of cancer development are shortly introduced and discussed.

Macklin et al. (2012) describe an agent-based cell model of ductal carcinoma in situ (DCIS), which is a type of breast cancer. Each cell is modelled as a mobile agent that moves according to adhesive, repulsive, and motile forces. Heterogeneity is accounted for by associating each agent to a phenotypic state, and by means of rules that govern phenotypic transitions depending on the internal state of the cell and the local microenvironment. Cellular volume is also reproduced accounting for variations during proliferation and necrosis processes. Such a model is coupled with a reaction-diffusion model for modelling substrate dynamics, namely for describing how cells affect the concentration of molecules in their microenvironment and the dynamic of intracellular molecular signalling. The model predicts that DCIS tumours grow, in agreement with mammographic data.

Olsen and Siegelmann (2013) design an ABM of various cancer types, focussing in particular on cancer growth. It has three levels of representations: the cellular, tissue, and molecular ones.

Cells are agents, and a notion of environment is explicitly represented: it mainly models blood vessels that bring nutrients to cells, thus modelling the molecular level. Agents are organised into a three-dimensional grid: the tissue level emerges from the agents and the environment dynamics and interactions. Their behaviour is represented via flow-charts.

Finally, it is worth mentioning a number of other multicellular systems whose mechanics has been investigated in terms of ABM. Gorochowski et al. (2012) analyse the relationship between single-bacterial-cell dynamic and the population level characteristics for studying the collective behaviours observable in many bacterial populations. In Rouillard and Holmes (2014) a case of ABM application to myocardial cells has been reported. It studies the long-term consequences of therapies that alter infart mechanics. These therapies aim at reducing complications of heart attack by altering heart structure, mechanics, and function over time. Walker, Wood, Southgate, Holcombe, and Smallwood (2006) present an extension of the *Epitheliome*, an agent-based model of epithelial cells. There, agents represent individual cells that, in the original *Epitheliome* model, iteratively change their state according to a set of rules based purely on observations and responsible for injecting behaviours into the agents, such as proliferation, intercellular adhesion, migration, and apoptosis. In the last version presented by Walker et al. (2006), a mathematical model is integrated within each agent, to directly model intracellular pathways that cause cell proliferation, thus making a connection between molecular-level events and growth on a tissue level.

Given this big picture of related works supporting the crucial role that ABM has in multicellular systems biology, hereafter we provide details on the computational model of a specific platform and we then show how it applies to modelling a case of embryo development. The goal is to provide readers with an example of what we described until now, an example that can possibly guide interested readers in this fascinating field. Among the tools described before, we choose ALCHEMIST because it is particularly suited to model the case study, and also because, being developed by our research group[2], we are particularly competent on the platform. The example is extracted from a literature paper (Montagna, Pianini, & Viroli, 2012).

AN ILLUSTRATIVE EXAMPLE: THE ALCHEMIST PLATFORM FOR SIMULATING *DROSOPHILA MELANOGASTER* DEVELOPMENT

The ALCHEMIST platform[3] was specifically developed for running models composed by interacting compartments whose internal behaviour is defined by a set of chemical-like rules (Pianini, Montagna, & Viroli, 2013). The basic idea was to overcome the limits we found in other platforms, where the computational model is hard to understand. There, the representation of biological phenomena is far from the real mechanisms actually occurring in biological systems, where everything is a reaction.

The engine is based on two optimised versions (Gibson & Bruck, 2000; Slepoy, Thompso, & Plimpton, 2008) of the popular and successful Gillespie's stochastic simulation algorithm (SSA), originally defined for the simulation of chemical systems (Gillespie, 1977). The computational model was then extended to shift the paradigm from the pure chemistry towards the word of ABM, by specifying the agent (internal and interacting) and environment behaviours through the concept of reactions, properly extended from the original concept of pure chemistry found in Gillespie.

Thus, while ALCHEMIST makes it possible to simulate multiple, separate, interacting, and possibly mobile entities, its approach to ABM is still structured around the concepts of *reaction* and *compartment*.

As a result, ALCHEMIST computational model, based on these two key concepts, is ideal for modelling multicellular systems, where the basic mechanisms are reactions, and the basic entities are compartments.

In the following the notions on top of which the ALCHEMIST computational model is grounded are described. Then we show how it is possible to develop a model of a multicellular system using the abstraction provided by ALCHEMIST by using a literature example, the *Drosophila Melanogaster* development shown in a previous work of ours (Montagna, Pianini, & Viroli 2012).

The ALCHEMIST Computational Model

In ALCHEMIST a compartment can represent an agent, and all the possible events in the model should, in the end, break down to a set of reactions (see Figure 3). As such the notion of agent we can deal with, modelling a system with ALCHEMIST is a soft one, *i.e.*, meaning that we can hardly give to agents cognitive capabilities –as it can be required in other simulation domains such as social sciences. However it is perfectly suited for cellular systems. Accordingly, in the following discussion the terms compartment and agent will be used interchangeably as synonyms.

To equip ALCHEMIST with all the abstractions proper of ABM, as first step, the notion of environment, missing in chemistry-derived SSAs, is introduced as a first-class abstraction (Weyns, Omicini & Odell, 2007). The environment has the responsibility to provide, for each compartment, a set of compartments that are its neighbours. The function of the current environment state that determines whether or not two compartments are neighbours can be arbitrarily complicated. Also, it is responsible of exposing possible physical boundaries, namely, to limit movements of compartments situated within the environment.

Figure 3. In the ALCHEMIST computational model, a system is composed of interacting agents whose state is defined by the molecules it contains. The agent state changes over time for the execution of a set of reactions that model the agent behaviour

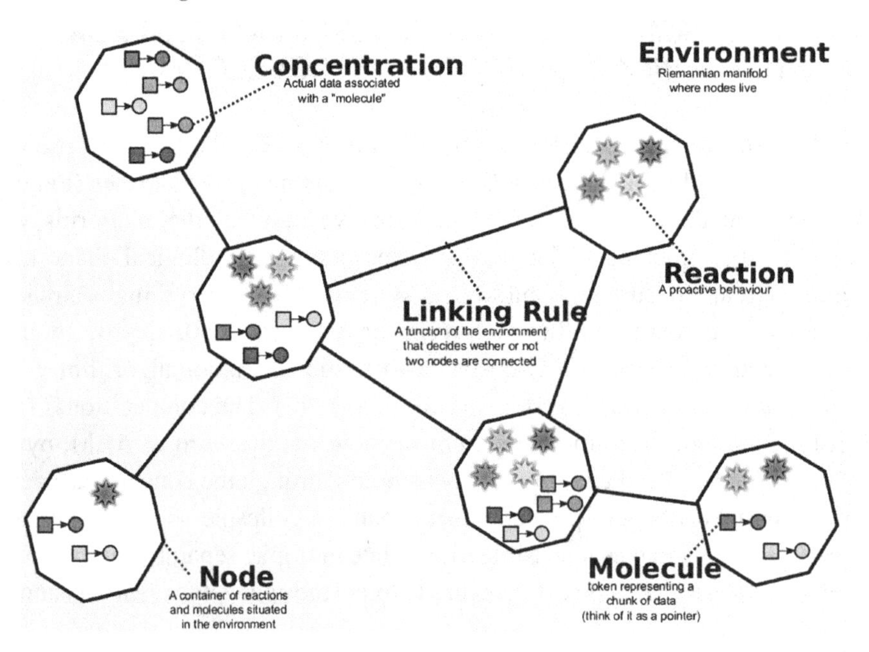

The fact that reactions are the only abstraction the modeller can rely upon in order to specify the agent behaviour and to let the simulated system progress does not actually hinder expressivity. In fact, the ALCHEMIST model does not use a strictly-chemical agent behavioural model (White & Pagurek, 1998), rather it generalises the concept of reaction as: *"a set of conditions that, when matched, trigger a set of actions on the environment"*. Accordingly, a condition is a function that associates a numeric value ranging from zero to positive infinity to each possible state of the environment and of agent internal state. Conditions can be everything: for instance, in computational biology, with conditions the modeller can verify the number of neighbouring compartments, the presence of specific molecules inside the agent or in the surrounding environment. If the value associated to condition is zero, the event cannot be scheduled; otherwise, it is up to the reaction to interpret the number. In this framework, actions are arbitrary changes to the environment and/or to agent state. Examples can be the movement of agents, the creation or destruction of molecules, the creation of new agents.

Reactions compute their own expected execution time. Such a putative time is generated by a function ("rate equation" in the following) taking as input all the values generated by conditions and a time distribution. The engine may require putative time to be updated in case that:

1. The reaction has been executed, or
2. Another reaction that the reaction depends on has been executed.

The injection of a time distribution as a parameter for the rate equation makes it possible to model a whole set of events that are not exponentially distributed: imagine, for instance, a burette dropping some quantity of reactant in a compartment every fixed time interval. Such an event is clearly not memoryless, and can be modelled in the proposed framework with a reaction having no condition and a single action increasing the reactant concentration in the compartment. Such a reaction can be fed with a Dirac comb, and the resulting system would seamlessly mix exponential and non-exponential events.

A Case Study in Developmental Biology

To show how to build a model of a multicellular system on top of ALCHEMIST, hereafter we present a case of embryo development. We inspire our discussion to a model of *Drosophila Melanogaster* development presented in Montagna et al. (2012). Drosophila is one of the best-known multicellular organisms. The model reproduces the process of *Drosophila* development from the fertilised egg until the pattern formation of the products of a set of crucial genes called gap genes—*hb*, *kni*, *gt*, and *Kr*.

A schematic representation of what happens in the embryo in the first hours after fertilisation is shown in Figure 4. The schema sketches (from top to bottom) how the concentration of diverse gene products (mRNA and proteins) varies: starting with a unique cell with a polarised distribution of maternal proteins, the downstream transcription of different genes divides the embryo in regions that are stripe-like patterns of expression that strongly characterise the *Drosophila* embryo—as shown in Figure 2. The formation of this pattern is crucial since these segments are closely related to the final anatomical structure. At the same time, fast compartment divisions happen so that the embryo – at that stage – consists of about 6000 separate cells. Understanding how it is possible to create such a precise spatial distribution of gene products, thus generating an exact arrangement of different cell types, is the fascinating challenge of developmental biology, for which the help of modelling techniques is crucial.

Figure 4. Hierarchy of genes establishing the anterior-posterior body plan of the Drosophila Melanogaster embryo

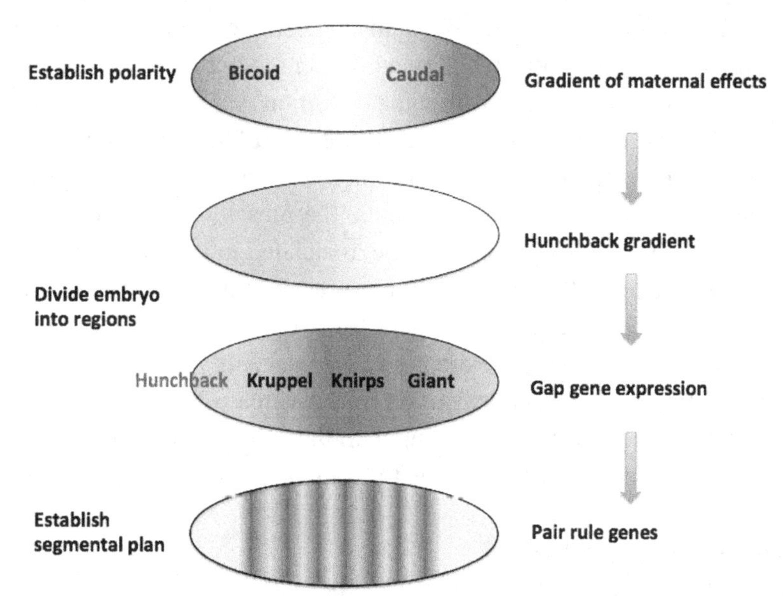

The Model

The whole embryo is modelled as a unique elliptic environment. Each cell is a compartment/agent situated inside the environment. It carries different types of molecules, can move and divide. The 2D continuous environment regulates the position of agents, specifically during movement and growth that must occur within its boundaries. Moreover it is filled with diffusing proteins. The mechanisms explicitly modelled are:

1. Molecule interactions,
2. Molecule diffusion,
3. Agent divisions, and
4. Agent migration.

Molecule Interactions

According to the literature, the gene regulatory network involving gap genes and maternal protein is modelled as a set of reactions where each gene is activated or inhibited by the others. In particular the four gap genes are the molecular actors of this model. They are modelled in two different states, active and inactive (for instance, for the *hb* gene, we have respectively gHb0 and gHb1). Their products are also explicitly modelled. For instance, for the *hb* gene, we have pHb. To exemplify this rule we provide here an example through the reaction that models the inhibition of the *hb* gene by the *Kni* protein: the conditions is that the agent comes with at least one *Kni* protein and the *hb* gene active; the action changes the state *hb* gene from active to inactive. The reaction looks like:

[pKni + gHb1] --> [pKni + gHb0]

The whole network, including all the interactions among gap genes, and their activation by maternal proteins is fully described by Montagna et al. (2012).

Molecule Diffusion

Diffusion occurs from / to agents and inside the environment along the x and y axis. Molecules move from one agent into the environment, or, the other way round, from one location of the environment into a neighbouring agent. Depending on the type of diffusion (passive or active transport), conditions – for this reaction to be scheduled – are diverse, such as the presence of membrane proteins or of energy in the form of ATP. A simple example follows:

[pHb in cell] --> [pHb in environment]

Agent Division

It is modelled as a chemical-like reaction, whose condition is given by the maximum number of other agents in the neighbourhood and whose action is the creation of a new agent. The new agent is situated close to the dividing one in a casual direction, provided that an other agent does not occupy the new position, and owns half of its molecular content. The rate of compartment division is determined according to the rate observed in the real system. In particular we refer to data reported in Montagna et al. (2010). Since experimental data show a given synchronisation between dividing cells, the phenomena was modelled through a non-Markovian reaction. The reaction looks like:

NumberOfNeighbors < 6 --> Division

Agent Migration

Movement of agents is based on biomechanical forces of repulsion among neighbouring agents: if two agents are closer than a distance – given as a parameter – that models agent surface (spatial extension), a repulsive force is generated and a new position for them is computed. The nearer two agents are, the stronger is the repulsion. Moreover they are forced to remain within the membrane-delimited area so as to filling pretty homogeneously the available space.

Simulation Results

The initial set up for simulations resembles the initial state of the embryo, where only one compartment is present, and where maternal proteins form a gradient in the environment as from experimental data (Pisarev et al., 2009). Simulations results are shown in Figure 5-6-7. They are evaluated observing the time evolution of the compartment number and of the gene expression pattern—see Figure 8. Figure 5 shows the initial condition with only one agent and the egg polarised by maternal effects localised in

Figure 5. The initial configuration of the simulation is characterised the fertilised egg, charactersied by only one compartment and by the gradient of maternal product, bcd (grey) and cad (light blue), in the environment

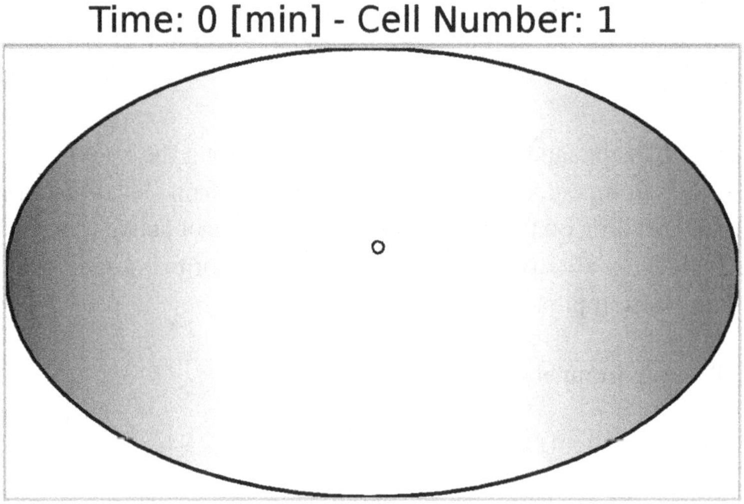

Time: 0 [min] - Cell Number: 1

Figure 6. At time 72, after 9 nuclear divisions, the number of compartments is around 2^9. The image shows half of them, being the environment 2D.

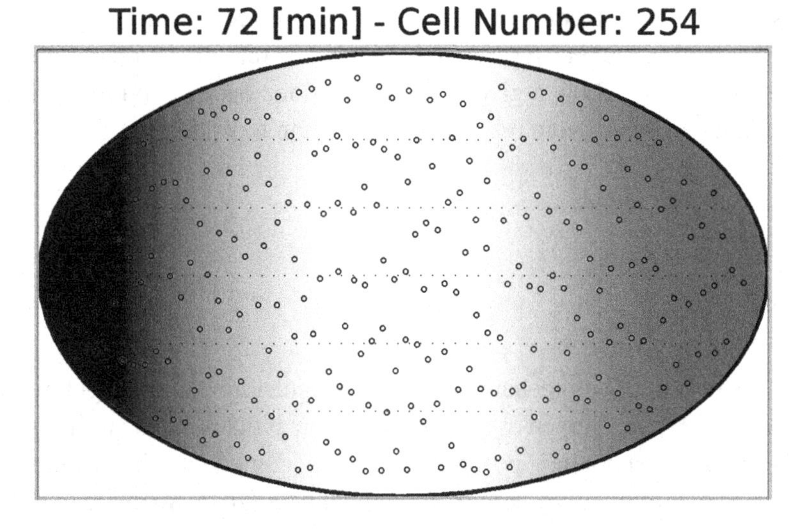

the extreme pole: *bcd* on the left and *cad* on the right. During the first minutes of simulation only agent divisions occur so as to fill the whole maternal cell at the end of cleavage cycle 9, *i.e.*, after 9 compartment divisions, with around 250 agents (half of the total 500), as shown in Figure 6. Finally, Figure 7 reproduces the expression pattern of the four gap genes, whose spatial organisation is compared with experimental data of Figure 8. Cells are coloured of yellow, red, blue, and green—if, in order, they express *hb*, *kni*, *gt* and *Kr*, and their size is proportional to the protein concentration.

Figure 7. Simulation results for the four gap genes hb (yellow), kni (red), gt (blue), Kr (green) at a simulation time equivalent to the eighth time step of cleavage cycle 14A

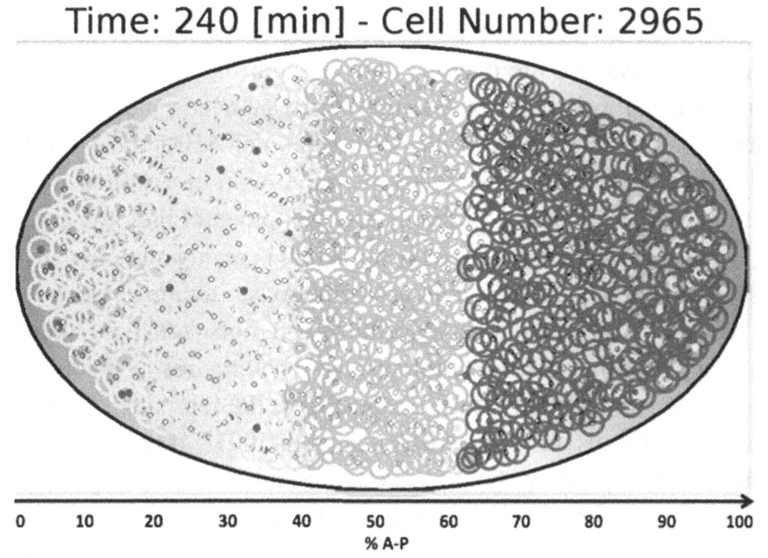

**For a more accurate representation of this figure, please see the electronic version.*

Figure 8. The experimental data for the expression of (from the top) hb, kni, gt, Kr at the eighth time step of cleavage cycle 14A
Source: Surkova et al., 2008 ©Maria Samsonova and John Reinitz

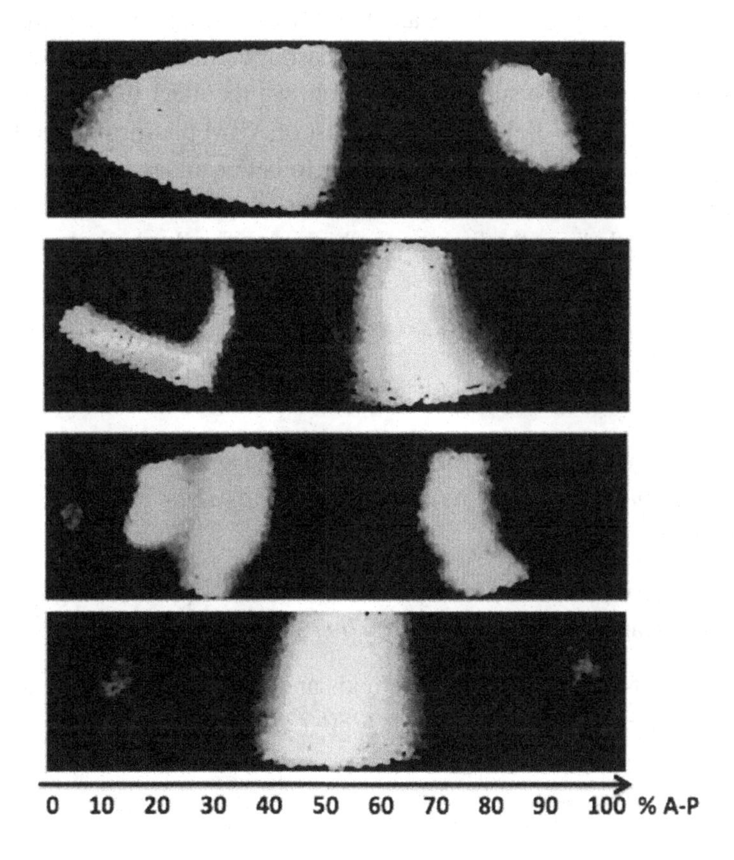

FUTURE RESEARCH DIRECTIONS

Directions for future research can be summarised into two main issues. The first theoretical issue is to refine the way multicellular systems are mapped into an agent-based model and the way simulations are executed, analysing the algorithms adopted to make time proceeds and events execute accordingly.

Secondly – to demonstrate the feasibility and effectiveness of ABM and MABS – our next focus is to apply the ALCHEMIST framework to the analysis of the main morphogenetic signalling pathways and of their interactions. This is a crucial problem in developmental biology, given that they are also involved in tumour growth. Since the integrated phenomenon is highly complex, involving a large set of mechanisms mediating cell-to-cell and intracellular interactions, literature evidences (Dada & Mendes, 2011; Deisboeck et al., 2011) show that novel algorithms and tools from computer science can significantly contribute in supporting experimental biologists to ravelling out such complexity.

To make the model description more clear, and to support the reproducibility and evaluation of the model, we plan to adopt protocols and standards, such as the ODD (Overview, Design concepts, and Details) one (Grimm et al., 2010), which should make agent-based models easier to write and read—and, more generally, the ABM approach easier to embrace for the multicellular systems biology community.

CONCLUSION

The goal of this chapter is to present a promising and challenging application domain for ABM: multicellular systems biology. The constituting characteristics of multicellular systems in fact highlight a clear mapping between the real system entities and the ABM abstractions. This makes building models and interpreting simulation results intuitive. For this reason we think that ABM is particularly suited for multicellular systems biology. To support our vision, we recollect the literature reporting on successful applications of ABM in this field, as well as a list of ABM platforms specifically developed for building models of interacting cells. To help the reader to better understand better the big picture, we finally choose one platform, ALCHEMIST, and motivate its adoption for ABM of multicellular systems. Finally we describe an example from the literature, which specifically shows how to model a system of interacting cells.

REFERENCES

An, G., Mi, Q., Dutta-Moscato, J., & Vodovotz, Y. (2009). Agent-based models in translational systems biology. *Wiley Interdisciplinary Reviews: Systems Biology and Medicine*, *1*(2), 159–171. doi:10.1002/wsbm.45 PMID:20835989

Anderson, A. R. A., Chaplain, M. A. J., & Rejniak, K. A. (Eds.). (2007). *Single-cell-based models in biology and medicine*. Birkähuser Basel; doi:10.1007/978-3-7643-8123-3

Bonabeau, E. (2002). Agent-based modeling: Methods and techniques for simulating human systems. *Proc. of the National Academy of Sciences*, *99*(3), 7280-7287. doi: 10.1073/pnas.082080899

Cannata, N., Corradini, F., Merelli, E., Omicini, A., & Ricci, A. (2005). An agent-oriented conceptual framework for Systems Biology. In E. Merelli, P. P. González Pérez & A. Omicini (Eds.), Transactions on Computational Systems Biology III (Vol. 3737, pp. 105–122). Springer. 8 doi:10.1007/11599128_8

Cickovski, T. M., Huang, C., Chaturvedi, R., Glimm, T., Hentschel, H. G. E., Alber, M. S., Glazier, J. A., Newman, S. A. & Izaguirre, J. A. (2005). A framework for three-dimensional simulation of morphogenesis. *IEEE/ACM Transactions on Computational Biology and Bioinformatics, 2*(4), 273–288. doi: 10.1109/TCBB.2005.46

Cooper, J., & Osborne, J. (2013). Connecting models to data in multiscale multicellular tissue simulations. *Procedia Computer Science, 18*, 712–721. doi:10.1016/j.procs.2013.05.235

Dada, J. O., & Mendes, P. (2011). Multi-scale modelling and simulation in systems biology. *Integrative Biology, 3*(2), 86–96. doi:10.1039/c0ib00075b PMID:21212881

Deisboeck, T. S., Wang, Z., Macklin, P., & Cristini, V. (2011). Multiscale cancer modeling. *Annual Review of Biomedical Engineering, 13*(1), 127–155. doi:10.1146/annurev-bioeng-071910-124729 PMID:21529163

Gelfand, A. (2013). The biology of interacting things: The intuitive power of agent-based models. *Biomedical Computation Review, 1*, 20–27.

Gibson, M. A., & Bruck, J. (2000). Efficient exact stochastic simulation of chemical systems with many species and many channels. *The Journal of Physical Chemistry A, 104*(9), 1876–1889. doi:10.1021/jp993732q

Gillespie, D. T. (1977). Exact stochastic simulation of coupled chemical reactions. *Journal of Physical Chemistry, 81*(25), 2340–2361. doi:10.1021/j100540a008

González Pérez, P., Cárdenas, M., Camacho, D., Franyuti, A., Rosas, O., & Lagúnez-Otero, J. (2003). Cellulat: An agent-based intracellular signalling model. *Bio Systems, 68*(2–3), 171–185. doi:10.1016/S0303-2647(02)00094-1 PMID:12595116

González Pérez, P. P., Omicini, A., & Sbaraglia, M. (2013). A biochemically-inspired coordination-based model for simulating intracellular signalling pathways. *Journal of Simulation, 7*(3), 216–226. doi:10.1057/jos.2012.28

Gorochowski, T. E., Matyjaszkiewicz, A., Todd, T., Oak, N., Kowalska, K., Reid, S., & di Bernardo, M. et al. (2012). 08). BSim: An agent-based tool for modeling bacterial populations in systems and synthetic biology. *PLoS ONE, 7*(8), 1–9. doi:10.1371/journal.pone.0042790 PMID:22936991

Grimm, V., Berger, U., DeAngelis, D. L., Polhill, J. G., Giske, J., & Railsback, S. F. (2010). The ODD protocol: A review and first update. *Ecological Modelling, 221*(23), 2760–2768. doi:10.1016/j.ecolmodel.2010.08.019

Hoehme, S., & Drasdo, D. (2010). A cell-based simulation software for multi-cellular systems. *Bioinformatics (Oxford, England), 26*(20), 2641–2642. doi:10.1093/bioinformatics/btq437 PMID:20709692

Hohm, T., & Zitzler, E. (2009). Multicellular pattern formation. *IEEE Engineering in Medicine and Biology Magazine, 28*(4), 52–57. doi:10.1109/MEMB.2009.932905 PMID:19622425

Holcombe, M., Adra, S., Bicak, M., Chin, S., Coakley, S., Graham, A. I., & Worth, D. et al. (2012). Modelling complex biological systems using an agent-based approach. *Integrative Biology*, *4*(1), 53–64. doi:10.1039/C1IB00042J PMID:22052476

Kitano, H. (2002). Systems Biology: A brief overview. *Science*, *295*(5560), 1662–1664. doi:10.1126/science.1069492 PMID:11872829

Luke, S., Cioffi-Revilla, C., Panait, L., Sullivan, K., & Balan, G. (2005). MASON: A multiagent simulation environment. *Simulation*, *81*(7), 517–527. doi:10.1177/0037549705058073

Macal, C. M., & North, M. J. (2010). Tutorial on agent-based modelling and simulation. *Journal of Simulation*, *4*(3), 151–162. doi:10.1057/jos.2010.3

Macklin, P., Edgerton, M. E., Thompson, A. M., & Cristini, V. (2012). Patient-calibrated agent-based modelling of ductal carcinoma in situ (DCIS): From microscopic measurements to macroscopic predictions of clinical progression. *Journal of Theoretical Biology*, *301*, 122–140. doi:10.1016/j.jtbi.2012.02.002 PMID:22342935

Merelli, E., Armano, G., Cannata, N., Corradini, F., dInverno, M., Doms, A., & Luck, M. et al. (2007). Agents in bioinformatics, computational and systems biology. *Briefings in Bioinformatics*, *8*(1), 45–59. doi:10.1093/bib/bbl014 PMID:16772270

Montagna, S., Donati, N., & Omicini, A. (2010). An Agent-based Model for the Pattern Formation in Drosophila Melanogaster. In *Proceedings of the 12th International Conference on the Synthesis and Simulation of Living Systems*. The MIT Press.

Montagna, S., Omicini, A., & Pianini, D. (2016). Extending the Gillespie's stochastic simulation algorithm for integrating discrete-event and multi-agent based simulation. In B. Gaudou & J. S. Sichman (Eds.), Multi-Agent Based Simulation XVI (Vol. 9568, pp. 3–18). Springer. 1 doi:10.1007/978-3-319-31447-1_1

Montagna, S., Pianini, D., & Viroli, M. (2012). A model for Drosophila Melanogaster development from a single cell to stripe pattern formation. In D. Shin, C.-C. Hung & J. Hong (Eds.), *27th annual ACM Symposium on Applied Computing* (SAC 2012) (pp. 1406–1412). Riva del Garda, Italy: ACM. doi:10.1145/2245276.2231999

Montagna, S., Ricci, A., & Omicini, A. (2008). A&A for modelling and engineering simulations in Systems Biology. *International Journal of Agent-Oriented Software Engineering*, *2*(2), 222–245. doi:10.1504/IJAOSE.2008.017316

North, M. J., Collier, N. T., Ozik, J., Tatara, E. R., Macal, C. M., Bragen, M., & Sydelko, P. (2013). Complex adaptive systems modeling with Repast Symphony. *Complex Adaptive Systems Modeling*, *1*(1), 3. doi:10.1186/2194-3206-1-3

Olsen, M. M., & Siegelmann, H. T. (2013). Multiscale agent-based model of tumor angiogenesis. *Procedia Computer Science*, *18*, 1016–1025. doi:10.1016/j.procs.2013.05.267

Omicini, A., & Zambonelli, F. (2004). MAS as complex systems: A view on the role of declarative approaches. In J. A. Leite, A. Omicini, L. Sterling & P. Torroni (Eds.), Declarative Agent Languages and Technologies (Vol. 2990, pp. 1–17). Springer. 1 doi:10.1007/978-3-540-25932-9_1

Pianini, D., Montagna, S., & Viroli, M. (2013). Chemical-oriented simulation of computational systems with Alchemist. *Journal of Simulation, 7*(S3), 202–215. doi:10.1057/jos.2012.27

Pisarev, A., Poustelnikova, E., Samsonova, M. & Reinitz, J. (2009). FlyEx, the quantitative atlas on segmentation gene expression at cellular resolution. *Nucleic Acids Research, 37*(1), 560–566. doi: 10.1093/nar/gkn717

Pogson, M., Smallwood, R., Qwarnstrom, E., & Holcombe, M. (2006). Formal agent-based modelling of intracellular chemical interactions. *Bio Systems, 85*(1), 37–45. doi:10.1016/j.biosystems.2006.02.004 PMID:16581178

Rouillard, A. D., & Holmes, J. W. (2014). Coupled agent-based and finite-element models for predicting scar structure following myocardial infarction. *Progress in Biophysics and Molecular Biology, 115*(2–3), 235–243. doi:10.1016/j.pbiomolbio.2014.06.010 PMID:25009995

Setty, Y. (2012). Multi-scale computational modeling of developmental biology. *Bioinformatics (Oxford, England), 28*(15), 2022–2028. doi:10.1093/bioinformatics/bts307 PMID:22628522

Slepoy, A., Thompson, A. P., & Plimpton, S. J. (2008). A constant-time kinetic Monte Carlo algorithm for simulation of large biochemical reaction networks. *The Journal of Chemical Physics, 128*(20), 205101. doi:10.1063/1.2919546 PMID:18513044

Starruß, J., de Back, W., Brusch, L., & Deutsch, A. (2014). Morpheus: A user-friendly modeling environment for multiscale and multicellular systems biology. *Bioinformatics (Oxford, England), 30*(9), 1331–1332. doi:10.1093/bioinformatics/btt772 PMID:24443380

Surkova, S., Kosman, D., Kozlov, K., Manu, , Myasnikova, E., Samsonova, A. A., & Reinitz, J. et al. (2008). Characterization of the Drosophila segment determination morphome. *Developmental Biology, 313*(2), 844–862. doi:10.1016/j.ydbio.2007.10.037 PMID:18067886

Uhrmacher, A. M., Degenring, D., & Zeigler, B. (2005). Discrete event multi-level models for Systems Biology. In C. Priami (Ed.), Transactions on Computational Systems Biology I (Vol. 3380, pp. 66–89). Springer. 6 doi:10.1007/978-3-540-32126-2_6

Uhrmacher, A. M., & Weyns, D. (Eds.). (2009). *Multi-agent systems: Simulation and applications* (1st ed.). Boca Raton, FL: CRC Press.

Walker, D., Wood, S., Southgate, J., Holcombe, M., & Smallwood, R. (2006). An integrated agent-mathematical model of the effect of intercellular signalling via the epidermal growth factor receptor on cell proliferation. *Journal of Theoretical Biology, 242*(3), 774–789. doi:10.1016/j.jtbi.2006.04.020 PMID:16765384

Wang, Z., Butner, J. D., Kerketta, R., Cristini, V., & Deisboeck, T. S. (2015). Simulating cancer growth with multiscale agent-based modeling. *Seminars in Cancer Biology, 30*, 70–78. doi:10.1016/j.semcancer.2014.04.001 PMID:24793698

Weyns, D., Omicini, A., & Odell, J. J. (2007). Environment as a first class abstraction in multi-agent systems. *Autonomous Agents and Multi-Agent Systems, 14*(1), 5–30. doi:10.1007/s10458-006-0012-0

White, T., & Pagurek, B. (1998). Towards multi-swarm problem solving in networks. In *3rd International Conference on Multi Agent Systems (ICMAS '98)* (pp. 333–340). doi:10.1109/ICMAS.1998.699217

Wilensky, U. (1999). *NetLogo. Center for Connected Learning and Computer-Based Modeling*. Northwestern University. Retrieved from http://ccl.northwestern.edu/netlogo/

Zambonelli, F., & Omicini, A. (2004). Challenges and research directions in agent-oriented software engineering. *Autonomous Agents and Multi-Agent Systems*, *9*(3), 253–283. doi:10.1023/B:AGNT.0000038028.66672.1e

KEY TERMS AND DEFINITIONS

Cellular Potts Model: A lattice model evolved as a time-discrete Markov chain, where the transition probabilities are specified with the help of a *Hamiltonian* or *energy* function.

Cleavage Cycle: A cell cycle occurring during the cleavage stage of embryo development and resulting in the synchronous division of the cells in the early embryo.

Gene: DNA is composed of thousands of genes (the number varies with the organism). A gene contains the set of instructions needed to code a specific protein.

Gene Expression: The process that, beginning with the information contained in a gene and passing through different steps, end up in the synthesis of a functional gene product, that in most of the cases is a protein.

Lattice Model: A physical model where space is defined on a lattice, *i.e.*, on a structure with a regular periodic unit.

mRNA: During the process of gene expression, the first step synthetizes a mRNA molecule that contains all the genetic information stored in the gene and necessary for the next step. From the mRNA cell is then able to code for a specific protein.

Ordinary Differential Equation: One of the most common modelling approaches in Systems Biology is mathematical and adopts the ordinary differential equations. They involve derivative of one or more dependent variables with respect to a single independent variable.

Protein: The basic functional unit of a cell is the protein. Depending on their tridimensional structures, they perform most of the activities inside a living cell.

Reaction Diffusion Systems: They are differential equations with two independent variables that normally account for the change in space and time of molecules.

ENDNOTES

http://www.nature.com/subjects/multicellular-systems

2 http://apice.unibo.it

3 http://alchemist.apice.unibo.it

Section 2

Applications in Biological and Environmental Systems

Chapter 8
Architecture with Multi-Agent for Environmental Risk Assessment by Chemical Contamination

Sergio Fred Ribeiro Andrade
UESC, Brazil

Lilia Marta Brandão Soussa Modesto
UESC, Brazil

ABSTRACT

Risk assessment for human health and ecosystems by exposure to chemicals is an important process to aid in the mitigation of affected areas. Generally, this process is carried out in isolated spots and therefore may be ineffective in mitigating. This chapter describes an architecture of a multi-agent system for environmental risk assessment in areas contaminated as often occur in mining, oil exploration, intensive agriculture and others. Plan multiple points in space-time matrix where each agent carries out exposure assessment and the exchange of information on toxicity, to characterize and classify risk in real time. Therefore, it is an architecture model with multi-agent that integrates ontology by semantic representation, classifies risks by decision rules by support vectors machines with multidimensional data. The result is an environment to exchange information that provides knowledge about the chemical contamination, which can assist in the planning and management of mitigation of the affected area.

INTRODUCTION

It is an important environmental problem the chemical contamination in compartments soil, air, surface and groundwater, mainly originating waste industries, either by operations or accidents. One way of assessing these contaminations is the risk assessment methodology to human health and ecosystems, originated from the National Research Council [NRC] (1983), which is an important process for decision-making in the search for solutions for the management of risk and mitigation of contaminated areas.

DOI: 10.4018/978-1-5225-1756-6.ch008

In general, for this assessment are considered integrated information from the fields of physics, chemistry and biology, and also the energy resulting from human actions that can directly or indirectly affect the health and welfare of the population (Lioy, 1990).

Much of the industry, among them, the mining, petroleum and intensive agriculture, there is exposure to toxic substances that affect a wide variety of serious acute and chronic, reversible or irreversible diseases, which prevent or disable the exposed individual and others, which cause health damage in several generations. This complexity of the contamination in hazardous locations requires a specific methodology for evaluation and indicators for knowledge and to aid in correcting the ills caused.

The tools used for operation and information, are summarized and evaluated in isolated spots without considering the extent of the affected area and the integration of the points analyzed in the space-time matrix. Knowledge of the level of contamination and exposure in the area affected, in real time, enables better conditions to planning and mitigation. For example, when a certain location that receives tributaries coming from other aquifers contaminated local have advance knowledge of the risk of toxicity, can plan actions for human protection and preservation of biota. Likewise, happen in subsequent sites that may take similar measures.

The motivation to develop this study was the lack of tools, criteria for analysis in the space-time matrix and simulators for decision-making. In addition, the absence of a system of multi-agent or similar, quantitative and qualitative risk visual characterization usefully both for the technical planning of prevention and mitigation of risks, as for communication to various social actors.

This proposal is based on the development of an integrated architecture and system for Multi-agent that will provide real-time information, in level of exposure, toxicity and risk classification to human health and to individuals of biota, with a simulation on the entire contaminated area and not only in isolated site, providing viewing appropriate information to decision-making.

The multi-agent system employment according to Wooldridge (2009) presented in this chapter, it is an adequate architecture, because it treats each computational entity as an autonomous decision-making through information sharing, enabling real-time risk assessments and in every area involved.

The objectives are: 1. Develop architecture with multi-agent simulation of environmental risks for chemicals in extensive area represented by a space-time matrix; 2. Develop system for environmental risk assessment based on ontological and multi-agent knowledge with support vector machine application and multidimensional data for decision-making; and 3. Simulation the process with the system developed for the environmental risk assessment in environmental impact activity.

BACKGROUND

According to Rebelo et. al (2014), decisions on risk management mainly depend on their evaluation on a scientific basis. This question is quite emphasized in the literature because it deals with the importance of the issue, and in particular on the use of toxicology to aid the environmental management systems of the productive sector.

The principal methodologies for risk assessment to health and the environment are available for (United States Environmental Protection Agency [USEPA] 1986, 1989, 2004), (Agency for Toxic Substances and Disease Registry [ATSDR], 2016), (World Health Organization [WHO], 2010), (European Union [EU], 2003), (Government of Canada [Canada] 2004, 2012) and (Government of Netherland

[Netherland], 2009). In Brazil, the (Environmental Company of The State São Paulo [CETESB], 2001), remains important procedures for the evaluation process of the risk to human health.

These methods, generally, based on a quantitative approach and numerical patterns that apply mathematical equations by route of exposure in isolated and not integrated points. Its procedures offer advantages that facilitate the measurement and comparison of contaminated scenarios and determining safety limits. However, two aspects in literature draw attention lack, both in the methods and procedures as the evaluative tools: a) the integrated exposure assessment in the space-time matrix; and b) visual qualitative risk characterization by categories of severity of adverse events (Andrade, 2015).

On the first topic, the methodologies used an approach in the evaluation of exposure to consider the concentration of the contaminant constant over long exposure periods. This approach facilitates the calculations and scenarios comparison, but prevents a more careful analysis in shorter exposure periods to contaminated site. This happens when environmental contamination in extensive areas that undergo chemical provision for accidents where there may be a little accurate knowledge of the temporal evolution of risk.

Mainly due to an emphatic assessment in isolated sites contaminated matrix without integrating data with other points to a holistic view of risk, which may impair the effectiveness of mitigation specific area affected at intermediate intervals or the all period.

On the second point, the evaluative tools available not characterizes visually the risk in the entire affected area and reduces the interpretation of the process as a whole and understanding of data for more efficient communication with society, which is the main interested in the topic.

More specifically, in these tools the risk assessed only based on quantitative limits, which can sometimes reduce the evaluative information to an apparent binary evaluation system that concludes: "risk" or "no risk". This situation may compromise the understanding of the evaluation of different degrees of risk, without proper understanding of qualitative differences in the types of risk and efficiency in its communication to society.

Some paper related to this work were found like, Xie and He (2014) on development of decision support systems in response to social risk catastrophic; Jiang et al. (2011) for a review on application of decision support systems for contaminated site; Dzemydine and Dzindzalieta (2010) on architecture of decision support systems for risk evaluation of transportation of hazardous materials. As well, Linard, Fontenille and Lambin (2009) on multi-agent simulation to assess the risk of malaria in France; Torrelas (2004) on the framework for multi-agent system engineering using ontology modeling for risk assessment in e-commerce; Stroeve, Blom and Van der Park (2003); and on multi-agent situation awareness error evolution in accident risk.

THE CHEMICAL RISK ASSESSMENT PROCESS

The environmental risk assessment process in world follows a procedure known as "risk assessment paradigm" (NRC, 1983), which has guided the main methods of risk assessment to health and ecosystems (USEPA, 1989) (WHO, 2010), (ATSDR, 2016) and (EU, 2003). The chemical risk assessment paradigm has four stages: hazard identification, exposure assessment, dose-response assessment and risk characterization, described in two topics.

Hazard Identification and Exposure Assessment Chemical

In step of hazard identification is needed describe the contributing factors what provoke damage to the environment. They are identified the chemicals of interest, potential toxicity, environmental concentrations, the migration routes, transport, fate and persistence of contaminants. They are also characterized the chemical properties, physics, climate and biota of the affected location and description of the activity is carried out, the generator contamination event, the population and the exposure of receptors sensitive groups of chemical compounds.

The concentration data in environmental compartments and their intake dose are organized by spatial point of sample collection in order of increasing distance from a central point chosen. This central point can be the supposed generator of pollutants with a higher level of chemical concentration. For example, the mining tailings basin or oil spill in an estuary. From this matrix can perform analysis georeferenced spatial and temporal of the levels and risk categories for the scenario contaminated under study, designated of space-time matrix.

Traditionally, the theoretical model for evaluation of chemical exposure as basic formulation in USEPA (1992) follows the Equation 1, which defines the equations for exposure route to calculate the dose potential intake.

$$D_{pot} = \int_{t_1}^{t_2} C(t)IR(t)dt \qquad (1)$$

Where D_{pot} is the potential dose $C(t)$ is the concentration of exposure as a function of time $t_2 - t_1$ is the estimated exposure time, and $IR(t)$ is the intake rate, inhalation or dermal contact of the contaminant, depending on the nature of the latter and the pathway in the intake.

The Equation 1 is the general expression of the dose, whereas the chemical concentration is in time function. The use of this equation imposes the need to know the time function of the concentration in day in exposure periods that can extend over decades. Therefore, in practice, many studies applying the methods of risk (USEPA, 2004) (WHO, 2010), (ATSDR, 2016) and (EU, 2003), in place of this formulation adopts the function of time concentration as a constant, which can express the greatest value of this concentration during the study period or average value expressed in the approach of Equation 2.

$$\mathrm{ADD} = \frac{\mathrm{CxEDxEF}}{\mathrm{BWxAT}}\left(\mathrm{IR}\right) \qquad (2)$$

Where *ADD* is the intake average daily dose in the exposure time in mg kg^{-1} day^{-1}. *C* is the average concentration in the exposure period in mg kg^{-1} mg L^{-1} or mg cm^{-3}. *ED* is the duration of exposure in years. *EF* is the average frequency of exposure in days by year, *IR* is the ingestion rate, inhalation or dermal contact by day, *BW* is the average body weight of the population in kg, and *AT* is the median time in days of life receivers.

The constant *C* average concentration in the period, the result leads to average values of dose and risk, considering the study period (USEPA, 1989).

The formula shown in Equation 3 corresponds to a formulation of complementarity in Equation 2, which intended to measure the intake dose with greater accuracy and real expression of exposure situations for medium and long multiplied term.

This Equation 3 in right side of the equality divided into two parts by a multiplicative factor. The first part of the right side of the equation is general and corresponds the exposure of the concentration or contaminant dose for a period of time in a sample located in the environmental compartment (soil, water, air, natural foods) multiplied by frequency of varied exposure by a period of receiver life time while exposed to the contaminant for non-carcinogenic, or the total time of the recipient's life for carcinogenic contaminants. This product divided by the average body weight of an adult or child of the population involved in the exhibition.

The second part of the equation on the right side, separated from the first by a product corresponds to absorption factors or retention in the receiver body, specific time rate of ingestion, inhalation or dermal contact, bioaccumulation factor or bio magnification, as measure converter or other important parameters agreed by environmental health authorities. These parameters are found in USEPA (1989, 2004), WHO (2010) and ATSDR (2016).

$$DCI(t)_{\exp} = \frac{\overline{C_{\exp}}(\Delta t)EF_f}{\overline{BW} x AT} x \left(factor_1 x\ factor_2\ x\ \ldots x\ factor_n\right) \quad (3)$$

For $\Delta t = t_{i+1} - t_i$ and i = (0, 1, 2,..., n) .

Where $DCI(t)_{exp}$ is the intake dose in a route of organic absorption in mg kg^{-1} dia^{-1}, C_{exp} is the concentration in the exposure route for a period of time determined in mg kg^{-1}, mg L^{-1} or mg cm^{-3}, Δt is the duration of exposure per sample period collected in day where t_{i+1} is the end time of the intermediate period and t_i is the initial time of the same period, EF_f is the variable frequency of exposure for each intermediate period in days/year, $\overline{BW}$ is the body weight average in kg for the affected subpopulation, AT is the total exposure time in days and $fator_n$ are the parameters of absorption or converters measures.

For the employment this formulation were taken as example the models (Equation 4 to 9) presented in Table 1, referenced by (USEPA, 1989, 2004) (WHO, 2010) and (ATSDR, 2016), which serves as an example for practical application.

These are equations that represent the exposure models for contamination scenarios in space-time matrix, whose subpopulations receptors are generally composed of industrial workers and agricultural, residents and their families, who are daily in these places because of work and housing.

Applications of equations for exposure assessment are adjusted to environmental parameters (as shown in Table 2), site evaluation data from the scenario and the affected population, according to the specification in the Hazard Identification phase.

Dose-Response Assessment and Risk Characterization

Dose-Response Assessment is done through the use of reference doses for contaminants not carcinogenic by route of exposure. The *RfD/RfC* (Equation 10) is a value originated bioassays toxicological study,

Table 1. Equations and parameters for exposure assessment

Model	Equation	Equation
Inhalation of particulate matter	$DCI\left(t\right)_{inal-air} = \frac{C_{ar}\left(\Delta t\right)xEF_{f}}{\overline{BW}xAT}\left(IPxFRxHExAF\right)$	(4)
Accidental intake of soil	$DCI\left(t\right)_{intake-soil} = \frac{C_{s}\left(\Delta t\right)xEF_{f}}{\overline{BW}xAT}\left(IR_{s}xCF_{s}\right)$	(5)
Contact dermal soil	$DCI\left(t\right)_{derm-soil} = \frac{C_{s}\left(\Delta t\right)xEF_{f}}{\overline{BW}xAT}\left(SAxFAxABSxEVxCF_{s}\right)$	(6)
Water intake	$DCI\left(t\right)_{\mathbf{intake-water}} = \frac{C_{a}\left(\Delta t\right)xEF_{f}}{\overline{BW}xAT}\left(IR_{a}\right)$	(7)
Contact Dermal with water	$DCI\left(t\right)_{derm-water} = \frac{C_{w}\left(\Delta t\right)xEF_{f}}{\overline{BW}xAT}\left(SAxPCxETxCF_{a}\right)$	(8)
Food intake	$DCI\left(t\right)_{food} = \frac{C_{food(i)}\left(\Delta t\right)xEF_{f}}{\overline{BW}xAT}\left(IR_{an}xFI_{an}\right)$	(9)

Source: USEPA, 1989, 2004; WHO, 2010

Table 2. Applied parameters in the equations for exposure assessment.

Variable	Measure	Parameters
$DCI\left(t\right)_{exp}$	mg kg^{-1} day^{-1}	Dose chronic daily intake by route of exposure
C_{air}	mg m^{-3}	Contaminant concentration in the air
EF_{f}	day-year	Variable frequency for the exposure period range
$\overline{BW}$	kg	Mean body weight
AT	day	Total period of exposure during life
IP	k gh^{-1}	Particles inhalation rate
FR	-	Factor particle retention in the lung
HE	h day^{-1}	Time particle inhalation (in hours)

continued on next page

Table 2. Continued

Variable	Measure	Parameters
AF	-	Absorption factor for inhalation
IR_s	mg day^{-1}	Daily rate of accidental ingestion of soil
CF_s	10^{-6} kgmg $^{-1}$	Conversion factor for soil
IR_a	kg/food	Food intake rate
SA	cm^2	Skin surface available (in contact)
FA	mg cm^{-2} - event	Soil adhesion to the skin
ABS	-	Fraction absorbed by the skin
EV	event-day	Exposure time
IR_a	mg day^{-1}	Water intake rate
C_s	mg kg^{-1}	Chemical concentration in soil
C_a	mg L^{-1}	Contaminant concentration in water
PC	cm h^{-1}	Constant dermal permeability
ET	h day^{-1}	Time in hours of daily exposure
CF_a	10^{-3}L / 1000cm^3	Conversion factor for water
$C_{food(i)}$	mg k g^{-1}	Concentration in food
IR_{an}	L day^{-1}	Daily rate of water intake by the animal
FI_{an}	-	Food fraction from the contaminated source

Source: USEPA, 1989, 2004; WHO, 2010

which states the highest level of chemical concentration for non-carcinogenic contaminants not that cause adverse health effects (USEPA 1989):

$$RfD = \frac{\text{NOAEL or LOAEL}}{\left(UF_1 x UF_2 \ldots x MF\right)} \quad (10)$$

Where *RfD* expressed in mg kg^{-1} day, *NOAEL* (No Observed Adverse Effect Level) and *LOAEL* (Lowest Observed Adverse Effect Level) determined by bioassays, UF_i are uncertainty factors and *MF* are modifying factor own of scientific study.

Can reply in the same equivalence the following indicators: DAI (Dose Acceptable Intake) or TDI (Daily Intake Tolerable) - (WHO, 2010), MRL (Minimum Risk Level) - (ATSDR, 2016), ACI (Acceptable Dose for Exposure Chronicle) - (USEPA, 1989).

In the assessment of the toxicity of carcinogenic substances is used *SF* (slope factor) in (USEPA, 1989, which is the confidence limit greater than 95% of the dose-response curve in mg kg^{-1} day. Also, for carcinogenic toxicological classifications are used groups by weight-of-evidence, published by (USEPA, 2016) and (IARC, 2016).

According to this methodology the risk characterization procedure is subdivided into the following steps:

1. Determination of the Hazard Index (*HI*) and the Carcinogenic Risk (*CR*) for contaminant and spaces and sampling periods;
2. Application of the USR (Unified Scale Risk);
3. Risk classification categorical ordinal kSVM (Multi-Class Support Vector Machine) of (Weston and Watkins, 1998) and (Hsu and Lin, 2002) for the qualitative risk characterization.

Characterization for Substances not Carcinogenic

The main indicator for non-carcinogenic substance is the *HI*, which is rate between intake dose and *RfD*. When the *HI* is greater than 1 there is a risk with any adverse health effects, when *HI* is equal or less than 1 there is no risk to health.

After obtaining the $DCI\left(t\right)_{exp}$ for not carcinogenic contaminants, the next step is determine the Hazard Quotient - *HQ*(*t*) of a specific contaminant by Equation 11 and *HI*(*t*) both as a function of time using the additive approach exposure for all routes by Equation 12 (USEPA 1989).

$$HQ_i(t) = DCI\left(t\right)_{exp} / RfD \quad (11)$$

$$HI(t) = \sum_{i=1}^{n} HQ_i\left(t\right) \quad (12)$$

Where *HI*(*t*) is the hazard index for cumulative exposure in time *t* for exposure to multiple routes; $DCI\left(t\right)_{exp_i}$ is the intake dose obtained by Equation 3 for the ith route of exposure in the time *t*, estimated at (mg kg^{-1} day^{-1}); and *RfD* is the dose reference (mg kg^{-1} day^{-1}) for the substance and found route.

The simultaneous action of exposure to various contaminants (*j*) for the additive ($HI^{tot}(t)$) of approximate effect calculated in Equation 13 at a point of the space-time matrix.

$$HI^{tot}(t) = \sum_{j=1}^{m} HI_j\left(t\right) \tag{13}$$

Characterization for Carcinogenic Substances

The main indicator for carcinogens is the Carcinogenic Risk (*CR*), which is the probability that an individual in given population developing cancer over a lifetime, considering exposure to contaminants classified as carcinogenic.

For each route of exposure to carcinogenic substances, in approximating to *HI*(*t*) is calculated by the *CR*Equation 14 for low dose daily intake ($CR \leq 0.01$) (USEPA, 1989).

$$CR\left(t\right) = DCI\left(t\right)_{exp} \text{ x } SF \tag{14}$$

In scenarios with high dose of daily intake ($CR > 0.01$), the *CR* must be estimated by Equation 15.

$$CR\left(t\right) = 1 - \exp\left(-DCI\left(t\right)_{exp} \text{ x } SF\right) \tag{15}$$

Where *CR*(*t*) is the probability of an individual developing cancer over a lifetime of exposure to contaminants to the time *t*; *SF* is the carcinogenic potential factor measured in mg $kg^{-1}day^{-1}$; and $DCI\left(t\right)_{exp}$ is the dose of daily intake to the contaminant given, considering exposure to the time *t* for a specific route.

In characterizing the Risk Multiple Carcinogenic (CR^{mult}) for various routes are combined odds of the *CR*, according to Equation 16.

$$CR^{mult}\left(t\right) = \sum_{i=1}^{n} CR_i\left(t\right) \tag{16}$$

In Equation 17 are considered multiple simultaneous substances (*j*) (USEPA, 1989).

$$CR^{tot}\left(t\right) = \sum_{j=1}^{m} CR_j^{mult}\left(t\right) \tag{17}$$

Table 3. Unified risk scale

Criteria Decision	Concept	Risk Class
$HI \leq 1$.	No effect	1
$HI > 1$ and $DCI(t)_{exp} < LOAEL$[e].	Adverse effect	2
$HI > 1$ and $DCI(t)_{exp} \geq LOAEL$; or RfD chronic [a]; and $DCI(t)_{exp} < DL_{50}$ (DC_{50}) or Group B2 - USEPA[c]; or $CR < 10^{-6}$.	Irreversible effect [d]	3
Carcinogenic: Group 2B – IARC[b]; or B1 or C - USEPA[c]; and/or $10^{-4} \leq CR \leq 10^{-6}$.	Serious effect	4
$HI > 1$ and $DCI(t)_{exp} \geq LOAEL$; or RfD chronic [a] and $DCI(t)_{exp} \geq DL_{50}$ (DC_{50}) Carcinogenic: Group 1 or 2A – IARC[b]; or A or B - USEPA[c]; or $CR > 10^{-4}$.	Very serious effect	5

[a]RfD/RfC – (USEPA, 2016), (WHO, 2016), (ATSDR, 2016); [b] (IARC, 2016); [c] (USEPA, 2016); [d] irreversible effects, mutagenic or carcinogenic evidence; [e]LOAEL - Lowest Observed Adverse Effect Level, (USEPA, 1989).

Unified Risk Scale

The risk is qualitatively evaluated by ordinal classes which represent categories (1 to 5), as URS as shown in Table 3 of Andrade (2015). The URS allows the assignment of ordinal classes of adverse health effects, according to a composition between the *HI* and *CR* of (USEPA, 1989), the carcinogenic groups (USEPA, 2016) and (IARC, 2016) and doses/ concentration lethal toxic (WHO, 2016).

The application of the URS is justified by the need to ordinal categorization of risk to data training step in the environment supervised by support vector machine, which will be used for Risk Classification (*CL*) in the final stage of qualitative characterization.

The criteria for defining the categories in URS are described following.

1. The risk category (1) corresponds the case where *HI* is less than or equal to *RfD*, and corresponds to the risk characterization where no adverse effect on human health was observed.
2. The risk category (2) corresponds to *HI* higher 1.0 and $DCI(t)_{exp}$ less than the dose corresponding to the LOAEL of WHO (2016), which was observed any moderate adverse effects, to health.
3. The category (3) risk corresponds to a higher HI 1.0 and $DCI(t)_{exp}$ greater than or equal to the corresponding dose to LOAEL; or the level of chronic *RfD* which was reported some important adverse effects, such as chronic poisoning, irreversibility, reproductive disorders, neurological or vital organs, mutagenicity or other congener adversity; It is $DCI(t)_{exp}$ less than DL_{50} / CL_{50}; or the contaminant is classified as B2 by (USEPA, 2016) or *CR* less than 10^{-6}.
4. The risk category (4) corresponds the chemical substance in 2B by (IARC, 2016), or B1 or C of (USEPA, 2016) - considered as a serious effect for relating contaminants that are likely to be carcinogenic to human health or animals; and/or *CR* between 10^{-4} and 10^{-6} inclusive.
5. The risk category (5) corresponds substance chemical in group 1 or 2A by (IARC, 2016); or A or B by (USEPA 2016), considered contaminants with very serious or lethal effect, related as carci-

Table 4. Rules generated by USR

Class	Rules
class 1	If: criterion in {HI<=1} then effect in {no_effect}
class 2	If: criterion in {HI>1 and $DCI(t)_{exp}$ <LOAEL} then effect in {adverse_effect}
class 3	If: criterion in {HI>1 and $DCI(t)_{exp}$ >=LOAEL; RfD ="Chronic"; $DCI(t)_{exp}$ <DL_{50}; GROUP="B2"; CR<1e-6} then effect in {irreversible_effect}
class 4	If: criterion in {GROUP="2B"; GROUP="B1" or GROUP="C"; CR>=1e-4 and CR<=1e-6} then effect in {serious_effect}
class 5	If: criterion in {HI>1; $DCI(t)_{exp}$ >=LOAEL; RfD="Chronic" and $DCI(t)_{exp}$ >= DL_{50}; GROUP="1" or GROUP="2A"; GROUP="A" or GROUP="B"; CR>1e-4} then effect in {very_serious_effect}

nogenic or carcinogenic capacity to human health; and/or *CR* greater than 10^{-4}; or noncancerous contaminants when *HI* is greater than 1 and $DCI(t)_{exp}$ greater or equal to DL_{50} (CL_{50}).

These criteria resulted in the definition of decision rules for logical tests, which were built as models for recognition by machines in Table 4.

Qualitative Characterization of Risk for Multiple Contaminants

With the obtainment of quantitative characterization of risk indicators (HI, HI^{tot}, CR, CR^{tot} and CR^{mult}) and the categorization of risk by URS, carried out the supervised training data with the primary results of the evaluations and the model preparation for risk classification. The qualitative risk characterization is performed for each specific point of space-time matrix.

Therefore, it is used Support Vector Machine (SVM) of Vapnik (1992), which is originally a binary classification technique. For execution of binary classes is carried out in two basic steps: assignment of binary classes and a set of data that are linked to these classes with determination of an optimal hyperplane that separates the points in much as possible on the same side and determines maximum distance each class in the hyperplane (Haykin, 1999).

It considered a set of training data $(x_{i,}, y_{i,})_{i=1}^{N}$, where $x_i \in R^n$ is a data input to sort and $y_i \in \{-1,+1\}$ are desired binary classes to exit the operation. The equation of hyperplane decision surface (equation 18) is given by:

$$w^T \varphi(x) + b = \quad (18)$$

Where: *w* is an adjusting weight vector, *b* is a bias that allows the hyperplane to be positioned to distance data. From this classification is given according to the following restrictions as (Equation 19):

$$W^T\varphi(X_i) + b \geq 1 \text{ for } Y_i = +1$$
$$W^T\varphi(X_i) + b \leq -1 \text{ for } Y_i = -1 \quad (19)$$

To find the values of term *w* and the bias *b*, the optimization problem is solved by the method of Lagrange multipliers (α_i) function and Green's in (Haykin, 1999), which minimizes with reference to the *w* and *b* and maximizes with reference to $\{\alpha_i\}_{i=1}^N$, the following construction as Equation 20 and 21.

$$Q(\alpha) = \sum_{i=1}^{N} \alpha - \frac{1}{2}\sum_{i=1}^{N}\sum_{j=1}^{N} \alpha_i \alpha_j y_i y_j x_i^T x_j \quad (20)$$

$$W = \Sigma_{(i=1)} N\square a_i K(x, x_i) + 1 \rightarrow \begin{pmatrix} w_m \\ b \end{pmatrix} = \sum_{i=1}^{1} \alpha_i^m \begin{pmatrix} \varphi(x_i) \\ 1 \end{pmatrix} \quad (21)$$

The *φ(.)* is mapping for the separation of spaces between data and represents a linear kernel that is a dot product between two vectors, denoted by Equation 22:

$$k(x, x_i) = \varphi^T(x).\varphi(x_i) \text{ for } i = 1,2, \ldots, N \quad (22)$$

The result of the SVM is obtained by applying Equation 23:

$$f(x) = \sum_{i=1}^{N} w_i y_i K(x, x_i) + b \quad (23)$$

For application of this method in risk categories 1 to 5, the multi-class technique kSVM was used by the methods "one-against-one" or "one-against-all." These algorithms allow expanding the basic binary classification to classification in multiple categories. An example of geometric hyperplane used for maximum separation between classes, by the decision function shown in Equation 17 is show in Figure 1.

Where *x* is the vector derived attributes *HI*, *CR*, and within the sample period representing the parameters of a particular contaminant, and HI^{tot}, CR^{tot} and CR^{mult} for multiple contaminants in action multiple. The *w* is the vector of adjustable weights; *b* is the bias term. The $b/\|w\|$ is the dislocation of the hyperplane until origin; and $2/\|w\|$ is the distance between the classes in the hyperplane. The axis x_1 of the graph is vector projection *x* represents contaminants risk for the individual and location of the sample to multiple risk; and axis x_2 is classes of *x* separated by geometric hyperplane.

The example in Figure 1 shows a binary classification that occurs when $f(x) \geq 1$ is determined class to HI (HI^{tot}) > 1 (USEPA, 1989) or CR (CR^{tot}) > 10^{-5} (WHO, 2010), and when $f(x) \leq 1$ for HI represents another class HI (HI^{tot}) ≤ 1 or CR (CR^{tot}) $\leq 10^{-5}$.

The algorithms chosen for multi-class application can be SMO (*Sequential Minimal Optimization*) of (Platt, 1998), the *MultiClassClassifierUpdateable* of (Mark et al., 2009) and the LibSVM (*Library for Support Vector Machines*) of (Chang & Lin, 2011). The choice of these algorithms is justified for being

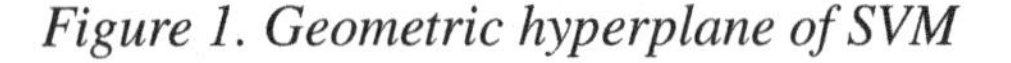

Figure 1. Geometric hyperplane of SVM

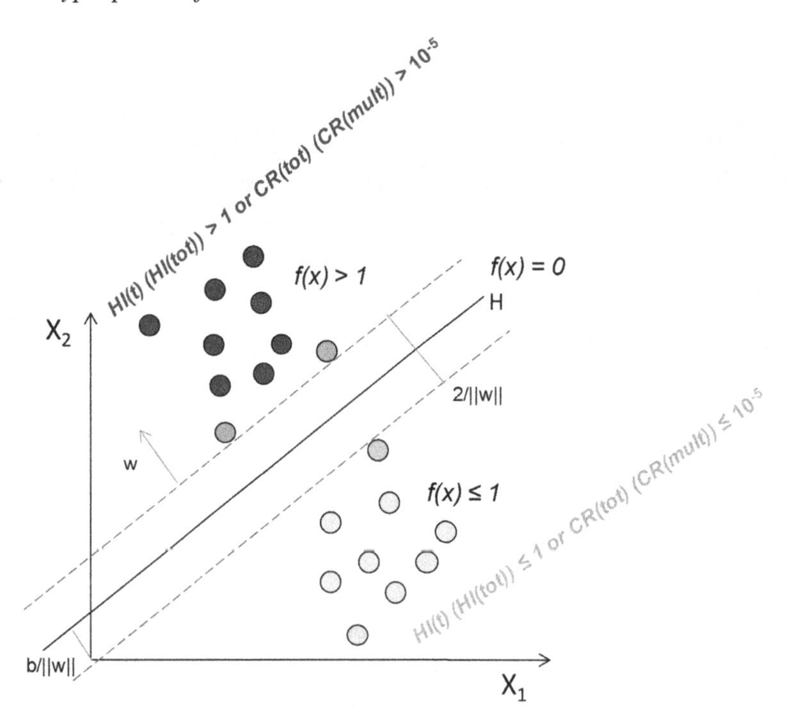

appropriate techniques for data classification with few categories or classes, and applying decomposition method, which resolves binary sub problems to combination multi-class, (Faceli et al., 2011).

Test of statistical significance between the SVM classifiers are applied like the index Kappa (k), absolute error and relative mean, mean square error and confusion matrix (Mark et al., 2009) (Wood, 2007). Using the Weka software, (Mark et al., 2009), or equivalents.

METHODOLOGY FOR DEVELOPMENT OF ARCHITECTURE AND SYSTEM

Environmental risk assessment is a relatively new activity, interdisciplinary and presents with several open fields and what still depends on research and studies. The risk assessment for chemical contamination is processed by the aid of multi-agent autonomous and integrated simulation in decision making is an area of distributed artificial intelligence that can contribute to the extensive knowledge of environmental impacts. The tools available currently make the calculation of the risk of isolation mode and without integration, of exposure and toxicological data.

This chapter investigates methods congeners and development an architecture to fill this gap, contributing to the computing applied to the environment and natural resources.

The feasibility of the project also is in the latent demand for applications for aid in planning for mitigation areas affected by large environmental impacts from mining, oil industry, intensive agriculture and others. It also considers the possibility of employment in other areas such as toxicology, industrial chemistry, bioinformatics and medicine, opening opportunities for further discoveries.

As previously mentioned, this work is based on the representation of chemical risk assessment process through the interoperability of multi-agent for visual simulation to support the decision. Therefore, it will use the method of communication semantic agents through ontology based on toxicological data, classified by support vector machines committee, training multidimensional cubes in data warehouse and information visualization for online processing tools. The basic tasks to achieve the system architecture are:

1. Adaptation to chemical risk assessment methodology according to evaluation paradigm of environmental risk (NRC, 1983);
2. Development of risk assessment ontology with use of the OWL language (Web Ontology Language), of (McGuinness et al, 2004), by PROTEGE software (2016).
3. Construction of multi-agent class, according to the ontological terms, based on the protocol (Foundation for Intelligent Physical Agents - Agent Communication Language [FIPA-ACL], 2002), by ontological interoperability;
4. Multi-agent system development with implementation of the framework (Java Agent Development Framework [JADE], 2016), and interfaces for application of SVM for decision making;
5. Classification of risk based in decision rules and Unified Risk Scale - URS, of Andrade (2015);
6. Development system for exposure assessment and risk characterization for single and multiple contaminant in synergistic action in space-time matrix, as described in approach Andrade (2015);
7. Final classification of risk for SVM committee (Vapnik, 1998), with application of strategies one-against-one and one-against-all;
8. Development of Data Warehouse (DW) with the load of public databases of toxicology (USEPA 2016) and (IPCS, 2016), with reference doses to non-cancerous contaminants and slope factor for carcinogenic contaminants, by exposure route; and
9. Visualization tools development for Online Analytical Processing (OLAP) on the DW database, (Codd, Codd and Salley, 1993).

ARCHITECTURE SYSTEM FOR RISK SIMULATION CHEMICAL

The system architecture for chemical risk assessment simulation for multi-agent is illustrated in Figure 2. This process use a the ontological taxonomy of terms that represent the risk assessment steps to health and the environment (risk identification, exposure assessment, toxicology and risk characterization, and their evaluation procedures) as integrated agents (Noy & McGuinness, 2001).

The aspect for recognition and communication between cooperative agents and ontology is provided by BeanGenerator library (Cancedda & Caire, 2008, 2010), which contains methods for agents operation through semantic ontology using the OWL. The entire process conforms to the (Foundation of Intelligent Physical Agent [FIPA], 2002). Each message between agents and on the procedures of the process is sent following the hierarchy in the process, according to risk assessment paradigm (NRC, 1983). Specifically, the risk characterization what is the integrating the results of the exposure assessment the equations and toxicological information.

This dynamic process is managed by *moderator* agents driving the process and each step indicating the procedures for implementing the *evaluator* agent according to the concentrations of analyzed chemical agents, or by entering concentrations predicted routes of exposure, considering the evaluation dose response to non-cancerous and cancerous contaminants.

Figure 2. Architecture of multi-agent system for risk assessment

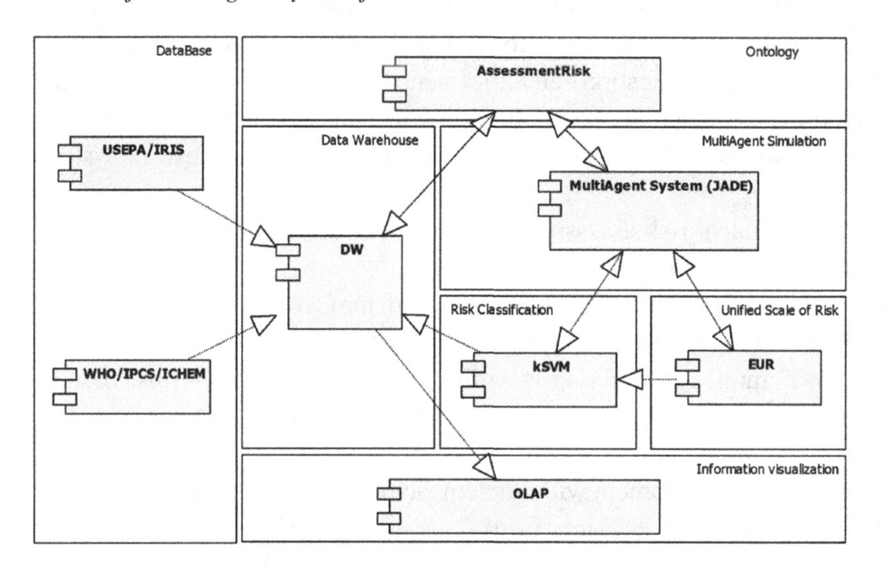

For the evaluation of toxicology, *moderator* agents make checks each other through access to DW basis for knowledge about the availability of reference dose and slope factor for the inventoried contaminants, and the existence of risk characterization to the location study.

They are employed toxicological databases (UEPA, 2016) and (IPCS, 2016) containing references to studies of doses and carcinogenic potential indicator for exhibitions and contaminant entering the human organism orally, inhalation, dermal. These databases along with the concentrations of chemicals will serve as input for the formation of multidimensional cubes based on the data warehouse. If necessary the evaluator agent should perform the calculations of the Hazard Index and Carcinogenic Risk and enable the method to multi-class risk by kSVM.

The application of support vector machine committee is justified by high performance ratings of ordinal categories. The multi-class is made via decomposition of a binary classification process through the use of strategies "one-against-one" and "one-against-all", by machine learning. Through training supervised in order to reuse in future risk prediction or decision making for planning the affected area mitigation actions.

The multidimensional cubes facilitate access to visual information for the characterization and classification of environmental risk matrix for spatiotemporal through OLAP tools with drill down capabilities, roll up, drill across, and slice and dice and others.

ONTOLOGY OF RISK ASSESSMENT ENVIRONMENTAL

The second phase is the definition of ontology that will serve to establish the criteria of multidimensional modeling, the formation of the star schema for the data warehouse and the taxonomy model used as a parameter for communication between agents, for messages and operation behaviors of action.

The justification for the application of data warehouse technology instead of relational database is the need to maintain the historical record of dose-response data sourced from toxicological studies and serve for comparison with other risk assessments.

The following is description each of these operations.

Modeling Multidimensional Based on Ontology

It was considered the Methontology approach, Fernandez et al. (1997), used for modeling ontologies combined with the construction roadmap of ontologies by conceptual integration between these methods and the application of (Noy and McGuinness, 2001).

The life cycle for the ontology presented by Methontology from a script that should be run during the construction process. The necessary steps for in development of ontology are illustrated in Figure 3.

The model illustrated in Figure 4 was followed to the schemes for *AssessmentRisk* ontology that obeyed the Content Reference Model (RCM) that is referenced by the ACL language rating, specified by (FIPA, 2016). This scheme for ontology served to guide the risk assessment process carried out by the assessment agents and the management of moderator's agents, through the messages and behaviors share.

Each ACL message contains the semantic properties according to the terms and *predicates*. The *predicates* are expressions that can be a INFORM, a REQUEST or other, with states behaviors or indicators that refer to the terms as *Concept*.

The *Concepts* are terms that indicate entities. Here are expressed as Risk Identification, Exposure Assessment, Dose-Response Assessment and Risk Characterization. As the risk assessment is a structured activity as a process, the first step Risk Identification is important for the implementation of Exposure Assessment that is related to the dose-response Assessment to result in Risk Characterization, by functions of actions defined as *AgentAction*.

Figure 3. Roadmap of definition of ontology

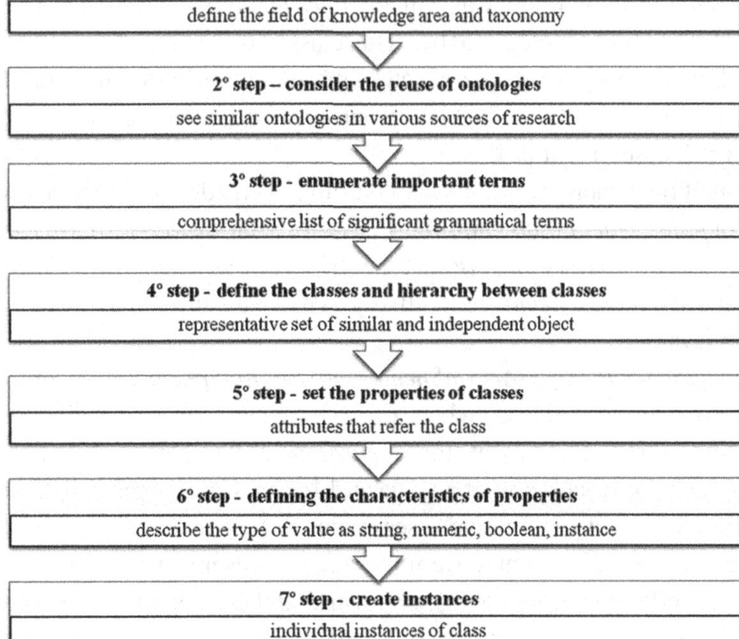

Figure 4. Model for definition of ontology

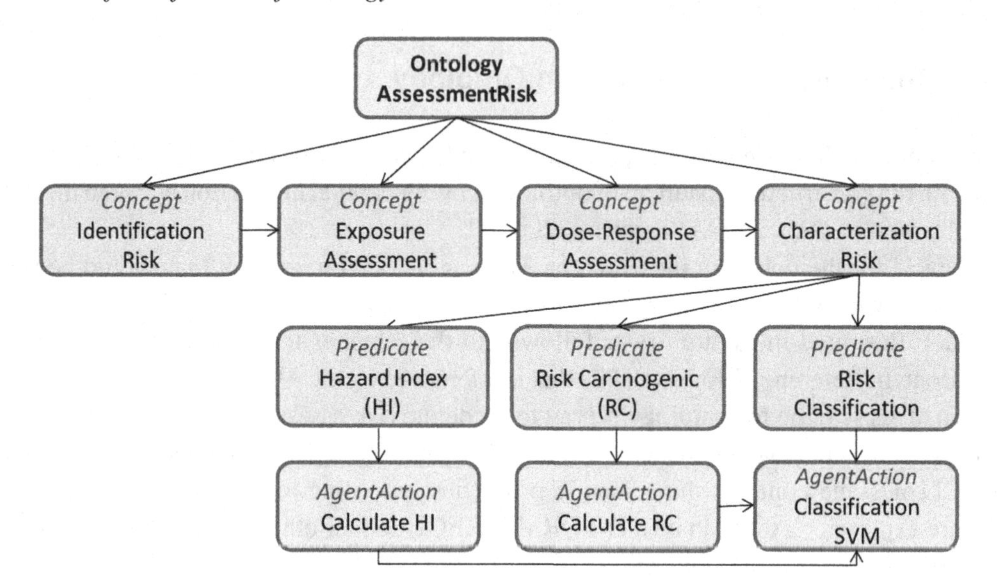

The *precicates* Hazard Index, Carcinogenic Risk and Risk Classification are status of the term *Concept* Risk Characterization, as the risk characterizations are result of the risk assessment process and expressed quantitatively by *HI* and *CR*, and qualitatively by the *RC*. These *predicates* are the result of behaviors that are represented by the evaluation calculation, according to the Exposure Assessment and Dose-Response Assessment.

The *AgentAction's* are terms of *Concepts*. In this scheme the *AgentAction's Calculate HI*, *Calculate RC* and *Classification* SVM are methods for calculation of the indicators of Risk Characterization correspond the states of *Predicates*.

With the result originated of ontology was generated logical representation in Java classes, by applying the beanGenerator plug-in of (Protégé, 2016). This class was developed in accordance with standard specified for the JADE for the purpose of appropriate communication between agents, favoring a semantic environment between the terms and relations in the ontology and behaviors necessary to implement the methods in chemical risk assessment process.

The representation of the general recognition of ontology is made by *AssessmentRiskOntology* class, which is an instance of *jade.content.onto.Ontology* that carries the various terms of *Concept*, *AgentAction* and *Predicate*, declared in vocabulary of constant type string, and as instances of classes *ConceptSchema*, *AgentActionShema* and *PredicteSchema*, respectively, which are in *jade.content.shema* package (like shows code partial in java).

The classes *ConceptSchema*, *AgentActionShema* and *PredictSchema* represent objects that define the terms of the entire evaluation process, where *ConceptSchema* class describe the entities that make up the four main stages of risk assessment, the *AgentActionSchema* class specify the parameters and mathematical formulation for calculation of risk hazard Index, Carcinogenic risk, single and multiple, and Risk Classification to provide qualitative assessment.

In the initial stage of the multi-agent system the agents present in the ontology are loaded of the formatted file as OWL and are mapped to objects for use and connection to the specific agents of the evaluation process of chemical, which are *evaluator* agents and *moderator* agents of the process.

```
//AssessmentRiskOntology.java generated by bean generator.
import jade.content.onto.*;
import jade.content.schema.*;
public class AssessmentRiskOntology extends jade.content.onto.Ontology  {
public static final String ONTOLOGY_NAME = "AssessmentRisk";
private static ReflectiveIntrospector introspect = new ReflectiveIntrospec-
tor();
private static Ontology theInstance = new AssessmentRiskOntology();
 public static Ontology getInstance() {
     return theInstance;}
 //VOCABULARY
 public static final String HAZARDINDEX="HazardIndex";
 public static final String RISKCARCINOGENIC="RiskCarcinogenic";
 public static final String CLASSIFICATIONRISK="ClassificationRisk";
 public static final String CALCULATEASSESSMENT_CHARACTERIZATIONRISK="characte
rizationRisk";
 public static final String CALCULATEASSESSMENT="CalculateAssessment";
 public static final String EXPOSUREASSESSMENT_HAZARDIDENTIFICATION="hazardIde
ntification";
 public static final String EXPOSUREASSESSMENT="ExposureAssessment";
 public static final String HAZARDIDENTIFICATION_CONTAMINATEDPERIOD="contamina
tedPeriod";
 public static final String HAZARDIDENTIFICATION_CONTAMINATEDSITE="contaminate
dSite";
 public static final String HAZARDIDENTIFICATION_POPULATIONINVOLVED="populatio
nInvolved";
 public static final String HAZARDIDENTIFICATION="HazardIdentification";
 public static final String RISKCHARACTERIZATION="RiskCharacterization";
 public static final String DOSEREPONSEASSESSMENT_
DOSEREFERENCE="doseReference";
 public static final String DOSEREPONSEASSESSMENT="DoseReponseAssessment";
//Constructor
private AssessmentRiskOntology(){
    super(ONTOLOGY_NAME, BasicOntology.getInstance());
 try {
   //Concept(s), AgentAction(s), Predicate(s)
    ConceptSchema doseReponseAssessmentSchema = new ConceptSchema(DOSEREPONSEA
SSESSMENT);
    add(doseReponseAssessmentSchema, assessmentrisk.ontology.DoseReponseAs-
sessment.class);
    AgentActionSchema calculateAssessmentSchema = new AgentActionSchema(CALCUL
ATEASSESSMENT);
    add(calculateAssessmentSchema, assessmentrisk.ontology.CalculateAssess-
ment.class);
```

```
    PredicateSchema classificationRiskSchema = new PredicateSchema(CLASSIFICAT
IONRISK);
    add(classificationRiskSchema, assessmentrisk.ontology.ClassificationRisk.
class);
```

STRUCTURAL MATRIX FORMATION OF DATA WAREHOUSE

The following factors to define the multidimensional logical model were adopted: risk evaluation process, data granularity, dimensions and facts that guide the Structural Matrix and the formation of multidimensional cubes for data mart (DW derived for each evaluation process).

To answer the questions of each evaluation process requires information from the dimensions (classes and subclasses of ontology) linked to it in the structural matrix.

The structural matrix is a composition that relates the risk assessment processes in each row, and in the classes of the ontology in each column, resulting in dependence intersections. To the questions about chemical risk assessment process, its numerical data, totals and statistics, and on the dimensions necessary for the formation of dimensional cubes, this can obtain crucial information for decision making.

The granularity of data corresponds to reduce structural information that cannot be subdivided and indicating danger or risk resulting chemical risk assessment process.

The facts are the main entities of the multidimensional model, as guard of interim evaluations data, such as exposure assessment and dose-response assessment, which have direct relationships with all dimensions, providing the granularity of data. The dimensions are qualitative or descriptive entities of the model that relate to the facts also form graininess data.

This composition is caused by the structural matrix and form multidimensional star schema to match an evaluation process of risk or in combination, defining multidimensional cubes to serve the OLAP tools and visualizations of information.

Algebraic Notation for Dimensions and Facts of Schema Star

The model of ontology provided the conceptual basis for multidimensional modeling. In this context, the entities were selected that formed the dimension tables and fact, which resulted in the DW architecture. Then, the notation applied below and its structural matrix for the logical model.

The formation of entities facts and dimensions of multidimensional modeling are originated in the ontology classes and their properties. To represent this definition, was considered the model of algebra (Datta & Thomas, 1999), which proposes to quintuple notation ***DW = {D,M,A,f,R}***, where the integrated elements indicate the features that resemble a multidimensional cube. These characteristic are:

1. $\boldsymbol{D} = \{\boldsymbol{d}_1, \boldsymbol{d}_2, \ldots, \boldsymbol{d}_n\}$, a set of ***n*** dimensional entities that represent descriptive domain attributes, where each d_i is the name of a dimension of the domain;
2. $\boldsymbol{M} = \{\boldsymbol{m}_1, \boldsymbol{m}_2, \ldots, \boldsymbol{m}_k\}$, a set of ***k*** facts entities that represent numeric data and domain measurements, where each m_i is the name of an extracted fact of the domain;

3. $A = \{a_1, a_2, \ldots, a_t\}$, is a set of attributes where each t is extracted property dimension or fact of the domain;
4. The function f is mapped as: $f: D \rightarrow A$, i.e., for each dimension entity exist a set of attributes. Each f is disjoint for attributes, i.e., $\forall\, i, j,\, i \# j,\, f(di) \cap f(dj) = 0$; and,
5. $R = \{r_1, r_2, \ldots, r_m\}$, it is a set of one-to-many relationship between dimensions and facts through surrogate keys *primary key (pk)* and *foreign key (fk),* in following algebraic relationship: d_i |X| $(d_n.pk = m_k.fk)$ m_i. The set of all *fk's* concatenated into facts becomes *pk* composed.

Modeling Multidimensional Cube

At this stage, it was examined every function elected indicating whether the term was a class, subclass, property or relationship. This list was ordered and served as the basis for the establishment of the ontology hierarchy.

The terms of ontology was originated of the risk assessment paradigm, according to the NRC (1983), whence the candidate terms that were selected as classes and subclasses of ontology, because they are higher hierarchical references that did not admit superposition. Some terms were not considered because they did not represent the generality of the concept of entities or their functions could be represented as behavior or operational methods. Example: contaminant intake dose, total accumulated dose in the period, reference dose, multiple risk and others.

Other terms of the list were not used in determining the classes and were assigned concepts as attributes of these classes. For example: *contaminantedPeriod, contaminantedSite, populatonInvolved, referenceDose* and others. The Figure 5 shows the definition of classes and subclasses forming the toxonomia to *AssessmentRisk* ontology.

The achievement of the multidimensional model is presented in four phases:

Phase 1 - Business Processes: Is defined for structural matrix (Table 5) the evaluation processes to characterize the chemical risk. Evaluative processes are formed by a set of methods represented by assessment agents that make calculations to quantify the non-carcinogenic risk, carcinogenic risk and chemical risk class as the USR;

Phase 2 – Granularity: The more granular data corresponds to an item that is the minimum to be used as an indicator of risk by exposure to a chemical contaminant, or hazard quotient that is an required index for characterization risk;

Phase 3 – Dimensions: By convention, the entities were named using the prefix "Dim_" to better identify their functions. These dimensional and derived entities are:

Dim_HazardIdentification, Dim_ExposureAssessment (Sub_RouteWater, Sub_RouteSoil, Sub_RouteAir, Sub_RouteFood), Dim_DoseResponseAssessment, Dim_RiskCharacterization, Dim_HazardIndex and Dim_RiskCarcinogenic.

As mentioned, the size *Dim_Data* is a special entity that saves history data record keeping of evaluative events. His attributes also serve to assist in the granularity of DW; and,

Figure 5. Classes and subclasses of the ontology AssessmentRisk

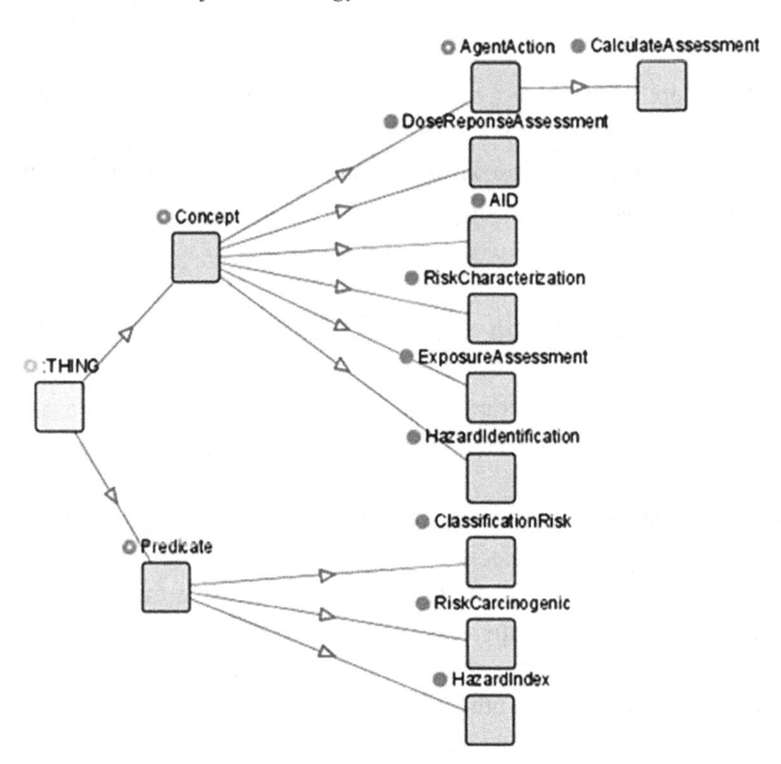

Table 5. Structural matrix for formation of the DW

	Class:	Hazard Identification	Exposure Assessment	Dose Response Assessment	Risk Characterization	Hazard Index	Risk Carcinogenic	Risk Classification
Evaluation processes	Daily dose intake $DCI(t)_{exp}$	X	X					
	Hazard Quotient $HQ_i(t)$	X	X	X				
	Hazard Index accumulated $HI^{tot}(t)$	X	X	X	X	X		
	Carcinogenic Risk Total $CR^{tot}(t)$	X	X	X	X	X	X	
	Carcinogenic Risk Mutiple $CR^{mult}(t)$	X	X	X	X	X	X	
	Risk Classification Whole Period CL	X	X	X	X	X	X	X

Phase 4 – Facts: The fact table is named the "*Fat_*" prefix, for easy identification in the star schema. Its name is *Fat_Assessment*.

As an example of application of the methodology for the formation of a derivative multidimensional cube DW, in a period of contamination of a carcinogenic substance, the following requirements and definitions for viewing by OLAP tool are presented:

It is thought that users are interested in the values of multiple carcinogenic risk and its quantitative. To support the decision needs to measure data in the facts entity corresponding to *M = {individual_ carcinogenic _ risk, total_ carcinogenic _ risk, multiple_ carcinogenic_ risk};*

Users want to analyze the information considering the dimensions: Dim_HazardIdentification, Dim_ ExposureAssessment, Dim_DoseResponseAssessment, Dim_RiskCharacterization, Dim_HazardIndex and Dim_Data. In other words, they seek to know:

"What is the total cancer risk contaminating *c1* in a given space *e1* at time *t1* of the space-time matrix *m1*?" or "What is the multiple carcinogenic risk in a given space *e1* at time *t1* of the space-time matrix *m1*?"

The Dim_Data dimension is described by the day, month and year properties. The dimension Dim_ HazardIdentification has the properties contaminatedSite, contaminatedPeriod, populationInvolved. The Dim_ExposureAssessment dimension has properties routeAirEvaluation routeWaterEvaluation, route-SoilEvaluation, routeFoodEvaluation. The dimension Dim_Dose-Response Assessment has reference-Dose, SlopeFactor properties. The Dim_RiskCharacterization dimension has individualHazardQuotient properties, and individualHazardIndex individualRiskCarcinogenic. The Dim_HazardIndex dimension has hazardQuotient properties, and intakeDose referenceDose noncarcinogenic, and the dimension Dim_RiskCarcinogenic has doseIntakeCarcinogenic and doseSlopeFactor properties.

Thus there was obtained: *A= {contaminatedSite, contaminatedPeriod, populationInvolved, routeAirE-valuation, routeWaterEvaluation, routeSoilEvaluation, routeFoodEvaluation, referenceDose, SlopeFactor, individualHazardQuotient, individualHazardIndex, individualRiskCarcinogenic, doseIntakeCarcinogenic and doseSlopeFactor}*. Each dimension consists of mutually disjoint properties.

The multidimensional cube was defined as illustrated in Table 6.

Table 6. Relationships between entities mapped to form a multidimensional cube

Function Mapping *(f: D → A)*	**Related Entities by Key** *pk* **and** *fk* **in** *(R)*
f(Dim_Data)	*{pk1, day, month, year};*
f(Dim_HazardIdentification)	*{pk2, contaminatedSite, contaminatedPeriod, populationInvolved};*
f(Dim_ExposureAssessment)	*{pk3, routeAirEvaluation, routeWaterEvaluation, routeSoilEvaluation, routeFoodEvaluation};*
f(Dim_DoseResponseAssessment)	*{pk4, referenceDose, SlopeFactor};*
f(Dim_RiskCharacterization)	*{pk5, individualHazardQuotient, individualHazardIndex, individualRiskCarcinogenic};*
f(Dim_HazardIndex)	*{pk6, hazardQuotient, intakeDose, referenceDoseNonCarcinogenic};*
f(Dim_RiskCarciongenic)	*{pk7, intakeDoseCarcinogenic, doseSlopeFactor};*
f(Fat_Assessment)	*{fk1, fk2, fk3, fk4, fk5, fk6, fk7, _ carcinogenic _ risk, total_ carcinogenic _ risk, multiple_ carcinogenic_ risk}.*

The formation of this multidimensional cube meets the definition date previous, because it expresses a mapping $D \rightarrow A$ for each relationship (R) "one-to-many", which allows to structure proper information views to the decision-making process and therefore responds to questions from users.

MULTI-AGENT SYSTEM DEVELOPMENT

For the implementation of multi-agent system, the agents were defined by the ontology structure. The main classes implemented for use of the agents required for operation of the evaluation process are *Authority* and *Assessment*. The *Authority* class corresponds to the process manager who demands services, is responsible for conducting the evaluation system and the replacement data memory and authorizes the operation that perform through the *Assessment* class. The *Assessment* class performs the action, receives messages on demand, checks the absence steps and the possibility of risk assessment in accordance with the provision of data on toxicological studies of dose response in the considered databases, and finally responds on the progress and process time, calculations and evaluation of classification or reported on the refusal or inability evaluation.

Specifically, as shown code in language java where the *Access* class is used to instantiate the required objects for assessment, considering a point of the contaminant event in space-time matrix. The *Assessment* class was instantiated as assessment agents (*Evaluator*), sends messages REQUEST type, which will be managed by instantiated agents for moderation (*Moderator*), the *Authority* class checking if exist or not data on exposure and dose-response, responds as appropriate, with messages like AGREE or REFUSE.

The first statement in the main method contains a String parameters matrix that carries the parameter "-gui" which launches a instance Remote Monitoring Agent (RMA) to enable the use of graphical tools, especially the *Sniffer Agent* to view the exchange of messages type FTA between the selected agents and the use of *Agent Directory Facilitator (DF)*. These parameters are evidenced by the *AgentSpecifier, AgentName's* (*Moderator*) and the instantiated class (*Authority*).

The second statement on main method contains a String matrix *Local1* in container as *Local_1* alias. These parameters are loaded *Evaluator1* the evaluator agent instantiated object of Assessment class, which takes as arguments *Moderator1* moderators to Moderator4.

```
import jade.Boot;
public class Access {
 public static void main(String[] args) {
   String[] parameters =
   {"-gui", "-local-host", "127.0.0.1", "Moderator1: Authority; Moderator2:
Authority; Moderator3: Authority; Moderator4: Authority",};
   jade.Boot.main(parameters);
   String[] Local1 =
         { "-container",
           "-container-name", "Local_1",
           " Evaluator1:Assessment(Moderator1,Moderator2,Moderator3,Moderat
or4)"};
   jade.Boot.main(Local1);
```

In general, the mechanism for chemical risk simulation for the decision-making process begins with the demand of the evaluators that loading the moderators for process management. Each evaluator has its georeferenced location in space-time matrix that operates through moderators who make the dialogue for the exchange of messages and the feasibility of actions for the process to meet logged data and update information.

This process is illustrated in Figure 6 which show an evaluation for four processes through the agents *Evaluator* ($E_{(1)}$ to $E_{(4)}$) and agents *Moderator* ($M_{(1)}$ to $M_{(4)}$). The moderation agents have a key role in the evaluation process because they lead the execution of the tasks according to the need for exposure assessment of chemical compounds and the existence of important toxicological studies to calculate the characterization of chemical risk.

Moderating agents interact with the DW proceeding to update the entities that allow such transaction and verifying data originating available databases (USEPA, 2016) and (IPCS, 2016), and results of other reviews for possible use . After that, it informs the assessment agent who ordered and waiting the time for such an assessment. If the data is available, read and load in the DW the assessment process, sharing messages, confirms operation, performs new evaluation and records new results.

Respecting the example illustrated in Figure 7, the interaction between four evaluators agents and four moderators agents, was simulated the performance of the *evaluator* agent $E_{(1)}$ and their interactions with the four present moderators $M_{(1)}$ to $M_{(4)}$ using the framework JADE. Each *evaluator* agent represented a location in space-time matrix of the contaminant event and was established in peripheral container and *moderator* agents were established in the main container to better organization on the basis of messages and actions shares.

It is observed in the example of Figure 6, the first interaction takes place between the *evaluator* agent *Evaluator1* and the agent known as *Yellow Page* DF, where *Evaluator1* makes a REQUEST for simple execution of the publication of his intentions in the evaluation process, to the knowledge of other assessors and moderators agents. The DF agent responds stating that the publication was successfully committed.

Figure 6. Communication scheme between modeling agents and appraisers

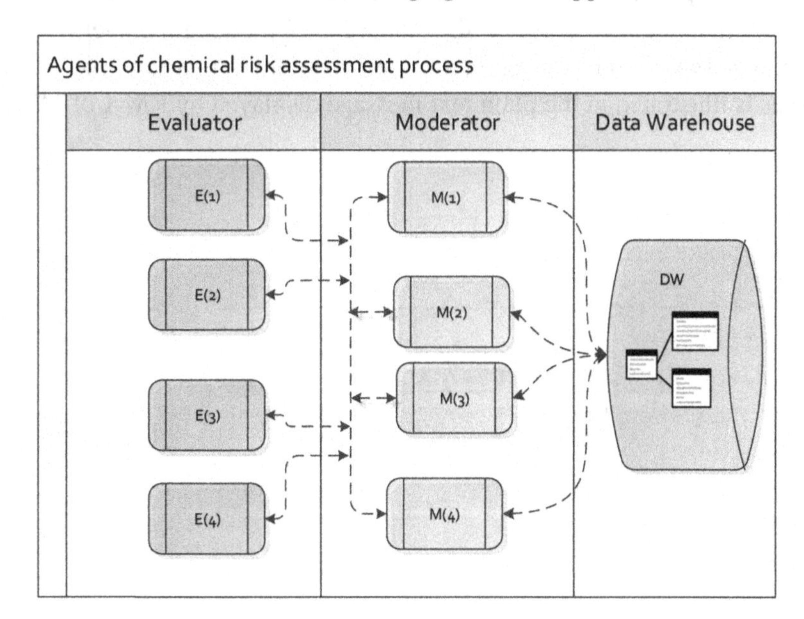

Figure 7. Communication scheme between modeling agents and assessment agents

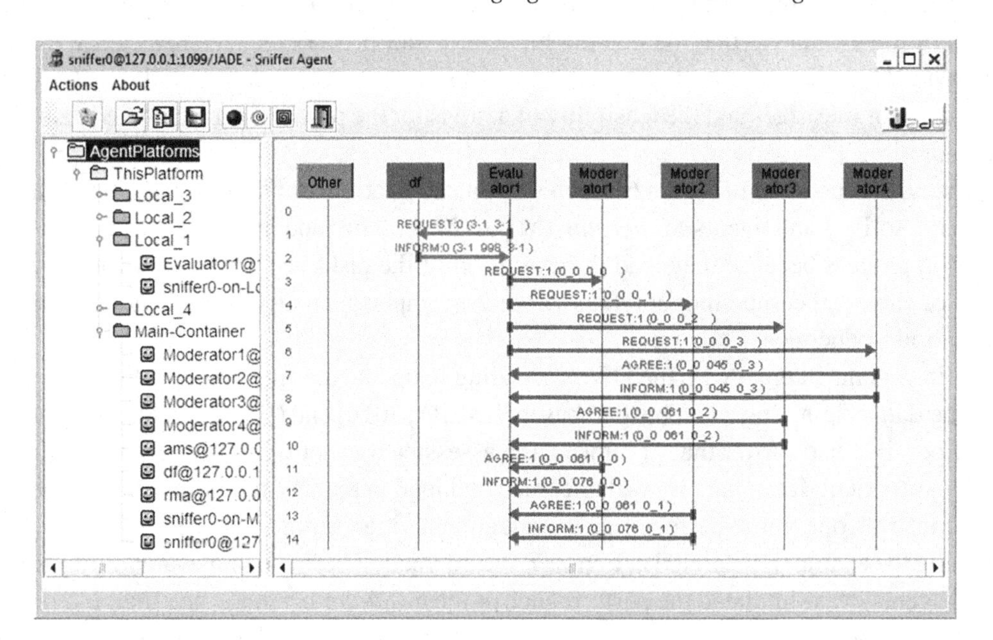

Then the *Evaluator1* agent perform interaction with each of moderating agents ($M_{(1)}$ to $M_{(4)}$), sending REQUEST messages on the purpose of knowledge about the need to follow the steps of the risk assessment process. Each *moderator* agent provides checks the available databases, represented by DW, and responds to the *evaluator* agent with *INFORM, REFUSE* or *AGREE*, as the status and the need for the evaluation process.

In the example in code shown in the *Evaluator1* agent initiates request for checks per *Moderator* agent ($M_{(1)}$ to $M_{(4)}$) on the chemical risk assessment process, specifically on the hazard identification step, exposure assessment and risk characterization, constant in terms of *Concept AssessmentRisk* ontology. The identification parameters are contaminated site, the population involved and contaminated period, making use of the language in FIPA-SL standard. The moderator's agents adopt the necessary procedures, including verification in DW and return information. In this example given the structure of the posted messages is illustrated in the plain text message displayed by RMA of JADE:

REQUEST

```
((action
    (agent-identifier
    :name Evaluator1@127.0.0.1:1099/JADE
    :addresses (sequence http://PC:7778/acc))
    :language fipa-sl
    :ontology AssessmentRisk
    (CalculateAssessment
    :characterizationRisk
```

```
        (ExposureAssessment
        :hazardIdentification
            (HazardIdentification
            :populationInvolved 2000
            :contaminatedSite 10-Oil_Platform_Accident
            :contaminatedPeriod 110915)))))
```

INFORM

```
((RiskCarcinogenic
    (ExposureAssessment
    :hazardIdentification
        (HazardIdentification
        :populationInvolved 2000
        :contaminatedSite 10-Oil_Platform_Accident
        :contaminatedPeriod 110915))))
```

Data Warehouse and Data Visualization Tools

This topic is discussed on the data warehouse and its OLAP tools for information visualization, such result of the application of the methods described previously in the ontology development stages, multidimensional modeling of DW and integration with multi-agent system. The purpose is to show the development of a prototype model illustrated in Figure 8, which served to systemic simulation architecture presented here.

The prototype was built with the IDE C ++ Builder (Embarcadero, 2015), with resources for access to DW, composition of multidimensional cubes and OLAP resource provide for simulations and systemic test.

The star schema entity-relationship that contain the DW was defined by multidimensional algebraic application and is represented by a cube with three dimensions of data, where the formulation of information views able to answer several questions about the assessment processes of chemical risk and fundamental stages.

It is observed in the DW input the information originate from databases (USEPA, 2016) and (IPCS, 2016), available on toxicological references and other studies on assessment of chemical exposure and especially data on the processes of risk assessments conducted by multi-agent. The extract, transform and load of data are occult in this work because of editorial space.

The dimensional cube example (Figure 8) shows information output and historical records, which are shown in three dimensions, attributes of *Population Involved* (hazard Identification dimension), *Chemical_Esp* (exposure assessment dimension) and *Hazard Index* (characterization risk dimension), corresponding to evaluation processes.

The interfaces have dynamic objects in grid with hierarchy of resources, association and group data, slice and dice methods, drill-down, bottom-up support and descending master-detail, function calculations, and graphic indicators.

Figure 8. Scheme exemplifying the architecture with input and output in DW

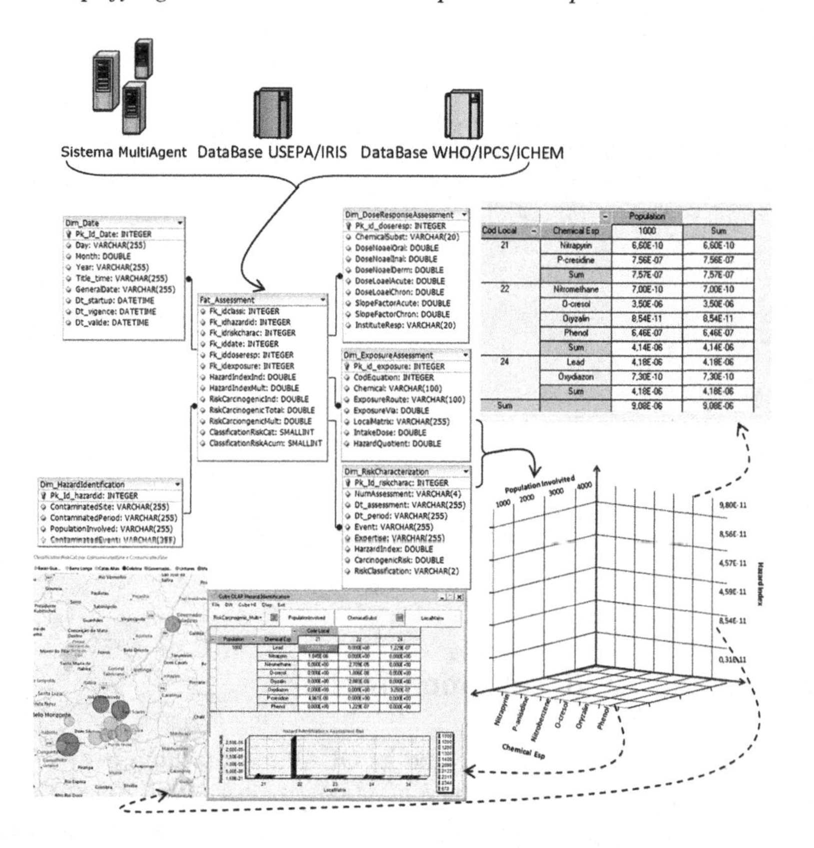

The visualization slice and dice is generated significant data matrix where the rows and columns may be changed to facilitate understanding the information with position change of the dimensions an abstraction of a dynamic cube. In the example can have a change in the lines between *contaminatedSite*, *populationInvolved* or by *contaminatedPeriod* hazard or risk classification index with different aspects for appreciation by users. *Risk Carcinogenic* lines can be extracted or compressed in accordance with the vision risk for single or multiple risks by chemical compound.

The drill down and drill up correspond to an increase or decrease of the granularity of the data as the granularity is subject to *quotient hazard*, which is the lowest risk assessment level, especially for a route of exposure and a chemical element, it can be raise the amount of chemical agents or simply affected period with opening tree view, with change in the degree of detail, subject to totalizations of multiple risks.

Alerts or ranking through risk ratings are useful in space-time matrix with georeferenced view a map, giving conditions for the user to better understand space and time affected with chemical contamination.

FUTURE RESEARCH DIRECTIONS

In the future, this architecture shown can easily be adapted to apply to contamination that causes acute and sub chronic exposure, just the time differential treatment that incorporates. It will also be interesting to work on developing a different approach to additive, more enhanced, to estimate the aggregate and cumulative risk involving multiple routes and multiple contaminants, an issue that is one of the most current open problems in the subject.

It is also important to improve the resources for services with interoperability with georeferenced information systems, cartographic base for application maps with more detailed information on the space-time matrix.

Having a software artifact for web and mobile platforms thus expanding the possibility of practical application and expand the channels of communication with those involved.

CONCLUSION

The environmental risk assessment is a relatively new activity, interdisciplinary and presents with several open fields, still depending of studies and research. The risk assessment for chemical contamination by employ of autonomous multi-agent for assisted decisions by machine learning and shared memory through multidimensional data, integrated simulation in decision-making, corresponding an activity that includes the distributed artificial intelligence and which can contribute to the vast knowledge of environmental impacts.

The proposal unlike the methods commonly applied in the literature allows analysis in geographic space contaminated and its temporal evolution, including a qualitative risk classification which considers the complex systems of multiple exposure routes and multiple contaminants. These attributes present in the architecture provide a new type of information risk analysis, which should contribute to better communication of risks to different social actors and therefore, greater efficiency in the risk management.

The integration of the set of methods and techniques employed showed possible from a technological point of view the data interoperability between methods, which depend on semantic recognition capabilities by machines and sharing information in common purpose.

This work provides help in this area of research, and can encourage congeners systems, as well as develop architectures to fill this gap, contributing to the computing applied to the environment and natural resources.

The feasibility of this general model have a focus on the demand for applications for assist in planning for mitigation areas affected by environmental impacts by chemicals, originating from the mining, oil industry, intensive agriculture and others. Other feasibility also considers the possibility of using this technology in other areas such as toxicology, industrial chemistry, bioinformatics and medicine, opening opportunities for further discoveries.

REFERENCES

Agency for Toxic Substances and Disease Registry (ATSDR). (2016). *Public Health Assessment Guidance Manual, 2005*. Available from http://www.atsdr.cdc.gov/HAC/PHAmanual/ch2.html

Andrade, S. F. R. (2015). *Risk assessment to health applied to nickel mining activity in southern Bahia* (Unpublished doctoral thesis). State University of Santa Cruz.

Bellifemine, F. B., Caire, G., & Greenwood, D. (2016). *JAVA Agent Development Framework* (Version 4.3.3) [Software]. Available from http://jade.tilab.com/

Cancedda, P., & Caire, G. (2008). *Creating ontologies by means of the bean-ontology class. JADE 4.0.* Telecom Italia S.A.

Cancedda, P., & Caire, G. (2010). *Application-defined content languages and ontologies. JADE 4.0. Ontology beanGenerator for JADE (Version 3.5)* [Software]. Telecom Italia S.A.

Chang, C. C., & Lin, C. J. (2011). LIBSVM: A library for support vector machines. *ACM Transactions on Intelligent Systems and Technology*, *2*(3), 27. doi:10.1145/1961189.1961199

Codd, E. F., Codd, S. B., & Salley, C. T. (1993). Providing OLAP (on-line analytical processing) to user-analysts: An IT mandate. *Codd and Date, 32*.

Datta, A., & Thomas, H. (1999). The cube data model: A conceptual model and algebra for on-line analytical processing in data warehouses. *Decision Support Systems*, *27*(3), 289–301. doi:10.1016/S0167-9236(99)00052-4

Dzemydiene, D., & Dzindzalieta, R. (2010). Development of architecture of embedded decision support systems for risk evaluation of transportation of dangerous goods. *Technological and Economic Development of Economy*, *16*(4), 654–671. doi:10.3846/tede.2010.40

Environmental Company of the State São Paulo (CETESB). (2001). *Manual management of contaminated areas*. São Paulo, Brazil: Author.

European Union - EU. (2003). *Technical Guidance Document (TGD) Part I. EUR20418 EN/1*. European Commission-Joint Research Center, Directive 98/8EC, 14 JRC Publication N°: JRC23785, EUR 20418 EN, 2003.

Faceli, K., Carvalho, A., Lorena, A., & Gama, J. (2011). *Inteligência Artificial – uma abordagem de aprendizado de máquina*. Rio de Janeiro: LTC.

Fernández-López, M., Gómez-Pérez, A., & Juristo, N. (1997). *Methontology: from ontological art towards ontological engineering.AAAI Symposium on Ontological Engineering*, Stanford, CA.

Foundation for Intelligent Physical Agentes (FIPA). (2002). Communicative Act Library Specification. Document number SC00037J - FIPA TC Communication. Geneva, Switzerland: Author.

Government of Canada. (2004). Canadian Handbook on Health Impact Assessment: The Multidisciplinary Team (vols. 1-3). Ottawa, Canada: Author.

Government of Canada. Health Canada's Social Media Tools. (2012). Guidance *Documents related to Human Health Risk Assessment.* Revised 2012, ISBN: 978-1-100-17671-0, Cat.: H128-1/11-632E-PDF. Retrieved from http://www.hc-sc.gc.ca/ewh-semt/contamsite/docs/index-eng.php#a1

Government of Netherland. National Institute for Public Health and the Environment (RIVM). (2015). *Risks of Chemicals - Proast*. Retrieved from http://www.rivm.nl/en

Hall, M., Frank, E., Holmes, G., Pfahringer, B., Reutemann, P., & Witten, I. H. (2009). The WEKA data mining software: an update. Waikato Environment for Knowledge Analysis (Version 3.7.5) [Software]. *ACM SIGKDD Explorations Newsletter, 11*(1), 10-18.

Haykin, S. (1999). Neural Networks. A comprehensive Foundation (2nd ed.). Prentice-Hall.

Hsu, C. W., & Lin, C. J. (2002). A comparison of methods for multiclass support vector machines. *Neural Networks. IEEE Transactions on, 13*(2), 415–425.

International Agency for Research on Cancer (IARC/WHO). (2016). *List of Classifications. Agents Classified by the IARC.* Retrieved from http://monographs.iarc.fr/ENG/Classification/index.php

Jiang, D., Lu, M. X., Li, F. S., Zhou, Y. Y., & Gu, Q. B. (2011). Review on current application of decision support systems for contaminated site management. *Huanjing Kexue yu Jishu, 34*(3), 170-174.

Linard, C., Ponçon, N., Fontenille, D., & Lambin, E. F. (2009). A multi-agent simulation to assess the risk of malaria re-emergence in southern France. *Ecological Modelling, 220*(2), 160–174. doi:10.1016/j.ecolmodel.2008.09.001

Lioy, P. J. (1990). Assessing total human exposure to contaminants. *A multidisciplinary approach. Environmental Science & Technology, 24*(7), 938–945. doi:10.1021/es00077a001

McGuinness, D. L., & Van Harmelen, F. (2004). OWL web ontology language overview. *W3C Recommendation, 10*(10), 2004.

National Research Council (NRC). (1983). *Risk Assessment in the Federal Government: Managing the Process.* Washington, DC: The National Academies Press. doi: 10.17226/366

Noy, N. F., & McGuinness, D. L. (2001). *Ontology development 101: A guide to creating your first ontology.* Palo Alto, CA: Stanford University Press.

Platt, J. (1988). *Sequetial minimal optimization: A fast algorithm for training support vector machines.* Technical Report MST-TR-98-14. Microsoft Research.

Protégé Software. (2016). *Protégé is a core component of The National Center for Biomedical Ontology* (Version 4.3.3) [Software]. Stanford Center for Biomedical Informatics Research. Available from http://protege.stanford.edu/

Rebelo, A., Ferra, I., Gonçalves, I., & Marques, A. M. (2014). A risk assessment model for water resources: Releases of dangerous and hazardous substances. *Journal of Environmental Management, 140*, 51–59. doi:10.1016/j.jenvman.2014.02.025 PMID:24726965

Stroeve, S. H., Blom, H. A. P., & Van der Park, M. N. J. (2003). Multi-agent situation awareness error evolution in accident risk modelling. In *Proceedings of the Fifth USA/Europe ATM R&D Seminar* (pp. 1-25).

Technologies, E. (2015). *Integrated development environment for language C++.* Retrieved from https://www.embarcadero.com/br/products/cbuilder

Torrellas, G. S. (2004). A framework for multi-agent system engineering using ontology domain modelling for security architecture risk assessment in e-commerce security services. In *Network Computing and Applications, 2004. (NCA 2004). Proceedings. Third IEEE International Symposium on* (pp. 409-412). IEEE. doi:10.1109/NCA.2004.1347810

United States Environmental Protection Agency (USEPA). (1986). Guidelines for Carcinogen Risk Assessment. EPA/630/R-00/004. Author.

United States Environmental Protection Agency (USEPA). (1989). Risk Assessment Guidance for Superfund. Vol I: Human Health Evaluation Manual (Part A). Washington, DC: Author.

United States Environmental Protection Agency (USEPA). (1992). Guidelines for Exposure Assessment. EPA/600Z-92/001. Author.

United States Environmental Protection Agency (USEPA). (2004). Supplemental Guidance for Dermal Risk Assessment. Washington, DC: Author.

United States Environmental Protection Agency (USEPA/IRIS). (2016). *Integrated Risk Information System - USEPA-IRIS* [Data file]. Retrieved from http://www.epa.gov/IRIS/

Vapnik, V. N., & Vapnik, V. (1998). *Statistical learning theory* (Vol. 1). New York: Wiley.

Weston, J., & Watkins, C. (1998). *Multi-class support vector machines*. Technical Report CSD-TR-98-04. Department of Computer Science, Royal Holloway, University of London.

Wood, J. M. (2007). Understanding and Computing Cohen's Kappa: A Tutorial. *WebPsychEmpiricist*. Retrieved November 8, 2015, from http://wpe.info/vault/wood07/wood07ab.html

Wooldridge, M. (2009). *An introduction to multiagent systems* (2nd ed.). John Wiley & Sons.

World Health Organization (WHO). (2010). Risk Assessment Toolkit: chemical hazards. WHO/IPCS.

World Health Organization (WHO/IPCS/ICHEM). (2016). *International Programmed on Chemical Safety - Chemical Safety Information from Intergovernmental Organizations - INCHEM* [Data file]. Retrieved from http://www.inchem.org/

Xie, Z., & He, Z. (2014, June). The Study on the Development of Decision Support Systems in Response to Catastrophic Social Risks. In *Third International Conference on Computer Science and Service System*. Atlantis Press. doi:10.2991/csss-14.2014.108

KEY TERMS AND DEFINITIONS

Data Warehouse: It is a database stored by dimensional subject, with detailed historical records, composed of descriptions and numerical facts, own to aid decision making.

Dose-Response Assessment: It is the toxicological evaluation of chemicals by exposure route, the dose-response effect, to estimate the severity of adverse actions.

Exposure Assessment: Quantitative assessment of the magnitude, duration and frequency of exposure to chemicals, for contaminated routes and pathways.

Hazard Identification: It is characterization of the contaminant event, the contaminated site, exposure routes, the affected population and inventory of chemical compounds.

Ontology: It is a model representing taxonomy of terms, classes, subclasses and attributes corresponding to a domain and employed for recognition of semantics by machines.

Risk Assessment: It is the technical-scientific process to estimate the probability and indicators of adverse effects on human health and biota exposed to chemicals.

Risk Characterization: It is the integrating of the exposure processes and toxicity, integrated to the probability and risk assessment indicators for non-carcinogen and carcinogenic effects.

Support Vector Machine: Machine-learning technique supervised for generalization through separator geometric hyperplane, applied for classification and regression.

Chapter 9
Microbial Fuel Cells Using Agent-Based Simulation:
Review and Basic Modeling

Diogo Ortiz Machado
FURG, Brazil & IFRS, Rio Grande, Brazil

Diana Francisca Adamatti
FURG, Brazil

Eder Mateus Nunes Gonçalves
FURG, Brazil

ABSTRACT

Microbial Fuel Cells (MFC) could generate electrical energy combined with the wastewater treatment and they can be a promising technological opportunity. This chapter presents an agent-based model and simulation of MFC comparing it with analytical models, to show that this approach could model and simulate these problems with more abstraction and with excellent results.

INTRODUCTION

A lot of research has been done with the energy and environment issues. About this subject, the environmental damages mitigation and the research for renewable energy solutions are some of the mains topics. Studies indicate that the bacteria could generate electrical energy combined with the wastewater treatment, resulting in a promising technological opportunity (Logan, 2008).

In this way, computational modelling and simulation bring tools that could expand the understanding of systems and processes with economic and time advantages in face to build pilot plants, prototypes and experimental planning and execution. It is known that just the modelling and the simulation could not bring all the real aspects and variables. Nevertheless, its uses can offer more confidence to the planning and development of strategies in order to fulfill experimental tests (Maria, 1997).

DOI: 10.4018/978-1-5225-1756-6.ch009

In Brazil, according to the last Basic Sanitation Census done in 2008, 34,8 million people live without sewerage system and treatment. It represents 44,8% cities without this basic service. And worst, the number of treatment is much less than the disposal presented, about 10%. This reflects in costs to the public health systems and in environmental impacts in medium and long time. There are some technical difficulties reasons to offer basic sewage treatment. In one hand, there is the economic viability to build and operate wastewater treatment plants, since exist a minimum population that turn the economic trade-off positive in a city. In other hand, there is the energy availability. In a huge country with geographical difficulties, the cities with more necessity of basic sanitation probably do not have energy required to run the plant. And, in small cities with electrical energy, the cost of the energy to use pumps and aerators turns the implantation of water treatment economically negative. If these costs could be mitigated, the construction of small profitable treatment plants could be made in remote places.

Analyzing the progress of energy matrix in Brazil, it is possible to see that it has good developed in terms of renewable energies. According to the last report of Ministério de Minas e Energia (Ministry of Mines and Energy - Brazil), Brasil has 39% of its energy matrix composed by renewable energy, against 10% of OECD[1]. The CO2 emission of Brazil are less than OECD, where Brazil has 1,59 tco2/tep versus 2,31 tco2/tep to OECD. These characteristics reflect a country with energy diversity and adapted to the new energies and environmental world scenario concerns . However, with the economic growth, the energy demand increase 3,1%, more than the GNP (Gross National Product) of 0,1%. This shows an economic issue due to not supported Brazilian development. Therefore, potential of new fonts of renewable energy and technology show promising roles, and the possibilities of a sustainable growth with MFC technology could be reached.

The use of MFC technology in wastewater treatment plants could support the build of sanitation plants in place that nowadays are difficult. The sewage contains great amounts of energy in form of organic matter. This matter can feed the bacteria present in the anodic chamber. Copper et al apud Logan (2008) state that the energy content of wastewater is up to 5,9 times greater than the power used in wastewater process. Therefore, the metabolic activity of these bacteria has ability to lower the BOD and COD of wastewater and the same time that generate energy.

Logan (2008) exemplify this potential, using a scenario of a town with 100 000 people. The save of the plant in one year, considering the typical efficiencies of MFC, could reach 5 million dollars. In order to calculate it, he shows the energy supply of 1700 homes, considering The United States consumption level. This potential combined with other energy recuperation, as methane combustion in a generator, turn the wastewater plant from a consumer of energy to a net producer. This application shows a special use of MFC technology and point out to the new energy paradigm that is to implement distributed energy saving technology spread industrial and residential process. The economic and environmental benefits appear through fitting application characteristics with the technology advantages.

One objective of this chapter is review some technological potentials of Microbial Fuel Cells (MFC) combined with wastewater treatment and gives some perspectives about the multidisciplinary of this problem. The main goal of this chapter is to explore, in a qualitative way, the possibilities of agent model approach to model and simulate a MFC. One of the issues of this paradigm is the abstraction transformation between the analytical models to the agent-based models. To do this study, we present a review about some technological potentials of Microbial Fuel Cells (MFC) combined with wastewater treatment and gives some perspectives about the multidisciplinary of this problem. Finally, we will select some basic bacterial behaviors and, qualitatively, confront them with widely used analytical models of MFC consumption and biofilm growth.

The chapter is structured in 4 stages. Firstly, a section introduces the MFC area. After, the agent-based model is presented as well as the simulated scenarios and results. Finally, the conclusions are presented.

MICROBIAL FUEL CELLS

In this section, is presented some theoretical concepts about Microbiological Fuel Cells (MFC). This cell is a way to generate energy using organic matter that could be contained in sewage, effluents and waste, brings economic advantages in both energy and sanitary aspects. There are other features of MFC that could be used, for example, as an economical way to hydrogen generation and biosensors production (Logan, 1998).

MFC are cells with representative developments since 90s, but their bioelectricity discovered was done by Potter in 1911. The bioelectricity is a quality of some bacteria to generate electrons in the metabolic cycle through the chemical reduction of some substrate. The idea behind a MFC is to build an energy cell using those microorganisms. Rahimnejad et al (2015) presents a complete review about MFC bioelectricity generation.

In a conventional electrochemical cell, there are two chambers with two electrodes and solutions. In one side, an anode and a reductive solution, and in the other side a cathode and an oxidative solution. The electrodes are connected and an electric load is put in the system. The arrange resulted is a closed circuit that generates current through the electrical potential between electrodes and chemical species. One specie oxides and loses electrons, that migrate to the anode and travel through the sludge, until reach the cathode, evidencing electrical work. There occurs the reduction of chemical species in the second chamber. This architecture explains the end of power and the need of recharging. The phenomenon occurs because of the consumption of all reductive species and the end of electrons transfer. And, because of the natural reversibility of the reactions, the cell has a performance lifetime, needing its charge or change. These electrochemical principles and their uses in batteries were developed initially by Luigi Galvani and Alessandro Volta, in the century18^{th}. This technology is basically the same of the modern lithium cells battery presents in many cellphones. The knowledge developed is used in advanced studies of electrochemistry as in MFC.

The MFC uses the same electrochemical principles. However, there are no electrochemical planed solutions in cathodic and anodic chambers, but microbiological planned substrate and bacteria. The nutriment that reduces and oxides is consumed by the living bacteria, and fulfill a complex and cycled metabolic mechanism inside of the living cell that uses NADH/NAD+ as electron donors and acceptors in biochemical routes (Rittman & McCarty, 2001). As a result of the effort to live and metabolic versatility, the work of the cell brings both reduction of any carbonic nutriment and release electrons. The electron generation in macroscale can be abstracted as a conventional electrochemical cell and, with an external sludge, three advantages emerges: the possibility of continuous feed cells, the long last electric production and the capability to feed of almost any substrate. Since the cell do not "want" die and for live it must feed, consequently, results in energy generation in a robust system. A scheme of this fuel cell is represented by Figure 1.

The energy production potential of these cells are thermodynamically calculated giving theoretical electrical potentials between 0,3 and 0,7 V. These potentials are not reached in real testing. The causes reported by Logan (1998) are:

Figure 1. Microbial Fuel Cell
Source: https://commons.wikimedia.org/wiki/File:Emefcy-MFC.JPG

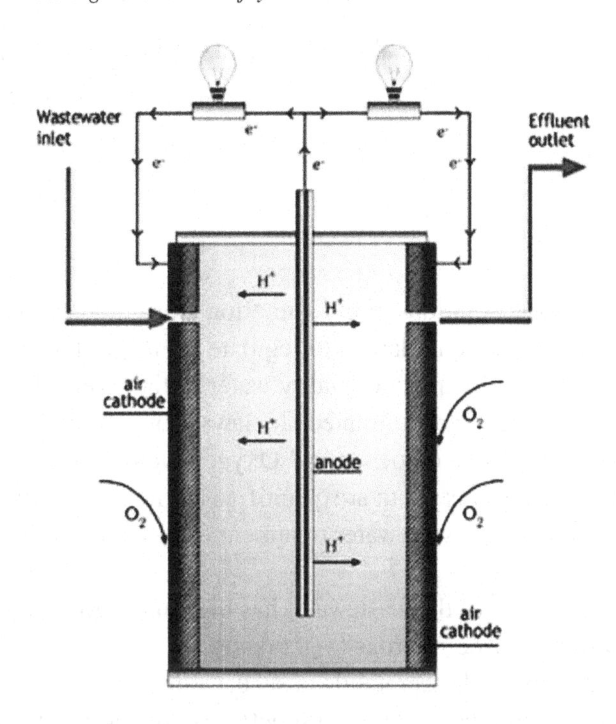

- Different stages of microbial colonization and life (lag phase, log phase, stationary phase, death phase).
- Species interactions (competition, symbiosis).
- Manufacture of enzymes or structures (EPS - Extracellular Polymeric Substance, nanowires, pilli).
- Activation (heat, electron transfer, reactions).
- Metabolism (pumping protons through cell membrane).
- Mass transfer (pH gradient, flux of reactants and products).
- Ohmic losses (Ion diffusion, flow of electrons through biofilm, architecture dependent).

Some studies have been done in order to answer many question about its functioning, and some tools are used in electrochemical analysis of the cell and they have important role in the simulation. Important parameters are External resistance (load), Coulombic and energy efficiency, and Polarization and Power density curves. These tools are used in order to generate meaningful data as: Open Circuit Voltage, Power density profile, Current density profile, and bring some of the cell capacities of sustain electrical potential through voltage drop varying external resistance.

As a living cell, behaviors bring a new layer of complexity to the system. The model of a MFC becomes greatly complex, because these cells are dynamic. We will present a review and a basic modelling to this problem using the Agent-based computational paradigm, and we expect that the advantages of this programming fit to this problem and their difficulties when mathematical modelling are used. The objective is to the develop a tool that could extrapolate and investigate the behavior of a MFC system initiating with simple validations.

Sewage Treatment and Energy Generation

There are three features to the MFC technical potential and they explain the great effort presented by researchers in the last 15 years:

1. The continuous feed work;
2. The long life possibilities; and
3. A wide variety of substrate uses.

If a wide analysis will be done, solutions in the sanitation area could be pointed.

The basic work of a sewage treatment plant is to regulate chemical, biochemical and microbiological parameters of water in order to dump good quality water in the rivers. This quality is standardized by environmental companies and they are guaranteed by laws. Some basic parameters of water is COD (Chemical Oxygen Demand) and BOD (Biochemical Oxygen Demand) and its quantify the pollution potential of sewage or industrial effluent. Both are quantified by the mass of oxygen per volume of solution. There are basically 3 steps in the wastewater treatment (Metcalf & Eddy, 2003):

- **Primary Treatment:** In this part the wastewater has big solids, gross, and suspensions materials. A mechanical separation is made in order to get ready for the next step. The first battery of settling tanks is used to precipitate bigger solids in suspension. This part of sewage treatment is also called mechanic treatment and secure the right functioning of chemical and microbiological tanks.
- **Secondary Treatment:** In this step the reduction of BOD and COD takes place through the uses of oxidizing bacteria concentrated solution called activated sludge. This part of wastewater treatment needs aeration and mixing, presenting expensive part of treatment and generating great amounts of sludge. However, in this part up to 90% of BOD is removed and water characteristics as turbidity, smell, color are enhanced toward quality standards levels.
- **Tertiary Treatment:** In this phase the rest of particles are filtrated and disinfection procedures are made. In this phase, living organisms, including bacteria and pathogens, are neutralized with chlorine, ozone or UV radiation, in order to integrate the water to nature.

The MFC uses living bacteria that already are used in sewage and effluent treatment as explained in secondary treatment of a wastewater plant. So, this is an opportunity of industrial application of the cell whereas a similar use of bacteria could be made. Still, the use of microorganisms is applied only to lower COD and BOD water levels. So, this synergy opportunity is other feature that could be used to water treatment and energy generation together. According to Logan (1998), there are four advantages of the use of MFC with wastewater treatment:

1. Production of electricity.
2. Less need of aeration and pump, because just cathode needs aeration.
3. Reduce solids production (sludge).
4. Potentially control for bad smell (odor).

Bacterial Behavior and Community (Biofilm)

Bacterial communities are very important in many processes in nature. The capability of self-regeneration in rivers, lagoons and swamps. And, in artificial wastewater treatment plants, they are used to restore water parameters to minor environmental impacts. In our bodies, these communities maintain populations that secure correct operation of the digestive system. In food industry, a plenty of products are made with the support of bacteria as brew, wine, bread, yogurt, and all fermented food. In pharmaceutical area, the antibiotics and other drugs are synthesized by these microorganisms.

However, some strains are harmful or unwanted in these processes. Some of these species can cause diseases as meningitis, botulism, cholera, tetanus or tuberculosis. In industrial process, some strains can compete with desirable communities and kill or difficult growing and production leading in efficiency problems.

In order to produce energy, the MFC needs to consolidate a community of specific bacteria in one place that is suitable to electron transfer. As already exposed, the device used to this function is the electrode. A positive one, called Anode, and a negative one, called Cathode. The anodic chamber is the place where exoelectric bacteria is inoculated and substrate is fed. Studies show that the bacteria will attach to better place to live and, in order to easily left electrons, will attach in the place with more electrical potential to discharge (the anode). This phenomenon is carried out by the population of bacteria presents in solution resulted by the crowding of the cathode surface. This formation is called biofilm and it has important role in MFC energy generation. Figure 2 shows a micrograph of a biofilm.

There are many interactions, phenomena and applications researched and developed in the area of biofilms. The main relations presented in biofilm formation and fuel cell performance are:

Figure 2. Microbial biofilm micrograph
Source: Rakosy (2004) - http://global.britannica.com/science/Staphylococcus

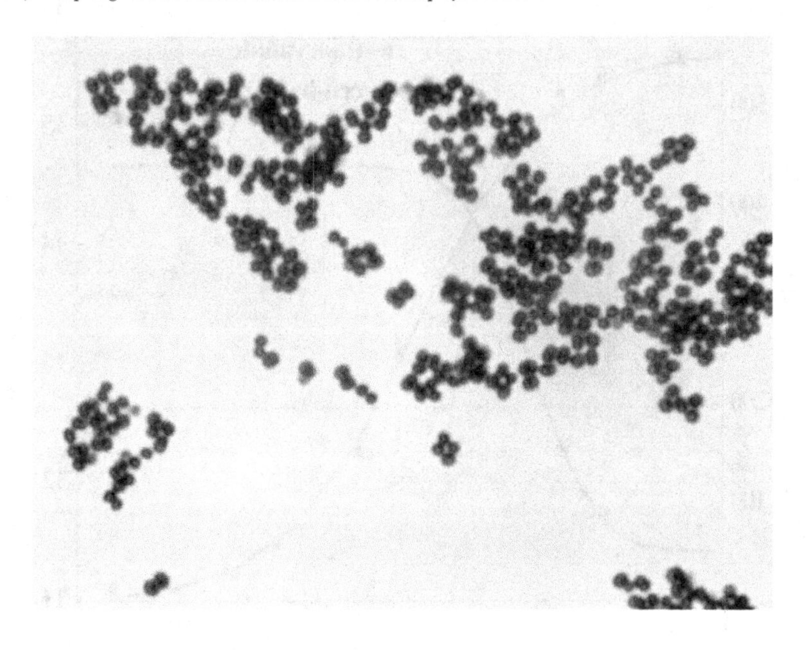

- Relation between biofilm thickness and fuel cell power density.
- Bacteria activity as a function of electron transmission easiness - function of external resistance.
- Relation between internal resistance and external resistance in order to reach optimal steady state.
- Community formation and different interactions between different bacteria species.
- Transport phenomena in attaching and detaching of the electrode surface.
- Electron transport through the biofilm layer.
- Relation between transport phenomena.
- Attachment and detachment of bacteria.
- Surface phenomena.
- Bacteria motility.

The basic analytic model used to express biofilm growth is the Monod's (1949)Equation 1.1.

$$\mu = \mu_{max} * \frac{S}{K_s + S} \tag{1.1}$$

Where μ e μ_{max} are the specific and maximum gowth rate (h^{-1}), respectively. S is the substrate concentration (mol/L) and K_s the Monod's constant (mol/L). The Monod's constant represents the concentration reached with the half of the maximum growth rate. The integration of this equation for one substrate is graphically shown in Figure 3.

Figure 3. Baterial growth profile
Source: Huang et al., 2013

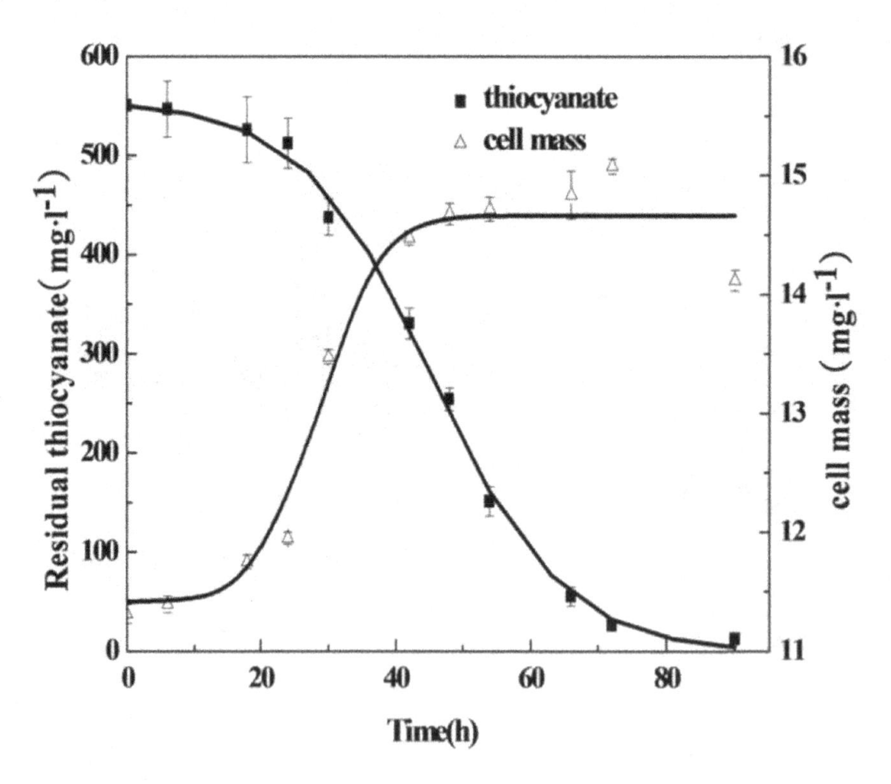

MFC AGENT-BASED MODEL

There are wide type of models and computational approaches of tools for tests and evaluation of MFC operation. Some of these works discourse about modelling and performance of microbial cell growth and dynamics (Esser et al, 2015), (Calder, 2007) and (Boghani et al, 2013); others in types of modelling as cellular automata, agent or analytical (Byrne & Drasdo, 2009), (Bandini et al, 2009), (Pizarro et al, 2005), (Gruner, 2010); anothers discuss a hybrid approach, analytical and multi-agent, in modelling as in (Picioreanu et al, 1997) and (Laspidou et al, 2010). In this work, the idea is to verify and validate representations, in order to evaluate MFC computational abstraction to use these tools. There are several reasons to do this type of study, since to extrapolation analysis, researches in different levels of detail and their interaction, going from microsimulation to macrosimulation of the system.

The model is basically a form to represent something through the abstraction of the object of study. These abstractions can be mathematical, computational, graphical, tridimensional, schematic and so on. The idea is to understand some interested aspects of the thing that is been modeled. For example, in the MFC design: the bacteria colonization, the stress with different external resistances, substrate concentration, geometry, motility, nanowires, EXP or electron transfer through biofilm.

The classical approach to model a system is a mathematical approach. The advantages are the well-known mathematical relations of variables. For example, in a transport phenomenon as reaction-diffusion nature. And the disadvantages are the difficulties to give understanding with the use of equations for abstractions and to fit equations (many variables) in very complex systems. This is the main problem of the analytical approach to develop a model of a MFC, the system has autonomous and dynamic units, a.k.a. bacteria.

According to Walker and Southgate (2009), differently of chemistry, physics or engineering, biology is a data-driven science, that generates estimated 1 terabyte of information per day for proteomics. This fact shows the difficulties to use differential equations to model this processes. For example, in health area, these authors put the bottom up and top down approaches to interpret a system. And they conclude that an interesting alternative for the data interpretation is a middle-out approach. Considering the biological cell as the most important level to model living system because the cells are the most basic representative of life.

In this section, an agent-based model and simulation is proposed to MFC problem, using NetLogo tool (Wilensky, 2016). The main concept is to develop agents with simple behavior, and then, to run qualitative assessment of the model against the mathematical model of bacterial growth of Monod (Equation 1). The proposed model shows the geometric organization of the biofilm in the anode surface against the micrographs of bacteria colonies.

Model and Simulation Environment

The Figure 4 presents the interface of the developed model.

Netlogo is an agent-based programming platform, where agents are called turtles, and patches are the places in the surface or space. The ticks are execution turns of the algorithm that could be attributed to time. All developed applications use the *setup* and *go* commands to give the initial step and to execute the program through the ticks, respectively.

Figure 4. Developed simulation interface

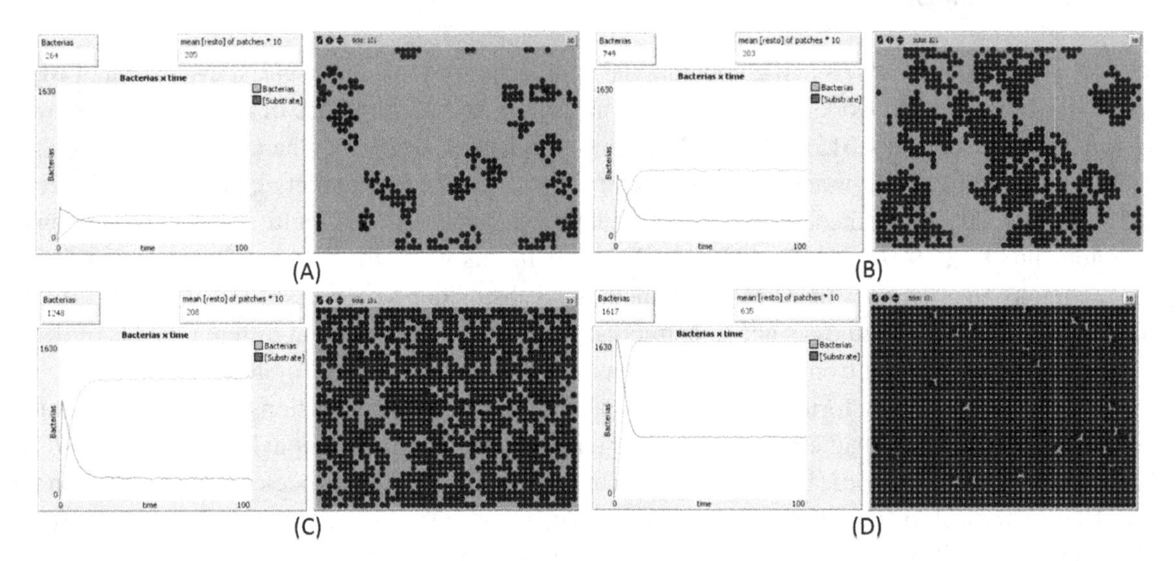

To the creation of the model was declared a set of agents called bacterias, with the individual called bacteria, represented by the dark circles in Figure 4. The attributes of these agents are the internal energy reserve, maximum internal energy and initial internal energy. The world of these agents is the area displayed in Figure 4, and it represents the anodic surface. This surface is divided by patches that represent the space where some bacteria can occupy. The attributes of patches are substrate concentration and the rest of substrate after the metabolism.

Also in Figure 4, it is possible to visualize some names utilized in the parameters like *SubstrateQt*, *Eficiency*, *numberbacterias*, *metabolismenergy*, that represent, respectively, the substrate concentration, the metabolic efficiency, the initial number of bacteria in the biofilm and initial internal energy. These variables can be configured with different values, for different scenarios to simulate.

The output of the algorithm is the number of bacteria or the biofilm size. The algorithm executes, in a simplified way, the following functions: substrate diffusion, biofilm growth, metabolism and bacterial decay.

The diffusion of substrate is based on a mass balance, considering steady state of substrate feed, namely, at each tick the patches concentration is updated, the bacteria metabolize and a rest is calculated as follows.

The biofilm growth was modelled starting with the following rule: if the internal bacterial energy is bigger than 40% of the maximum energy, then this bacteria could reproduce. This action leads the drop of a quarter of the initial mother energy before the reproduction. The reproduction results in the born of four new individuals, that have the quarter of the progenitor energy. In fact, the bacterial reproduction generates just one individual. In this way, for each tick, approximately two cellular divisions are made, and in the end of the tick, five bacteria are alive.

The internal energy was modelled starting of the metabolism, that following the rule: the internal energy from the substrate is stored in function of the metabolism efficiency of the bacteria. In this way, it is dependent of the bacteria capability to absorb nourishment, and of its concentration, that the bacteria fulfill its internal material balance.

Therefore, the simplified model of biofilm growth is reached. The input quantities were normalized from 0 to 100%, and the internal energy quantities and the substrate concentrations vary from zero to saturation. In the case of internal energy equals to zero, the cell dies.

SIMULATED SCENARIOS AND RESULTS

The Table 1 presents the values of variables to simulate two scenarios of tests. The number of bacteria and the metabolism energy do not change in the two scenarios. In the first scenario, we have changed the substrate concentration. In the second scenario, we have changed the perceptual of efficiency.

Scenario Test 1: Substrate Concentration Variation Simulation

For the first scenario test, we have varied the microbial growing from the initial quantity until the steady state. With concentrations from 0 to 100%, with variations of 10% and efficiency with a constant value of 50%.

The Figure 5 shows the population growth in function of the nutrient availability. A pattern is evidenced, with aggregates and steady state values. This occurs because the reproduction of cells and the adhesion of new cells around the progenitor. In A and B, part of the cathode is empty, but in D is shown a full surface occupation. This figure (5-D) shows the physical limit of energy production of the system, since just the cells that are in the biofilm will have suitable conditions to flow its electrons to the anode and generate used energy.

The results of population versus time shows an exponential bacterial growth and exponential decay of substrate concentration. It is possible to view noise in the model and the growth and stabilization phases. When we vary the substrate concentration, there is no changes in the stabilization time of the system. This shows that raising the substrate the cell develops a bigger growing rate to, in the global result, the substrate metabolization reach the steady state in the same time. Finally, it is possible to evaluate the behavior of the agents in function of the Monod equation, comparing the growth and the real data values.

In the the simulation presented in Figure 5-D, the concentration value were set at 40%, evidencing the elevation of the rest in concentration, not metabolized. This is due the fact of the system is fed in a continuous form, as the Figure 6 presents (the behavior of the cells).

Figure 6 evidences the bacterial growth in the increase of substrate. The response of this increase is instantaneous. However, there is a time delay of the reproduction to reach a new steady state. This behavior varies in a microorganism to other, basing to an adaptation time, a new lag, to the new environment

Table 1. Test variables and variations

Variable	Description	Test 1	Test 2
substrateQt	Substrate concentration (%)	10, 20, 30, 40, 50, 60, 70, 80, 90, 100	20
efficiency	Metabolic Eficiency(%)	50	25, 50, 75, 100
numberbacterias	Initial bactéria population	25	25
metabolismenergy	Initial internal Energy (%)	50	50

Figure 5. Scenario Test 1- Substrate concentration 10%(A), 20%(B), 30%(C) and 40%(D)

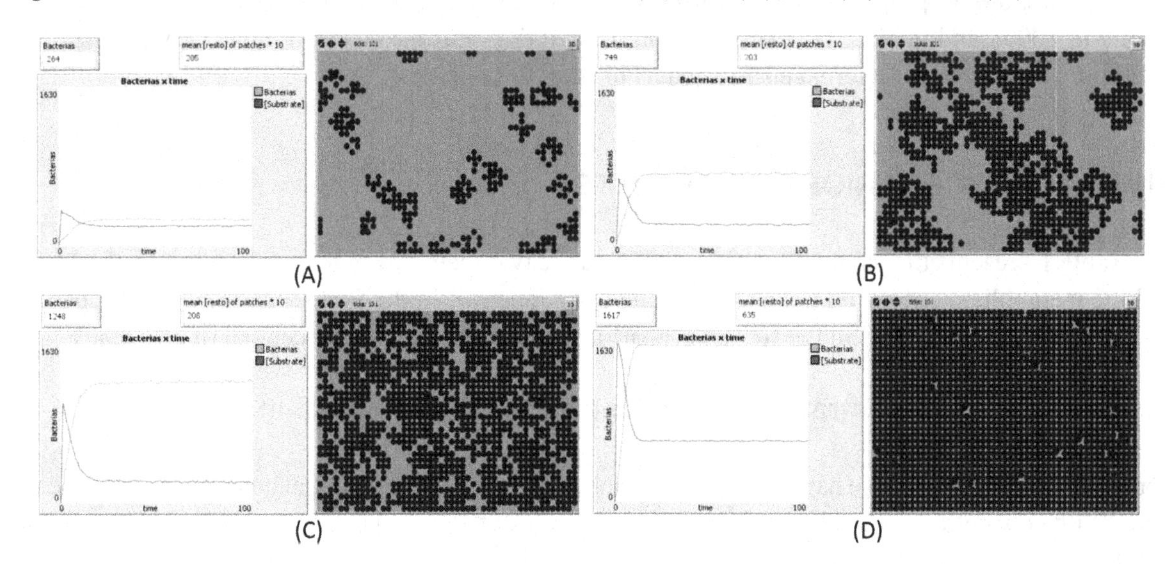

Figure 6. Test - Continuous substrate variation: 15%(1), 30%(2), 45%(3), 60%(4), 75%(5), 90%(6)

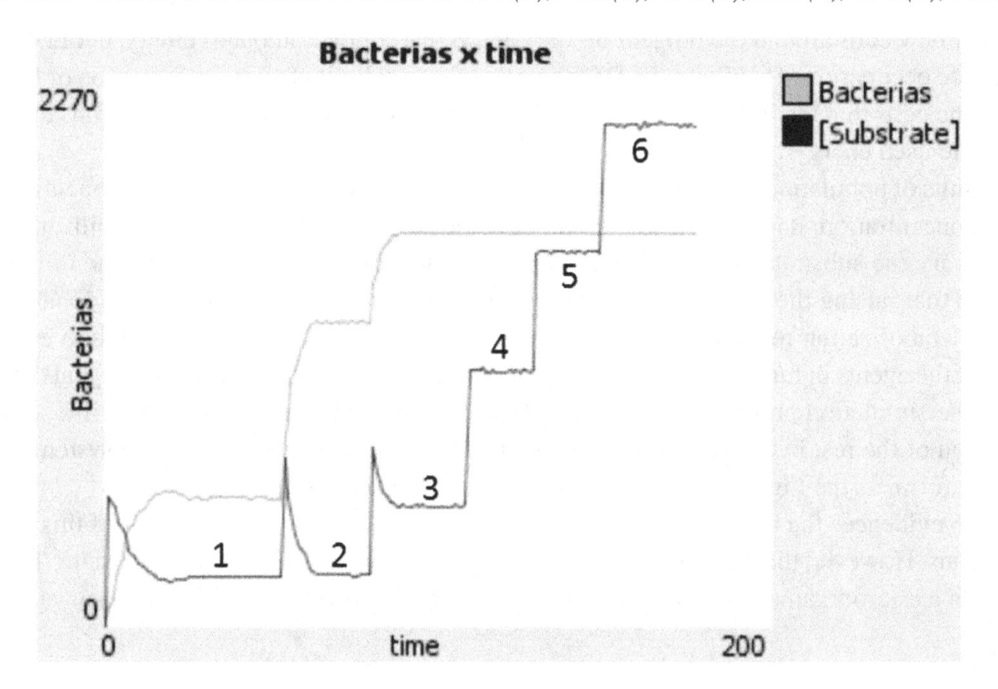

conditions. In the third change (Figure 6-3) it is possible to visualize the saturation trend, because the increase of the residual substrate (the substrate concentration in solution). From the fourth to the sixth changes, the increase of substrate do not result in the increase of biofilm population, fact explained by the maximum possible population on the anode that is limited by its surface area.

Scenario Test 2: Metabolic Efficiency variation simulation

In this scenario test, we havechanged the metabolic efficiency of bacteria for a constant substrate feed and evaluated the impact in the biofilm population. We have used a constant concentration of substrate with 20%, and changes of 25% in efficiency (see Figure 7).

Over again, it is possible to evaluate the biofilm growth through the increase of metabolic efficiency. Meanwhile, apparently, for the same period, the cells are dispersed. Over the efficiency increase, there is a stabilization time difference between the scenario tests. This occur because of the necessary time to the bacteria reach the needed internal energy for its reproductions. This parameter, as exposed in the Monod equation, is called "Monod constant". This scenario test shows a relationship between the bacteria metabolic efficiency with this parameter of the analytical model.

The Figure 8 present the reduction of residual substrate concentration after the equilibrium. There was the variability increase in bacterial population in steady state. This fact that apparently did not occurred in the tests with the concentration changes. The test indicate that, for the same substrate concentration, different bacterial concentrations and different final substrate concentration are reached. The oscillation frequency in the population values is bigger in the values (3) and (4). This implies that, with higher efficiencies, the system become noisy.

Figure 7. Test 2- Efficiency variation: 25%(A), 50%(B), 75%(C) and 100%(D)

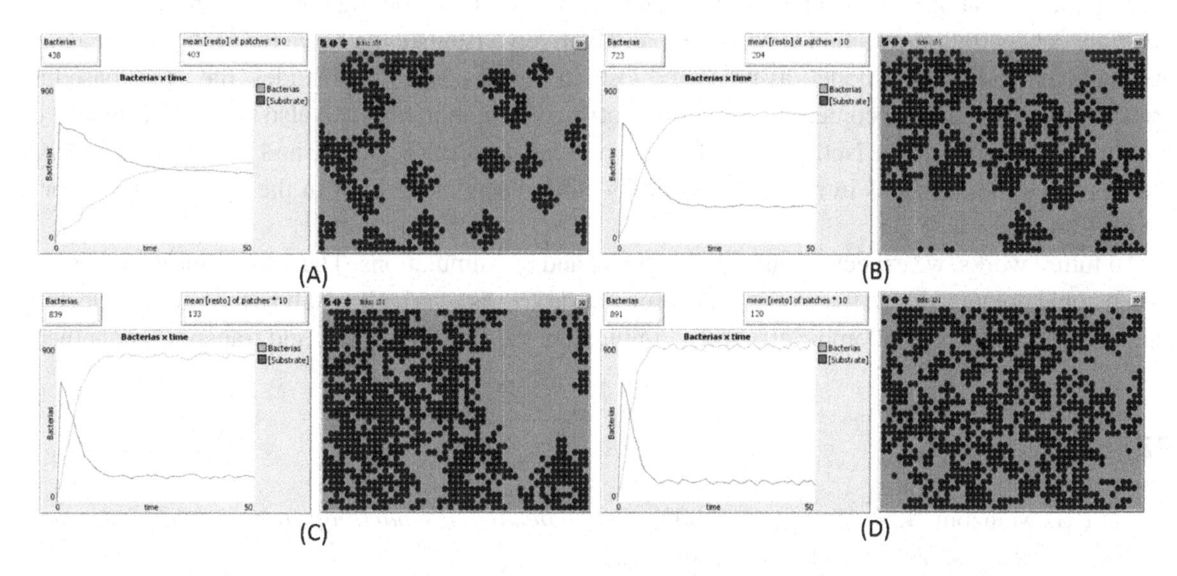

Figure 8. Test 2- Continuous efficiency variation: 25%(1), 50%(2), 75%(3) and 100%(4)

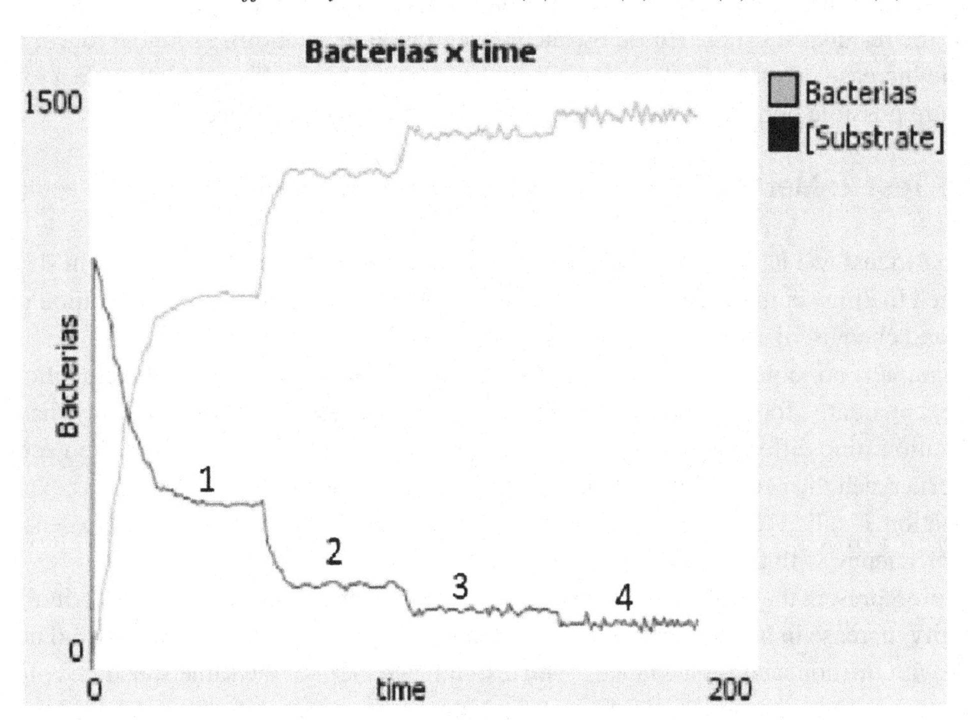

CONCLUSION

The main goal of this chapter was, through simple abstractions and basic bacterial behavior programming, to prove that agent-based models are suitable tools in the modeling the MFC problem. Over the substrate concentration variation is possible to qualitatively visualize the behavior of the model against the mathematical Monod model (as presented tests in Figures 3 and 5). Besides, the agent-based approach provides valuable geometric and spatial biofilm information, that can also be verified against the biofilm micrograph and the NetLogo interface (as presented tests in Figures 2 and 7). This characteristic can answer some questions in researches, thereby could bring some light in the dynamic behavior of the biofilm.

To future works, we expect to improve the model and our simulations. The main improvements are a dimensional organization of the model; use the electrode size as a variable; analyze the variable electrons balance; exploring the coulombic efficiency; implement the fluid dynamics and transport phenomena.

REFERENCES

Bandini, S., Manzoni, S., & Vizzari, G. (2009). *Crowd behaviour modelling: from cellular automata to multi-agent systems*.Taylor and Francis Group, LLC.

Boghanim, H. C., Kim, J. R., Dinsdale, R. M., Guwy, A. J., & Premier, G. C. (2013). *Analysis of the dynamic performance of a microbial fuel cell using a system identification approach. Journal of Power Sources*.

Byrne, H., & Drasdo, D. (2008). *Individual-based and continuum models ofgrowing cell populations: a comparison. Journal of Mathematical Biology.*

Calder, M. A. (n.d.). *Modelling of a Microbial Fuel Cell* (Master thesis). Norwegian University of Science and Technology.

Esser, D. S., Leveau, J. H. J., & Meyer, K. M. (2015). *Modelling microbial growth and dynamics. In Applied Microbiology Biotechnology.* Springer-Verlag Berlin Heidelberg.

Gruner, S. (2010). *Mobile agents systems and cellular automata. Autonomous, Agent Multi-agent System Journal.*

Huang, H., Feng, C., Pan X., Wu, H., Ren, Y., Wu, C., & Wei, C. (2013). Thiocyanate Oxidation by Coculture from a Coke Wastewater Treatment Plant. *Journal of Biomaterials and Nanobiotechnology, 4*(2A).

Instituto Brasileiro de Geografia e Estatística (IBGE). (2008). Pesquisa Nacional de Saneamento Básico. Rio de Janeiro: IBGE.

Laspidou, C. S., Kungolos, A., & Samaras, P. (2010). *Cellular-automatta and individual-based approaches for the modelling of biofilm structures: Pros and cons. Desalination Journal.*

Logan, B. E. (2008). *Microbial Fuel Cells.* John Wiley & Sons.

Maria, A. (1997) Introduction to modeling and simulation.*Proceedings of the 1997 Winter Simulation Conference.*

Metcalf & Eddy. (2003). *Wastewater Engineering treatment Disposal Reuse.* New York: McGraw - Hill Book.

Ministério de Minas e Energia. (2015). *Resenha Energética Brasileira – Exercício de 2014.* Retrieved from http://www.mme.gov.br/web/guest/publicacoes-e-indicadores Acessed in 15/03/2015

Monod, J. (1949). The growth of bacterial cultures. *Annual Review of Microbiology, 3*(1), 371–394. doi:10.1146/annurev.mi.03.100149.002103

Piccioreanu, C., Loosdrecht, M. C. M., & Heijnen, J. J. (n.d.). Mathematical modeling of biofilm structure with a hybrid differential-discrete cellular automaton approach. *Biotechnology and Bioengineering, 58.* PMID:10099266

Pizarro, G. E., Teixeira, J., Sepúlveda, M., & Noguera, D. R. (2005). Bitwise implementation of a two-dimensional cellular automata biofilm model. *Journal of Computing in Civil Engineering, 19*(3), 258–268. doi:10.1061/(ASCE)0887-3801(2005)19:3(258)

Potter, M. C. (1911). Electrical effects accompanying the decomposition of organic compounds. *Proceedings of the Royal Society of London. Series B, Containing Papers of a Biological Character, 84*(571), 260–276. doi:10.1098/rspb.1911.0073

Rahimnejad, M., Adhami, A., Darvari, S., Zirepour, A., & Oh, S. (2015). Microbial fuel cell as new technology for bioelectricity generation: A review. *Alexandria Engineering Journal, 54*(3), 745–756. doi:10.1016/j.aej.2015.03.031

Rittmann, B. E., & McCarty, P. L. (2001). *Environmental Biotechnology: Principles and Applications*. New York: McGraw-Hill Book Co.

Walker, D. C., & Southgate, J. (2009). The virtual cell – a candidate co-ordinator for 'middle-out' modelling of biological systems. Briefings in Bioinformatics, 10(4), 450-461.

Wilensky, U. (n.d.). *The NetLogo 5.3.1 User Manual*. Retrieved from https://ccl.northwestern.edu/netlogo/

ENDNOTE

1 OECD - Organization for Economic Co-operation and Development - Composed by 34 countries: Germany, Australia, Austria, Belgium, Canada, Chile, South Korea, Denmark, Slovenia, Spain, United States, Estonia, Finland, France, Greece, Holland, Hungary, Iceland, Ireland, Israel, Italy, Japan, Luxembourg, Mexico, Norway, New Zealand, Poland, Portugal, United Kingdom, Slovakia, Czech Republic, Switzerland, Sweden and Turkey .

Chapter 10
Use SUMO Simulator for the Determination of Light Times in Order to Reduce Pollution:
A Case Study in the City Center of Rio Grande, Brazil

Míriam Blank Born
Universidade Federal do Rio Grande (FURG), Brazil

Diana Francisca Adamatti
Universidade Federal do Rio Grande (FURG), Brazil

Marilton Sanchotene de Aguiar
Universidade Federal de Pelotas (UFPel), Brazil

Weslen Schiavon de Souza
Universidade Federal de Pelotas (UFPel), Brazil

ABSTRACT

Nowadays, urban mobility and air quality issues are prominent, due to the heavy traffic of vehicles and the emission of pollutants dissipated in the atmosphere. In the literature, a model of optimal control of traffic lights using Genetic Algorithms (GA) has been proposed. These algorithms have been introduced in the context of control traffic. In order to search for possible solutions to the problems of traffic lights in major urban centers. Thus, the study of the dispersion of pollutants and Genetic Algorithms with simulations performed in Urban Mobility Simulator SUMO (Simulation of Urban Mobility), seek satisfactory solutions to such problems. The AG uses the crossing of chromosomes, in this case the times of the traffic lights, featuring the finest green light times and the sum of each of the pollutants each simulation cycle. The simulations were performed and the results compared analyzes showed that the use of the genetic algorithm is very promising in this context.

DOI: 10.4018/978-1-5225-1756-6.ch010

INTRODUCTION

Currently, the vehicle fleet in major urban centers grows increasingly every year, according to the Brazilian National Traffic Department (DENATRAN), causing traffic problems for drivers, pedestrians and the environment. Besides, about 500,000 pedestrians are killed at intersections traffic lights in China and Spain (TURKY; AHMAD; YUSOFF, 2009). However, pedestrian mobility problem affects cities around the world, causing the population become vulnerable along the walk activity. In the conventional traffic lights controllers, the lights change at constant cycle time, and these systems calculates the cycle time based on the average cargo traffic, disregarding the natural dynamics of it, aggravating congestion and contributing immeasurably to the dispersion of pollutants in the atmosphere.

In this research area, this type of solution is clearly not the optimal solution. Additionally, performance is need and the Genetic Algorithms (GAs) obtained promising results. The GAs proved to be an interesting heuristic with fast and simple implementation. However, the study of pollutant dispersion in the vehicle traffic context in addition to the control of traffic lights becomes extremely important, therefore justifying the purpose of our work. Thus new techniques and intelligent models of traffic lights systems can improve the flow of both vehicles and pedestrians, so that the transit could be feasible at all users and reduce the emission and dispersion of pollutants into the environment improving the air quality.

Thus, the main goal of this work is to develop a GA, which simulates the joints from the city center of Rio Grande - Brazil, where vehicle traffic is intense, as the municipality's population has a high growth due to the concentration of companies of the naval area. The SUMO simulator (Simulation of Urban Mobility) (KRAJZEWICZ et al, 2012) will be used for the simulation scenarios, as well as an analysis of pollutant dispersion by controlling traffic lights in this particular area. SUMO is a microscopic simulation of traffic that was developed in 2001 by the German Aerospace Center (DLR), in order to assist the vehicle traffic research community with a tool where algorithms could be implemented and evaluated, without the need for get a complete traffic simulation. It is open source, licensed under the GPL (General Public License), portable and designed to simulate modeling of large road networks. In recent years, SUMO toolkit achieved considerable progress in traffic modeling including: tools to import route networks in different formats; tools to generate demand and routing; and, a high-performance simulator. The environment of SUMO simulator is a resource that can be seen as a multi-agent system and vehicles and traffic lights included in this are the implementation of agents. Therefore, this work has relevance to society, searching the integration of multi-agent systems with evolutionary algorithms in order to manage traffic lights and to reduce the pollution concentration, with Rio Grande/Brazil as a case study.

BACKGROUND

At 60's John Holland and students from the University of Michigan created Genetic Algorithms (GAs) The purpose of Holland was to study the phenomenon of "evolution" and play it somehow in computing (AGUIAR, 1998). According to (GOLDBERG, 1989) it was possible to get a computer version of the process of evolution and that it would be able to solve similar problems to the evolution of characteristics.

According to Holland work, the system was composed of a bit string (0's and 1's), called individuals. These individuals evolved to find a better chromosome that meets a specific problem. The solution was found automatically and unsupervised way and the information was provided in the settings of each chromosome.

The natural evolution explored by Genetic Algorithms, in this proposed system, was obtained based on chromosome ability to analyze and match the best result. That is, the GAs were created based on genetic processes of living organisms reproducing what happens in real life, seeking to solve problems of high computational complexity in the development environment (AGUIAR & TOSCANI, 1997).

In each chromosome of individuals is encoded knowledge themselves. There are reproductive mechanisms that alter these training, among which the most used are: Mutation. The mutation operator is required for the introduction and maintenance of genetic diversity of the population analyzed. Inversion is the inversion on chromosome code, and the Crossing or Crossover performs an exchange with the genetic material of chromosomes generators.

Considering the principle of evolution, through which the most appropriate individuals have a greater chance of survival, the crossover greatly increases the chances of individuals possessing ideal characteristics, able to perpetuate itself in the process, since the level of adaptation to the environment is utmost importance to playing more frequently (AGUIAR, 1998; REZENDE, 2003). Among some characteristics of the GAs can highlight, according to (LINDEN, 2008):

- **Parallelism:** The population of solutions is assessed simultaneously;
- **Global:** One of the most important features of the GAs is it, as these algorithms not only use local information, ie, they are not tied to maximum local. For this reason, it is considered a very appropriate technical solution in search for complex and real problems;
- **Randomness:** Genetic algorithms are not considered totally random, considering that they have random components because they use the public information in question to determine the next search condition;
- **Discontinuities in the Function:** GAs do not use derived from information in their evolution and even neighborhood information to perform your search. It is suitable for functions with discontinuities or those who can not calculate a derived;
- **Discrete and Continuous Functions:** These algorithms have the ability to deal with real functions, discrete, Boolean and categorical (non-numeric), and such functions are mixed without any damage to the resolution of capacity problems by the Gas.

In the literature, several studies using the techniques of Artificial Intelligence (AI), as Multi-Agent Systems (MAS) and Evolutionary Computation (EC), in the context of natural resource management, because the strong influence on the population's quality of life. The dispersion of pollutants is studied, especially in large urban and industrial centers, such as the city of Rio Grande (Brazil), where there are heavy pollutant emission sources and vehicle traffic (NUNES, 2013).

In the work of Silva and Mandredini (2010) can be seen the application of AI in the control of road traffic. The objective of this project was the use of concepts of genetic algorithms (GA), along with the Tango traffic simulator to optimize the times of the traffic lights and thus improve vehicle traffic in the city of São Paulo (Brazil).

In Heinen et al (2013) was proposed to control traffic lights using Artificial Neural Networks with Radial Basis Function, which consists of the use of existing resources in the current traffic system, seeking to improve the performance of the traffic. They used the concepts of Reinforcement Learning (AR), Artificial Neural Networks (ANN) and Multi-Agent Systems (MAS).

According to Retore et al (2006), the use of MAS in urban traffic control becomes quite feasible. Their paper proposes the use of the theory of Distributed Artificial Intelligence (IAD) to analyze the relationship between the opening order of traffic lights of an intersection and the average waiting time by the drivers. The results obtained from the simulation of the proposed scenarios can analyze the degree of impact that seemingly simple choices can result in a traffic system. According to the simulated scenarios, two lights were set to open at the same time, causing the average waiting time of vehicles were reduced considerably. Thus, it can be said that the replication of this configuration in real traffic situations brings significant advantages to users of traffic lanes.

In the work of Melo (2005) was promoted the mobility and accessibility of pedestrians in urban areas considering the traffic environment and the components that are part of the same as people, roads, vehicles, planning of public policies, structural, operational and legislation. Thereby identifying the problems related accidents with pedestrians and develop a set of actions to contribute to reducing the number of pedestrian accidents in the same transit routes in the city center of Fortaleza (Brazil).

The work of Turky, Ahmad and Yusoff (2009) is intended to provide an optimized traffic control study using GA. It consists in four-way junction with two lanes and a pedestrian crossing. Initially, the authors created a project template for light, which were five sensors in the track, four of these detect the number of vehicles on the track and the fifth row of pedestrians. This system calculates the time a vehicle takes to travel a route with the source and a destination D. The model has input and output variables to compose the algorithm and it is simulated in two modes: static and dynamic.

To design the vehicle driving rules used an Automaton Cellular (AC), as this enables represent significant events that occur during congestion, traffic stoppage, movement of return and so on. An algorithm was developed to represent these rules and a model of the genetic algorithm.

Another work recently developed by Dessbesell (2015) presents where intelligent agents were employed in traffic control study in a multi-agent simulation environment, in order to reduce the problems caused by congestion, the case study occurred in a region of Porto Alegre (Brazil).

For the development work, the author used the SUMO traffic simulator; real data on the traffic signal of local programming and the volume of some crossings specific area traffic. In this context, an intelligent traffic light was developed, which applies a heuristic in the form of control algorithm for decision-making. The results obtained from the simulations were compared with and without the intelligent traffic, where the results were considered significant.

About the related works cited, we can conclude that the techniques mentioned bring many benefits to the study of the control of road traffic lights. Thus, our proposal was interdisciplinary, searching techniques and methods that help in the study of natural resources such as the dispersion of pollutants. Note the importance of this study as every year, in large urban centers, pollution becomes a concern of society. The vehicle traffic contributes significantly to the increase of this pollution, this study developed a genetic algorithm that seeks to assist in the management of traffic lights, aiming to reduce the dispersion of pollutants in the downtown area of the city of Rio Grande (Brazil).

SUMO (Simulation of Urban Mobility) is a microscopic simulation of traffic, which was developed in 2001 by the German Aerospace Center (DLR), in order to assist the vehicle traffic research community with a tool where algorithms could be implemented and this, without the need for a comprehensive traffic simulation. It has open source and licensed under the GPL (General Public License), portable and designed to simulate modeling of large road networks.

In recent years, SUMO achieved considerable progress in its utilities for modeling traffic, such as roads network importer with reading ability in different formats, demand generation utilities and routing and high-performance simulation (KRAJZEWICZ et al, 2012) and the application simulation environment is a resource that can be seen as a multi-agent system where vehicles and traffic lights included in this are the agents of this application.

Among the main features of SUMO include:

- It has all the applications needed to simulate a traffic network,
- Different types of vehicles,
- Vehicle movements occur in continuous space and discrete time,
- Streets with multiple lanes and lane change,
- GUI for users,
- Speed of execution in the simulations, and
- Interoperability with other applications at runtime, among others.

The software provides a pollutant emission library set based on the HBEFA database (Handbook Emission Factors for Road Transport) for all categories of vehicles. The pollutants are inserted in the simulator CO2 (carbon dioxide), CO (carbon monoxide), HC (hydrocarbons), NOx (nitrogen oxide), PMx (Particles or aerosols suspended in the air) and fuel consumption, as shown in Figure 1.

The tool, in addition to performing the simulation of traffic from any region also provides a set of applications that assist in implementation of these simulations in various traffic situations. The SUMO applications are divided into: Network Generation, Generation of Vehicles and routes, and Simulation (KRAJZEWICZ, et al, 2012).

Figure 1. Graphical representation the SUMO and of the pollutants for each vehicle
Source: SUMO, 2015

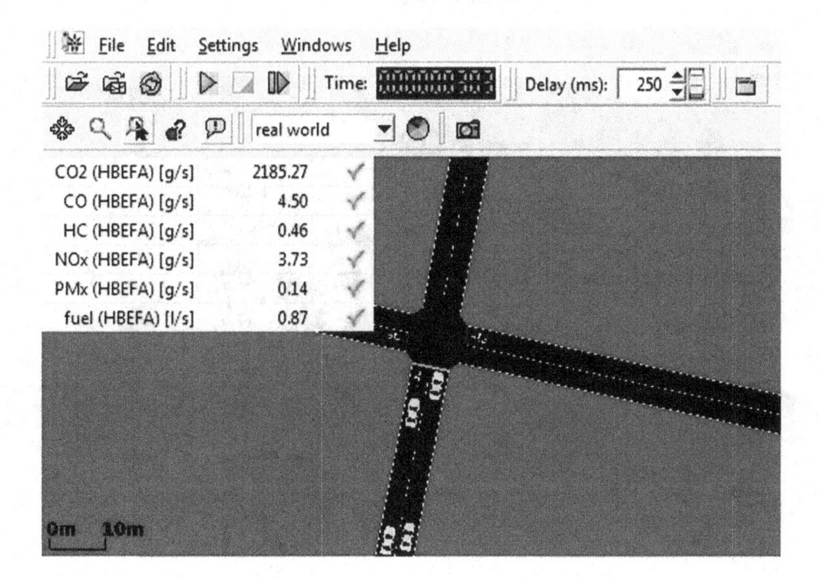

PROPOSED MODEL

For the development of this proposal, we prospected related work to traffic light control and dispersion pollutant. From this detailed analysis, we observe the need to assess the amount of dispersion of pollutants emitted by vehicles.

Thus, the management of traffic lights, aided by the use of GAs aimed at reducing pollution as a contribution to vehicle traffic problem and the emission of pollutants. The chosen area analyzed in this case study were the main streets of the city center of Rio Grande/Brazil, as shown in Figure 2, because the vehicle traffic is intense, as the municipality's population have a high growth due to the concentration of companies of the naval area.

The methodology used in this study is divided into three stages, which are shown as follows:

- **Stage 1:** Generation of data in SUMO and implementation of GA, understood in the sections:
    - Generate simulation in SUMO;
    - Capture the time of traffic lights in SUMO;
    - Set the time of traffic lights at the GA and
    - Perform the GA operations and choose individuals with better fitness.
- **Stage 2:** Simulations, divided into sections:
    - The input file with the best individuals;
    - Generate new simulations in SUMO; and
    - Comparing the levels of pollution.
- **Stage 3:** Results Analysis: this step the results obtained from the simulations developed in the previous steps are showed.

Figure 2. Graphical representation of the map of the city of Rio Grande (Brazil)

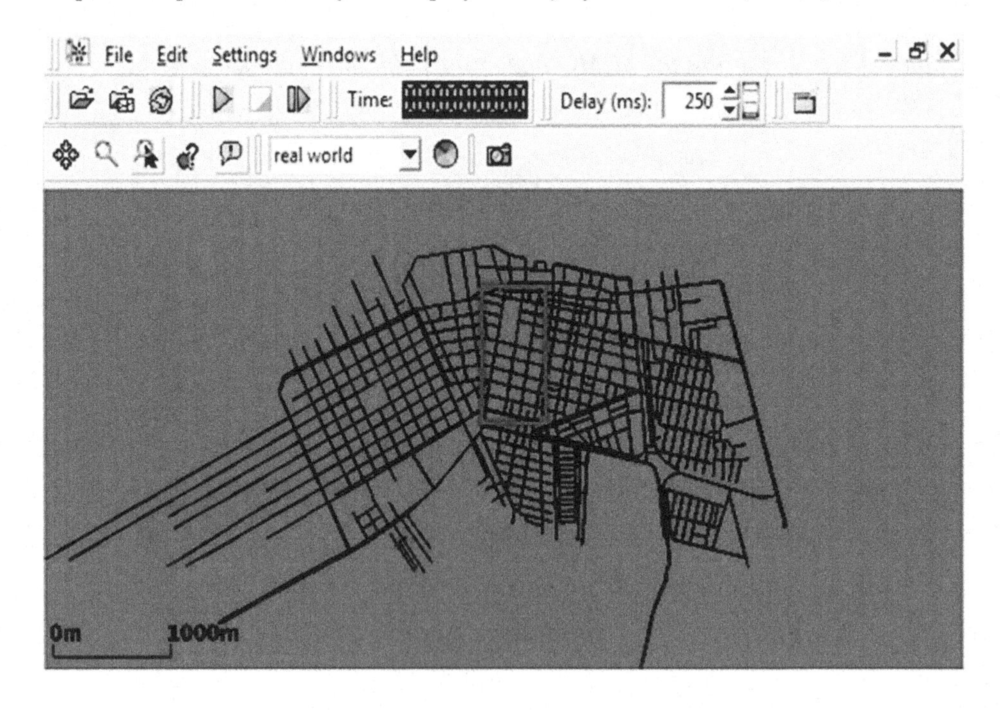

The genetic algorithm developed for this case study is shown in Figure 3. In the Figure 4, it is the flowchart representing the operation of the genetic algorithm developed for this study.

The GA developed for this study uses two classes: Chromosome and Cell. The Chromosome class has the vector attribute of type Boolean and two attributes (timeGreenLight and timeOfCycle), both with integer data type. After, classic methods the following is defined:

Figure 3. Functional flowchart proposal

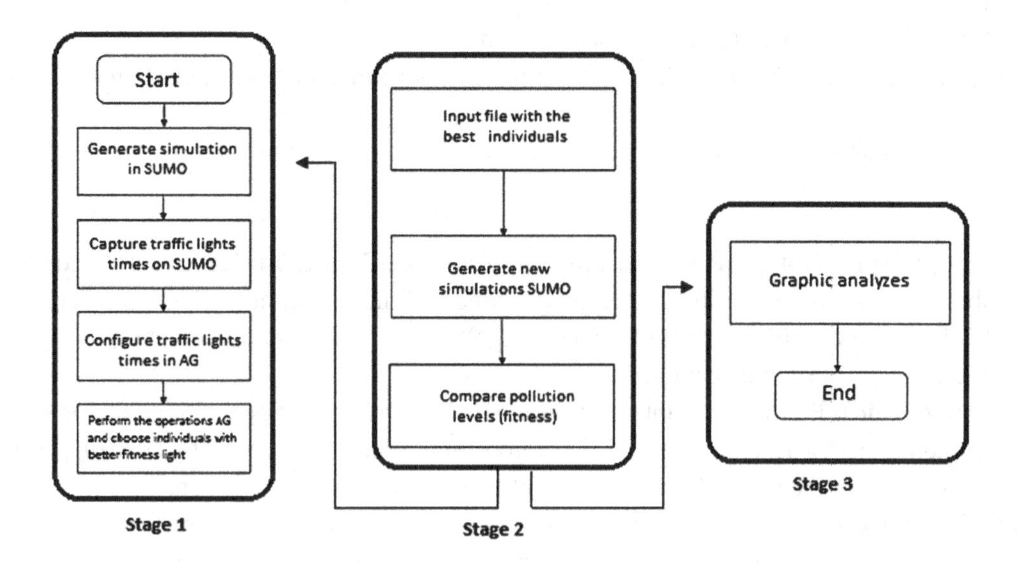

Figure 4. AG functional flowchart

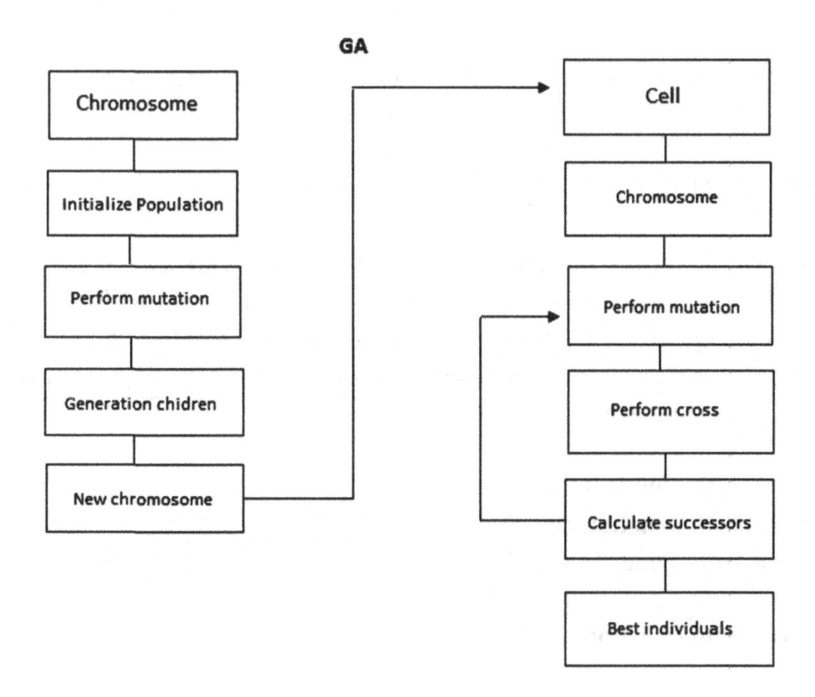

- **FillsTime (Void):** Initially fills time in vector as random;
- **Mute (Void):** This method performs operations mutations in random bits of the chromosome, wherein and chosen randomly one position vector, with the even boolean which receives one (1) to green light and zero (0) to red;
- **Generatesson (Chromosome DadA, Chromosome DadB) (Void):** Transforms the chromosome one of the children, passed as parameter of function;
- **countTimeOfGreen (Void)**: Sum of the total time of green traffic light seen that the same change the operations mutation;
- **getTimeGreenLight (Int):** This method get green signal time;
- **getTimeOfCycle (Int):** Method get the cycle time;
- **Get Chromosome (Boolean):** Returns the new chromosome with operations de mutation applied.

The cell class has a vector attribute that is a list of Chromosome type and its classic methods are constituted by:

- **addCromossomo (int numChromosome, int timeGreenLight, int timeCycle):** Add the desired number of chromosomes, with the green light time and the total time of the light cycle;
- **muteChromosome (int mutationOfPercentage):** Carries the mutation on chromosomes changing the cell with certain percentage;
- **Crossing (Void):** Performs the intersection of all chromosomes of generating cell his successors. This function has two son from A and B chromosomes which are formed by crossing the first half of the A dad with the first half B dad, and so forth to generate new children.
- **calculatesSuccessors (int numberOfSucessors) (Void):** Shows a file time green signal for ten (10) traffic light, who obtained less polution for certain time setting themselves. It is important to note that in this sum of pollutions are considered the following pollutants: CO, CO2, HC, NOx and PMX.

The Figure 4 is the GA flowchart proposed, as well as its classes and attributes.

RESULTS AND ANALISYS

For this case study was performed one hundred (100) simulations for each mutation value (10, 20, 30, 40 and 50), automatically, and in each simulation these SUMO performs 100 time steps (cycles), where the vehicles run through the main streets of the city center of Rio Grande / RS.

The results obtained in this study considered for analysis of genetic algorithm behavior, different values of mutation (rates) to perform the simulations:

- 100 simulations with mutation rate 10;
- 100 simulations with mutation rate 20;
- 100 simulations with mutation rate 30;
- 100 simulations with a mutation rate 40; and
- 100 simulations with mutation rate 50.

With mutation rates cited above, seek to verify the times of the traffic light and the total pollution charges (sum of pollutants: CO, CO2, HC, NOx and PMX). It is changed during the simulations without stabilization.

The simulations were carried out in an automated way for this process, and the configuration of the traffic lights every generation simulation corresponds to the files generated by SUMO tool. All files generated by the simulator provide the input settings for the GA. Furthermore, the execution of each generation time (involving 100 time steps/SUMO cycles) and a certain processing time demand (between 11 and 16 minutes each).

In Figures 5 and 6 shows the data generated from the simulations, with the number of generations, mutation rates applied and the amount of pollution. It is observed that in all mutation rates between generation 45 and 49, there 'peaks' in the values (SUMO) pollutions and earlier and later generations these pollution values stabilize.

In Figures 7, 8, and 9 it is noted that the simulations did not achieve significant peaks as in the previous simulations, pollution values for each mutation rates 30, 40 and 50 obtained total pollution values and approximate, and that simulations mutation rates 40 and 10 had the same total amount of pollution, representing the lower rates according to the tests.

Table 1 represents, for each ten (10) past traffic lights respective times in each mutation rate (10, 20, 30, 40 and 50) and lastly, Table 2 summarizes the total pollution values in each simulation with corresponding mutation rates.

In this way, we conclude that with the process automation and a greater number of simulations AG can stabilize showing what might be considered a good heuristic for this case study.

Thus, it is concluded that, with automation of the process and a greater number of simulations can stabilize AG showing that the same can be considered a good heuristic for this case study. Considering a simulation, only with SUMO, with 100 (cycles) generated is obtained at the end a total of 281,322 of the sum of the pollutants, performing the multiplication of this value by 100 (generation considered in this

Figure 5. Total pollution values for all the settings mutation - Mutation 10 and 100 generations

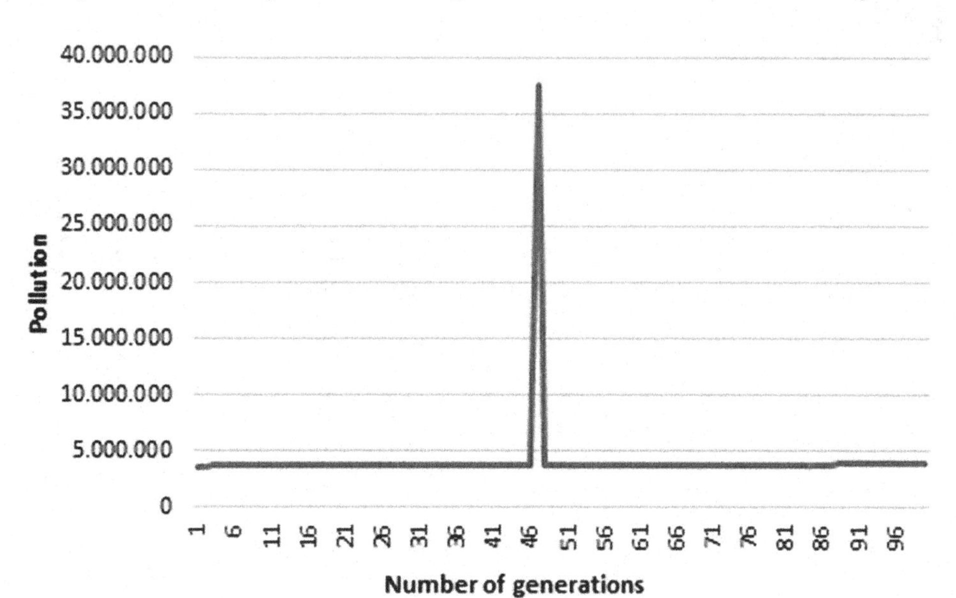

Figure 6. Total pollution values for all the settings mutation - Mutation 20 and 100 generations

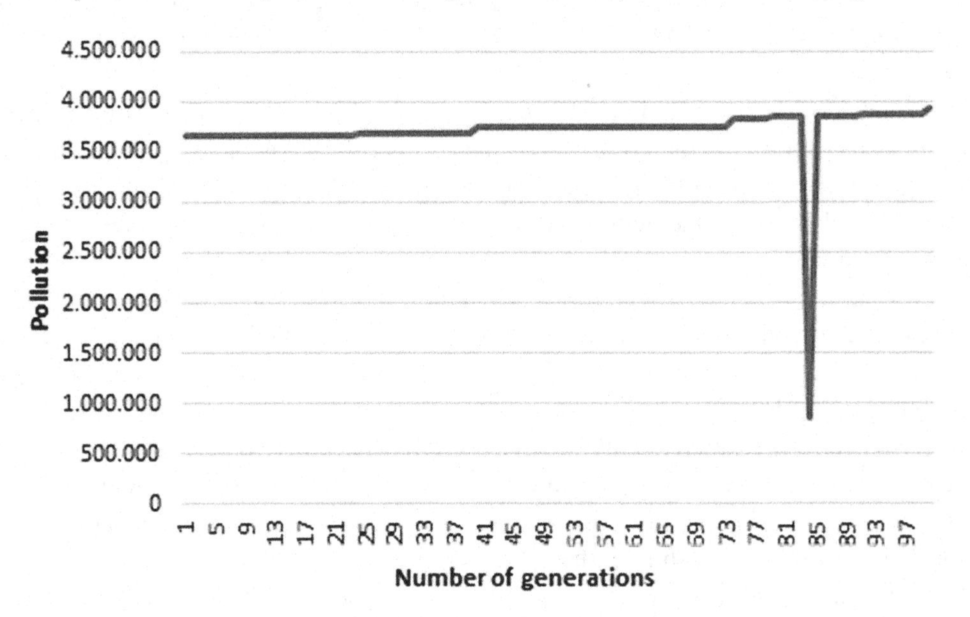

Figure 7. Total pollution values for all the settings mutation - Mutation 30 and 100 generations

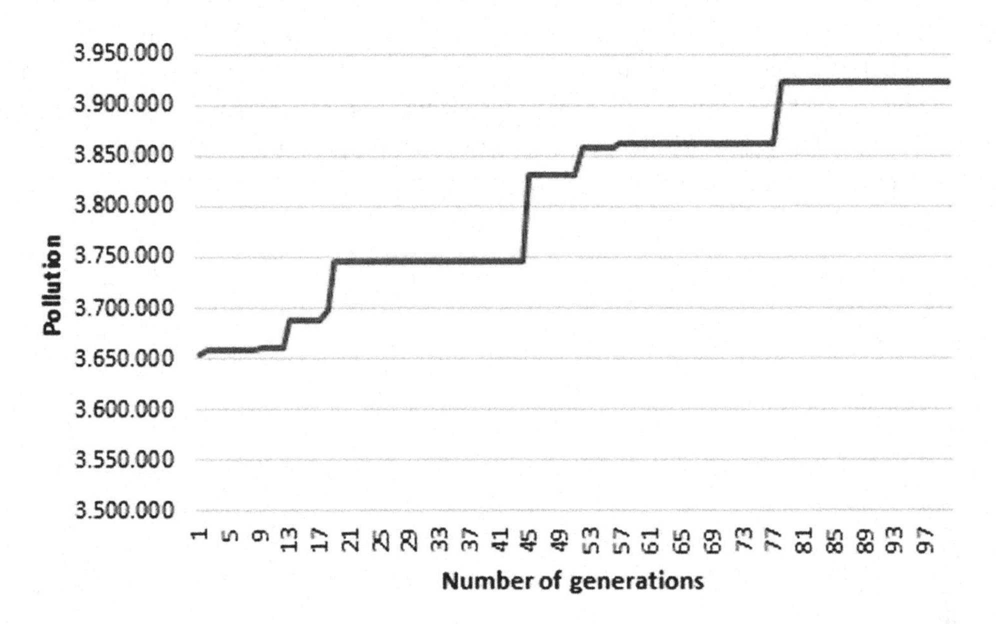

Figure 8. Total pollution values for all the settings mutation - Mutation 40 and 100 generations

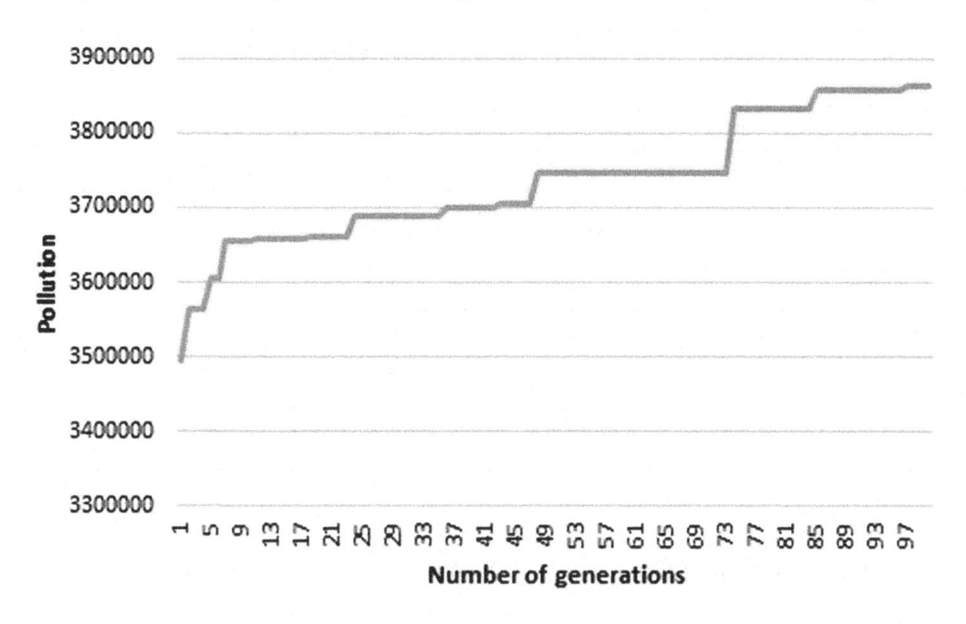

Figure 9. Total pollution values for all the settings mutation - Mutation 50 and 100 generations

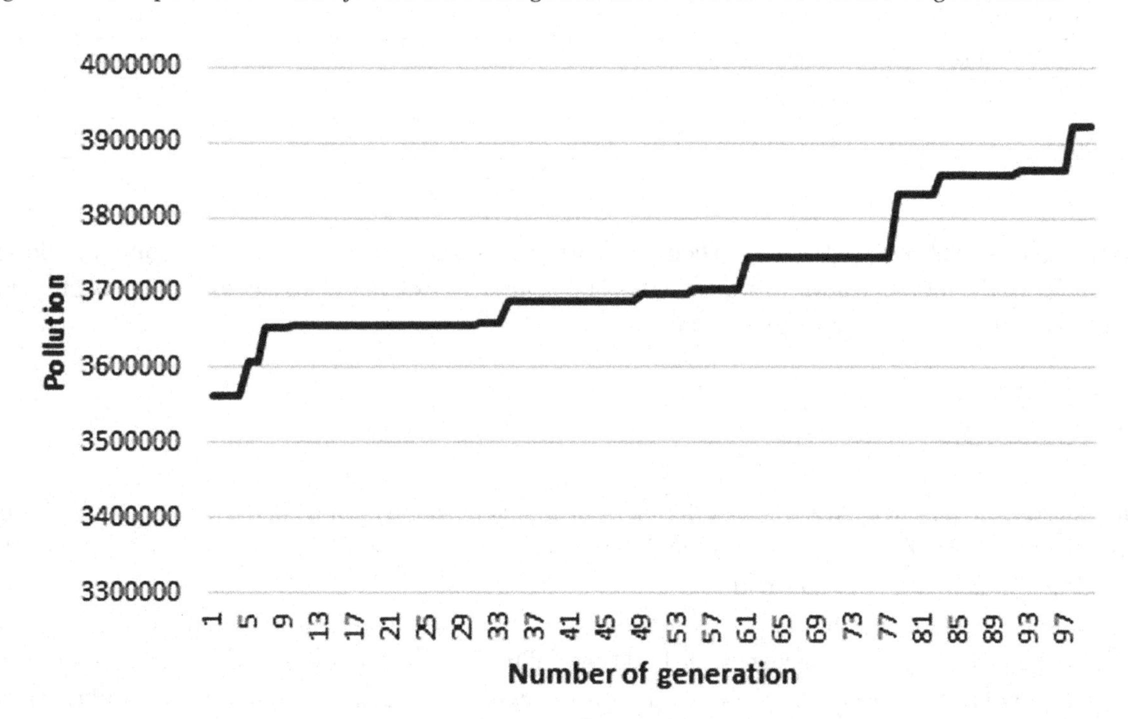

Table 1. Times of each traffic lightsl (10) for each of the mutations

Traffic Lights	Mutation 10	Mutation 20	Mutation 30	Mutation 40	Mutation 50
S0	29	26	28	29	29
S1	29	26	28	29	26
S2	29	26	28	29	26
S3	29	26	28	29	26
S4	29	26	28	29	26
S5	29	26	28	29	26
S6	29	26	28	26	26
S7	29	26	28	26	28
S8	26	26	28	26	28
S9	26	28	28	26	28

Table 2. Total pollution for each of the simulations with the respective changing rates

	Total Pollution
Simulation 1 – 100 generations of 10 mutation	3.862.363
Simulation 2 – 100 generations of 20 mutation	3.922.451
Simulation 2 – 100 generations of 30 mutation	3.922.451
Simulation 4 – 100 generations of 40 mutation	3.862.363
Simulation 5 – 100 generations of 50 mutation	3.922.451

study) obtains a value of 28,092,219. Then, comparing these data with the greatest amount of pollution generated by the reorganization of traffic lights and heuristics AG, has been a considerable decrease in pollution of the region targeted for this study.

CONCLUSION

The purpose of this work was the development of a genetic algorithm to help manage traffic lights, through the reorganization of the same, to minimize the dispersion of pollutants in the city center of Rio Grande / RS, using the SUMO simulator.

GA implemented has mutation and selection operations to choose of the best individuals. The data generated by these operations generate the best times of traffic lights. For this study, the best individuals are considered those who have the lowest values of fitness function, and this function is calculated from the sum of the pollutants emitted by the vehicle simulation. We hope that a reduction of pollution in the study area, aimed at automating the process with reasonable computational cost.

The proposal presented here was inspired by the work of Turky; AHMAD; Yusoff (2009), where they used GA for managing 1 (one) signal in a hypothetical scenario. However, in our study, we have used ten (10) lights in a real situation (the center of city Rio Grande (Brazil). Furthermore, we have used a traffic simulator based in multi-agent systems, the SUMO simulator, which is a reference for simulations in this area.

The data entry system so that the GA perform a satisfactory number of generations for the fitness calculation, resulting in the reduction of pollution in the study area. In this case, there were set a number of one hundred (100) generations to one hundred (100) cycles time step (each iteration) SUMO.

We believe that the use of GA, even with the preliminary results reported, is very promising. It is important highlight that the computation of all times of possible traffic lights is quite large and the times of traffic lights are interrelated. Thus, the use of this heuristic enables a reduction in the scope of search for possible solutions. Thus, it can obtain good solutions in a reasonable computational time, but in some situations, the heuristic remains in a local minimum, having no to the optimal solution. This behavior is possibly due to the use of mutation operator, which may have reduced the variability of solution candidates.

There are in the literature different heuristics which can be used in this context as neural networks and other types of evolutionary algorithms. However, due to the characteristics of the GAs, as before mentioned, they showed that generate satisfactory results for this approach.

From the results until this moment with this study, we have as future work perform simulations by changing other parameters of the GA as crossing. Another important thing to do is changing some SUMO parameters during the simulations, as as amount of time step or examine each of the pollution levels of the components (SO2, CO and NO2) and compare these values with CONAMA (National Environment Council) resolution.

REFERENCES

Aguiar, M. S. (1998). *Análise Formal da Complexidade de Algoritmos Genéticos (Dissertação de Mestrado em Ciência da Computação).* Porto Alegre, RS, Brazil: PPGC/UFRGS.

Aguiar, M. S., & Toscani, L. V. (1997). Algoritmos Genéticos. In *I Workshop sobre métodos formais e qualidade de software* (pp. 78-87). Porto Alegre/RS.

Dessbesell, G. J. (2015). *Simulação de Controle adaptativo de tráfego urbano através de sistema multiagentes e com base em dados reais. Dissertação (Mestrado em Sistemas e Processos Industriais).* Santa Cruz do Sul/RS.

Goldberg, D. (1989). *Genetic algorithms in search, optimization, and machine learning*. Addison-Wesley. *Artificial Intelligence*.

Heinen, M., Sá, C., Silveira, F., Cesconetto, C., & Sohn, G. (2013). *Controle Inteligente de Semáforos Utilizando Redes Neurais Artificiais com Funções de Base Radial*. Frederico Westphalen/RS.

Krajzewicz, D., Eedmann, J., Behrisch, M., & Bieker, L. (2012). Recent Development and Applications of SUMO - Simulation of Urban MObility. *International Journal On Advances in Systems and Measurements*, *3-4*(5), 128–138.

Linden, R. (2008). *Algoritmos Genéticos*. Rio de Janeiro: RJ Brasport.

Melo, F. B. (2005). *Proposição de Medidas Favorecedoras à Acessibilidade e Mobilidade de Pedestres em áreas Urbanas. Estudo de Caso: O Centro de Fortaleza. Dissertação (Mestrado em Engenharia de Transportes).* Fortaleza, CE: Universidade Federal do Ceará.

Nunes, G. (2013). *Estudo e análise da Dispersão de Poluentes: um estudo de caso para a cidade de Rio Grande/RS. Dissertação (Mestrado em Modelagem Computacional).* Rio Grande, RS: Universidade Federal do Rio Grande.

Retore, P., Santos, R., Marietto, M., & Sá, C. (2006). *Sistemas Multi-Agentes Reativos Modelando o Controle de Tráfego Urbano.* Uruguaiana/RS.

Rezende, S. O. (2003). *Sistemas Inteligentes: fundamentos e aplicações.* Editora Manole Ltda.

Silva, E., & Manfredini, V. (2010). *Aplicação de Conceitos da Inteligência Artificial no Controle de Tráfego Rodoviário. Trabalho de Conclusão de Curso* (Graduação em Ciência da Computação). São Paulo/SP.

Turky, M., Ahmad, S., & Yusoff, M. (2009). The Use of Genetic Algorithm for Traffic Light and Pedestrian Crossing Control. *International Journal of Computer Science and Network Security*, *9*(2).

KEY TERMS AND DEFINITIONS

DENATRAN: National Traffic Department, the agency that regulates the traffic laws in Brazil.

Fitness: Evaluation of individuals in a population in genetic algorithm, used as objective function value of the simulations.

Genetic Algorithms: Algorithms based on the evolution of species.

HBEFA: Handbook Emission Factors for Road Transport, standard pollutants library SUMO.

Mutation Rate: Values to be applied to the population for applying the mutation operation in Genetic Algorithms.

Pollutants Dispersion: Chemical transmitters that pollute the atmosphere and proliferate.

Rio Grande / RS: The interior of Rio Grande do Sul City - Brazil, where the case study for this work occurred.

Sumo: Simulation of Urban Mobility.

Chapter 11
Multi-Agent Systems in Three-Dimensional Protein Structure Prediction

Leonardo de Lima Corrêa
Federal University of Rio Grande do Sul, Brazil

Márcio Dorn
Federal University of Rio Grande do Sul, Brazil

ABSTRACT

Tertiary protein structure prediction in silico is currently a challenging problem in Structural Bioinformatics and can be classified according to the computational complexity theory as an NP-hard problem. Determining the 3-D structure of a protein is both experimentally expensive, and time-consuming. The agent-based paradigm has been shown a useful technique for the applications that have repetitive and time-consuming activities, knowledge share and management, such as integration of different knowledge sources and modeling of complex systems, supporting a great variety of domains. This chapter provides an integrated view and insights about the protein structure prediction area concerned to the usage, application and implementation of multi-agent systems to predict the protein structures or to support and coordinate the existing predictors, as well as it is advantages, issues, needs, and demands. It is noteworthy that there is a great need for works related to multi-agent and agent-based paradigms applied to the problem due to their excellent suitability to the problem.

INTRODUCTION

Bioinformatics consists in the study of biological problems and their intrinsic properties through the development of theoretical models and computational techniques (Chou, 2004; Gibas & Jambeck, 2001). Examples of bioinformatics studies include analysis and integration of -omics data, prediction of protein structure and function, and development of computational strategies to identify the affinity binding of drugs to a receptor and its effects. Depending on the biological question to be answered, the research in bioinformatics can be classified into two main lines: sequence analysis and structural bioinformatics.

DOI: 10.4018/978-1-5225-1756-6.ch011

The first is focused on the study and analysis of biological sequences, e.g., nucleotides and amino acid residues, over data mining and computational methods, such as sequence alignment, the inference of pathways from metabolic networks, morphometrics, and evolution (Chou, 2004). The second addresses biological questions from a three-dimensional point of view, covering most of the techniques included in computational chemistry or molecular modeling. It is related to researchers based on the three-dimensional (3-D) structures of molecules, including the computational prediction of protein/polypeptide structures, protein docking with different molecules (RNA, DNA, and proteins), simulation of dynamic behaviors of proteins, protein structure characterization and classification, and study of structure-function relationships (Chou, 2004; Xu, Xu, & Liang, 2007).

In this chapter, we provide an introduction and overview of the current state of multi-agent systems developed for study and prediction of three-dimensional structures of proteins. The study of protein structure and the prediction of their three-dimensional structures is one of the key research problems in structural bioinformatics. Over the last years, many computational methods, systems, and algorithms have been developed for the purpose of solving this complex issue. However, the problem still challenges biologists bioinformaticians, chemists, computer scientists, and mathematicians because of the complexity and high dimensionality of the protein conformational search space. Experimentally, the generation of a protein sequence is considerably easier than the determination of its 3-D structure. The 1990's GENOME projects resulted in a significant increase in the number of protein sequences. Unfortunately, the number of identified 3-D protein structures did not follow the same trend. Currently, the number of known sequences is far higher than the number of known 3-D structures; there is a large gap between the number of protein sequences we can generate and the number of new protein folds we can determine by experimental methods such as X-ray diffraction and Nuclear Magnetic Resonance (NMR). We intend to give an integrated view and insights about the protein structure prediction area concerned with the development and application of multi-agent systems to predict 3-D protein structures, as well as its advantages, issues, current needs, and demands. This chapter is useful for bioinformaticians, computer scientists, mathematicians, and biologists interested in beginning research in this field or to improve their current research.

This chapter is organized as follows. Section 2 provides some background knowledge relevant to proteins, its representation models, energy functions and molecular forces, and an overview of the protein structure prediction problem. In Section 3, prediction methods based on agents and multi-agent are introduced. The chapter concludes, and further research is outlined in Section 4.

BACKGROUND

Proteins

From a structural perspective, a protein or polypeptide is an ordered linear chain of building blocks known as amino acids. An amino acid residue is a small molecule containing an amino group (H2N+), a carboxyl group (COOH−), and a hydrogen atom attached to a central alpha carbon (Cα) (Figure. 1). Besides, each amino acid has also an R organic group (side-chain) connected to the Cα. The group R distinguishes one amino acid from another and confers the chemical properties of each amino acid residue. In nature, there are 20 distinct amino acid residues, each one with its chemical properties (Lodish et al., 1990). The side-chains of amino acids can differ in size; electric charge; and polarity. Also, depending on the polarity of the side chain, amino acids vary in their hydrophilic/hydrophobic character. Each protein is defined by its unique sequence of amino acid residues that causes the protein to fold into a particular three-dimensional shape. This shape or fold gives the protein its specific biochemical properties (Lesk, 2010; Liljas et al., 2009). A peptide is a molecule composed of two or more amino acid residues chained by a chemical bond called the peptide bond. This peptide bond is formed when the carboxyl group of one residue reacts with the amino group of the other residue, thereby releasing a water molecule (H2O) (Figure 1). Two or more linked amino acid residues are referred to as a peptide, and larger peptides are called polypeptides or proteins (Creighton, 1990; Lesk, 2002).

Proteins can be studied into four levels of abstraction (Lehninger, Nelson, & Cox, 2005; Lodish et al., 1990):

1. Primary structure;
2. Secondary structure;
3. Tertiary structure; and
4. Quaternary structure.

This hierarchy facilitates the description and the understanding of proteins. However, it does not aim at describing precisely the physical laws that produce protein structures; it is an abstraction that aims at making protein structure studies more tractable (Scheef & Fink, 2003). The primary structure only describes the sequence of amino acid residues in a linear order (Branden & Tooze, 1998). Each amino

Figure 1. Chemical representation of two amino acid residues and schematic representation of a model peptide. The carboxyl group of one amino acid 1 reacts with the amino group of the amino acid 2. A molecule of water is removed from two amino acids to form a peptide bond. N is nitrogen, C is carbon, O oxygen, and H hydrogen

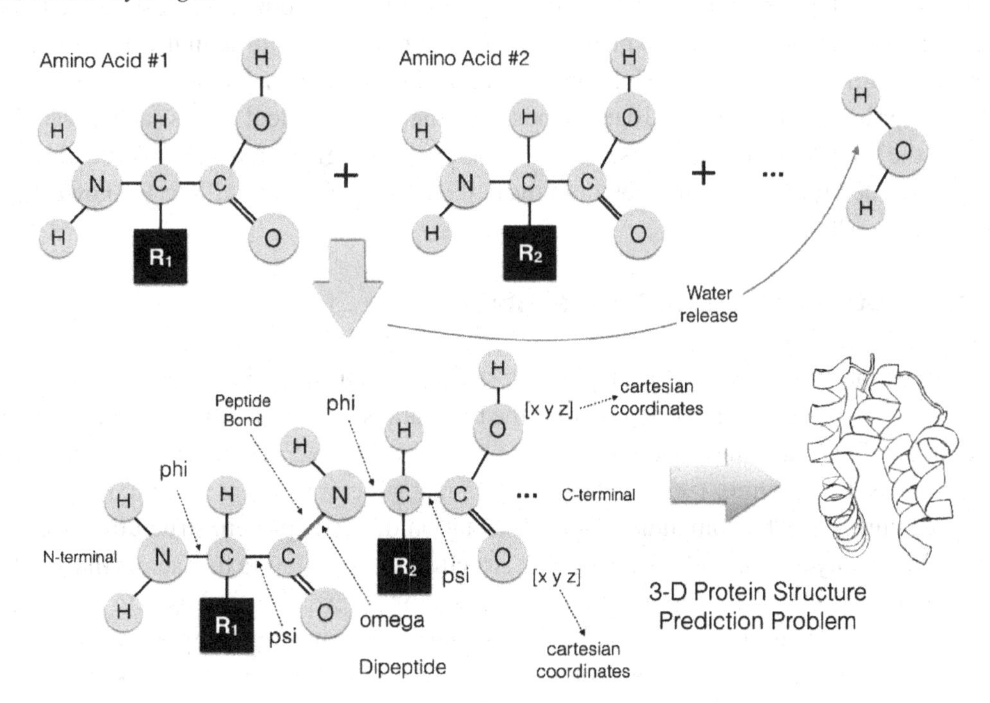

acid residue binds to another amino acid through a peptide bond. The beginning of the primary structure corresponds to its N-terminal region, and the end of its primary structure is the C-terminal region. The secondary structure is defined by the presence of hydrogen bond patterns between the hydrogen atoms of the amino groups and the oxygen atoms of the carboxyl groups in the polypeptide chain. Regularity in the spatial conformation is maintained through these intermolecular interactions (Richardson, 1981). There are two most commonly secondary structures: α-helices (Pauling, Corey, & Branson, 1951) and β-sheets (Pauling & Corey, 1951). There are other periodic conformations (coils and turns), but the α-helices and β-sheets are the most stable and can be considered as the main elements present in 3-D structures. The tertiary structure of a protein is related to its topology (or fold). The topology of a protein is given by the type of succession of secondary structures that are connected to and from the shape in which these structures are organized in a 3-D space. The three-dimensional shape assumed by a protein is also called the native or functional structure. The native structure of a protein is formed by the variation of thermodynamic factors, i.e., covalent interactions, hydrogen bonds, hydrophobic interactions, electrostatic interactions, van der Waals, and repulsive forces (Gibas & Jambeck, 2001; Richardson, 1981). The quaternary structure of a protein is the arrangement of various tertiary structures. This structure is maintained by the same forces that determine the secondary and tertiary structures (hydrogen bonding, hydrophobic interactions, hydrophilic interactions) (Lehninger et al., 2005; Lodish et al., 1990).

Understanding the protein structure allows the investigation of biological processes more directly, with higher resolution and finer detail. The sequence-protein-structure paradigm (also known as the "lock-and-key " hypothesis) says that the protein can achieve its biological function only by folding into a unique, structured state determined by its amino acid sequence (Anfinsen, 1973). Nevertheless, currently, it has been recognized that not all protein functions are associated with a folded state (Dunker et al., 2001; Dunker, Silman, Uversky, & Sussman, 2008; Tompa, 2002; Tompa & Csermely, 2004; Uversky, 2001; Wright & Dyson, 1999). For some cases, proteins must be unfolded or disordered to perform their functions (Gunasekaran, Tsai, Kumar, Zanuy, & Nussinov, 2003). These proteins are called intrinsically disordered proteins (IDP) and represent around 30% of the protein sequences. Despite the presence of IDP proteins, an important aspect for explaining the function of a given protein involves the analysis of complex molecular interactions. These interactions can be intramolecular (ionic bonds, covalent bonds, metallic bonds) or intermolecular (hydrogen bonds and other non-covalent bonds such as van der Waals forces). Thus, the knowledge of the 3-D structure of polypeptides gives researchers crucial information to infer the performed function of the protein in the cell (Branden & Tooze, 1998; Laskowiski, Watson, & Thornton, 2005a, 2005b).

Models for Protein Structure Representation

Protein structures can adopt a variety of shapes. The structure of one protein is defined by its amino acid sequence that folds spontaneously during or after the biosynthesis. The relation between the amino acid sequence of a protein and its conformation was first proven by Anfinsen's experiments (Anfinsen, 1973; Anfinsen, Haber, Sela, & White, 1961) and depends on many factors such as solvent, the concentration of salts, temperature, etc. The computational representation of a 3-D protein structure is a challenging task due to the difficulty in representing the protein structure and simulating the factors that contribute to the native structure stability. This representation is related to the level of detail used to describe the 3-D protein structure. The higher the number of features, higher is the capacity of representing the protein in its native state. The most detailed representation includes all atoms of the proteins and solvent molecules.

The geometric representation is one of the most important elements of 3-D protein structure prediction methods and is directly related to the reduction or increase of the protein conformational search space. Using the all atoms model to describe the protein is computationally expensive, and thus, simplified representations are often used (Chivian, Robertson, Bonneau, & Baker, 2003). There are two most common representations of polypeptides structures found in the literature. The first model represents the 3-D protein structure through the Cartesian position of the atoms. In this case, a polypeptide chain can be described as a set P of atoms in the three-dimensional space (R3). The second model represents the polypeptide structure using the set of dihedral torsion angles and is based on the fact that bond lengths are nearly constant in a polypeptide chain (Neumaier, 1997). The specific characteristics of the peptide bond have significant implications for the 3-D fold that can be adopted by proteins. The peptide bond (C-N), namely omega (ω) torsional angle, has a partial double bond and is not allowed rotation of the molecule around this bond. The rotation is only permitted around the bonds N-Cα and Cα-C. These bonds are known as phi (φ) and psi (ψ) dihedral angles and are free to rotate (Lesk, 2002; Lodish et al., 1990), varying from -180o to +180o. This freedom is mostly responsible for the conformation adopted by the polypeptide backbone. However, the rotational freedom around the φ (N-Cα) and ψ (Cα-C) angles is limited by steric hindrance between the side-chain of the amino acid residue and the peptide backbone (Branden & Tooze, 1998; Scheef & Fink, 2003). As a consequence, the possible conformation of a given polypeptide is quite limited and depends on the amino acid chemical properties. The peptide bond itself tends to be planar, with two allowed states: trans, ω=180o (usually) and cis, ω=0o (rarely) (Branden & Tooze, 1998; Lesk, 2002). The sequence of φ, ψ and ω angles of all residues in a protein defines the backbone conformation or fold (Hovmoller & Ohlson, 2002). Similar to the polypeptide backbone, side-chain also have dihedral angles, and its conformation contributes to the protein structure stabilization and packing. The number of angles Chi (χ) of the side-chain depends on the amino acid type (Liljas et al., 2009), ranging from 0 to 4 angles.

The use of dihedral angles has the advantage over the Cartesian model by having the degree of freedom reduced. For the backbone representation of a polypeptide, this gives rise to $3m$ degrees of freedom, where m is the number of amino acid residues. The main disadvantage of the usage of dihedral angles is that a small change in one dihedral angle can cause drastic changes in the polypeptide structure. There are other simplified models for protein molecules, such as the lattice protein model (Kolinski & Skolnick, 2004) and the off-lattice AB toy (Stillinger, Head-Gordon, & Hirshfeld, 1993). The most known type is the HP model (Hydrophobic-polar protein folding model) (Dill et al., 1995; Lau & Dill, 1989), which is based on the fact that native protein folds tend to form very compact cores driven by dominant hydrophobic interactions. In this model, each amino acid residue is classified either as hydrophobic (H) or hydrophilic (P). Two hydrophobic amino acids are in contact if they are adjacent in the fold but non-adjacent in the primary sequence. The protein structure prediction in the HP model can be described as a maximization problem, where the goal is to maximize the number of contacts between hydrophobic atoms (H-H contacts) (Jiang, Cui, Shi, & Ma, 2003). In general, lattice representations are highly simplified models for protein-folding phenomena that consider only the interactions between hydrophobic and polar amino acid residues distributed in a 2-D or 3-D spatial orientation. The off-lattice AB model (Stillinger et al., 1993) is simpler than the off-lattice 3-D torsional angles representation. Such scheme consists of a highly simplified model for protein-folding phenomena, which is also based only on the interactions between hydrophobic (A) and polar (B) amino acid residues.

Residues are linked together by rigid bonds with preset distance to form linear unoriented segments of amino acids that reside in two dimensions. In AB model, for any protein structure composed by *n*-residues, 2*n* bend angles will be needed. These angles are defined in the range of -180o to +180o.

Energy Functions and Molecular Forces

Energy functions are used in Molecular Mechanics (MM) simulations (Jorgensen & Tirado-Rives, 2005), Protein Design (Gordon, Marshall, & Mayo, 1999) and Protein Structure Prediction (Lazaridis & Karplus, 2000). There are two categories: MM potentials and protein structure-derived potential functions (scoring functions) (Zhang & Skolnick, 2004a). The first category aims at modeling the forces that determine protein conformations using physically based parameterized functional forms from small molecule data or in vacuo quantum mechanics (QM) calculations. The second category is empirically derived from experimentally determined structures from the Protein Data Bank (PDB) (Chivian et al., 2003; Hao & Scheraga, 1999; Koppensteiner & Sippl, 1995; Lazaridis & Karplus, 2000). These two classes of potentials represent the forces that determine the macromolecular conformation:

- Solvation,
- Electrostatic,
- Van der Waals interactions,
- Covalent bonds,
- Angles, and
- Torsions (Boas & Harbury, 2007; Chivian et al., 2003; Park, Huang, & Levitt, 1997; Pokala & Handel, 2000).

The main advantage of using a knowledge-based energy function is that it can model any behavior observed in known protein crystal structures, even when there is no real physical understanding of their behavior (Boas & Harbury, 2007). The disadvantage is that these functions cannot predict new behaviors absent in the training set obtained from the PDB. A potential energy function incorporates two types of terms: bonded and non-bonded (MacKerell, 2004). The bonded terms (bonds, angles, and torsions) are covalently linked. The bonded terms constrain angles and bond lengths near their equilibrium values. The bonded terms also include a torsional potential that models the periodic energy barriers encountered during bond rotation. The non-bonded potential comprises ionic bonds, hydrophobic interactions, hydrogen bonds, van der Waals forces, and dipole-dipole bonds. There are some potential energy functions used in computational molecular biology. AMBER (Cornell et al., 1995), CHARMM (Brooks et al., 1983) and ECEPP (Momany, McGuire, Burgess, & Scheraga, 1975) are the most widely used potential energy functions in the Protein Structure Prediction problem. A review of potential energy functions is found in Halgren (1995).

Protein Structure Prediction Problem

The prediction of the three-dimensional structure of proteins (PSP) is one of the most important research areas in Structural Bioinformatics, and can be explained as the efforts to predict the unknown 3-D structures of polypeptides, that may or may not have similar known structures in some protein data bank (Dorn, e Silva, Buriol, & Lamb, 2014). The most accurate way to determine the 3-D structure of proteins remains attached to experimental methods. Nowadays, X-Ray Crystallography (McRee, 1999) and Nuclear Magnetic Resonance (NMR spectroscopy) (Cavanagh, Fairbrother, Palmer III, & Skelton, 1995) are the most common experimental methods. However, such kind of method present some critical limitations, such as the expensive cost and also in several cases significant amount of time is required to find out the tertiary structure of a protein molecule. The difficulty in determining the 3-D structure of proteins has generated a large discrepancy between the volume of data (number of sequences of amino acid residues) and the number of 3-D structures of proteins which are currently known and stored in the Protein Data Bank (PDB — http://www.rcsb.org) (Berman et al., 2000). Therefore, only $\approx 0.11\%$ of non-redundant protein sequences stored in the NCBI Reference Sequence Database (RefSeq — http://www.ncbi.nlm.nih.gov/refseq/) (Pruitt, Tatusova, & Maglott, 2005) have representatives on the PDB. RefSeq and PDB are the most accessible databases of publicly available non-redundant and well- annotated sequences of proteins, genomic DNA and transcripts, and 3-D structures of proteins and other complex biomolecules experimentally determined, respectively.

The protein structure prediction problem is currently one of the challenging problems in Structural Bioinformatics (Tramontano & Lesk, 2006) and has been challenging Biochemists, Biologists, Computer Scientists and Mathematicians over the last decades (Baxevanis & Ouellette, 2004). The PSP problem can be classified according to the computational complexity theory as NP-hard problem as a result of the high dimensionality and complexity of the conformational protein search space (Crescenzi, Goldman, Papadimitriou, Piccolboni, & Yannakakis, 1998; Guyeux, Côte, Bahi, & Bienia, 2014; Unger & Moult, 1993). The PSP challenge arises due to the combinatorial explosion of plausible shapes, where a long amino acid chain ends up in one out of a vast number of 3-D conformations (Lesk, 2002; Levinthal, 1968). In the literature, one can find several classifications of the 3-D protein structure prediction methods.

In a simplified way, protein structure prediction can be viewed as the application of a search engine (genetic algorithm, tabu search, memetic algorithms, Monte Carlo, Monte Carlo with replica exchange, integer programming, etc.) to a physics-based or knowledge-based energy function. In the literature, it can be observed a significant number of works using constraints techniques to model the PSP as a Constraint Optimization Problem (COP). Thus, we briefly describe in the next sections some issues about the application of constraining techniques to the PSP, as well as a general description of the categorization of methods used in the area.

Constraint Programming

Constraint programming (CP) is a technique that aims to filter the most feasible solutions within a large set of candidate solutions through the planning of arbitrary constraints to solve a determined problem. Constraints can be defined as conditions or properties of a particular challenge. The use of CP eases the modeling of a problem and allows to integrate different constraints to it. Problems defined with constraints are named Constraint Satisfaction Problems (CSPs). CSPs are mathematical models with constraints where the goal is to find suitable solutions assigning values to the variables of the problem such that every constraint is satisfied (Rossi, Van Beek, & Walsh, 2006). In this case, the solver is a complete method, which systematically explore the entire search space (Cipriano, 2008). Although CP is based on feasibility of solutions rather than optimization of an objective function, it can be jointly used with a heuristic algorithm to reduce the possibilities and prune the size of the search space, prioritizing the method efficiency and dealing with the approximate solution concept (Cipriano, 2008; Hentenryck

& Michel, 2009). This strategy characterizes incomplete methods focused on Constraint Optimization Problems (COPs).

The search space is defined according to the problem formalization, composed by the set of variables and their domain. So, it is expressed by the constraints among the variables of the model and their assigned values. Also, in COPs beyond of the set of constraints that define the problem, there exist an optimization function used to guide the search procedure to reach good-enough solutions. A CSP can be defined as a 3-tuple of values $CSP = (X, D, C)$, where $X = \{x_1, ..., x_n\}$ is a finite set of variables that describes the theoretical model, $D = \{d_1, ..., d_n\}$ is the corresponding set of variable domains, and $C = \{c_1, ..., c_m\}$ represents the constraints on set X. So, a COP $P = (CSP, E)$ can be formalized by the joining of a CSP and an optimization cost function E. A feasible solution $S = \{s_1, ..., s_n\}$ for P must be restrained on D and satisfy the constraints defined on C. To be the best solution for the problem, $E(S)$ must be minimal.

A general definition of the PSP problem as a COP $P = ((X, D, C), E)$ can be done through the mapping of the PSP components to the CP space. Let ps be a primary sequence of a protein of length n, the variable $x_i \in X$ is associated with the i–th amino acid of the ps, and $d_i \in D$ is a set of values defined according to the protein representation corresponding to the i–th amino acid, e.g., pairs of dihedral angles (backbone and side-chain) or Cartesian coordinates. The constraint set C defines spatial geometric properties that a candidate solution must satisfy to be physically admissible. E is a fitness function representing the free energy of the protein structure.

The use of CP scheme allows describing spatial properties (defined by the protein representation) of the unknown protein regarding geometric constraints. The protein structure is represented on a discretized manner of the 3-D space where each amino acid is assigned to an autonomous agent of the MAS. Such kind of strategy facilitates the deal of physically forbidden conformations, accelerating the evaluation process.

Classification of Protein Structure Prediction Methods

Computational methods for protein structure prediction can be classified into four groups (Dorn et al., 2014; Floudas, 2007):

1. First principles (ab initio) methods without database information;
2. First principles methods with database information;
3. Comparative homology; and
4. Fold recognition.

First principles methods without database information characterize the ab initio methods. These methods aim at predicting new protein folds based only on the primary amino acid sequence (Osguthorpe, 2000). Ab initio methods are guided by thermodynamics concepts and physicochemical properties of the folding process of proteins in nature, and also that the native structure of a protein corresponds to the global minimum of its free energy (Tramontano & Lesk, 2006). Pure ab initio methods do not use any structural templates information from a database such as the PDB. In some cases, structural information is only used to empirically derive an energy function, for example in the parametrization of some all-atom potential energy functions, such as Rosetta scoring function (Rohl, Strauss, Misura, & Baker, 2004).

Groups ii, iii and iv are classified as knowledge-based methods and are capable of making predictions when template structural information's are available from experimentally determined protein structures. First principles methods with database information represent a hybrid class of methods that makes use of template information combined with a first principle ab initio approach (Srinivasan & Rose, 1995). In these methods, general characteristics of protein structures are extracted from a protein database and used to build starting point of the protein structure (Dorn et al., 2014). Ab initio with database information strategies do not compare the whole protein sequence to a known protein structure, but they compare only short fragments of the amino acid sequence as an attempt to get relevant information that would help in the prediction (Rohl et al., 2004). Fold recognition or protein threading methods rely on the theory that structure is more evolutionary preserved than sequence, where proteins with different amino acid sequences could have similar folds (Floudas, 2007). These methods aim to fit an amino acid sequence correctly against a structural model (Jones, Taylor, & Thornton, 1992). To choose similar models from a protein database, the method considers structural information such as residue-residue contact patterns, secondary structure, and solvent accessibility, since the similarities between the amino acid sequences are not enough (Bryant & Altschul, 1995). On the other hand, in comparative modeling methods, the primary goal is to align a target sequence of amino acid residues against to the sequence of another protein with a known 3-D structure, stored in the PDB, considering an evolutionary relationship factor and a similarity level between the target protein and the template protein. If both sequences are similar, the structural information obtained from the known protein structure is used to model the target protein (Martí-Renom et al., 2000; Sánchez & Sali, 1997). However, the application of empirical approaches to protein structure prediction is entirely dependent on experimental databases.

Regardless of the group classification, all developed 3-D protein structure prediction methods have to be tested for the ability to predict new protein structures. Every two years since 1994, a worldwide experiment called CASP (Critical Assessment of Structure Prediction – http://predictioncenter.org/) (Moult, Fidelis, Kryshtafovych, Schwede, & Tramontano, 2014) is performed to test protein structure prediction methods. Structural biologists who are about to publish novel protein structures are asked to submit the corresponding sequences in the CASP to be structurally predicted. The predictions are then compared with the newly experimentally determined structures (by NMR or X-Ray crystallography methods). CASP provides to research groups an opportunity to test their structure prediction methods objectively and delivers an independent assessment of the state-of-the-art in protein structure modeling to the research community and software users.

These methods can be classified into two broad categories according to the protein target models to be predicted: 1. free modeling and 2. template-based modeling.

1. **Free Modeling - (FM):** In free modeling category, there are either no usefully related structures, or the relationship is so distant that it cannot be detected. This is the most difficult category in the CASP, however, as fewer and fewer new folds are discovered experimentally, targets in the FM have become increasingly difficult to obtain. To address this problem, starting in December 2011, CASP introduced a mechanism by which FM targets are continuously solicited from the experimental community and immediately presented to the prediction community, in a procedure known as CASP ROLL (Moult et al., 2014).
2. **Template-Based Modeling - (TBM):** In template-based modeling category, a relationship to one or more experimentally determined structures could be identified, providing at least one modeling template and often more (Moult et al., 2014).

According to the latest CASP editions (Moult et al., 2014), the best results are being achieved by knowledge-based methods (Kryshtafovych, Fidelis, & Moult, 2014). Among the methods that have been tested on the CASP, QUARK (Xu & Zhang, 2012) and Zhang-Server (Zhang et al., 2015) can be pointed out as "reference methods" in the PSP area due to the best results achieved. QUARK is a first principle method with database information based on a fragment assembly approach (Simons, Kooperberg, Huang, & Baker, 1997). It uses fragments of known protein structures of length 1-20 amino acid residues to generate the initial structure templates. The models coming from the fragment arrangement are optimized over a set of Monte Carlo (MC) simulations known as Replica Exchange Monte Carlo (REMC). Such kind of optimization algorithm generates several replicas from the initial structures through the process of exchange of fragments. The Zhang-Server combines QUARK and I-TASSER (Iterative Threading ASSEmbly Refinement) (Zhang, 2008, 2009) systems. I-TASSER is a fold recognition interactive implementation of the TASSER method (Zhang & Skolnick, 2004b, 2004c). In the first stage, the target sequence is threaded through the PDB to identify appropriate local fragments. Such fragments will incur further structural reassembly using a parallel MC sampling for assembling/refinement (Zhang, Kihara, & Skolnick, 2002). In the second stage, the trajectories obtained by the simulation in the first stage are clustered (Zhang & Skolnick, 2004d); the clusters' centroids are obtained, and an MC simulation is applied starting up with the centroids' conformations. The conformation with the lowest energy is selected, and the backbone atoms are added by PULCHRA (Feig, Rotkiewicz, Kolinski, Skolnick, & Brooks, 2000) and the side-chains are added and optimized by SCWRL (Canutescu, Shelenkov, & Dunbrack Jr., 2001). Zhang-Server takes advantage of the ab initio approach of QUARK and the threading modeling technique of I-TASSER. Because of such combination of methods, Zhang-Server has been achieving the best positions in CASP both for the Free Modeling and Template Based Modeling categories.

Protein structure prediction is an immense field that cannot be thoroughly surveyed in this chapter. A complete review of protein structure prediction methods is found in Dorn et al. (2014). In the next section, some relevant PSP methods based on agents and multi-agent are presented and discussed in terms of the main characteristics of a MAS, such as general architecture, types of agents, protein representation, and search heuristic.

AGENT-BASED AND MULTI-AGENT SYSTEMS FOR THE 3-D PROTEIN STRUCTURE PREDICTION PROBLEM

Nowadays, to determine the 3-D structure of protein, a wide range of approaches are being applied to find approximated solutions. Machine Learning Algorithms, Data Mining Techniques, Optimization strategies, and Meta-heuristics are used in the context of the PSP problem (Dorn et al., 2014). In the same way, multi-agent systems are applied to support the prediction of three-dimensional structures of unknown protein sequences. Multi-agent systems (MAS) are being used to face complex problems, devising tasks among the agents of the system and exploring a more distributed approach. An agent is an autonomous computer process that runs in an independent way, where under some given circumstances, interacts and cooperates with the other agents to solve a major problem (Jennings, Sycara, & Wooldridge, 1998; Wooldridge, 2009). The agent-based paradigm has been shown a useful technique for the applications that have repetitive and time-consuming activities, knowledge share and management, such as integration of different knowledge sources and modeling of complex systems, supporting a great variety of domains (Merelli et al., 2007). Therefore, the multi-agent approach is a suitable strategy to model and deal with the complex PSP problem. It can focus on first principles methods, where the agents devise tasks and goals, interact and compete in an attempt to explore the search space in a more efficient way, as shown in Campeotto, Dovier, and Pontelli (2013), Lipinski-Paes and De Souza (2014), Muscalagiu, Iordan, Osaci and Panoiu (2012), Muscalagiu, Popa, Panoiu and Negru (2013), and Pérez, Beltrán, Rojo-Domínguez, Eduardo and Gutiérrez (2009). Or it can figures as a framework with the goal to predict the protein 3-D structure using and integrating the results of different existing predictors based on comparative modeling and fold recognition techniques (Bates, Kelley, MacCallum, & Sternberg, 2001; Garro, Terracina, & Ursino, 2004; Jin & Kim, 2004). According to the literature, many computational methods have been developed to the protein structure prediction problem. Nevertheless, it can be seen that there is a lack of works related to multi-agent and agent-based paradigms applied on PSP problem, principally when restricting the search to methods based on concepts of the first principle methods. Multi-agent systems offer models for representing complex real-world problems. Agents can be classified into different types based on their characteristics: agents have distinct features such as integration, cooperation, knowledge, simulation, etc. (Nwana, 1996) (Figure 2). A collaboration agent helps users to solve problems, especially in complex or unfamiliar domains by correcting errors, suggesting what to do next, and taking care of low-level details. In this category, there is more emphasis on cooperation and autonomy than on learning (Nwana, 1996). Interface agents employ machine learning techniques to provide assistance to a user dealing with a particular problem/application (Lashkari, Metral, & Maes, 1994). Commonly, interface agents take an amount of time to understand and learn human behavior before they are onto work. Smart agents interface with others in a concept that they do not necessarily need to be entirely intelligent. But by working together in a smart way, the agents form a type of emergent intelligence

Figure 2. Agent Topology originally proposed by Nwana (1996)
Source: Adapted from Nwana, 1996

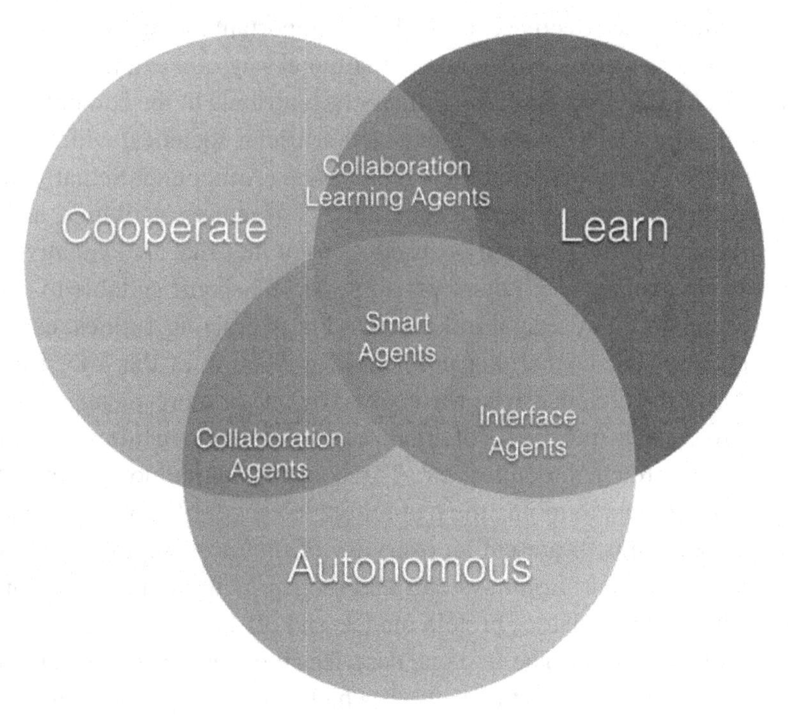

that may appear to exhibit intelligence. Collaboration learning agents is a mixed up of autonomous and learning agents; this group has to deal with significant or inherently distributed problems where they have to work as a group but, at the same time need to be flexible to respond to the changes from the environment. However, Nwana (1996) stated that is unaware of the existence of any such agents which collaborate and learn, but are not autonomous.

Therefore, we note most of the agents that will be presented in this review comprise the classes of Collaboration and Smart agents since they just follow a set of steps to reach a predetermined goal or use some intelligence behavior (search heuristic) to complete a task, respectively. One example of the collaboration agents are the worker agents described in Campeotto et al. (2013), which just play the role of assigning values to the variables attached to it. But, in general, the coordinator agents in the system or even the intermediaries, which are responsible for intermediate the interactions among the lower and higher level agents, can be classified as Smart Agents, since they assume the responsibility of control and coordinate the agents under to it through intelligent techniques, such as plan-based schemes, algorithmic search, and reinforcement learning. To illustrate, the CODY agent presented in Jin and Kim (2004) is a mediator and coordinator agent that efficiently controls the steps of the protein prediction process and changes the prediction process autonomously, when necessary, using a plan-based strategy. Additionally, the strategic agent described in Bortolussi, Dovier, and Fogolari (2005) controls the environmental properties through a global coordination of the simpler agents activities, based on a heuristic search, in an attempt to obtain a more effective exploration of the configuration space.

Multi-agent systems for the 3-D protein structure prediction problem can be studied according to its intrinsic properties related to cooperation, autonomy and learning capacity. In this section, some relevant publications in the field of MAS to deal with the PSP problem are presented and discussed.

Multi-Agent Systems: Prediction Methods

The multi-agent paradigm can be effective exploited as an important tool to investigate the properties of biological systems that are difficult to study in more traditional ways, for example with in vivo or in vitro experiments. Especially, MAS' concepts can significantly contribute to the computational biology field regarding conceptual frameworks to model agent-based artificial societies, which can easily simulate and implement the complex behaviors of biological processes (Amigoni & Schiaffonati, 2007). Focusing on the critical aspects biological systems, concerning to structures, activities, and interactions, the use of agent-based strategies can be modeled as abstractions of this, that are kept alive during the whole process from design to simulation. These concepts allow MAS become suitable to simulate biological systems that can be decomposed in several independent but interacting entities, each one represented by an agent. For example, in the concerning problem of prediction of the 3-D structure of proteins, each amino acid residue of the protein can be represented by one or more agents with different types of knowledge about the protein, e.g., properties of amino acids and their structures, protein templates from experimentally determined protein structures, etc. These agents autonomously interact, interchanging specific knowledge, in an attempt to mimic the real folding behavior of proteins in nature.

The purpose of this section is to present an overview of the most suitable frameworks for protein structure prediction regarding multi-agent systems and their principles. That attempt of the survey is divided into methods that comprise lattice protein models and off-lattice protein models. The following subsections are structured as follows. The first one describes the methods that use lattice protein representation, whereas the second represents the methods which focus on off-lattice protein representation.

Lattice Protein Models

Multi-Agent Systems Applied in the Modeling and Simulation of Biological Problems: A Case Study in Protein Folding (Pérez et al., 2009)

Pérez et al. (2009) proposed an iterative multi-agent framework for modeling biological systems heavily based on heterogeneous behavior, complex nature, localization, distribution and interaction of the components of the system. The agent-based approach, called as Evolution, aims to mimic a virtual laboratory to explore minimalist models of protein folding, encompassing distinct lattice protein models, evolutionary algorithms, and other computational techniques. These models are described by the coarse-grained 2-D and 3-D HP lattice representation.

The main components of the Evolution system are the agents and the blackboard communication scheme. The communication by blackboard provides both data sharing and coordination artifacts in a simple way. The blackboard levels are used to isolate and detail the solution elements needed for the resolution of the current problem. So, the MAS architecture encompasses three types of agents that interact by indirect communication, through a communication abstraction technique. The interactions among agents are defined by a set of semantics assigned to them according to the different blackboard levels and types of agents.

In this work, such strategy was divided into five distinct levels of abstraction of the protein folding prediction process:

1. The amino acid sequence level represents the solution elements related to protein sequences introduced by the user or read from a sequence file;
2. The HP(HPN) sequence level records HP(HPN) sequences that can be generated by the model generator agent or introduced by the user;
3. The initial conformational space level creates the initial conformational space. It is performed when the HP model, being discrete or continuous, is executed on 2-D triangular, 2-D square, 3-D diamond or 3-D cubic lattices. So, the solution elements on this level correspond to 2-D/3-D protein conformations defined regarding three main parameters: energy, the radius of gyration and maximal diameter allowed.;
4. The workspace algorithm level records all the states (conformations) visited by two iterative processes, the heuristic search, and the optimization, for seeking good solutions, providing the full landscape (evolutionary story) of the conformational space. These iterations are performed by the actions of the genetic agents; and
5. The plausible conformation level stores the solution elements corresponding to the optimal and suboptimal conformations found by the heuristic search and the optimization process. So, the conformations on this level represent the best solutions found by the genetic agents.

Evolution implements three different agents acting on the blackboard levels. From a bottom-up view of the abstraction levels, the HP(HPN) sequence generator agents are responsible for the translation of an amino acid sequence to an HP sequence, composed of 2 (HP) or 3 (HPN) symbols. Further the typical HP model, an extended HPN model is used in this work, which classifies the amino acids as Hydrophobic (H), Polar (P) or Neutral (N). The 2-D/3-D conformation generator agents have the task to generate a 2-D/3-D starting conformational space of lattice protein models. The agent implements the HP (HPN) model on 2-D, triangular and square lattices, or 3-D, diamond and cubic lattices. The agent also calculates the energy of each conformation thus generated. Finally, the higher level agents are the genetic agents. Each of them can be represented by a simple genetic algorithm (GA). Among these agents is performed the heuristic search and optimization processes on the search space for searching good-enough protein conformations. So, the genetic agents implement a wide variety of genetic operators and techniques, such as selection, reproduction, crossover, mutation and elitism. There is another type of agent, the interface agents, which are defined as abstract entities in the MAS architecture and later implemented for their specific purpose when predicting the protein structure. These agents provide, through a graphical user interface (GUI), the possibility of visualization of the 2-D/3-D initial conformational space and interaction with a particular conformation. This interaction allows the user discover more details of the conformation through the zoom, rotation, and translation of it. Additionally, the interface agents also allow the user to change the characteristics of a particular visited iteration by the optimization processes, recording and using as a seed to reset the whole optimization process from this state.

The authors refer the most remarkable characteristics of Evolution are the vast number of GUI and graphical tools provided for the user-system interaction. Evolution offers the user a wide variety of 2-D/3-D charts, e.g., diagrams, graphics, and tables, to interpret the partial and final results obtained during experiment execution. Also, new graphic outputs of indexes can be added, according to the user problem. All GUI and graphical tools have been implemented as interface agents. Hence, during the processes execution, a GUI is an interface agent that provides personalized assistance to users with their tasks. However, we note the absence of details about programming languages and technologies used in the development of this work.

Multi-Agent Systems Applied in the Modelling and Simulation of the Protein Folding Problem Using Distributed Constraints (Muscalagiu et al., 2013)

Muscalagiu et al. (2013) proposed an agent-based framework for the PSP problem, composed of autonomous agents, which collaborate to find good protein conformations. This work is an extension of a previous version described in Muscalagiu et al. (2012).

The authors have employed the Distributed Constraint Programming strategy (Yokoo, Durfee, Ishida, & Kuwabara, 1998). As already explained in Section 2.4.1, constraint programming (CP) is a programming paradigm used to describe and solve large classes of problems such as searching, combinatorial and planning problems, where relations among variables are established in the form of constraints. Distributed Constraint Satisfaction (DisCSP) and Distributed Constraint Optimization (DisCOP) is an extended version of the CP paradigm for describing a problem regarding constraints that are known and can comprise a set of distinct agents. Such distributed scheme is suitable for many problems for which the information was distributed to many agents. In DisCSP, variables of the problem corresponding to each agent are connected by constraints; the agents must assign values to their variables so that all constraints among them are satisfied. The agents must assign values to a set of variables so that the cost of a set of constraints over the variables is either minimized or maximized.

Most of the methods to solve DisCSP are distributed and asynchronous. To execute such searching technique, an agent has to be able to send messages to any other agents of its neighborhood. DisCOPs can be solved either using distributed search or by using distributed inference. In this MAS, the ADOPT algorithm (Modi, Shen, Tambe, & Yokoo, 2005) for distributed search was employed. According to the authors, ADOPT is a backtracking based bound propagation algorithm and was the first decentralized method to guarantee optimality, while at the same time allowing the agents to operate asynchronously. The process of optimization based on constraints aims to find any conformation which is feasible, i.e., respects all constraints imposed in the variables of agents and their domain, and minimizes the sum of the agents utilities, i.e., the minimization of an energy function related to the lattice protein model.

On this system, each amino acid is viewed as an autonomous agent that communicates with others by transmitting messages. The input protein is represented by lattice-based models with distributed constraints. The lattice protein representation used is the 2-D/3-D HP model. The architecture of the MAS was structured on two levels, corresponding to the two stages of implementation. The first level characterizes the external level and refers to the way of representing the surface of the multi-agent system. The second level is the internal level and defines the way in which asynchronous techniques will be programmed so that the agents will run concurrently and asynchronously. To this level are simulated the agents associated with the amino acids and are implemented the DisCSP/DisCOP algorithms for finding an optimal conformation. From a top-down (external to internal) perspective:

1. The HP sequence level is responsible for select HP protein sequences, which can be either generated by the NetLogo module HP generator or loaded from a file;
2. The initial Conformational space level corresponds to the initial conformational space created when the HP model, being discrete or continuous, is executed on 2-D triangular, 2-D square or 3-D cubic lattices; and
3. The algorithm workspace level represents the DisCSP/DisCOP algorithm and optimization for finding good protein solutions. We note that this architecture is similar to that described in Pérez et al. (2009) and already summarized in this review.

The MAS was implemented using NetLogo environment (https://ccl.northwestern.edu/netlogo/) (Tisue & Wilensky, 2004). Even about the internal level, the multi-agent system needs to simulate the agents of amino acids type. In NetLogo, the representation of amino acids, bonds, and lattice was done through a grid with some cells where each cell can be empty or occupied by an amino acid. The authors publish a first working examples in open-source NetLogo for the PSP problem. These set of examples can be accessed from the website http://discsp-netlogo.fih.upt.ro/. Still, they have developed a methodology to run the system in a cluster computing environment using the Java API of NetLogo as well as LoadLeveler; that is a job scheduler to control scheduling of batch jobs (Muscalagiu, Popa, & Vidal, 2013).

Off-Lattice Protein Models

Multi-Agent Simulation of Protein Folding (Bortolussi et al., 2005) and Agent-Based protein structure prediction (Bortolussi, Dovier, & Fogolari, 2007)

Bortolussi et al. (2005, 2007) developed an ab initio agent-based system following the general architecture, namely MAGMA, presented by Milano and Roli (2004) for agent-based optimization systems. MAGMA comprises four levels of agents, which interact to perform the optimization tasks:

- Level 0 deals with the initialization of the solutions space;
- Level 1 is focused on the stochastic search in the search space;
- Level 2 is responsible for global strategic tasks; and
- Level 3 is concerned with cooperation strategies.

This work is an extension of Bortolussi, Dal Palu, Dovier and Fogolari (2004), where the authors have only implemented just level 1 agents, whereas in this work level 2 and 3 were also included. Level 0 agents are trivial as the solutions are generated following one of four strategies, i.e., straight line, zig-zag, random or based on known protein templates.

Thus, the proposed work uses a hierarchical structure of agents, which are divided into three different layers by the knowledge and power of the agents. The first one is focused on the exploration of the conformational search space; the second layer implements global strategies to guide the optimization process and the latter cooperation tasks.

The protein representation adopted by Bortolussi et al. (2005) is a simplification of the all-atom model (De Mori, Micheletti, & Colombo, 2004), which consists in an off-lattice abstraction that represents each amino acid as a single sphere centered in the Cα atom. The protein is modeled with small fragments or local structures that can be obtained by the fulfillment of rigid constraints. The employed energy function comprises three terms, related to interaction (pairwise term), cooperation (secondary structure packing term) and chirality (chiral term). Using CP technique, the authors have also included some hard constraints implemented via energy barriers to prevent non-physical configurations. However, the system is independent both in the representation of the protein and on the force field model employed.

According to the MAS architecture, each amino acid of the protein target is seen as an independent agent called as searching agents or amino agents. Each one is represented only by the Cartesian coordinates of the Cα atom. The amino agents are responsible for performing the exploration of the conformational state space. The optimization of the problem is carried out by the agents' interactions and also through the information exchange among them within a Parallel Simulated Annealing (SA) (Kirkpatrick, Gelatt, & Vecchi, 1983) scheme that controls all the process. The agents' movements are guided by the knowledge of the position of neighboring agents, whereas the communication is restricted to the agents' neighborhood defined by a radius with a length of 14Å. Beyond the exploration task of the amino agents, the system includes more two types of agents allocated in the higher level layers. The strategic agent controls the environmental properties through a global coordination of the basic agents' activities in an attempt to obtain a more effective exploration of the configuration space. It also regulates the SA optimization parameters. The strategic agent has total knowledge of the current configuration since searching agents do not. Furthermore, the cooperative agent exploits some external knowledge to improve the folding process. It combines concurrency and previous knowledge about the protein folding to force the amino agents to assume a particular configuration, which is supposed to be favorable. This additional information can be extracted from an Oligomers database, proposed by Micheletti, Seno, and Maritan (2000), which encompass 40 different protein fragments with a length of 5 amino acids. According to Micheletti and co-authors, only these 40 protein pieces are needed to reproduce the whole ensemble of structures. Therefore, if 5 consecutive residues are in a configuration which is close to a particular oligomer, a biasing potential term was introduced in an attempt to force them to adopt this structure to check if this move is effective or not.

The communication scheme is based on the Linda tuple space (Carriero & Gelernter, 1989), that is a concurrent paradigm used to ease the handle of constraints among objects in the system. All the communications are performed through writing and reading logical atoms in the Linda tuple space. The MAS was implemented in SICStus Prolog (Carlsson & Mildner, 2012). An equivalent multithreading version of the system was performed, which is much faster than the sequential one. Details of the implementation can be found in Bortolussi et al. (2005). Additionally, in a most recent work, Bortolussi et al. (2007) give a more detailed description of the system. This work extends the first version to analyze the feasibility of the MAS with two different force fields.

MASTERS: A General Sequence-Based Multi-Agent System for Protein TER-Tiary Structure Prediction (Lipinski-Paes & De Souza, 2014)

Another multi-agent approach was proposed by Lipinski-Paes and Norberto De Souza (2014) and was called as MASTERS framework. MASTERS is a cooperative hierarchical MAS guided by search heuristics as Simulated Annealing and Monte Carlo (MC) simulations. This framework uses a coarse-grained protein representation, which is the off-lattice AB model as described above (Section 2.2).

MASTERS was developed using NetLogo (Tisue & Wilensky, 2004), which is a multi-agent programmable modeling environment. The authors developed the system based on the framework of Bortolussi et al. (2005), already presented in this review. The primary purpose of this work is to provide a system that matches to the user needs through an intuitive graphical user interface (GUI) and a simple modeling language. With the GUI, the users can setup and run simulations, edit the code, analyze the execution results (step and current energy) and generate 3-D plots (Energy vs. Time and Acceptance Ratio vs. Time). MASTERS allows the use of different energy functions and different abstraction level to guide the optimization of a protein, however, these fitness functions are simpler and designed specially to the AB model representation. It is interesting to note the framework can also be applied to a wide range of optimization problems that involve Cartesian coordinates (2-D or 3-D).

As in the works of Bortolussi et al. (2005, 2007), the MASTERS' agents are hierarchically organized according to the MAGMA architecture, proposed in Milano and Roli (2004), which includes three distinct types of agents. Describing from a bottom-up hierarchical view, the searching agents are the lower-level agents and have the goal of exploring the conformational search space. The set of all searching agents characterizes the optimization algorithm. Each amino acid of the protein is assigned to one or more searching agent, depending on the chosen abstraction level. The agents' position are represented by Cartesian coordinates, and the movements of each one are independent and local. All movements of the searching agents are controlled by a Monte Carlo criterion allied to a Simulated Annealing scheme. Such scheme defines a current conformational state S_1 and a possible new state S_2 with energy values E_1 and E_2, respectively. According to the MC criterion, the movement from S_1 to S_2 is done if $E_2 < E_1$, or if the probability of accepting a move is reached. The probability of accepting a move is described in Equation 1, where k is the Boltzmann constant and T is the system temperature.

$$e^{-\left(E2-E1\right)kT} \quad (1)$$

The Simulated Annealing is an MC improvement that performs a global optimization over the gradually cooled of the system. At the beginning of the simulation, the system is set to a high temperature, allowing it to escape from local minima. As the temperature decreases, the algorithm tends to concentrate the search in low energy regions of the conformational space and, consequently, converge to the global optimum though there is no guarantee. In MASTERS the temperature is decreased according to Equation 2, where $\alpha = 0.98$ and T_s is the temperature of step s.

$$T_{s+1} = T_s\alpha \quad (2)$$

The other two types of agents are coordinating agents that do not have representation in the Cartesian space. The director agent has a global view of the optimization process, and its role is to guide the optimization to a more efficient spatial exploration. It performs global moves on the searching agents. The higher-level environment agent is responsible for control all the simulation flow. It controls all other agents, as well as the optimization scheme, number of movements per time/temperature step, real-time plots, outputs and also the simulation ending.

Protein Structure Prediction on GPU: A Declarative Approach in a Multi-Agent Framework (Campeotto et al., 2013) and a Declarative Concurrent System for Protein Structure Prediction on GPU (Campeotto, Dovier, & Pontelli, 2015)

In Campeotto et al. (2013, 2015) was proposed a multi-agent system aiming to predict tertiary structures of proteins in a faster platform using a General Purpose Graphical Processing Units (GPU) architecture. They have created a set of agents with distinct functions to concurrently explore and assemble conformations of local fragments of the protein. Agents have the goal of retrieving, filtering, and coordinating local information about parts of a protein that is broken into smaller pieces, aiming to reach a global consensus after an optimization process. To model the PSP problem, the authors have used declarative techniques, i.e., Constraint Programming, which the interactions between fragments of the protein are represented as angle constraints, which are propagated via the communication among agents. Thus, the work uses an off-lattice structure representation, where a target protein is represented only by its dihedral angles. The scoring function employed encompass the sum of three components:

1. Contact Potential component that uses the statistical table of contact energies (Fogolari et al., 2007);
2. Torsional and Correlation potential component that is a statistical potential for the torsional angles and the correlation between amino acids in the protein's sequence that uses a pre-calculated table of energies; and
3. Hydrogen bond potential component.

The authors designed the MAS in a hierarchical multi-level organization covering four different types of agents. The supervisor agent is located at the top of the hierarchy, and are responsible for distributing sub-sequences of the primary protein sequence among the agents right below to it, i.e., the coordinator and the structure agents, and also guide the entire PSP optimization process until a stable global configuration. The primary sequence of a protein is broken according to the super secondary structures' location (α-helices and β-sheets) and computed by the structure agents that are specialized in the folding of secondary structures. Each structure agent receives a segment of secondary structure and applies a beam search strategy in an attempt to improve the given fragment. For each variable (amino acid) of the sub-structure, a worker agent is invoked to assign in parallel all the possible values in its domain to the corresponding variable, whereas the other variables of the fragment remain unchanged. These "multiple" tries performed by a worker agent, as placed by the authors, produce a set of new structures that are obtained by rotating the dihedral angles of the i–th amino acid of the target sequence. So, the best variables (minimum energy) found by the worker agents are assigned to the new current substructure. Then, the coordinator agent, specialized in loop modeling, is invoked to model the protein by moving the loop regions over a space sampling. The same idea of the structure agents is applied to the coordinator agent. For each variable of the loop segment, it invokes a worker agent to explore the search space in its domain and refine the fragment. In contrast to the structure agents, some random values to the remaining variables of the fragment are also concurrently explored. The energetically best among these structures is selected. At the end of the optimization process, a set of feasible and optimized solutions is returned to the higher level agent. After a global consensus achieved, all the best parts of the protein are grouped by the supervisor agent to reassemble the entire protein.

The MAS was implemented using the NVIDIA's Compute Unified Device Architecture (CUDA) for general purpose computing. The evaluation of the structures in the set of each worker agent is done by a parallel implementation of the energy function as well as the assignments and consistency checking of constraints. The communication between agents is based on memory sharing data structures and components of the own CUDA architecture. Previous knowledge about known protein structures is used to reduce the conformational search space, where the structure and coordinator agents use different sets of fragments as well as different weights for the energy function, according to their expertise. We note the CP formalization is similar to that described in Section 2.4.1. The set of angle constraints and implementation details can be found in Campeotto et al. (2013). In a later version of this work, Campeotto et al. (2015) provide the first complete description of the proposed multi-agent system and a precise analysis of its performance and properties.

Multi-Agent Simulated Annealing Algorithm Based on Differential Perturbation for Protein Structure Prediction Problems (Zhong, Lin, Du, & Wang, 2015)

In Zhong et al. (2015), the authors proposed a multi-agent system combined with a Simulated Annealing algorithm (MSA) to predict the tertiary structure of proteins using the 2-D off-lattice AB representation model (Section 2.2). Zhong et al. (2015) also incorporate to the method three differential perturbation (DP) operators taken from the learning ability of the mutation operators in differential evolution (DE) algorithm.

Simulated Annealing (SA) is a local search technique that provides a means to escape from local optima by allowing hill-climbing moves in the hope to find a global optimum. However, the main disadvantage is it may be extremely slow and require much more processing time to convergence than other algorithms. To overcome these issues, the authors implemented a system based on DE algorithm with a population of agents that run SA algorithm collaboratively in parallel, namely MSA. The DE operators are employed to adjust the neighborhood structure adaptively and improve the exploiting of the promising search space.

This work is based on the previous work of Zhong, Wang, Wang and Zhang (2012), where mutation operator formulas were directly used to generate a solution. In Zhong et al. (2012), as the randomly selected individuals to take part in the operation may be far from each other, this strategy may not be able to search the state space around of a particular basin finely enough. So, this paper defines three differential perturbation (DP) operators to generate candidate solutions collaboratively within a limited neighborhood. For further details of the DP operators we refer to Zhong et al. (2015). The SA scheme controls the movements of candidate solutions, which may accept worse results based on a probability of acceptance (Equation 1) to overcome local minima. The cooling schedule adopted to adjust the temperature of the SA follows the same equation presented in Equation 2. The energy function used in the method encompass two kinds of molecular interactions that compose the intermolecular potential energy for each molecule: backbones bend potentials and non-bonded interactions.

The method is structured with a set of agents that run independently and perform the optimization process over a candidate solution. So, each agent is assigned to a possible solution, and thc optimization process consists of a sequence of steps involving the initialization, mutation, evaluation and replacement of a solution. We note that no details about programming languages and technologies were given.

Comparison Prediction Methods

In this section, we present a comparison of the prediction methods described in the above section regarding the key features included in a multi-agent system. Table 1 comprises the comparisons of the methods according to the exploration of distinct architectures, constraint programming techniques, language and implementation technologies, communication protocols, protein representation models, external knowledge from a protein database, and search heuristics. It is noteworthy the all fundamental concepts of each definition in Table 1 were previously described in the discussion section of the methods (Section 3.1). According to the distinct schemes of architecture addressed in this chapter, in general, the methods use

Table 1. Comparison of the prediction methods regarding the main characteristics included in a multi-agent system. The first column represents the identification of the methods described in the above section

Method ID	Architecture	*CP* Strategies	Language/Implementation	Communication Protocol	Protein Model	External Knowledge	Search Heuristic
M-1	5-abstraction levels of the protein folding process	-	-	Blackboard scheme	2-D/3-D HP lattice model	-	Genetic Algorithm
M-2	2-abstraction levels structure	Distributed constraint programming (DisCSP/ DisCOP)	NetLogo	Transmitting messages	2-D/3-D HP lattice model	-	DisCSP/ DisCOP algorithms
M-3	MAGMA	Energy function with hard constraints	SICStus Prolog	Linda tuple space	Off-lattice (C-α/ Cartesian coordinates)	Oligomers database (40 fragm. with length 5)	SA
M-4	MAGMA	-	NetLogo	NetLogo environment	Off-lattice (AB model)	-	SA with Monte Carlo simulations
M-5	Hierarchical multi-level organization	Declarative techniques (angle constraints)	GPU architecture (NVIDIA's CUDA)	CUDA environment	Off-lattice (dihedral angles)	Known protein fragments	Beam search strategies
M-6	Independent runs of SA	-	-	-	Off-lattice (AB model)	-	SA with differential perturbation operators

M-1 (Perez et al., 2009); **M-2** (Muscalagiu et al., 2013); **M-3** (Bortolussi et al., 2005, 2007); **M-4** (Lipinski-Paes & De Souza, 2014); **M-5** (Campeotto et al., 2013, 2015); and **M-6** (Zhong et al., 2015). "-" indicates the item is not applicable to the method or the information was not informed by the authors

a hierarchical organization of agents. They tend to vary regarding the abstraction levels of the protein folding process and the agents' roles performed in the optimization processes. The M-6 method is an exception; it uses a single-layer structure of agents where each of them runs an independent execution of the SA algorithm with differential perturbation operators. Each execution comprises some steps of the optimization process, such as initialization, mutation, evaluation and replacement of a solution.

The constraint programming techniques in the PSP can be modeled as restrictions imposed in the protein conformations generated by the prediction methods. According to Table 1, the constraints can be treated as distributed constraints as in M-2, explicitly used in the scoring function like in M-3 or used as declarative techniques as in M-5. Still, one dealing with constraint programming can make use of a specialized search algorithm for CP, such as the DisCSP/DisCOP algorithms employed in M-3.

From Table 1, we can observe that the languages and implementations adopted by the methods can significantly differ. It is hard to point out what is the best choice, but we note that regardless of the technology used to build the system, it must favor the parallelism and the distribution of tasks among the entities of the system. Following this line, the communication protocol used follows the technology used to build the system as a way to facilitate the implementation. It can be seen in the protocol of transmitting messages employed by the M-2, provided by the NetLogo. M-4 also uses the NetLogo, but the authors do not explicit what type of protocol is applied. However, the protocol of communication is also provided by the NetLogo environment. M-3 uses the Linda tuple space from the SICStus Prolog, and M-4 uses the own CUDA environment to make the interactions. M-1 is an exception because it uses the blackboard communication protocol but any detail of the system implementation is provided.

The protein representation model used by each method is related to the level of detail used to describe the 3-D protein structure. Thus, we can group the described methods into two categories: lattice model (M-1 and M-2) or off-lattice model (M-3, M-4, M-5, and M-6). Still, the off-lattice model can represent a protein structure using the Cartesian coordinates of the C-α (M-3), dihedral angles (M-5), or from the AB model (M-4 and M-6). We note the higher is the number of features, higher is the complexity of the model and the capacity of representing the protein structure. Due to the complexity of the problem, the incorporation of previous knowledge about known protein structures is a common strategy used to reduce the size of the conformational search space in the PSP. In Table 1, we observe that only two methods explore such strategy. M-3 uses previous knowledge about the protein folding to force the amino agents to assume a particular configuration, which is supposed to be favorable. This external knowledge is extracted from an Oligomers database, which encompasses 40 different protein fragments with a length of 5 amino acids. M-5 uses different sets of fragments assigned to the agents to generate the protein structures, as well as different weights for the energy function, according to their expertise. Hence, we emphasize that the external knowledge about the problem should be better explored by the MAS prediction methods to improve their effectiveness. Regarding the search heuristics, from Table 1, we observe the Simulated Annealing is the most used algorithm due to its simplicity of implementation and accuracy. However, M-1 uses a genetic algorithm as its meta-heuristic and M-5 uses a beam search heuristic.

Table 2. illustrates the classification of different agents designed in the prediction methods: (i) the collaboration agents; (ii) the interface agents; and (iii) the smart agents.

Table 2. Classification of the prediction methods' agents according to the classification of agents proposed by Nwana (1996). The first column represents the identification of the methods described in the above section

Method ID	Collaboration Agents	Interface Agents	Smart Agents
M-1	*1) Sequence generator agents;* *2) Conformation generator agents*	*Abstract entity*	Optimization: *genetic agents* (each one implements a genetic algorithm)
M-2	*1)* Sequence level (implicit) *2)* Conformational space level (implicit)	NetLogo interface (implicit)	Optimization based on constraints: *agents of amino acids*
M-3	*Initialization of solutions* (implicit)	-	*1)* Exploration tasks: *searching agents*; *2)* Global exploration: *strategic agent;* *3)* External knowledge exploration: *cooperative agent*
M-4	-	Intuitive graphical user interface (NetLogo interface) (implicit)	*1)* Exploration tasks: *searching agents;* *2)* Global exploration: *director agent*; *3)* Global exploration: *higher-level environment agent*
M-5	*Worker agents*	-	*1)* Global coordination: *supervisor agent*; *2)* Optimization of the super secondary structures: *structure agents*; *3)* Optimization of the protein structures by modeling loop regions: *coordinator agents*
M-6	-	-	Optimization of solutions: *Independent agents* (Each one runs an SA to optimize a candidate solution)

M-1 (Perez et al., 2009); **M-2** (Muscalagiu et al., 2013); **M-3** (Bortolussi et al., 2005, 2007); **M-4** (Lipinski-Paes & De Souza, 2014); **M-5** (Campeotto et al., 2013, 2015); and **M-6** (Zhong et al., 2015). "-" indicates the item is not applicable to the method or the information was not informed by the authors.

This categorization is done using the classification of agents proposed by Nwana (1996) and is based on the distinct features of the agents, such as integration, cooperation, knowledge, and simulation. The definitions of these concepts were previously described in Section 3, as well as the description of each agent. It is noteworthy that some agents are not explicit declared in the works, but the systems encompass some levels that can be easily modeled by agents. For example, the Sequence level and the Conformational space level described in M-2 can be seen as agents that translate the amino acid sequence to the HP sequence and generate the initial protein conformations, respectively. Or the initialization of solutions step from M-3 can be modeled as an agent that has the task of initializing the solutions. Thus, this kind of module was noted in Table 2 with the "implicit" word.

Therefore, it is possible to note that the methods for the PSP present a great variation in terms of the system design and problem definition, consequently, despite the advances in the area there is not even a general method for the prediction of structures. Thus, in the next section we describe the meta-servers, which relies on the concept of consensus-based approach where independent predictions are combined to produce better results than the individual systems.

Multi-Agent Systems: Meta-Servers

A remarkable advance in the field of protein structure prediction is the meaning of meta-strategies or meta-servers (Bujnicki, Elofsson, Fischer, & Rychlewski, 2001a; Fischer, 2006). This idea relies on the concept of consensus-based approach (Lundstrom, Rychlewski, Bujnicki, & Elofsson, 2001) where independent predictions are combined to produce better results than the individual tools. Thus, errors in particular prediction can be revealed, leading to even better performance. For instance, in the recent CASP experiment, there is no general method which is always the best for all target proteins, which varies in size, class, etc. This occurs mainly because the quality of the predictions results depends on many specific target factors that are partially or entirely unknown when the prediction is run.

Currently, with the increasing number of available prediction methods, the chances of obtaining a reliable model also increases. Therefore, to generate a reliable hypothesis by computational analysis, one needs to consult many predictors and integrate their results, making comprehensive structural analysis and select the best-predicted 3-D protein model (Dorn et al., 2014; Fischer, 2006). Meta-servers have been developed to reduce the difficulty to combining various tools, integrating and displaying their results. In meta-servers, a set of protein structure predictors is applied to a target amino acid sequence (Jaskowski, Blazewicz, Lukasiak, Milostan, & Krasnogor, 2007). When compared with single protein structure prediction methods, meta-servers approaches have some advantages: meta-servers can produce good results, and they are better than the individual servers; 3-D protein structure prediction in meta-serves are more stable than those made when only a single prediction method is used. A complete review of meta-servers can be found in Fischer et al. (2006).

Meta-server approaches represent one of the most significant advances in the area of PSP. Due to the nature of the PSP problem, independent prediction strategies can be combined to produce better results than individual methods. This integration may be developed through multi-agent strategies. This section presents three meta-servers based on MAS and some other meta-servers that could be easily implemented by agent philosophies to provide more reliable prediction methods.

Plan-Based Coordination of a Multi-Agent System for Protein Structure Prediction (Jin & Kim, 2004)

Jin and Kim (2004) described the design architecture and implementation of a multi-agent system for the prediction of the tertiary structure of proteins, namely MAPS (Multi-Agent system for Protein Structure Prediction). Using intelligent techniques, MAPS can adjust its actions in reaction to the environment influences while performing different activities. According to the authors, the system also meets a variety of needs by connecting to *Agentcities network* (Willmott, Dale, Burg, Charlton, & O'Brien, 2001), accommodating a great deal of changes in biological databases both on quantitative and qualitative aspects, while sharing the data at the same time being operated as an integrated unit. Agentcities network (Willmott et al., 2001) is a project that provides connection services among various and heterogeneous agents following the FIPA specifications (http://www.fipa.org/).

In this work, the protein-related resources are presented as biological databases that directly support the protein structure prediction processes, and as software tools that facilitate such tasks as creating structure, searching similar sequences, aligning multiple sequences, or analyzing sequences. The most common online protein resources and databases are PDB (Berman et al., 2000), 3D-pssm (Kelley, MacCallum, & Sternberg, 2000), SWISS-PROT (Boeckmann et al., 2003), PSIPRED (Buchan, Minneci,

Nugent, Bryson, & Jones, 2013), SCOP (Andreeva et al., 2008), and HMMSTR (Bystroff, Thorsson, & Baker, 2000). The software tools include NCBI-Blast (http://blast.ncbi.nlm.nih.gov/Blast.cgi) which is used to find similar sequences, ClustalW (http://embnet.vital-it.ch/software/ClustalW.html) for multiple sequences alignment and SWISS-MODEL (homology-modelling server) (Guex & Peitsch, 1997) for the prediction of 3-D structures with querying sequences and templates. The system aims to facilitate several studies about proteins through transforming the existing protein resources into agents and coordinating these agents regardless of platforms or locations.

The authors classified the coordination mechanism of agents as cooperation or competition. The cooperation mechanism down that all agents should cooperate to reach one global goal, whereas in competition mechanism, conflict, and negotiation among the agents take place to secure benefits. In MAPS, the cooperation mechanism is used over a plan-based intelligent agent. It is a distributed planning method in which all the component agents together participate in making plans. The system has an agent, named CODY, which is a mediator and coordinator. Its role is to efficiently control the steps of the prediction process, changing the prediction process autonomously when necessary. Frequently, the data generated in the middle of processing stage need to be processed or changed so the data can be easily worked, or in some cases, the data should be processed through some specified software. The agents in MAPS decide whether or not to perform these jobs through interactions among agents.

The MAPS architecture is divided into four types of agent that can be classified according to their roles and functions:

1. Interface agents UI and PE;
2. Brokering agents AMS, DF and CODY;
3. Task agents NBLAST, PBLAST, CLUSTALW, MODELLER and SVIEWER; and
4. Resource agents PDB, 3DPSSM, SPROT, PPRED, SCOP, PDPROTEIN and HMMSTR.

Interface agents are the interface of the system and have to support users in using the entire system. Brokering agents control and coordinate the operations among agents based on plans, and save that information as needed. CODY agent is the most important component of the process, performing the planning, coordinating, communicating and mediating based on specific multi-agent plans. The coordinating process of the CODY agent is mostly placed by interactions with other agents. These interactions are a request and response-based message communications. AMS and DF agents perform administration and directory services that are essential in operating a multi-agent system. In MAPS, the whole process is divided into primitive plan units, and the global goal is achieved by carrying out these plans consecutively. However, these primitive plan units change along the prediction process, so it is necessary to specify the preconditions for plan execution and selectively choose the plans to execute in accordance with these conditions. Task agents compute the information produced by resource agents and analyze the results. Resource agents connect to the protein databases which are used for protein structure prediction, query, and find the information, replacing human browsing and searching. For further details regarding the agents and their tasks see Jin and Kim (2004).

To predict protein structures, MAPS handles with multiple agents that are distributed on many different platforms. Hence, the system cannot operate based on a complete, single plan that is predefined. The MAPS is connected to Agentcities network to support openness and decentralization of agent system utilizing JADE (Java Agent Development Framework) for openness, and JAM architecture for decentralization. JADE (Bellifemine, Caire, & Greenwood, 2007) is an FIPA-ACL (http://www.fipa.org/

repository/aclspecs.html) compliant middleware fully implemented in the Java language. It allows the development of multi-agent application system based on peer-to-peer network architecture. Agentcities network environment is used to facilitate the integration services of heterogeneous agent platforms utilizing a mechanism highly adaptable and useful due to its capability of interaction through ACL messages. Agentcities are compatible with several agent platforms, such as April, Comtec, FIPA-OS, JADE, and ZEUS. JAM is an agent architecture for supporting distributed plan-based reasoning. It achieves a goal by performing a hierarchy of small plan units, which leads to the fulfillment of one major global plan. Therefore, MAPS performs the protein structure prediction by coordinating CODY agents by plans, unlike the conventional agent-based systems. Also, MAPS is capable of joining and operating on heterogeneous platforms whenever the system requires, or wherever it is located, by connecting to agentcities network, increasing its openness, decentralization, and extensibility.

A Framework for Improving Protein Structure Predictions by Teamwork (Palopoli & Terracina, 2003)

Firstly, the authors formally defined the model used for representing the generic predictor. They modeled a predictor as a function *F* which takes as its input a protein sequence and returns the corresponding 3-D structure. *F* can be defined as follow:

$$F : SD \rightarrow TD \quad (3)$$

where *SD* is the source domain and *TD* is the target domain of *F*. The source domain is concerned with the representation of the protein sequences, whereas the target domain is related to the representation of the 3-D protein structures. In this work, the *SD* is defined regarding the amino acid sequences of the target proteins, and the *TD* represents the three-dimensional structure of protein as a sequence of 3-D relative positions of the Cα atoms composing the backbone. Such representation was chosen as several prediction tools are capable of predicting just the positions of the Cα atoms. The second step of the framework consists of the definition of measures relating pairs of protein structures. The goal is to evaluate the correctness of a protein structure prediction regarding its experimentally determined structure. This is the base step allowing to analyze the behavior of a predictor. The third step is a generalization of the previous one, where the interesting is to define a way to assess the behavior of a prediction method over a set of proteins. Then, after the mapping definition of the values of single predictors domains into the values of the reference domains *SD* and *TD,* and to express all the functions associated with the predictors regarding them, the next step is related to evaluating different protein structure predictors. It can be performed by means of an affinity coefficient of a pair of predictors. Finally, the last step of the framework is intended to define a technique for obtaining one, and as accurate as possible, the prediction from the set of predictions yielded by a group of methods. For a complete description of the system formalization, we refer the reader to Garro et al. (2004), and Palopoli and Terracina (2003).

This framework has been exploited for the definition of a multi-agent system called X-MACoP (XML Multi-Agent system for the Collaborative Prediction of protein structures). The main features of X-MACoP are the automatic selection of a team of predictors to be jointly applied to the prediction of protein structures, the integration of the predictions yielded by the predictors of the team for obtaining a unique and feasible prediction, and the mapping of the predictor inputs and outputs in such a way that a user handles a similar data format. The X-MACoP architecture has two kinds of agents, called as the user agent and the predictor agent. A generic user agent UA_i is associated with a user u_i and assists it in performing prediction tasks, such as the definition of a set of proteins *PS* to associate with the prediction task t_i; computation of the affinity coefficients of the predictors related to *PS*; construction of a predictor team for UA_i and t_i; and prediction of the 3-D structure of the input protein sequence *ips* obtained by integrating the results returned by the predictor agents of t_i when applied on the *ips*. A predictor agent is associated with a particular prediction tool and collaborates with both the user agent and other predictor agents for defining predictor teams and carrying out prediction tasks. A predictor agent PA_i supports a user agent UA_i in the execution of a prediction task t_i by predicting the 3-D structure of proteins comprising the set *PS* related to the UA_i assigned to it, and suggesting other predictor agents, unknown to the user agent UA_i, which might be considered for the composition of the predictor team associated with UA_i and t_i.

In this work, the authors have applied the framework to analyze the behavior of three methods; that is the SWISS-MODEL (Guex & Peitsch, 1997), the CPHmodels (Lund et al., 1997) and the DOE FOLD Server (Fischer & Eisenberg, 1996). SWISS-MODEL and CPHmodels are homology modeling based tools, whereas DOE FOLD Server is a threading-based approach. The three tools have been applied to a small set of protein sequences. After the formalization and translation of the methods to the framework definitions, the authors stated this system contributes to the context of collaborative protein structure prediction, but according to the results extracted from this test case, the exploitation of the DOE FOLD tool would be not helpful to improve the prediction quality. It was confirmed by the example where, by applying a team comprising the three methods on a protein of unknown structure, only the results yielded by SWISS-MODEL and CPHmodels are exploited to constitute the final team prediction. They also showed that the framework provided a formalism allowing to handle different predictors easily and has been exploited for the definition of the X-MACoP multi-agent system supporting users in the task of predicting the 3-D structure of proteins.

GeneSilico Protein Structure Prediction Meta-Server (Kurowski & Bujnicki, 2003)

Kurowski and Bujnicki (2003) developed a meta-server called as GeneSilico in an attempt to provide a convenient, secure and simple on-line structure prediction service. The system prioritizes the quality of the final results instead of the execution speed. For this, users need to submit manually refined sequence alignments (multiple sequence alignments) to obtain potentially more accurate predictions. GeneSilico is a WWW meta-server as a gateway to several protein structure prediction methods. It eases the access to several methods through a single, secure and user-friendly WWW interface. The system was implemented to allows easy web scripting to simplify automated request and retrieval of data by clients (user-agents) based on the XML-RPC serialization.

The user has several options for submission of the prediction query. The server accepts both single sequences and multi-alignment sequences. The system includes a set of components, such as: (i) the HMMPFAM tool (Mulder et al., 2003) which is used to identify and analyze the primary structure of the target; (ii) PSIPRED (Buchan et al., 2013), SAM-T02 (Karplus et al., 2001) and PROF (Ouali & King, 2000) can be used to predict the secondary structure; (iii) a local PDB-BLAST filter with the PSI-BLAST algorithm (Altschul et al., 1997) for identification of closely related sequences of known structures in PDB; (iv) the 3-D structure prediction core; and (v) the consensus server Pcons (Lundstrom et al., 2001) used to analyze the results returned by the prediction servers. The 3-D prediction core of the GeneSilico is the most important module of the method. It comprises the best methods available, according to the CAFASP (Fischer et al., 2001) and LiveBench (Bujnicki, Elofsson, Fischer, & Rychlewski, 2001b) experiments: RAPTOR (Xu, Li, Kim, & Xu, 2003), 3D-pssm (Kelley et al., 2000), FUGUE (Shi, Blundell, & Mizuguchi, 2001), GenTHREADER (Jones, 1999), SAM-T02 (Karplus et al., 2001) and BIOINBGU (Fischer, 2000).

3D-Jury: A Simple Approach to Improve Protein Structure Predictions (Ginalski, Elofsson, Fischer, & Rychlewski, 2003)

The primary purpose of the method is to create a simple but powerful method for generating an ensemble of potential solutions using variable sets of models extracted from different prediction server sources. The motivation of the method can be explained using the experience with ab initio prediction methods which lead to the conclusion that averages of low-energy conformations obtained most frequently by folding simulations are closer to the native structure than the conformation with the lowest energy. Currently, 3D-Jury (Ginalski et al., 2003) is one of the most simple and traditional meta-servers. It computes structural similarities between models generated by a set of servers using the MaxSub tool (Siew, Elofsson, Rychlewski, & Fischer, 2000). However, the authors placed any other similar program can be utilized as well. The system neglects the assigned confidence scores to the models. The final 3D-Jury score of a given model is the sum of all similarity scores of the considered model pairs divided by the number of the considered pairs plus one. At the end of the comparison step, among all the possible models for the protein target, it chooses the most realistic one (higher similarity score) as the predicted final result. According to the authors, the system follows a simple protocol that can be easily reproduced and incorporated into other fold recognition programs. Furthermore, the proposed protocol should help to improve the quality of structural annotations of novel proteins. They also note the system does not guarantee that the correct model will be selected from a set of preliminary models, especially if the right solution is an outlier and is provided by only a single server.

3D-Judge: A Metaserver Approach to Protein Structure Prediction (Jaskowski et al., 2007)

The proposed framework, namely 3D-Judge, is a meta-server that uses an artificial neural network (ANN) to select the best model from among models produced by individual servers. This decision is made basing on the mutual similarities between models produced by the servers and the knowledge obtained during the training. ANN is trained on historical data (models from CASP experiment). In contrast of the 3D-Jury (Ginalski et al., 2003), described above, 3D-Judge tries to find more sophisticated relationships between models and underlying methods. Historical knowledge about similarity patterns among models and real structures is incorporated in the prediction scheme using a powerful mechanism for neural networks. The method used to compare the protein structures was GDT (Global Distance Test) (Zemla, 2003), but any similarity comparison method could be used instead, or indeed a similarity meta-server. The system was implemented in C++ with the use of Fast Artificial Neural Network Library (FANN) (Nissen, 2003). The obtained results for the meta server show that it has a great potential. In comparison with the 3D-Jury, the authors have shown that for some set of servers, 3D-Jury can perform very poor whereas 3D-Judge outperforms all the individual servers. The practical problem with the 3D-Judge is that the good performance is intrinsically related to the quality of historical data, and the method cannot operate without them. Nevertheless, the performed experiments have shown that the 3D-Judge meta-server is competitive to the other successful methods, such as the 3D-Jury.

CONCLUSION

This chapter has provided an integrated view and insights about the protein structural area concerned with the project, application and implementation of multi-agent systems to predict the 3-D structure of protein (methods described in Section 3.1) or to support and coordinate the existing predictors (meta-servers outlined in section 3.2), as well it is advantages, issues, needs, and demands. Determining the 3-D structure of a protein is both experimentally expensive (due to the costs associated with crystallography or NMR), and time-consuming. Tertiary protein structure prediction is currently one of the challenging problems in Structural Bioinformatics, and a wide range of approaches are being applied to find approximated solutions. Multi-agent systems are also used to cover complex problems such as PSP, and MAS are developed to explore the wealth of information in experimental structures solved at atomic resolution; investigate the properties of biological systems that are difficult to study in more traditional ways; simulate and implement the complex behaviors of biological processes; etc.

Protein Structure Prediction is a tough problem. The development of new strategies, the adaptation, and investigation of new methods and the combination of existing and state-of-the-art computational methods and techniques to the PSP problem is clearly needed. Multi-agent systems can substantially contribute to this field regarding conceptual frameworks to model agent-based artificial societies, which can easily simulate and implement the complex behaviors of biological processes. Nevertheless, it can be seen that, currently, there is a lack of works related to multi-agent and agent-based paradigms applied to PSP problem, principally when restricting the search to methods based on concepts of the first principle methods.

Structural bioinformatics deals with problems where the rules that govern the biochemical processes and relations are partially known which makes hard to design efficient computational strategies for these problems. Especially in the PSP problem, there is a broad range of unanswered biological questions related to the protein folding process in nature. Predicting the structure of large proteins remains a challenge, with bottlenecks from the force field, conformational search methods, identification of correct templates, refinement of model structures closer to the native one, etc. Along the last decades, many methods and computational strategies have been proposed as a solution for the PSP problem. Each method incorporates different biological knowledge to model the PSP problem. These knowledge are related on how the polypeptide structure is represented, how the molecular forces are computed and how predicted protein structures are scored. As reviewed in this chapter some representations of the Protein Structure Prediction problem explores the hydrophobicity of atoms, the interaction between hydrophobic and polar amino acid residues, and others provide more detailed descriptions using all-atom models. Despite the advances in the development of methods for the PSP problem, especially in the application of techniques using multi-agent concepts, there is not even a general method for the prediction of protein structures. A remarkable advance in the field of protein structure prediction is the meaning of meta-strategies or meta-servers. This idea relies on the concept of consensus-based approach where independent predictions are combined to produce better results than the individual tools. Meta-servers could be easily implemented by agent philosophies to provide more reliable prediction methods. Considering the state of the art of prediction methods, where each method takes into account specific properties of the problem, the use of meta-servers employing concepts of multi-agents systems to guide the search process seems one of the best avenues for future research.

REFERENCES

Altschul, S. F., Madden, T. L., Schäffer, A. A., Zhang, J., Zhang, Z., Miller, W., & Lipman, D. J. (1997). Gapped blast and psi-blast: A new generation of protein database search programs. *Nucleic Acids Research, 25*(17), 3389–3402. doi:10.1093/nar/25.17.3389 PMID:9254694

Amigoni, F., & Schiaffonati, V. (2007). Multiagent-based simulation in biology. In *Model-based reasoning in science, technology, and medicine* (pp. 179–191). Springer Berlin Heidelberg. doi:10.1007/978-3-540-71986-1_10

Andreeva, A., Howorth, D., Chandonia, J. M., Brenner, S. E., Hubbard, T. J., Chothia, C., & Murzin, A. G. (2008). Data growth and its impact on the scop database: New developments. *Nucleic Acids Research, 36*(suppl 1), D419–D425. doi:10.1093/nar/gkm993 PMID:18000004

Anfinsen, C. (1973). Principles that govern the folding of protein chains. *Science, 181*(4096), 223–230. doi:10.1126/science.181.4096.223 PMID:4124164

Anfinsen, C., Haber, E., Sela, M., & White, F. H. J. (1961). The kinetics of formation of native ribonuclease during oxidation of the reduced polypeptide chain. *Proceedings of the National Academy of Sciences of the United States of America, 47*(9), 1309–1314. doi:10.1073/pnas.47.9.1309 PMID:13683522

Bates, P. A., Kelley, L. A., MacCallum, R. M., & Sternberg, M. J. (2001). Enhancement of protein modeling by human intervention in applying the automatic programs 3d-jigsaw and 3d-pssm. *Proteins: Struct., Funct. Bioinf., 45*(S5), 39–46.

Baxevanis, A. D., & Ouellette, B. F. (2004). *Bioinformatics: a practical guide to the analysis of genes and proteins*. John Wiley & Sons.

Bellifemine, F. L., Caire, G., & Greenwood, D. (2007). *Developing multi-agent systems with jade*. John Wiley & Sons. doi:10.1002/9780470058411

Berman, H. M., Westbrook, J., Feng, Z., Gilliland, G., Bhat, T., Weissig, H., & Bourne, P. E. et al. (2000). The protein data bank. *Nucleic Acids Research, 28*(1), 235–242. doi:10.1093/nar/28.1.235 PMID:10592235

Boas, F. E., & Harbury, P. B. (2007). Potential energy functions for protein design. *Current Opinion in Structural Biology, 17*(2), 199–204. doi:10.1016/j.sbi.2007.03.006 PMID:17387014

Boeckmann, B., Bairoch, A., Apweiler, R., Blatter, M.-C., Estreicher, A., & Gasteiger, E. et al.. (2003). The SWISS-PROT protein knowledgebase and its supplement TrEMBL in 2003. *Nucleic Acids Research, 31*(1), 365–370. doi:10.1093/nar/gkg095 PMID:12520024

Bortolussi, L., Dal Palu, A., Dovier, A., & Fogolari, F. (2004). Protein folding simulation in CCP. In Proceedings of bioconcur2004.

Bortolussi, L., Dovier, A., & Fogolari, F. (2005). Multi-agent simulation of protein folding. In *Proceedings of the first international workshop on multi- agent systems for medicine, computational biology, and bioinformatics.*

Bortolussi, L., Dovier, A., & Fogolari, F. (2007). Agent-based protein structure prediction. *Multiagent and Grid Systems, 3*(2), 183–197. doi:10.3233/MGS-2007-3204

Branden, C., & Tooze, J. (1998). *Introduction to protein structure* (2nd ed.). New York: Garlang Publishing Inc.

Brooks, R., Bruccoleri, R., Olafson, B., States, D., Swaminathan, S., & Karplus, M. (1983). Charmm: A program for macromolecular energy, minimization, and dynamics calculations. *Journal of Computational Chemistry*, *4*(2), 187–217. doi:10.1002/jcc.540040211

Bryant, S. H., & Altschul, S. (1995). Statistics of sequence-structure threading. *Current Opinion in Structural Biology*, *5*(2), 236–244. doi:10.1016/0959-440X(95)80082-4 PMID:7648327

Buchan, D. W., Minneci, F., Nugent, T. C., Bryson, K., & Jones, D. T. (2013). Scalable web services for the psipred protein analysis workbench. *Nucleic Acids Research*, *41*(W1), W349–W357. doi:10.1093/nar/gkt381 PMID:23748958

Bujnicki, J., Elofsson, A., Fischer, D., & Rychlewski, L. (2001a). Structure prediction meta server. *Bioinformatics (Oxford, England)*, *17*(8), 750–751. doi:10.1093/bioinformatics/17.8.750 PMID:11524381

Bujnicki, J., Elofsson, A., Fischer, D., & Rychlewski, L. (2001b). Livebench-2: large-scale automated evaluation of protein structure prediction servers. *Proteins: Struct., Funct. Bioinf.*, *45*(S5), 184–191.

Bystroff, C., Thorsson, V., & Baker, D. (2000). Hmmstr: A hidden markov model for local sequence-structure correlations in proteins. *Journal of Molecular Biology*, *301*(1), 173–190. doi:10.1006/jmbi.2000.3837 PMID:10926500

Campeotto, F., Dovier, A., & Pontelli, E. (2013). Protein structure prediction on gpu: a declarative approach in a multi-agent framework. In *Parallel processing (icpp), 2013 42nd international conference on* (pp. 474–479). doi:10.1109/ICPP.2013.57

Campeotto, F., Dovier, A., & Pontelli, E. (2015). A declarative concurrent system for protein structure prediction on gpu. *Journal of Experimental & Theoretical Artificial Intelligence*, *27*(5), 503–541. doi:10.1080/0952813X.2014.993503

Canutescu, A., Shelenkov, A., & Dunbrack, R. Jr. (2001). A graph-theory algorithm for rapid protein side chain prediction. *Proteins*, *12*(9), 2001–2014. doi:10.1110/ps.03154503 PMID:12930999

Carlsson, M., & Mildner, P. (2012). Sicstus prologâthe first 25 years. *Theory and Practice of Logic Programming*, *12*(1-2), 35–66. doi:10.1017/S1471068411000482

Carriero, N., & Gelernter, D. (1989). Linda in context. *Communications of the ACM*, *32*(4), 444–458. doi:10.1145/63334.63337

Cavanagh, J., Fairbrother, W., Palmer, A. III, & Skelton, N. (1995). *Protein nmr spectroscopy: principles and practice* (1st ed.). New York: Academic Press.

Chivian, D., Robertson, T., Bonneau, R., & Baker, D. (2003). Ab initio methods. *Methods of Biochemical Analysis*, *44*, 547–557. PMID:12647404

Chou, K. C. (2004). Structural bioinformatics and its impact to biomedical science. *Current Medicinal Chemistry*, *11*(16), 2105–2134. doi:10.2174/0929867043364667 PMID:15279552

Cipriano, R. (2008). On the hybridization of constraint programming and local search techniques: Models and software tools. In *Logic programming* (pp. 803–804). Springer Berlin Heidelberg. doi:10.1007/978-3-540-89982-2_81

Cornell, W., Cieplak, P., Bayly, C., Gould, I., Merz, K. Jr, Ferguson, D., & Kollman, P. et al. (1995). A second generation force field for the simulation of proteins, nucleic acids, and organic molecules. *Journal of the American Chemical Society*, *117*(19), 5179–5197. doi:10.1021/ja00124a002

Creighton, T. E. (1990). Protein folding. *The Biochemical Journal*, *270*(1), 1–16. doi:10.1042/bj2700001 PMID:2204340

Crescenzi, P., Goldman, D., Papadimitriou, C., Piccolboni, A., & Yannakakis, M. (1998). On the complexity of protein folding. *Journal of Computational Biology*, *5*(3), 423–465. doi:10.1089/cmb.1998.5.423 PMID:9773342

De Mori, G. M., Micheletti, C., & Colombo, G. (2004). All-atom folding simulations of the villin headpiece from stochastically selected coarse-grained structures. *The Journal of Physical Chemistry B*, *108*(33), 12267–12270. doi:10.1021/jp0477699

Dill, K., Bromberg, S., Yue, K., Fiebig, K., Yee, D., Thomas, P., & Chan, H. (1995). Principles of protein folding: A perspective from simple exact models. *Protein Science*, *4*(4), 561–602. doi:10.1002/pro.5560040401 PMID:7613459

Dorn, M., Silva, M. B., Buriol, L. S., & Lamb, L. C. (2014). Three-dimensional protein structure prediction: Methods and computational strategies. *Computational Biology and Chemistry*, *53*, 251–276. doi:10.1016/j.compbiolchem.2014.10.001 PMID:25462334

Dunker, A., Lawson, J., Brown, C., Williams, R., Romero, P., Oh, J., & Obradovic, Z. et al. (2001). Intrinsically disordered protein. *Journal of Molecular Graphics & Modelling*, *19*(1), 26–59. doi:10.1016/S1093-3263(00)00138-8 PMID:11381529

Dunker, A., Silman, I., Uversky, V., & Sussman, J. (2008). Function and structure of inherently disordered proteins. *Current Opinion in Structural Biology*, *18*(6), 756–764. doi:10.1016/j.sbi.2008.10.002 PMID:18952168

Feig, M., Rotkiewicz, P., Kolinski, A., Skolnick, J., & Brooks, C. (2000). Accurate reconstruction of all-atom protein representations from side-chain- based low-resolution models. *Proteins*, *41*(1), 86–97. doi:10.1002/1097-0134(20001001)41:1<86::AID-PROT110>3.0.CO;2-Y PMID:10944396

Fischer, D. (2000). Hybrid fold recognition: Combining sequence derived properties with evolutionary information. *Pacific Symposium on Biocomputing. Pacific Symposium on Biocomputing*, *5*, 119–130. PMID:10902162

Fischer, D. (2006). Servers for protein structure prediction. *Current Opinion in Structural Biology*, *16*(2), 178–182. doi:10.1016/j.sbi.2006.03.004 PMID:16546376

Fischer, D., & Eisenberg, D. (1996). Protein fold recognition using sequence- derived predictions. *Protein Science*, *5*(5), 947–955. doi:10.1002/pro.5560050516 PMID:8732766

Fischer, D., Elofsson, A., Rychlewski, L., Pazos, F., Valencia, A., Rost, B., & Dunbrack, R. L. et al. (2001). Cafasp2: the second critical assessment of fully automated structure prediction methods. *Proteins: Struct., Funct. Bioinf.*, *45*(S5), 171–183.

Floudas, C. (2007). Computational methods in protein structure prediction. *Biotechnology and Bioengineering*, *97*(2), 207–213. doi:10.1002/bit.21411 PMID:17455371

Fogolari, F., Pieri, L., Dovier, A., Bortolussi, L., Giugliarelli, G., Corazza, A., & Viglino, P. et al. (2007). Scoring predictive models using a reduced representation of proteins: Model and energy definition. *BMC Structural Biology*, *7*(1), 1–17. doi:10.1186/1472-6807-7-15 PMID:17378941

Garro, A., Terracina, G., & Ursino, D. (2004). A multi-agent system for supporting the prediction of protein structures. *Integr. Comput. Aid. E.*, *11*(3), 259–280.

Gibas, C., & Jambeck, P. (2001). *Developing bioinformatics computer skills*. Sebastopol, CA: O'Reilly Media, Inc.

Ginalski, K., Elofsson, A., Fischer, D., & Rychlewski, L. (2003). 3d-jury: A simple approach to improve protein structure predictions. *Bioinformatics (Oxford, England)*, *19*(8), 1015–1018. doi:10.1093/bioinformatics/btg124 PMID:12761065

Gordon, D., Marshall, S., & Mayo, S. (1999). Energy functions for protein design. *Current Opinion in Structural Biology*, *9*(4), 509–513. doi:10.1016/S0959-440X(99)80072-4 PMID:10449371

Guex, N., & Peitsch, M. C. (1997). SWISS-MODEL and the Swiss-Pdb Viewer: An environment for comparative protein modeling. *Electrophoresis*, *18*(15), 2714–2723. doi:10.1002/elps.1150181505 PMID:9504803

Gunasekaran, K., Tsai, C., Kumar, S., Zanuy, D., & Nussinov, R. (2003). Extended disordered proteins: Targeting function with less scaffold. *Trends in Biochemical Sciences*, *28*(2), 81–85. doi:10.1016/S0968-0004(03)00003-3 PMID:12575995

Guyeux, C., Côte, N. M.-L., Bahi, J. M., & Bienia, W. (2014). Is protein folding problem really a np-complete one? first investigations. *Journal of Bioinformatics and Computational Biology*, *12*(01), 1350017–1350041. doi:10.1142/S0219720013500170 PMID:24467756

Halgren, T. A. (1995). Potential energy functions. *Current Opinion in Structural Biology*, *5*(2), 205–210. doi:10.1016/0959-440X(95)80077-8 PMID:7648322

Hao, M., & Scheraga, H. (1999). Designing potential energy functions for protein folding. *Current Opinion in Structural Biology*, *9*(2), 184–188. doi:10.1016/S0959-440X(99)80026-8 PMID:10322206

Hentenryck, P. V., & Michel, L. (2009). *Constraint-based local search*. The MIT Press.

Hovmoller, T., & Ohlson, T. (2002). Conformation of amino acids in protein. *Acta Crystallographica*, *58*(5), 768–776. PMID:11976487

Jaskowski, W., Blazewicz, J., Lukasiak, P., Milostan, M., & Krasnogor, N. (2007). 3d-judge–a metaserver approach to protein structure prediction. *Found. Comput. Decis. Sci.*, *32*(1), 3–14.

Jennings, N. R., Sycara, K., & Wooldridge, M. (1998). A roadmap of agent research and development. *Autonomous Agents and Multi-Agent Systems, 1*(1), 7–38. doi:10.1023/A:1010090405266

Jiang, T., Cui, Q., Shi, G., & Ma, S. (2003). Protein folding simulations of the hydrophobich-hydrophilic model by combining tabu search with genetic algorithms. *The Journal of Chemical Physics, 119*(8), 4592–4596. doi:10.1063/1.1592796

Jin, H., & Kim, I. C. (2004). Plan-based coordination of a multi-agent system for protein structure prediction. In *International Conference on AI, Simulation, and Planning in High Autonomy Systems* (pp. 224–232). Springer Berlin Heidelberg.

Jones, D., Taylor, W., & Thornton, J. (1992). A new approach to protein fold recognition. *Nature, 358*(6381), 86–89. doi:10.1038/358086a0 PMID:1614539

Jones, D. T. (1999). Genthreader: An efficient and reliable protein fold recognition method for genomic sequences. *Journal of Molecular Biology, 287*(4), 797–815. doi:10.1006/jmbi.1999.2583 PMID:10191147

Jorgensen, W., & Tirado-Rives, J. (2005). Potential energy functions for atomic- level simulations of water and organic and biomolecular systems. *Proceedings of the National Academy of Sciences of the United States of America, 102*(19), 6665–6670. doi:10.1073/pnas.0408037102 PMID:15870211

Karplus, K., Karchin, R., Barrett, C., Tu, S., Cline, M., Diekhans, M., & Hughey, R. et al. (2001). What is the value added by human intervention in protein structure prediction? *Proteins: Struct., Funct. Bioinf., 45*(S5), 86–91.

Kelley, L. A., MacCallum, R. M., & Sternberg, M. J. (2000). Enhanced genome annotation using structural profiles in the program 3d-pssm. *Journal of Molecular Biology, 299*(2), 501–522. doi:10.1006/jmbi.2000.3741 PMID:10860755

Kim, D. E., Chivian, D., & Bȧker, D. (2004). Protein structure prediction and analysis using the Robetta server. *Nucleic Acids Research, 32*(S2), W526–W531. doi:10.1093/nar/gkh468 PMID:15215442

Kirkpatrick, S., Gelatt, C., & Vecchi, M. (1983). Optimization by simulated annealing. *Science, 220*(4598), 671–680. doi:10.1126/science.220.4598.671 PMID:17813860

Kolinski, A., & Skolnick, J. (2004). Reduced models of proteins and their applications. *Polymer, 45*(2), 511–524. doi:10.1016/j.polymer.2003.10.064

Koppensteiner, W. A., & Sippl, M. J. (1995). Knowledge-based potentials-back to the roots. *Biochemistry, 63*, 247–252. PMID:9526121

Kryshtafovych, A., Fidelis, K., & Moult, J. (2014). Casp10 results compared to those of previous casp experiments. *Proteins: Struct., Funct. Bioinf., 82*(S2), 164–174.

Kurowski, M. A., & Bujnicki, J. M. (2003). Genesilico protein structure prediction meta-server. *Nucleic Acids Research, 31*(13), 3305–3307. doi:10.1093/nar/gkg557 PMID:12824313

Lashkari, Y., Metral, M., & Maes, P. (1994). Collaborative interface agents. In *Proceedings of the 12th national conference on artificial intelligence* (p. 444-449). Elsevier.

Laskowiski, R., Watson, J., & Thornton, J. (2005a). Profunc: A server for predicting protein functions from 3d structure. *Nucleic Acids Research*, *33*, 89–93. doi:10.1093/nar/gki414

Laskowiski, R., Watson, J., & Thornton, J. (2005b). Protein function prediction using local 3d templates. *Journal of Molecular Biology*, *351*(3), 614–626. doi:10.1016/j.jmb.2005.05.067 PMID:16019027

Lau, K., & Dill, K. (1989). A lattice statistical mechanics model of the conformation and sequence spaces of proteins. *Macromolecules*, *22*(10), 3986–3997. doi:10.1021/ma00200a030

Lazaridis, T., & Karplus, M. (2000). Effective energy functions for protein structure prediction. *Current Opinion in Structural Biology*, *10*(2), 139–145. doi:10.1016/S0959-440X(00)00063-4 PMID:10753811

Lehninger, A., Nelson, D., & Cox, M. (2005). *Principles of biochemistry* (4th ed.). New York, USA: W.H. Freeman.

Lesk, A. M. (2002). *Introduction to bioinformatics* (1st ed.). New York, NY: Oxford University Press Inc.

Lesk, A. M. (2010). *Introduction to protein science: architecture, function, and genomics*. New York: Oxford University Press.

Levinthal, C. (1968). Are there pathways for protein folding? *Journal de Chimie Physique*, *65*(1), 44–45.

Liljas, A., Liljas, L., Piskur, J., Lindblom, G., Nissen, P., & Kjeldgaard, M. (2009). *Textbook of structural biology*. Singapore: World Scientific. doi:10.1142/6620

Lipinski-Paes, T., & De Souza, O. N. (2014). Masters: A general sequence- based multiagent system for protein tertiary structure prediction. *Electronic Notes in Theoretical Computer Science*, *306*, 45–59. doi:10.1016/j.entcs.2014.06.014

Lodish, H., Berk, A., Matsudaira, P., Kaiser, C. A., Krieger, M., & Scott, M. (1990). Molecular cell biology (5th ed.). New York: Scientific American Books, W.H. Freeman.

Lund, O., Frimand, K., Gorodkin, J., Bohr, H., Bohr, J., Hansen, J., & Brunak, S. (1997). Protein distance constraints predicted by neural networks and probability density functions. *Protein Engineering*, *10*(11), 1241–1248. doi:10.1093/protein/10.11.1241 PMID:9514112

Lundstrom, J., Rychlewski, L., Bujnicki, J., & Elofsson, A. (2001). Pcons: A neural-network based consensus predictor that improves fold recognition. *Protein Science*, *10*(11), 2354–2362. doi:10.1110/ps.08501 PMID:11604541

MacKerell, A. D. (2004). Empirical force fields for biological macromolecules: Overview and issues. *Journal of Computational Chemistry*, *25*(13), 1584–1604. doi:10.1002/jcc.20082 PMID:15264253

Martí-Renom, M., Stuart, A., Fiser, A., Sanchez, A., Mello, F., & Sali, A. (2000). Comparative protein structure modeling of genes and genomes. *Annual Review of Biophysics and Biomolecular Structure*, *29*(16), 291–325. doi:10.1146/annurev.biophys.29.1.291 PMID:10940251

McRee, D. (1999). *Practical protein crystallography* (1st ed.). London: Academic press.

Merelli, E., Armano, G., Cannata, N., Corradini, F., dInverno, M., Doms, A., & Schroeder, M. et al. (2007). Agents in bioinformatics, computational and systems biology. *Briefings in Bioinformatics*, *8*(1), 45–59. doi:10.1093/bib/bbl014 PMID:16772270

Micheletti, C., Seno, F., & Maritan, A. (2000). Recurrent oligomers in proteins: an optimal scheme reconciling accurate and concise backbone representations in automated folding and design studies. *Proteins: Struct., Funct. Bioinf.*, *40*(4), 662–674.

Milano, M., & Roli, A. (2004). MAGMA: A multiagent architecture for metaheuristics. *IEEE Transactions on Systems, Man, and Cybernetics. Part B, Cybernetics*, *34*(2), 925–941. doi:10.1109/TSMCB.2003.818432 PMID:15376840

Modi, P. J., Shen, W.-M., Tambe, M., & Yokoo, M. (2005). Adopt: Asynchronous distributed constraint optimization with quality guarantees. *Artificial Intelligence*, *161*(1), 149–180. doi:10.1016/j.artint.2004.09.003

Momany, F., McGuire, R., Burgess, A., & Scheraga, H. (1975). Energy parameters in polypeptides vii, geometric parameters, partial charges, non- bonded interactions, hydrogen bond interactions and intrinsic torsional potentials for naturally occurring amino acids. *Journal of Physical Chemistry*, *79*(22), 2361–2381. doi:10.1021/j100589a006

Moult, J., Fidelis, K., Kryshtafovych, A., Schwede, T., & Tramontano, A. (2014). Critical assessment of methods of protein structure prediction (CASP)-round x. *Proteins: Struct., Funct. Bioinf.*, *82*(S2), 1–6.

Mulder, N. J., Apweiler, R., Attwood, T. K., Bairoch, A., Barrell, D., & Bateman, A. et al.. (2003). The interpro database, 2003 brings increased coverage and new features. *Nucleic Acids Research*, *31*(1), 315–318. doi:10.1093/nar/gkg046 PMID:12520011

Muscalagiu, I., Iordan, A., Osaci, M., & Panoiu, M. (2012). Modeling and simulation of the protein folding problem in discsp-netlogo. *Global Journal on Technology*, *2*, 229–234.

Muscalagiu, I., Popa, H. E., Panoiu, M., & Negru, V. (2013). Multi-agent systems applied in the modelling and simulation of the protein folding problem using distributed constraints. In *Multiagent system technologies* (pp. 346–360). Springer Berlin Heidelberg. doi:10.1007/978-3-642-40776-5_29

Muscalagiu, I., Popa, H. E., & Vidal, J. (2013). Clustered computing with netlogo for the evaluation of asynchronous search techniques. In *Intelligent software methodologies, tools and techniques (somet), 2013 IEEE 12th international conference on* (pp. 115–120). IEEE. doi:10.1109/SoMeT.2013.6645651

Neumaier, A. (1997). Molecular modeling of proteins and mathematical prediction of protein structure. *Society for Industrial and Applied Mathematics Rev.*, *39*, 407.

Nissen, S. (2003). *Implementation of a fast artificial neural network library (fann).* Report, Department of Computer Science University of Copenhagen (DIKU).

Nwana, H. S. (1996). Software agents: An overview. *The Knowledge Engineering Review*, *11*(3), 205–244. doi:10.1017/S026988890000789X

Osguthorpe, D. J. (2000). Ab initio protein folding. *Current Opinion in Structural Biology*, *10*(2), 146–152. doi:10.1016/S0959-440X(00)00067-1 PMID:10753815

Ouali, M., & King, R. D. (2000). Cascaded multiple classifiers for secondary structure prediction. *Protein Science, 9*(06), 1162–1176. doi:10.1110/ps.9.6.1162 PMID:10892809

Palopoli, L., & Terracina, G. (2003). A framework for improving protein structure predictions by teamwork. In *Proceedings of the first asia-pacific bioinformatics conference on bioinformatics 2003* (vol. 19, pp. 163–171).

Park, B., Huang, E., & Levitt, M. (1997). Factors affecting the ability of energy functions to discriminate correct from incorrect folds. *Journal of Molecular Biology, 266*(4), 831–846. doi:10.1006/jmbi.1996.0809 PMID:9102472

Pauling, L., & Corey, R. (1951). The pleated sheet, a new layer configuration of polypeptide chains. *Proceedings of the National Academy of Sciences of the United States of America, 37*(5), 251–256. doi:10.1073/pnas.37.5.251 PMID:14834147

Pauling, L., Corey, R., & Branson, H. (1951). The structure of proteins: Two hydrogen-bonded helical configurations of the polypeptide chain. *Proceedings of the National Academy of Sciences of the United States of America, 37*(4), 205–211. doi:10.1073/pnas.37.4.205 PMID:14816373

Pérez, P. P. G., Beltrán, H. I., Rojo-Domínguez, A., Eduardo, M., & Gutiérrez, S. (2009). Multi-agent systems applied in the modeling and simulation of biological problems: A case study in protein folding. *World Acad. Sci. Eng. Technol., 3*(10), 497–506.

Pokala, N., & Handel, T. (2000). Review: Protein design - where we were, where we are, where were going. *Journal of Structural Biology, 134*(2-3), 269–281. doi:10.1006/jsbi.2001.4349 PMID:11551185

Pruitt, K. D., Tatusova, T., & Maglott, D. R. (2005). Ncbi reference sequence (refseq): A curated non-redundant sequence database of genomes, transcripts, and proteins. *Nucleic Acids Research, 33*(Database issueS1), D501–D504. doi:10.1093/nar/gki025 PMID:15608248

Richardson, J. S. (1981). The anatomy and taxonomy of protein structure. *Advances in Protein Chemistry, 34*, 167–339. doi:10.1016/S0065-3233(08)60520-3 PMID:7020376

Rohl, C. A., Strauss, C. E., Misura, K. M., & Baker, D. (2004). Protein structure prediction using rosetta. *Methods in Enzymology, 383*, 66–93. doi:10.1016/S0076-6879(04)83004-0 PMID:15063647

Rossi, F., Van Beek, P., & Walsh, T. (2006). *Handbook of constraint programming*. Elsevier.

Sánchez, R., & Sali, A. (1997). Advances in comparative protein-structure modeling. *Current Opinion in Structural Biology, 7*(2), 206–214. doi:10.1016/S0959-440X(97)80027-9 PMID:9094331

Scheef, E., & Fink, J. (2003). Fundamentals of protein structure. In P. Bourne & H. Weissig (Eds.), *Structural bioinformatics* (Vol. 44). Hoboken, NJ: John Wiley & Sons, Inc.

Shi, J., Blundell, T. L., & Mizuguchi, K. (2001). Fugue: Sequence-structure homology recognition using environment-specific substitution tables and structure-dependent gap penalties. *Journal of Molecular Biology, 310*(1), 243–257. doi:10.1006/jmbi.2001.4762 PMID:11419950

Siew, N., Elofsson, A., Rychlewski, L., & Fischer, D. (2000). Maxsub: An automated measure for the assessment of protein structure prediction quality. *Bioinformatics (Oxford, England)*, *16*(9), 776–785. doi:10.1093/bioinformatics/16.9.776 PMID:11108700

Simons, K. T., Kooperberg, C., Huang, E., & Baker, D. (1997). Assembly of protein tertiary structures from fragments with similar local sequences using simulated annealing and Bayesian scoring functions. *Journal of Molecular Biology*, *268*(1), 209–225. doi:10.1006/jmbi.1997.0959 PMID:9149153

Srinivasan, R., & Rose, G. D. (1995). Linus: a hierarchic procedure to predict the fold of a protein. *Proteins: Struct., Funct. Bioinf.*, *22*(2), 81–99.

Stillinger, F. H., Head-Gordon, T., & Hirshfeld, C. L. (1993). Toy model for protein folding. *Physical Review E: Statistical Physics, Plasmas, Fluids, and Related Interdisciplinary Topics*, *48*(2), 1469–1477. doi:10.1103/PhysRevE.48.1469 PMID:9960736

Tisue, S., & Wilensky, U. (2004). Netlogo: A simple environment for modeling complexity. In *International conference on complex systems* (vol. 21).

Tompa, P. (2002). Intrinsically unstructured proteins. *Trends in Biochemical Sciences*, *27*(10), 527–533. doi:10.1016/S0968-0004(02)02169-2 PMID:12368089

Tompa, P., & Csermely, P. (2004). The role of structural disorder in the function of RNA and protein chaperones. *The FASEB Journal*, *18*(11), 1169–1175. doi:10.1096/fj.04-1584rev PMID:15284216

Tramontano, A., & Lesk, A. M. (2006). *Protein structure prediction* (1st ed.). Weinheim, Germany: John Wiley and Sons, Inc.

Unger, R., & Moult, J. (1993). Finding the lowest free energy conformation of a protein is an np-hard problem: Proof and implications. *Bulletin of Mathematical Biology*, *55*(6), 1183–1198. doi:10.1007/BF02460703 PMID:8281131

Uversky, V. (2001). What does it mean to be natively unfolded? *European Journal of Biochemistry*, *269*(1), 2–12. doi:10.1046/j.0014-2956.2001.02649.x PMID:11784292

Willmott, S., Dale, J., Burg, B., Charlton, P., & O'Brien, P. (2001). Agentcities: a worldwide open agent network. *Agentlink News*, *8*.

Wooldridge, M. (2009). *An introduction to multiagent systems* (2nd ed.). John Wiley & Sons.

Wright, P., & Dyson, H. (1999). Intrinsically unstructured proteins: Re-assessing the protein structure-function paradigm. *Journal of Molecular Biology*, *293*(2), 321–331. doi:10.1006/jmbi.1999.3110 PMID:10550212

Xu, D., & Zhang, Y. (2012). Ab initio protein structure assembly using continuous structure fragments and optimized knowledge-based force field. *Proteins: Struct., Funct. Bioinf.*, *80*(7), 1715–1735.

Xu, J., Li, M., Kim, D., & Xu, Y. (2003). Raptor: Optimal protein threading by linear programming. *Journal of Bioinformatics and Computational Biology*, *1*(01), 95–117. doi:10.1142/S0219720003000186 PMID:15290783

Xu, Y., Xu, D., & Liang, J. (2007). *Computational methods for protein structure prediction and modeling*. Springer.

Yokoo, M., Durfee, E. H., Ishida, T., & Kuwabara, K. (1998). The distributed constraint satisfaction problem: Formalization and algorithms. *IEEE Transactions on Knowledge and Data Engineering*, *10*(5), 673–685. doi:10.1109/69.729707

Zemla, A. (2003). LGA: A method for finding 3d similarities in protein structures. *Nucleic Acids Research*, *31*(13), 3370–3374. doi:10.1093/nar/gkg571 PMID:12824330

Zhang, W., Yang, J., He, B., Walker, S. E., Zhang, H., Govindarajoo, B., & Zhang, Y. (2015). *Integration of QUARK and I-TASSER for Ab Initio Protein Structure Prediction in CASP11. Proteins: Struct., Funct. Bioinf.*

Zhang, Y. (2008). I-tasser server for protein 3d structure prediction. *BMC Bioinf.*, *9*(40), 1–8. doi:10.1093/bib/bbn041 PMID:18215316

Zhang, Y. (2009). I tasser: Fully automated protein structure prediction in casp8. *Proteins*, *77*(S9), 100–113. doi:10.1002/prot.22588 PMID:19768687

Zhang, Y., Kihara, D., & Skolnick, J. (2002). Local energy landscape flattering: Parallel hyperbolic Monte Carlo sampling of protein folding. *Proteins*, *48*(2), 192–201. doi:10.1002/prot.10141 PMID:12112688

Zhang, Y., & Skolnick, J. (2004a). Scoring function for automated assessment of protein structure template quality. *Proteins*, *57*(4), 702–710. doi:10.1002/prot.20264 PMID:15476259

Zhang, Y., & Skolnick, J. (2004b). Tertiary structure predictions on a comprehensive benchmark of medium to large size proteins. *Biophysical Journal*, *87*(4), 2647–2655. doi:10.1529/biophysj.104.045385 PMID:15454459

Zhang, Y., & Skolnick, J. (2004c). Automated structure prediction of weakly homologous proteins on a genomic scale. *Proceedings of the National Academy of Sciences of the United States of America*, *101*(20), 7594–7599. doi:10.1073/pnas.0305695101 PMID:15126668

Zhang, Y., & Skolnick, J. (2004d). Spicker: A clustering approach to identify near-native protein folds. *Journal of Computational Chemistry*, *25*(6), 20–22. doi:10.1002/jcc.20011 PMID:15011258

Zhong, Y., Lin, J., Du, Q., & Wang, L. (2015). Multi-agent simulated annealing algorithm based on differential perturbation for protein structure prediction problems. *International Journal of Computer Applications in Technology*, *51*(3), 164–172. doi:10.1504/IJCAT.2015.069330

Zhong, Y., Wang, L., Wang, C., & Zhang, H. (2012). Multi-agent simulated annealing algorithm based on differential evolution algorithm. *Int. J. Bio-Inspir. Com.*, *4*(4), 217–228.

Chapter 12
Biomass Variation Phytoplanktons Using Agent-Based Simulation:
A Case Study to Estuary of the Patos Lagoon

Diego de Abreu Porcellis
FURG, Brazil

Diana F. Adamatti
FURG, Brazil

Paulo Cesar Abreu
FURG, Brazil

ABSTRACT

The phytoplanktons are organisms that have limited locomotion about the current being drift in aquatic environment. Another characteristic of phytoplankton their growth and energy are result about photo-synthetic process. It is important to emphasize that the phytoplankton is the main primary producer of aquatic environment, it means that, it is the base the aquatic food chain . The organic material produced by phytoplankton is responsible in provide the material and energy which sustains the growth of fish, crustaceans and mollusks, in marine ecosystems. Because of this, it is important to know the factors that interfere with their accumulation in environments mainly in fishing regions. In this way, this study tries to demonstrate the importance of retention time, often caused by hydrological issues, in the variation of phytoplankton biomass in the estuary of the Patos Lagoon (ELP), in Rio Grande/RS. To do that, we created one model that simulates this environment, using techniques of multi-agent-based simulation and its implementation was done with the NetLogo tool.

DOI: 10.4018/978-1-5225-1756-6.ch012

INTRODUCTION

Plankton (in Greek – plâgchton) means wandering because of its power of locomotion be void against the current, and so it is "no moviment" in the aquatic environment. The Plankton can be divided into two distinct categories which are: the animals (zooplankton) and plants (phytoplankton) (Lalli and Timothy 1997). Thus, the focus of this study is to determine the variation of phytoplankton biomass in the estuary of the Patos Lagoon (ELP) in response to hydrologic conditions.

Many studies (Howarth et al. 2000, Jassby, 2008) have as the aim to know the factors that control the amount of phytoplankton biomass in aquatic environment. The tasks have been doing in this area show some factors as light nutrients and predation for the production of phytoplankton. There are other studies (Abreu et al. 2010; Lucas, Thompson and Brown, 2009) show the interference of waters movement and the process of accumulation, mainly when it refers to retention or transport time.

The previous mentioned studies are divided by types of aquatic environments and they are related to rivers, lakes, sea, estuaries and others. In our studies, we will work with estuaries. They are transitional environmental between fresh water and salty water (Pritchar, 1967). The estuary which will be used as an example for this proposal is the Estuary of the Patos Lagoon (ELP).

Estuaries are environments where mixing occurs between fresh water coming from the continent and marine waters (Pritchar, 1967). The ecological processes of estuaries are similar to other aquatic environments such as oceans, lakes and rivers. The ELP is located in the extreme south of Brazil and it is included by Patos Lagoon and its connection with Atlantic Ocean by a canal about 800 meters wide. That region has approximately 900 square kilometers and it is an environment of low depth 75% of this area has less than two deep meters (Seeliger & Odebrecht 1997). By having a close link with the sea and to be near an anfidromic point (it is a point where the tide is null), the tides are small amplitudes. Another feature about this estuary is that it is surrounded by two big cities, Pelotas and Rio Grande. Moreover, in this place is situated the Port of Rio Grande, one of the biggest of Brazil (Seelinger & Odebrecht 1997).

There is a lot of phenomena involved in the growth of phytoplankton and a great diversity of phenomena involved with estuaries and it becomes difficult to have conclusive data which determine the control factors of the variation of the biomass of phytoplankton. At present, the collection is carried out intensively in time and the space through sensors. However, these sensors must have periodic maintenance and the cost is high.

These factors are of great importance for the life of the species located in the estuary because of the low depth ensures that sunlight is not a limiting factor for the photosynthetic activity of the primary producers and the action of man ensures excess nutrient needed the development of life in this environment (Lalli & Timothy 1997). Because of this, studies have sought other factors influencing phytoplankton biomass accumulation in the ELP. These studies are directed mainly to abiotic factors such as the wind and rain, which act directly on the hydrology of this ecosystem (Odebrecht, Abreu, & Carstensen, 2015).

The multiagent system (MAS) can be an alternative to deterministic models, because with increased computational power and high flexibility can represent various behaviors of the system, using variable in close relationship to the phenomena observed in actual experiments thus making the simplest model to be improved and portray emergent phenomena (Tang, Parsons, & Sklar, 2006). Use Multi-Agent-Based Simulation (MABS) has expanded due to the increased computing power, because simulations based on this technique are characterized by encapsulating behavior of each agent and the interaction of these agents in an environment preestablished. The use of MABS to simulate complex environments can assist decision-making of the systems.

The phytoplankton biomass variation in ELP has been studied for years by several researchers (Odebrecht Abreu Carstensen, 2015) in the same way has been studied in many other estuaries around the world (Søballe & Kimmel, 1987; Reynolds, 2000; Allan & Benke, 2005; Howarth et al., 2000; Jassby, 2008). These studies, mostly, are used techniques of statistics to predict and determine the phenomena involved in phytoplankton biomass accumulation.

There is also a study of Lucas, Thompson and Brown (2009), which creates a deterministic model for general estuaries. This model was validated using statistical studies of estuaries located in the northern hemisphere. However, this model does not represent true results for the ELP. Thus, the main objective of this work focuses on the creation of an adequate model to simulate and predict the behavior of this complex system. To create this model, we used MABS. This model will assist research in regard to phytoplankton biomass accumulation in the estuary. Therefore, basing to the flexibility of MABS, it can be adapted for use in other estuaries. We are using the Netlogo tool to investigate our problem and to have a clear idea of the motivator to variation of biomass of phytoplankton in the ELP.

The main goal of our study is to clear the reasons of the biomass variation of phytoplankton in the estuary in the Patos Lagoon through a agent-based model.

We want to demonstrate the validity of studies carried out at the Institute of Oceanography of FURG (Federal University of Rio Grande) (Abreu Carstensen & Odebrecht, 2015) (Abreu et al., 2010), where the main motivator of the variation of biomass of phytoplankton is the retention time (it is the time where the estuary stays retained by the action of the wind) and we will refute the Lucas, Thompson, and Brown (2009) thesis, that affirm that the retention time does not influence on the accumulation of phytoplankton. This model will be validated using real data that were obtained in studies developed at the Institute of Oceanography of FURG.

We will show the importance of retention time this accumulation, to demonstrate that the Lucas, Thompson and Brown (2009) study it is not appropriate to our environment.

During the development of this work were followed four steps as shown in Figure 1. The Problem Study step involved the study of theoretical areas to obtain an understanding of the problem to be treated; In the second step, we performed the proposed model in NetLogo tool. At the end, with the completed model, we performed the validation from real data obtained in research of the Institute of Oceanography at FURG.

Figure 1. Adopted methodology

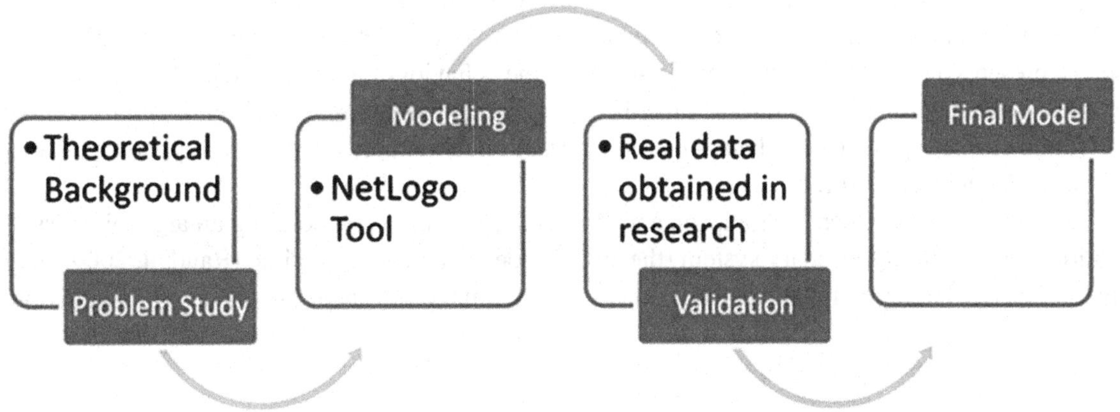

• Theoretical Background
Problem Study
Modeling
• NetLogo Tool
• Real data obtained in research
Validation
Final Model

THEORETICAL BACKGROUND

Phytoplankton

Plankton that has its nomenclature derived from the Greek word plăgchton, which means wanderer. This is a designation applied to all marine organisms that have limited mobility with respect to current and ends up drifting in the environment. The Plankton can be divided in zooplankton, which refers to small animals, and phytoplankton, which means all chlorophyll planktonic organisms (Lalli & Timothy 1997).

Phytoplankton is the main primary producer of the marine environment. Primary producers are responsible for transforming inorganic substances into organic matter through photosynthesis. This process has vital importance to the food chain of aquatic environments because phytoplankton provides the matter and energy that sustains the growth of fish, crustaceans and molluscs in marine ecosystems (Lalli & Timothy 1997).

Studies in estuaries, as Howarth et al. (2000) and Jassby (2008), have the focus of knowing the factors that control of the phytoplankton biomass accumulation. Such studies have shown the importance of light, nutrients and predation for the production of phytoplankton. In addition to these works, transport time, also known as retention time or circulation is, in some studies, considered the main factor influencing the phytoplankton biomass variability, considering the long and short-term studies (Abreu, Bergesch and Proenca, 2010; Howarth et al., 2000; Odebrecht, Abreu, & Carstensen, 2015). However, Lucas, Thompson and Brown (2009) mentions that the retention time has small influence on the variability of phytoplankton, but what really affects this variability are the duplication and loss taxes of phytoplankton.

In this way, to investigate and clarify the influence of transport time in phytoplankton variability in the ELP requires a study with large amount of data and samples. However, the cost, both in time and in financial terms, are high. Because of this, to have a computer model to simulate this environment could make accurate predictions about the phytoplankton biomass accumulation, with accurate and fast results.

Related Works

Real studies that are using this work are from Institute of Oceanography of FURG, which depict the importance of retention time in the short term phytoplankton biomass accumulation in the ELP. These studies report that the action of wind on the estuary of the input blocks the output of phytoplankton to the sea, making these phytoplankton reproduce in the estuary and, thus, are generated "Bloons" of phytoplankton (Abreu, Bergesch, & Proenca, 2010; Odebrecht, Abreu, & Carstensen, 2015).

A counterpoint to these studies is the research made by Lucas, Thompson and Brown (2009). This research presents a simple conceptual model to model the behavior of a water system. This model has as variables the loss, growth and transport time. It is also mentioned that several simplifications are made and that, even so, it provides explanations for different relations between the phytoplankton biomass accumulation and transport time.

The method used by Lucas, Thompson and Brown (2009) begins by defining an algal biomass distribution equation for a stationary system (the system does not vary in the time (Randall, 2009)) using only the variable loss, growth and distance downstream from the entrance. This is shown in Equation 1.

$$B(x) = B_{out} = B_{in} \exp\left(\frac{\mu_{growth} - \mu_{loss}}{u} x\right) \quad (1)$$

In studies developed in Institute of Oceanography at FURG were obtained from salinity and temperature data through a submersible sensor (RBR) of continuous monitoring that was implemented to 5m deep at the main pier entrance of the Navy of Brazil, in which the channel is narrower (800 m). Analyzing these data, it was identified patterns that demonstrate the importance of the retention time for the accumulation of biomass in the short term (Abreu, Bergesch, % Proenca, 2010; Odebrecht, Abreu, & Carstensen, 2015), unlike the studies by Lucas, Thompson, and Brown (2009), which use only the transmission time as a constant value. These studies demonstrate, by means of data, which are time periods contained wind action on the entry of the estuary causing the water flow be dammed, and it is the main factor of biomass accumulation (Abreu, Bergesch, & Proenca, 2010; Odebrecht, Abreu, & Carstensen, 2015).

PROPOSED MODEL

The proposed model in this work want to to demonstrate the importance of the retention time to the accumulation of phytoplankton as well as demonstrate by means of parameters, the relationship of each of them how to increase or decrease in phytoplankton in ELP. Other objective of the model is to be quite general model and it can be used in other environments.

Model Construction Steps

During the modeling steps, we have built a first model through which it was possible to verify the importance of the relationship between the retention time/phytoplankton biomass accumulation. This model had as input parameters the number of initial algae, the duplication time of phytoplankton (time, in days, that it takes for a seaweed become two) and retention time (time, in days, that the opening of the estuary is blocked due to wind action) in the ELP. This first model did not consider some variables that were considered irrelevant to this estuary but as a way to generalize the model he suffered an incremental process of building, and the final model that is shown in Figure 2.

The Simulation

The simulation in NetLogo is managed by ticks, which is the relative time interval within the simulation. And, this time is relative, because the real-time tick of a clock is not necessary due to the amount of agents and the computational power of the environment in which the simulation is running. In this model also has the agents that are called algae. Each agent has a different ID, what differentiates an alga to other. In each tick, each of these algae can perform one or some of these functions in the system:

- **MoveAlga:** This function is performed by all of the algae every tick. In this role, each of the agents "Alga" performs a spin in its own space at random between 0 and 360 °. The agent checks the space in front is a valid space to perform the movement . If it is a valid space, it performs the move walking in that direction; if space is not valid, it performs a new twist and again tests the space until it can perform a movement.
- **DuplicAlga:** This function is performed every tick, but is not performed by each agent. The special feature of this function is that it starts selecting some agents (algae), and these agents perform a duplication, ie, each of the selected agents creates a new one with its own ID and this agent will

Figure 2. Interface of final model

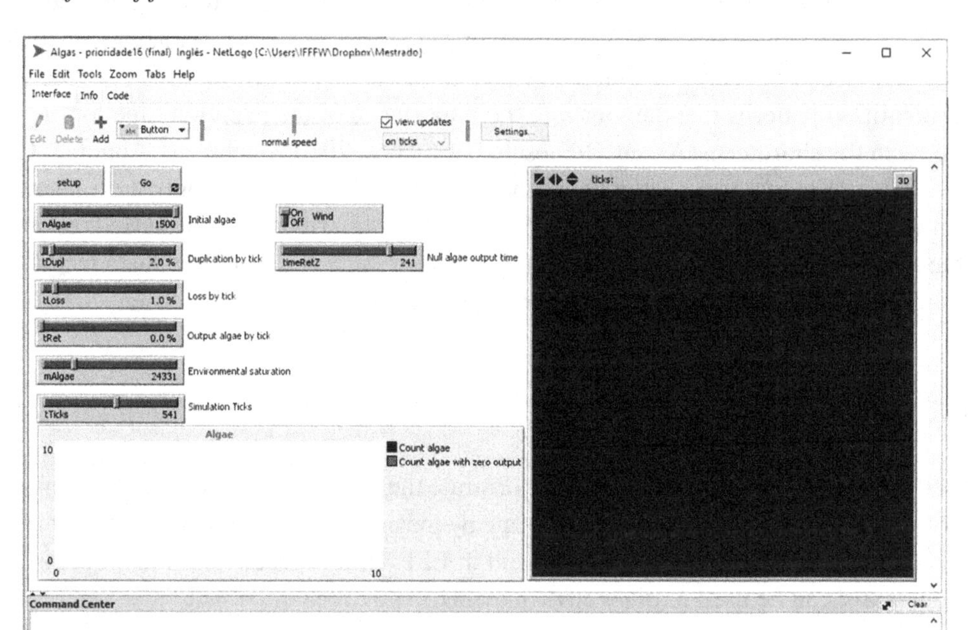

be available at tick next to perform all possible functions of any other algae. Its operation can be seen in Figure 3.

- **LossAlga:** This function, also the previous one, is held every tick and it is not performed by all agents. This function selects agents (algae) that will be deleted from the model. However, it differs from the previous function during this selection, as it performs a different selection of agents where there is a greater number of algae selected from the saturation parameter in the model interface. This parameter saturation means that the environment does not support the current number of algae to lack of nutrients and space, thereby generating a higher rate of loss of algae. This function simulates the loss of natural causes, predation and other, except for the removal of estuarine algae with the output of fresh water. Figure 4 shows the operation of the LossAlga function.
RetVol: This function is similar to duplicAlga function. This function is performed every tick, except the ticks related to retention time in which this function is not active. The function selects a number of algae that will be removed from the simulation. Thus, this function simulates the output mouth of the algae, which is zero for a retention time (wind action on the output estuary). We can see how this function works in the Figure 5.

Setting Up the Environment

Before do any simulation, the user must set the environment parameters, which can be realized through the configuration interface in Figure 6. The parameters to be set are as follows:

- **nAlgae:** This parameter inform the initial number of agents in the environment, ie, it can be understood that value as the initial number of algae on the environment.

Figure 3. Function operation DuplicAlga

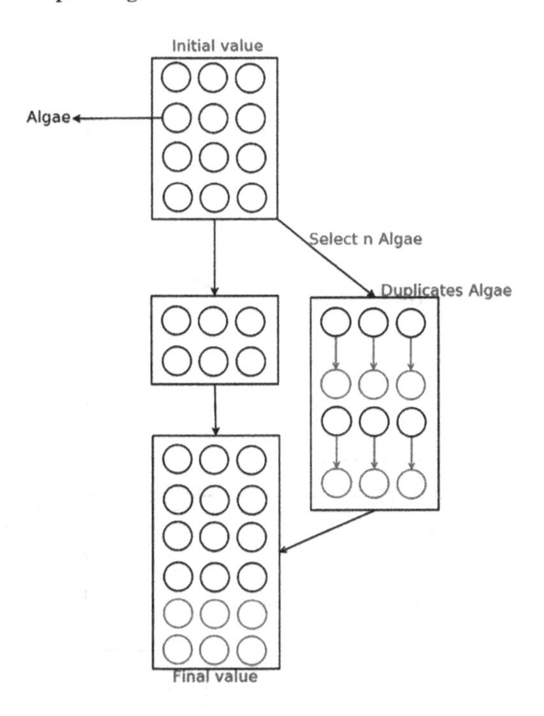

Figure 4. Function LossAlga operation

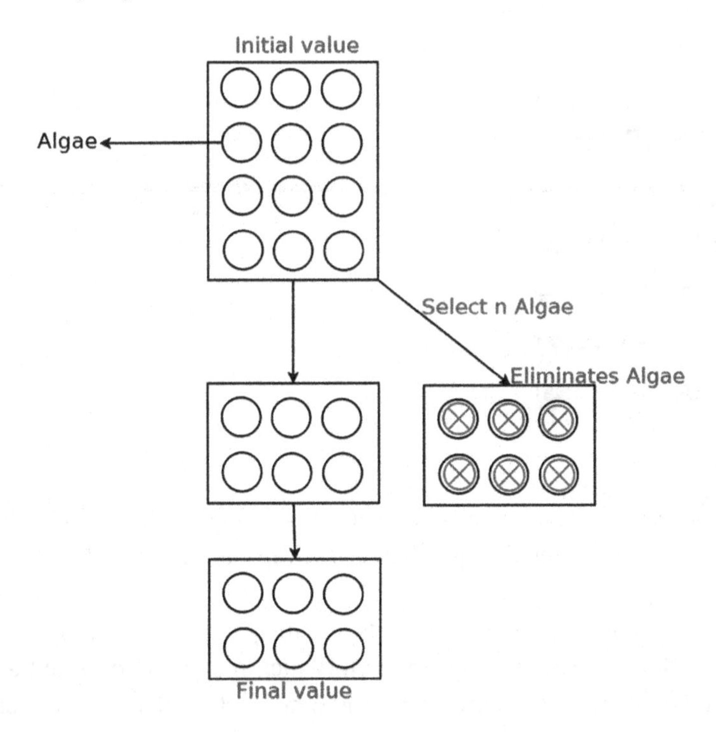

Figure 5. Function flowchart RetVol

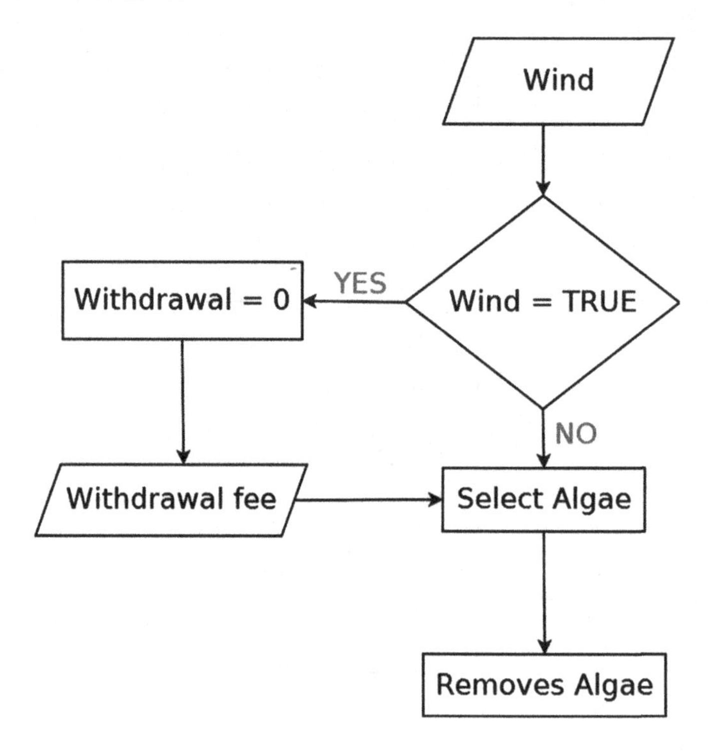

Figure 6. Interface of parameters configuration

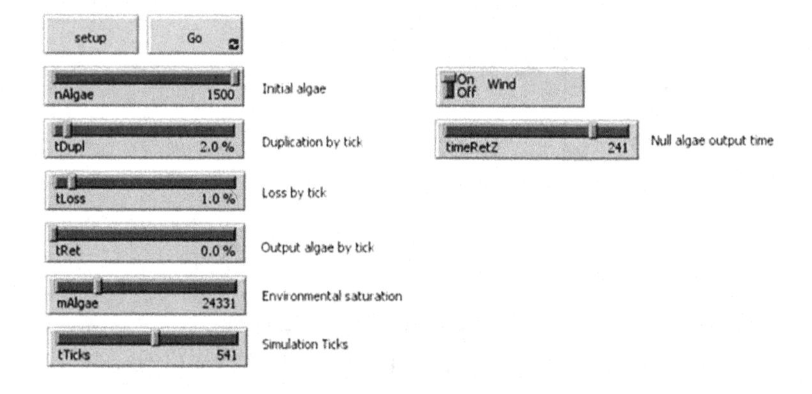

- **tDupl:** Inform the duplication rate, ie, what percentage of agents that will be duplicated every tick. Through this parameter defines the speed of growth in the number of algae on the environment every time unit.
- **tLoss:** Inform the rate of loss agents in each tick. This parameter controls the amount of algae to be excluded from the simulation each time unit, which may be interpreted as the loss rate compared to the death of algae, natural predation and other factors. This parameter can change during simulation when there is environmental saturation, which can increase the rate.

- **tRet:** The values that are selected for this parameter is related agents rate that will leave the simulation. This parameter simulates the rate at which algae out of the estuary. It is important to note that this rate may be changed during a simulation due to the retention time, ie, when there is retention time the rate is zero.
- **mAlgae:** If this value is reached, the tLoss parameter is changed,, so that the loss of algae is larger every time unit. This parameter is used to simulate the environment of the use of resources due to excess algae in the environment.
- **tTicks:** This parameter controls the amount of ticks of each simulation. It is important to note that the parameter name is due to the relative time unit in simulations using NetLogo is tick. During validation, this value will be transformed into a similar unit to the data observed in loco (real data).
- **Wind:** This checkbox is used to turn on/off the retention time during the simulation.
- **timeRetZ:** This is the parameter that controls the amount of ticks that there is retention time (withdrawal = 0). This parameter simulates the wind, performing an action on the entrance of the estuary and blocking the exit of the estuary of algae.

VALIDATION

Model validation was through the analysis of simulated data in this model comparing with the real data obtained in loco by the Institute of Oceanography at FURG. During this step, the real data obtained were analyzed. These data are used in the study of Odebrecht, Abreu, and Carstensen (2015). These data were obtained by sensors that were putted in the main pier entrance of the Navy of Brazil in the ELP, where the channel is narrower. This place was chosen because it is a mixed environment. However, there may be pycnocline (increased density) in some periods, especially in salt water into the estuary wedge-shaped, interfering with sensors. The sensors used were RBR (submersible continuous monitoring sensor) and a fluorescence sensor (which captures minute-to-minute data). It is important to note that the sensors had monthly maintenance, which were used anti-corrosive paints and made the removal of "barnacles", among other necessary maintenance.

Data from these sensors were transferred weekly to a computer. Finally, data were obtained comprising a period of nearly two years (March 2010 - December 2011), and the period of data collection is hour (in relation to the data that were obtained from minute to minute was held an average that was embedded in the time period), which generated a data table with about 13800 lines. The data have collection time, temperature, voltage, salinity and others. For this work, he considered the time of collection, the salinity and the voltage, that was transformed in Chlorophyll A (the pigment is found in the plant cells, there are 4 types of chlorophyll and A is the resulting green photosynthetic pigment, and it is used in this work due to the sensors used to capture the pigmentation of water to measure the amount of phytoplankton in the estuary).

Selection of Real Data

Due to the large amount of data obtained were selected during the validation step, some of the data from Abreu, Bergesch and Proenca (2010) and Odebrecht, Abreu, and Carstensen (2015) studies were used. This selection was chosen four periods that demonstrate the behaviors related to retention time and phytoplankton biomass accumulation. These data represent different growth behaviors at different times of the year. Also during this validation, the data had a normalization, to perform the average of the last 5 hours to demonstrate the value of the current time. In this way, values that were far outside the natural curve were corrected. The chosen periods have one X axis that represents the hours, and two Y axes, where the left axis ranging from 0 to 35 and on the salinity, and the right axis that goes from 0 to 100 is related to chlorophyll A.

In Figure 7-a presented the period from 20/11/2011 (15h) to 25/11/2011 (05h). In this figure, there is a relatively short retention period of approximately 60 hours (the period where salinity reaches values of about 30) increased by Chlorophyll A, which was in values close to 3μm and reached the end of the retention time, approximate values of 95μm. From this figure, we can do the relationship between the retention time and the increase of phytoplankton in the environment. And after leaving the salt water estuary, the Chlorophyll A tends to be drained, returning to initial parameters.

The Figure 7-b shows the period 26/10/2010 (00:00) to 30/10/2010 (22:00). It is possible to observe in this figure that even in the event of an approximate retention time Figure 7-a, no substantial growth of Chlorophyll A. However, it can be noted that the existence of a higher retention time, there could be further growth, since the tendency of the figure relative to the Chlorophyll is rising, and it was stopped by the retention time and the end has been drained.

Figure 7-c is related to the period from 20/04/2010 (15h) to 27/04/2010 (23h). It is possible to verify that in this figure, even though the retention time has exceeded the 110h, no real increase in chlorophyll

Figure 7. Selection real data

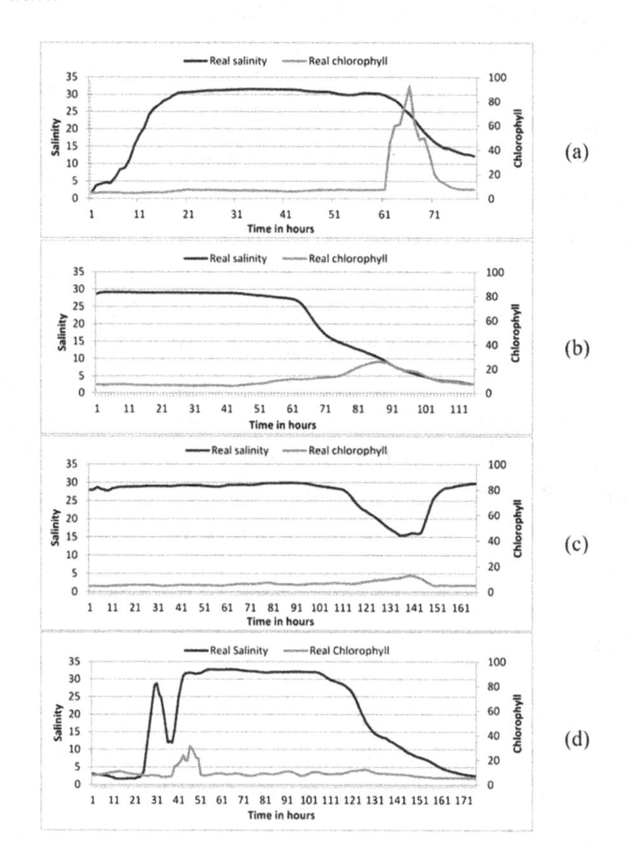

A. It could happen because it is an autumn period, where there the reduction of sunlight, especially in points further south, as is the case of the studied estuary.

Figure 7-d shows that the place was pycnocline, because the salt water into the wedge-shaped estuary, which alter the results obtained in the sensor. Because of this, there was an increase of Chlorophyll A for the increase of salinity in the estuary, which characterizes this wedge-shaped entrance made the estuary bottom sediment were suspended. Thus, there was a change in sensor input data and, after a period, the value of Chlorophyll A returned to normal because this sediment which was erected during the salt water into the estuary back to its natural state. This type of situation is not covered by the model created due to a data capture atypical data by the sensor.

Generation of Simulated Data

Simulated data were obtained from several tests with the model. From these tests, the intervals were chosen that were close to the maximum of the real data, to check that the model, when within a standard, can simulate a behavior similar to the behavior of the real data. For each of the three periods (Figure 7-a-b-c), we have done simulations. Before comparing the data obtained in loco to the simulated values, it was necessary that there was a normalization of the data obtained in the model so that the simulated data were consistent with real data. To this, we were standardized the following relationships to the proposed model:

- Each real time data is equivalent to 2 simulation ticks: this means that the real data have equivalent to 100 hours, the simulated data will be equivalent to 200 ticks (time within the simulation). We have done an average every 2 ticks, so that only one tick is generated.
- The model works with Algae nomenclature for the amount of algae system in order to work with equivalent data. The algae were divided by 300, so that values are approximate during the simulation and thus the simulated values and real values with the nomenclature employed Chlorophyll A.
- Salinity is a purely visual data and the model was considered value 30 whenever there retention time.
- An important factor to point out is that during the simulations the retention time (timeRetZ) was chosen to take the beginning of the fall of salinity in the estuary.

Figure 8 a simulates the actual results of Figure 7-a, with the retention time of approximately 60 hours. The figure generated in the simulation shows great similarity to the graph generated from actual data.

Figure 8 a was generated from the model, using the following input data:

nalgae = 1500.
tDupl = 8%.
tLoss = 4%.
tRet = 10%.
mAlgae = 55000.
tTicks = 350.
timeRetZ = 120.

Figure 8. Simulated data

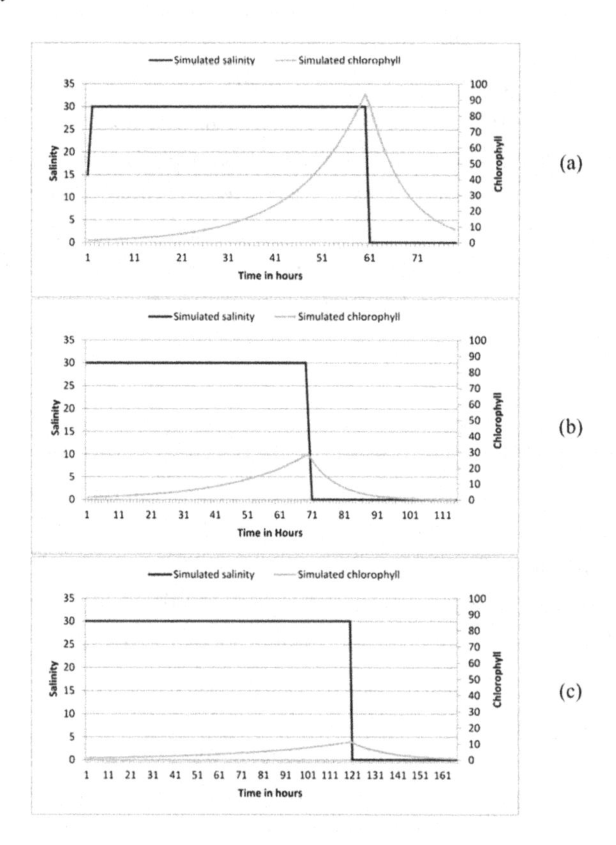

Importantly, these data have been entered in the simulation interface and time-related data as timeRetZ and tTicks must be normalized according to the normalization performed in the final results. Thus, the timeRetZ the end of the simulation equivalent to 60h and the tTicks is equivalent to 175h. Another value that is normalized to be the value of the nAlgae, which should be divided by 300 to Chlorophyll for the initial 5μm values. The value of mAlgae follows the same pattern, but in the current simulation this did not interfere.

Figure 8 b was generated from the model, using the following input data:

nalgae = 1500.
tDupl = 7%.
tLoss = 4,5%.
tRet = 8%.
mAlgae = 55000.
tTicks = 350.
timeRetZ = 140.

Importantly, the nAlgae, tTicks and timeRetZ values were normalized according to the aforementioned (data related to algae were divided by 300 and related tick were divided by 2), so that the standard remains between simulated graphics. Thus, the nAlgae value was equal to 5 and tTicks equal to 175h and tempRetZ being equivalent to 70 hours. By checking the real data, the retention time value is approximate to 70h, and thus is equivalent to simulation.

Figure 8. c was generated from the model, using the following input data:

nalgae = 1500.
tDupl = 2%.
tLoss = 1%.
tRet = 4%.
mAlgae = 55000.
tTicks = 541.
timeRetZ = 241.

To maintain the same pattern as the previous simulations, the same data normalization units were used for related algae values division was performed by 300 and ticks values related to a division by 2. Thus was performed, equivalent to 5µm for nAlgae, 270h for tTicks and 120h for timeRetZ.

Discussion of Data

Figure 9 a demonstrates the value of Real and Simulated ChlorophyllA and Chlorophyll to the first period of real data (Figure 7-a) and simulated data (Figure 8-a). In this scenario, it can be seen that with a retention time of approximately 70 hours, there was a marked increase in Chlorophyll A. This increase can be justified at the time of the year of data collection, in the southern hemisphere is late spring and early the summer period when the sun is predominant, what may have caused a duplication rate greater. Because of this, it was chosen to simulate an 8% duplication rate per hour. In this simulation were chosen tLoss values equal to 4% per hour. After the simulation and data normalization, we can check the similar behavior between the curves. Importantly, the X axis represents the time, and in this figure the retention time occurs between the time 1h to about 61h time.We can check the estuary water output curve (from 61h), there is a substantial drop of algae, demonstrating the importance of retention time in this accumulation.

When we analyze the Figure 9-b, we can see initially that there is a less substantial growth relative to Figure 9-a. The reasons for this lower growth, even with the amount of retention of 1h to 71h (approximately), it is due to a time less provides the solar interference in the environment, which may have reduced the efficiency of Chlorophyll A. We know that the sun has great influence on photosynthesis and a reduction of the sun can negatively influence the growth of algae, or even it can characterize this decrease by increasing loss of Chlorophyll A to predators or other causes. During simulation, there was a decline in tDuplic values and there was an increase in tLoss per hour. Even with these reductions, we can see that due to the retention of 71h, the accumulation occurred within this time, it instantly after the end of the retention, the amount of Chlorophyll A was reduced to default values.

Figure 9-c shows mainly a situation in which no substantial growth of Chlorophyll A, even with a relatively large retention time of approximately 140h. It is possible to check, also, the retention time generates a slight increase, indicating that there is a greater duplication than a loss of Chlorophyll A in the estuary, which generate a small increase (approximately 12µm), and that after the retention time, the system suffers with the flow.

Figure 9. Real and simulated Clorofila A

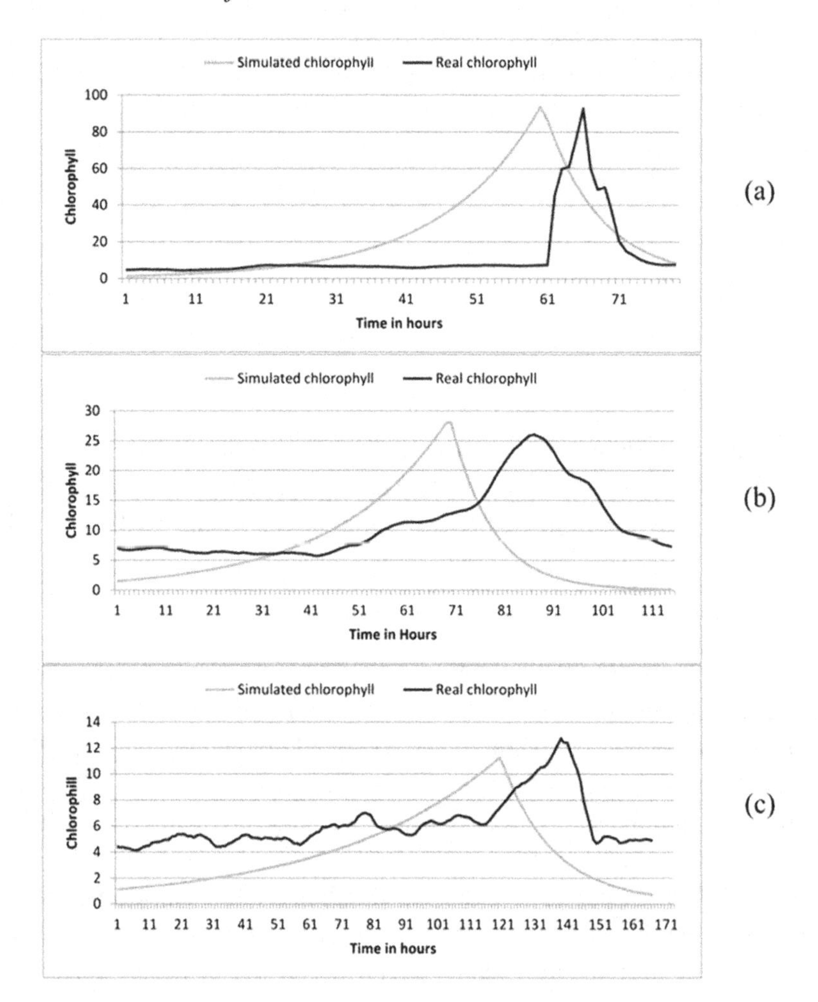

From the study of these three scenarios, we found that the simulation results are satisfactory because, managed with real results, creating similar behavior. It is important to note that through this simulation we can visually draw the importance of retention time to the existence of accumulation, which contradicts the Lucas, Thompson and Brown (2009) study.

Another important factor which we must consider is that the Chlorophyll A simulations and real data is their "peak" at different time interval, that may originate retention time of choice in the simulation. The time chosen to create the simulation in relation Figure 9-a was 61h, that is the time understood the real from the beginning of salinity rise until the beginning of the salinity decrease. In Figure 9-b and 6-co was based on the peak salinity (the input saltwater estuary was not considered), this choice was due to the different scenario in the hours before the salt water entering the estuary. In Figure 9-a, it had a longer period of low salinity as in Figure 9-b and 9-c was interspersed periods of input and output, a fact that may have left these graphics not as close as Figure 9-a.

CONCLUSION

This study aimed to demonstrate the importance of the retention time in the phytoplankton biomass accumulation. To to that, a model was developed to demonstrate the effect of retention time on biomass accumulation. Due to the complexity of models based on equations, we developed a model based on agents, which demonstrates the full potential of this technology in the creation of models directed to the area of emerging and biological phenomena, such as biomass accumulation model.

Other motivation of this work is to create a model that can effectively simulate possible situations, so that, to obtain relevant data for future research. The data generated by the model can be used in research and reduce data acquisition costs, which is an issue for this type of study, because the amount of temporal data and other influences that estuary and the sensors suffer during this time search.

Therefore, we developed the model with the help of experts in area and literature researches, creating the rules that the agent should follow, and the behavior of the system as a whole (environmental, agent and actions).

We can conclude that the studies developed during research were promising, because they could demonstrate in a simulated way the importance of retention time in the estuary of the the Patos Lagoon and through data generated changes that would require months or years of research, because of the difficulty to obtain data to sensor insert in the ELP.

Other important factor is the possibility of generating new scenarios that have not been found and check how these scenarios would affect the growth of Chlorophyll A in the estuary. This possibility may be important as a way to predict future situations and also perform reverse engineering. The model has also been designed to be as generic as possible so that it can be used in other environments and represent satisfactory results.

As future work, we will improve a refinement of the source code to have results more approximate to the real data. Also, we must understand the input values, seeking to find the actual values and verify the accuracy and influence of input values in the final result. Even as future work can be carried out studies in other environments so that we can verify if indeed the model was generalized to the most diverse environments.

REFERENCES

Abreu, P. C., Bergesch, M., Proença, L. A., Garcia, C. A. E., & Odebrecht, C. (2010). Short-and long-term chlorophyll a variability in the shallow microtidal Patos Lagoon estuary, southern Brazil. *Estuaries and Coasts*, *33*(2), 554–569. doi:10.1007/s12237-009-9181-9

Allan, J. D., & Benke, A. C. (2005). Overview and prospects. In Rivers of North America. Elsevier.

Howarth, R. W., Swaney, D. P., Butler, T. J., & Marino, R. (2000). Climatic control on eutrophication of the Hudson River Estuary. *Ecosystems (New York, N.Y.)*, *3*(2), 210–215. doi:10.1007/s100210000020

Jassby, A. D. (2008). *Phytoplankton in the Upper San Francisco Estuary: Recent biomass trends, their causes and their trophic significance*. Retrieved from http://repositories.cdlib.org/cgi/viewcontent.cgi?article51103&context5jmie/sfew

Knight, R. D. (2009). *Física: Uma Abordagem Estratégica* (Vol. 4). Bookman.

Lalli, C., & Parsons, T. R. (1997). *Biological Oceanography: An Introduction: An Introduction*. Butterworth-Heinemann.

Lucas, L. V., Thompson, J. K., & Brown, L. R. (2009). Why are diverse relationships observed between phytoplankton biomass and transport time. *Limnology and Oceanography*, *54*(1), 381–390. doi:10.4319/lo.2009.54.1.0381

Odebrecht, C., Abreu, P. C., & Carstensen, J. (2015). Retention time generates shortterm phytoplankton blooms in a shallow microtidal subtropical estuary. *Estuarine, Coastal and Shelf Science*, *162*, 35–44. doi:10.1016/j.ecss.2015.03.004

Pritchard, D. W. (1967). Observations of circulation in coastal plain estuaries. American Association for the Advancement of Science.

Reynolds, C. S. (2000). Hydroecology of river plankton: The role of variability in channel flow. *Hydrological Processes*, *14*(16-17), 3119–3132. doi:10.1002/1099-1085(200011/12)14:16/17<3119::AID-HYP137>3.0.CO;2-6

Seeliger, U., & Odebrecht, C. (1997). *Introduction and overview. Subtropical convergence environments: the coast and sea in the southwestern Atlantic*. New York: Springer. doi:10.1007/978-3-642-60467-6

Søballe, D. M., & Kimmel, B. L. (1987). A large-scale comparison of factors influencing phytoplankton abundance in rivers, lakes, and impoundments. *Ecology*, *68*(6), 1943–1954. doi:10.2307/1939885

Tang, Y., Parsons, S., & Skalar, E. (2006). Modeling human education data: From equation-based modeling to agent-based modeling. Hakodate.

Chapter 13
Participatory Management of Protected Areas for Biodiversity Conservation and Social Inclusion:
Experience of the SimParc Multi-Agent-Based Serious Game

Jean-Pierre Briot
LIP6/UPMC-CNRS, France & PUC-Rio, Brazil

Gustavo Mendes de Melo
EICOS/IP/UFRJ, Brazil

Marta de Azevedo Irving
EICOS/IP/UFRJ, Brazil

Isabelle Alvarez
IRSTEA-LISC/UPMC-LIP6, France

José Eurico Vasconcelos Filho
UNIFOR, Brazil

Alessandro Sordoni
University of Montréal, Canada

Carlos José Pereira de Lucena
DI/PUC-Rio, Brazil

ABSTRACT

The objective of this paper is to reflect on our experience in a serious game research project, named SimParc, about multi-agent support for participatory management of protected areas for biodiversity conservation and social inclusion. Our project has a clear filiation with the MAS-RPG methodology developed by the ComMod action-research community, where multi-agent simulation (MAS) computes the dynamics of the resources and role-playing game (RPG) represents the actions and dialogue between stakeholders about the resources. We have explored some specific directions, such as: dialogue support for negotiation; argumentation-based decision making and its explanation; technical assistance to the players based on viability modeling. In our project, multi-agent based simulation focuses on the negotiation process itself, performed by human players and some artificial participants/agents, rather than on the simulation of the resources dynamics. Meanwhile, we have also reintroduced the modeling of the socioecosystem dynamics, but as a local technical assistance/analysis tool for the players.

DOI: 10.4018/978-1-5225-1756-6.ch013

INTRODUCTION

A significant challenge involved in biodiversity conservation is the management of protected areas (ex: national parks, marine reserves, biosphere reserves… – the main focus of this work being national parks), which usually undergo various pressures on resources use and access, resulting in many conflicts. Methodologies intending to facilitate participatory management and conflict resolution are being addressed via bottom-up approaches that emphasize the role of social actors (stakeholders) involved in these conflicts. Examples of social actors are: park managers, representatives of local communities at the border area, tourism operators, public agencies and nongovernmental organizations (NGOs). Examples of inherent conflicts connected with biodiversity protection are: irregular occupation, inadequate tourism exploration, water pollution, environmental degradation and illegal use of natural resources. In Latin America, the process of management of protected areas for biodiversity conservation usually takes place through the mediation of institutional arenas of dialogue and conflicts, implemented though a management council (Irving 2006). This council, of a participatory nature, includes representatives of various social actors/stakeholders. They represent and express various perspectives and participate in dialogue, negotiations and decisions about protected areas management.

We are conducting a research project to explore how advanced computer support (distributed role-playing game, simulation, artificial agents, negotiation support, decision support…) can help participatory management of protected areas. The project is named SimParc (which stands in French for "Simulation de gestion participative de parcs") and is based on a serious game. Serious games and role-playing games have indeed received increased attention (Crookall, 2010) as effective approaches for exploration and training, in context but without high costs or risks (Michael, 2006). They are considered as promising pedagogical tools for complex collective decision making processes.

Note that, although our project has a very clear filiation with the well known MAS-RPG (Multi-Agent-Simulation – Role-Playing Games) methodology (Barreteau, 2003), developed by the companion modeling (ComMod, by Barreteau et al. (2003)) action-research community about participatory management of renewable resources, we have explored some specific directions, such as: dialogue support for negotiation; argumentation-based decision making and its explanation; technical assistance to the players based on viability modeling.

The objective of this chapter is therefore to discuss the rationales behind the specificities of our project and detail our design, results as well as prospects for future works.

RELATED WORK

The ComMod (Companion modeling, see (Barreteau et al., 2003)) community had (and still has) a profound impact on research and methods for participatory management of environmental resources and on multi-agent simulation. More specifically, ComMod is about methods for participatory management of renewable resources, with interaction between natural resources inner processes (e.g., hydrodynamics, animals population evolution…) and the human and social processes of their usage (consumption, control…). They have proposed the combination of multi-agent simulation (MAS) – to represent and compute the dynamics of the natural resources – with role-playing games (RPG) – to represent the dialogue between stakeholders, to explore individual and collective decision strategies about the resources (e.g., actual use, access control, conflict resolution, etc.), see (Barreteau, 2003)). The multi-agent simulation platform Cormas (Le Page et al., 2012) is used to implement the simulation part. In the MAS-RPG combination, simulation runs are interlaced with the different game steps, thus allowing players to understand the consequences of their decisions/actions and their interrelations with decisions/actions of other players. Initially, and still in most of ComMod projects, RPG is conducted manually or semi-manually. Pioneering works, such as JogoMan-ViP by Adamatti et al. (2007) and Simulación by Guyot et al. (2006), have aimed at integrating MAS and RPG, that is providing a support for distributed players (inspired by distributed video games) and interfacing it with the multi-agent simulation, thus leading to a more fluid integration between simulation steps and decision steps. They have also started introducing some artificial agents, as players or as assistants. We have followed on such directions, with some specific objectives and results, as we will see.

APPROACH

General Objectives

As explained in the introduction, the objective of our SimParc research project is to explore the use of various computer-related techniques, such as serious games, role-playing games, multi-agent systems, simulation, decision support systems, user interfaces, dialogue systems, argumentation-based systems, viability theory, to help at participatory management of protected areas. It is based on the observation of several case studies in Brazil. However, we chose not to reproduce exactly a real case, in order to leave the door open for broader game possibilities. Our focus (very much inline with ComMod) is indeed on improving the process and not the result. In other words, as we will see, we have an epistemic objective, to help stakeholders at understanding the nature of conflicts and negotiation in the management of protected areas. Current SimParc serious game is not (or at least not yet) aimed at decision support (i.e., we do not expect the resulting decisions to be directly applied to a specific park).

More precisely, the SimParc role-playing game is based on the negotiation process that takes place within the park management council. Each player plays a role (as for each council member). Depending on its role profile and the elements of concerns (e.g., tourism spot, traditional community, endangered species, etc.) in each of the landscape unit (sub-area) of the park, each player will try to influence the decision about the level of conservation for each unit. It is clear that conflicts of interest will quickly emerge (e.g., between a biodiversity conservation ONG and a tourism operator about controlling access to a specific landscape unit populated by endangered species), leading to various strategies of negotiation (and possible conflict resolution, e.g., an agreement about a regulated ecotourism policy). A special role in the game is the park manager. He is an arbiter and decision maker.

Specific Research Objectives

As we will see, the SimParc project somehow departed from standard ComMod/MAS-RPG approach and it has explored some complementary objectives. Let us go back to the characteristics of the domain. At first, in the Brazilian legislation about national parks, direct use of natural resources is forbidden. Only indirect use, such as e.g., tourism, may be allowed. Also, the park management council is of consultative and not decisionary nature. Thus, as opposed to ComMod cases, there is no direct action by the players on the resources. The focus of the players is thus on identification of conflicts and negotiation. At the end of the negotiation (which does not necessarily lead to a consensus), each player makes a final proposal and the park manager has the role of the final decision maker.

The conduction of the SimParc project and the nature of the serious game prototype that we have constructed reflect this specificity. We have initially focused on the support for dialogue and negotiation between players and not on the modeling and simulation of the evolution of ecosystem resources, as would have been the case in standard MAS-RPG projects. As a result, the project started working on a prototype Web-server-based architecture for distributed role-playing (similar to Simulación by Guyot et al. (2006) and JogoMan-ViP by Adamatti et al. (2007)). We have then explored the following research directions: 1) dialogue support for negotiation; 2) argumentation-based decision making and its explanation; 3) technical assistance to the players. In that sense, the SimParc serious game is a multi-agent based simulation of the negotiation process itself, performed by human players and some artificial participants. Meanwhile, as we will see below, we have also reintroduced the modeling of the socioecosystem dynamics, but as a local technical assistance/analysis tool for the players and not as a global system modeling. Let us now summarize these three directions:

1. We have designed and implemented a specialized user-interface for negotiation. It uses rhetorical markers and dialogue filtering/structuring mechanisms (detailed in (Vasconcelos et al., 2009a)). A personal assistant also provides help as well as information about the situation of the player within the discussion and proposals (e.g., affinity and discordance with other players proposals, dominance relations, etc.).
2. We have designed and implemented an artificial agent taking the role of the park manager (Sordoni et al., 2010). Its architecture is dual. The first part models the manager individual decision-making process, deliberating about conservation types for each landscape unit. The second part consists in combining the players votes and the park manager vote, based on an influence function. Note that we have decided to use an argumentation-based framework, as traces of arguments produced for the decision are a basis for the explanation of the decision (to the players).
3. We have designed a viability assistant agent (Wei et al., 2012), based on viability theory by Aubin (1991), which helps a player at analyzing the possible impact of his proposed constraints (desired level of conservation) on the park viability. Our rationale is that defining and negotiating about constraints seems more intuitive for players than expressing global optimization objectives. The viability assistant can provide the player with various viability analysis (viability kernel, capture basin, suggest relaxation of constraints) to help him analyzing the viability of his proposal and negotiating with other players.

Organization of the Chapter

We will start by introducing the design of the role-playing game, at the heart of our project. This role-playing game is an abstraction of the discussion and negotiation process taking place between stakeholders representatives within a park management council. We will then briefly introduce the principles of the server-based architecture to support the game. Then, we will successively detail our three specific contributions about: 1) negotiation support; 2) manager automated decision; 3) viability analysis. We then present some experimental evaluation of our proposals and discuss future prospects before concluding.

THE SIMPARC GAME

The Role-Playing Game

Current SimParc game has an epistemic objective: to help and support each participant discover and understand the various factors, conflicts and the importance of dialogue for a more effective management of parks. Note that this game is not (or at least not yet) aimed at decision support (i.e., we do not expect the resulting decisions to be directly applied to a specific park).

The game is based on a negotiation process that takes place within the park council. This council, of a consultative nature, includes representatives of various stakeholders (e.g., community, tourism operator, environmentalist, non governmental association, water public agency…). The actual game focuses on a discussion within the council about the "zoning" of the park, i.e. the decision about a desired level of conservation for every sub-area of the park (called landscape unit – unidade de paisagem: UP). See at Figure 1 a screen dump of the game showing the representation of the (imaginary) park (with different layers of information: geography, subdivision into landscape units, objects of interest – with the explanation of their pictograms…) and at Figure 2 a focus on one specific landscape unit (UP).

In the game, we consider nine pre-defined potential levels (that we will consider as types) of conservation, from more restricted conservation to more flexible indirect use of natural resources, as defined by the (Brazilian) legal system. Examples are: "Intangible", the most conservative type, "Primitive" and "Recuperation". Figure 3 shows the interface for a player to decide what type of conservation he will propose for each landscape unit.

Figure 1. SimParc interface for analyzing the park

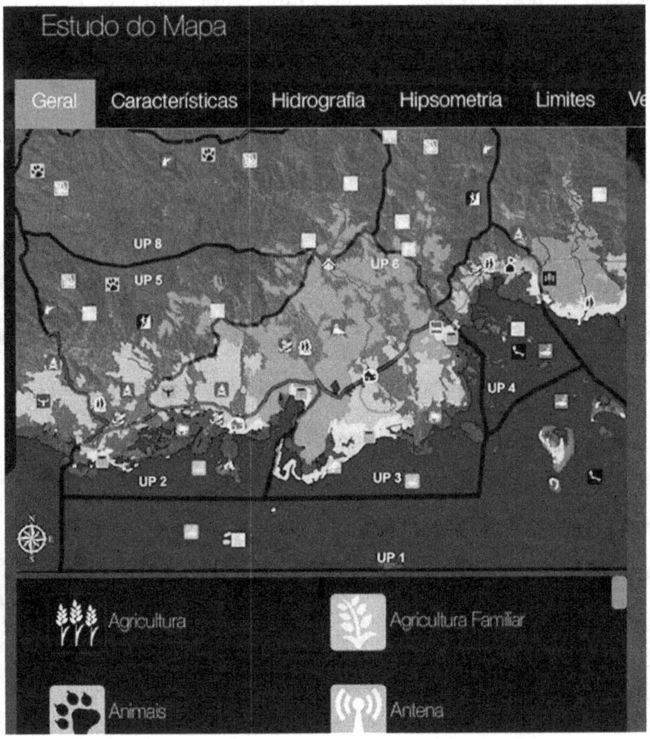

Figure 2. SimParc interface for analyzing a landscape unit (UP)

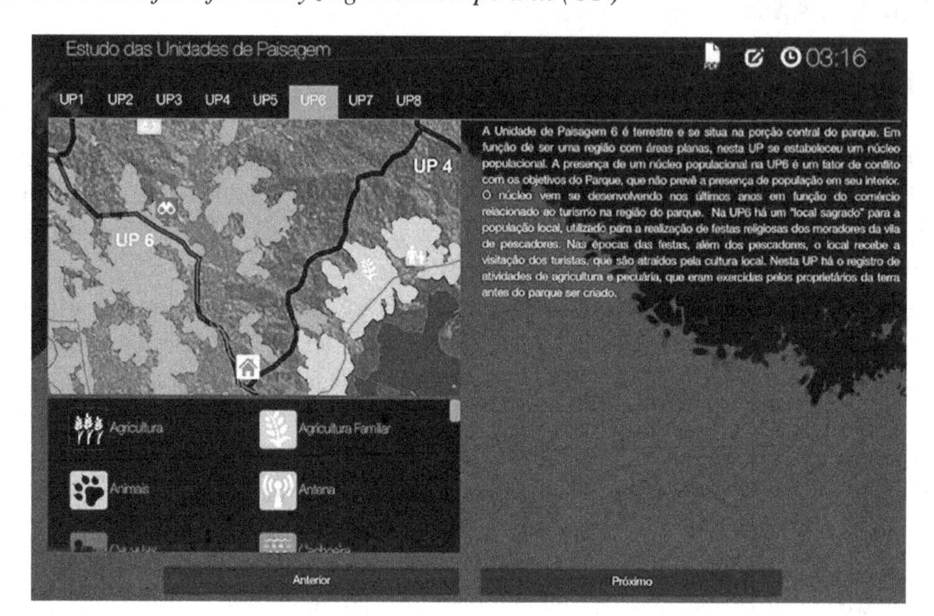

The game considers a certain number of players' roles, each one representing a certain stakeholder. Each player, as in any role-playing game, has to embody the designed/selected role with its respective background culture, postures and objectives. To facilitate the incorporation of the role by the player, SimParc offers a set of personas to represent him/her during the game. Depending on its profile and the elements of concerns (e.g., tourism spot, traditional community, endangered species, cultural attraction… – each type represented by a pictogram) in each landscape unit (see at Figure 2), each player will try to influence the decision about the level (type) of conservation for each landscape unit. Conflicts

Figure 3. SimParc interface for selecting conservation types

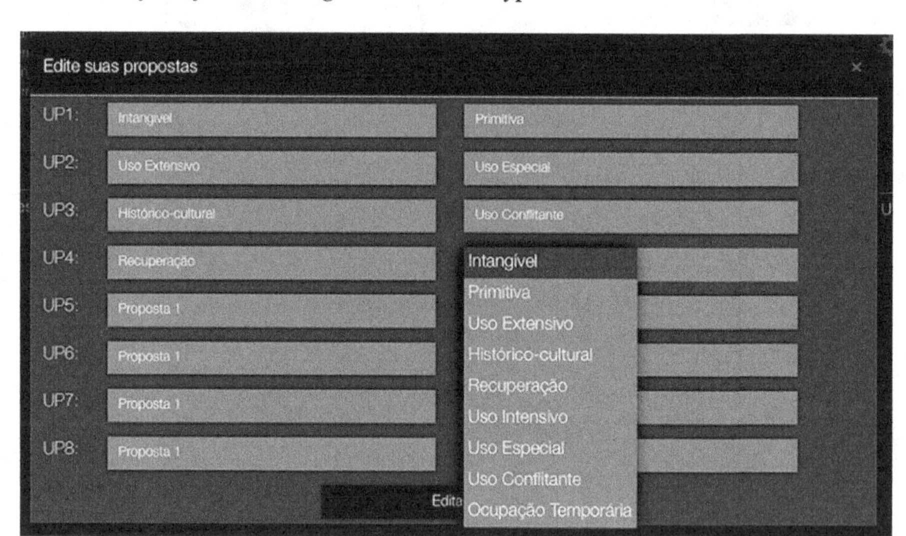

of interest will quickly emerge, leading to various strategies of negotiation (e.g., coalition formation, trading mutual support for respective objectives, etc).

A special role in the game is the park manager. He is a participant of the game, but as an arbiter and decision maker, and not as a direct player. He observes the negotiation taking place among players and takes the final decision about the levels of conservation for each landscape unit. (It is important to remind that this follows the situation of a real national park in Brazil, where the park management council is only of a consultative nature, thus leaving the final decisions to the manager.) Decision by the park manager is based on the legal framework, on the negotiation process among the players, and on his personal profile (e.g., more conservationist or more open to social concerns) (Irving 2006). He may also have to explain his decision, if the players so demand.

The Game Cycle

The game is structured along six steps, as illustrated at Figure 4.

At the beginning (step 1), each participant is introduced to basic information about the game general context (in an interactive and ludic way, through the metaphorical touring of a conceptual park, see at Figure 5). Then, he is associated with a role (ex: environmentalist, representative of a local community, tourism operator…). Then, an initial scenario is presented to each player, including the setting of the landscape units (see at Figures 1 and 2), the possible types of conservation and the general objective associated to his role. Then (step 2), each player decides a first proposal of types of conservation for each landscape unit (see at Figure 3), based on his/her understanding of the objective of his/her role and his analysis of the situation. Once all players have done so, each player's proposal is made public.

In step 3, players start to interact and to negotiate about their proposals. This step is, in our opinion, the most important one, because players collectively build their knowledge by means of an argumentation

Figure 4. The six steps of the SimParc game

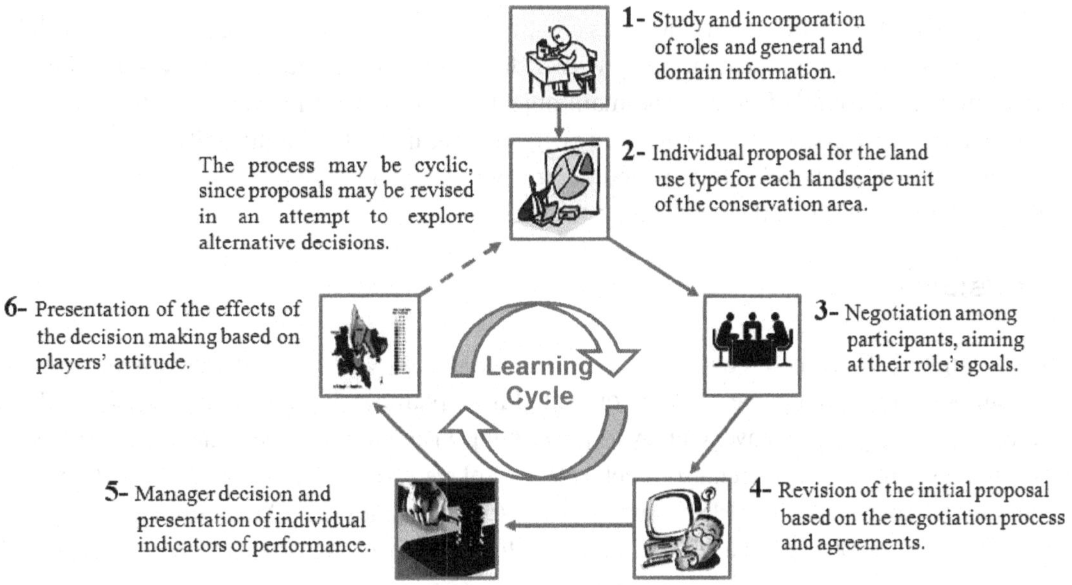

1- Study and incorporation of roles and general and domain information.

2- Individual proposal for the land use type for each landscape unit of the conservation area.

3- Negotiation among participants, aiming at their role's goals.

4- Revision of the initial proposal based on the negotiation process and agreements.

5- Manager decision and presentation of individual indicators of performance.

6- Presentation of the effects of the decision making based on players' attitude.

The process may be cyclic, since proposals may be revised in an attempt to explore alternative decisions.

Learning Cycle

Figure 5. Metaphorical tour for introducing the game

and negotiation process (the interface will be shown at Figure 12). In step 4, they revise their proposals and commit themselves to a final proposal for each landscape unit. In step 5, the park manager makes the final decision, considering the negotiation process, the final proposals and also his personal profile (e.g., more conservationist or more sensitive to social issues). Each player can then consult various indicators of his/her performance (e.g., closeness to his initial objective, degree of consensus, etc.). He can also ask for an explanation about the park manager decision rationales. The last step (step 6) "closes" the epistemic cycle by considering the possible effects of the decision. In the current game, the players provide a simple feedback on the decision by indicating their level of acceptance of the decision.

A new negotiation cycle may then start, thus creating a kind of learning cycle with possibly different allocation of roles to players (so that they play with different objectives and perspectives and thus reflect on the multiplicity of points of view). The main objectives are indeed for participants: to understand the various factors and perspectives involved in decision making and in biodiversity conservation challenges and how they are interrelated; to negotiate; to try to reach a group consensus; and to understand cause-effect relations based on the decisions.

Initial Version

The initial design of the game (version 1) was conducted during year 2007. It was tested, without any computer support, through a game session conducted in September 2007 (see at Figures 6 and 7, corresponding respectively to the players' analysis phase (step 1) and to the players' negotiation phase (step 3)). There were six roles in the scenario. Each role was played by a team of two players. Players were researchers and students of the APIS research group (which stands for "Áreas Protegidas e Inclusão Social", in English: "Protected Spaces and Social Inclusion"), at UFRJ (Rio de Janeiro), led by Marta Irving, and specialized in biodiversity participatory management.

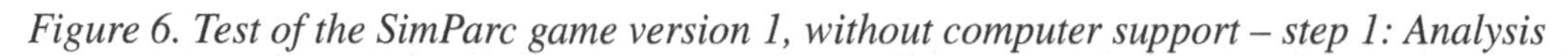

Figure 6. Test of the SimParc game version 1, without computer support – step 1: Analysis

Figure 7. Test of the SimParc game version 1, without computer support – step 3: Negotiation

A first version of the computer prototype was then designed and implemented on top of Simulación, the platform designed by Paul Guyot (2006), as a first proof of concept (Briot et al., 2007). Based on this, we then designed a second version of the prototype, that we will now describe.

THE SIMPARC GAME SUPPORT ARCHITECTURE

The second version of the prototype benefited from our previous experiences (game sessions and initial prototype) and has been based on a detailed design process. Based on the system requirements, we have adopted Web-based technologies (more precisely Java-based J2EE and JSF) that support the distributed and interactive character of the game as well as an easy deployment. After significant real size testing with specialists of park management issues (which will be described in a forthcoming section about evaluation), we have recently designed and implemented a third version, based on web applications frameworks JRaptor and Spring, as well as responsive Web design (using CSS and HTML handling libraries, in order to make the contents of the system adaptable to any display devices, such as computers, tablets, smartphones), more specifically frameworks such as: AngularJS, Html5, CSS3, Bootstrap and jQuery.

Thanks to the Web server approach, it is easy and quick to set up a game session: one just needs for each player a computer or tablet running a web browser and with access to Internet (or to a local network, not necessarily connected outside, as long as it can connect to the Web server). The SimParc Web server to be used may be running on a local computer or on our online server made continuously available.

Figure 8 shows the general architecture and communication structure of SimParc prototype. Distributed users (the players and the park manager) interact with the system mediated internally by communication broker agents (CBA). The function of a CBA is to abstract the fact that each role may be played by a human or by an artificial agent. A CBA also translates user messages in http format into the multi-agent KQML format and vice versa (KQML is a language for interoperability/intercommunication between agents, see (Finin et al., 1994)). For each human player, there is also an assistant agent offering assistance during the game session (see more details in (Vasconcelos 2009b)). During the negotiation phase,

Figure 8. SimParc prototype general architecture

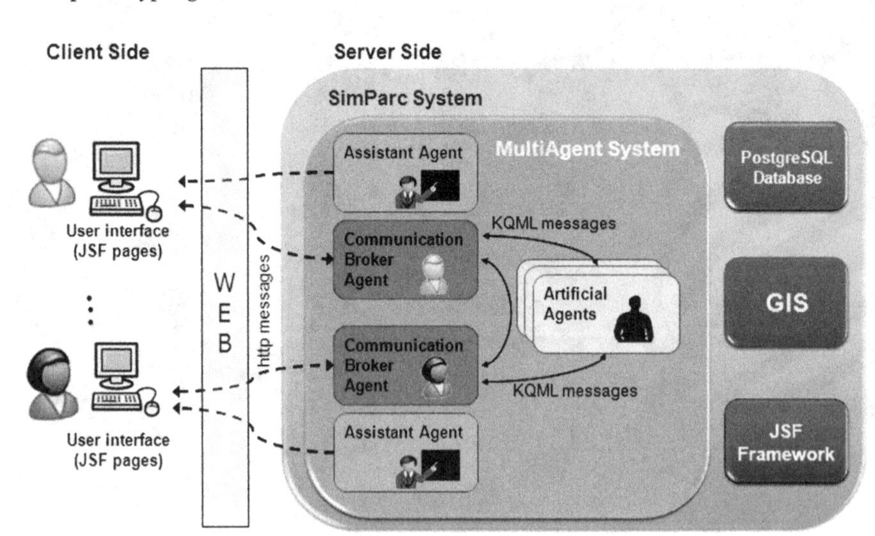

players (human or artificial) negotiate among themselves to try to reach an agreement about the type of use for each landscape unit of the park.

A Geographical Information System (GIS) offers to users different layers of information (such as flora, fauna and land characteristics) about the park geographical area (see at Figure 1). All the information exchanged during negotiation phase – namely users' logs, game configurations, game results and general management information – are recorded and read from a PostgreSQL database.

THE INTERFACE

Design of the Interface

The interface for negotiation includes specific support for negotiation (rhetorical markers and dialogue filtering/structuring mechanisms, see Vasconcelos et al. (2009a)); access to different kinds of information about other players, land, law; and the help of a personal assistant. Different interfaces are made available depending on the step (phase of the game, ex: analysis, proposal, negotiation) and the needs of the player, e.g., analyze the area based in its different layers (e.g., land, hydrography, vegetation..., see at Figure 1), select the desired types of conservation (see at Figure 3), access information, etc.

The process of design has been based on communication-centered design, and its more agile version, eXtreme Communication-Centered Design (Aureliano et al., 2006), design proposals based on the semiotic engineering theory of human-computer interaction. (According to it, both designers and users are interlocutors in an overall communication process that takes place through the interface of the system. Designers must tell users what they mean by the artifact they have created, and users must try to respond to what they are being told (de Souza, 2005)). We have adapted the application of the methodology to the characteristics of the SimParc project.

The output products of the analysis phase are the records from interviews with experts and users, scenarios (use cases), goals diagram and tasks model. The scenarios were constructed based on interviews, in a narrative form, to help at identify contextualized types of usages. The goals diagram (see at Figure 9) was constructed from the scenarios and interviews, with the aim of representing the goals (identified *a priori*) of the users. We believe that the task model represents an intermediary step, easing a conceptual transition from the analysis phase (what, why and by whom) to the design (how). Note that task models are also widely used and accepted in human-computer interaction (HCI). Overall, the goal of the diagrammatic representation of task models is to provide an overview of the design process for each goal and how these goals are decomposed into tasks and sub-tasks. This diagram provides a new set of information about the process, presenting the hierarchy and flow of tasks, preparing designers and users to an outline of the interaction. We used an adaptation of the Hierarchical Task Analysis (HTA) (Annett et al., 1967) for modeling tasks identified from the goals diagram and the scenarios.

Design of the Negotiation Language

We consider negotiation as a particular form of communication process between two or more parties, focused on mutual agreement(s) on a given conflict of interest or opinions (Putnam et al., 1992). We further believe that the adoption of a dialogue language, based on argumentation models and linguistics theory, can offer different ways of support to a computer mediated negotiation process. The main objec-

Figure 9. Diagram of final goals

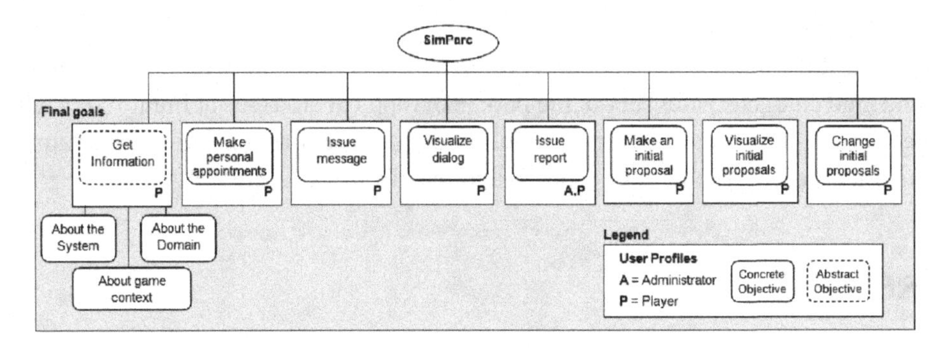

tive for that negotiation language is to find the inflection point between the necessary "framing" and the maintenance of fluidity and naturalness of the dialogue.

The structure of the dialogue is an important factor, because it helps at a better management of the history of the negotiations facilitating the inclusion of artificial agents in the process, increasing the focus on the process, on issues negotiated and on the clarity of dialogue. Many interaction protocols for negotiation between agents have been proposed (e.g., via the FIPA-ACL effort, see (FIPA, 2002)), but they privilege the agent-agent communication at the expense of human communication. Note also that computer mediated communication suffers from various types of impoverishment of the dialogue, particularly in relation to non-verbal communication, considering the body language (Ekman, 2007) and the vocal intonation. Thus, we are looking for an intermediate and simple way to promote both human-human and human-agent communication.

We have considered many proposals of notation for structuring and visualization of the argumentation, as, e.g., in (Kirschner, 2003). Among them: the Toulmin model, a reference for the majority of the posterior models; the Issue-Based Information System (IBIS), an informal model based on a grammar that defines the basic elements present in dialogues about decision-making; the "Questions, Options and Criteria" (QOC); the "Procedural Hierarchy of Issues" (PHI) and the "Decision Representation Language" (DRL) (Kirschner, 2003). Based on this analysis, we believe that it is possible to offer a pre-structure, adding to the informal and interpretative characteristic of prose, while maintaining the fluidity of dialogue. Our main inspirations for rhetorical markers is IBIS (Kirschner, 2003), as well as theories of negotiation, e.g., (Wall et al., 1991) (Raiffa, 1982) and Speech Act Theory (Searle, 1969). These markers are basically composed of rhetorical identifiers of intention (see at Figure 10), the object focus of the intention and of a free speech. These elements give the tone of the dialogue, making clear the illocution, and thus facilitating the expression of the desired perlocution (Searle, 1969).

We therefore provide the structure by threading from the dialogue, which minimizes risks of losing context, common in computer mediated communication (via chat) (Fuks et al., 2006). Figure 11 shows an example of threading based on the proposed structure.

In complement to this semi-structure applied to the text, we propose to model each speak from players as an object. These objects have the following attributes: identifier, sender, receiver(s), marker, focus, and a free text. This modeling eases at the management and indexing of dialogue by the system. For instance, filters may be applied to analyze the history of a dialogue, e.g., filtered along a given speaker, or a specific type of marker. But it also opens the way for its processing by software agents.

Figure 10. Semi-structure for the text based on rhetorical markers

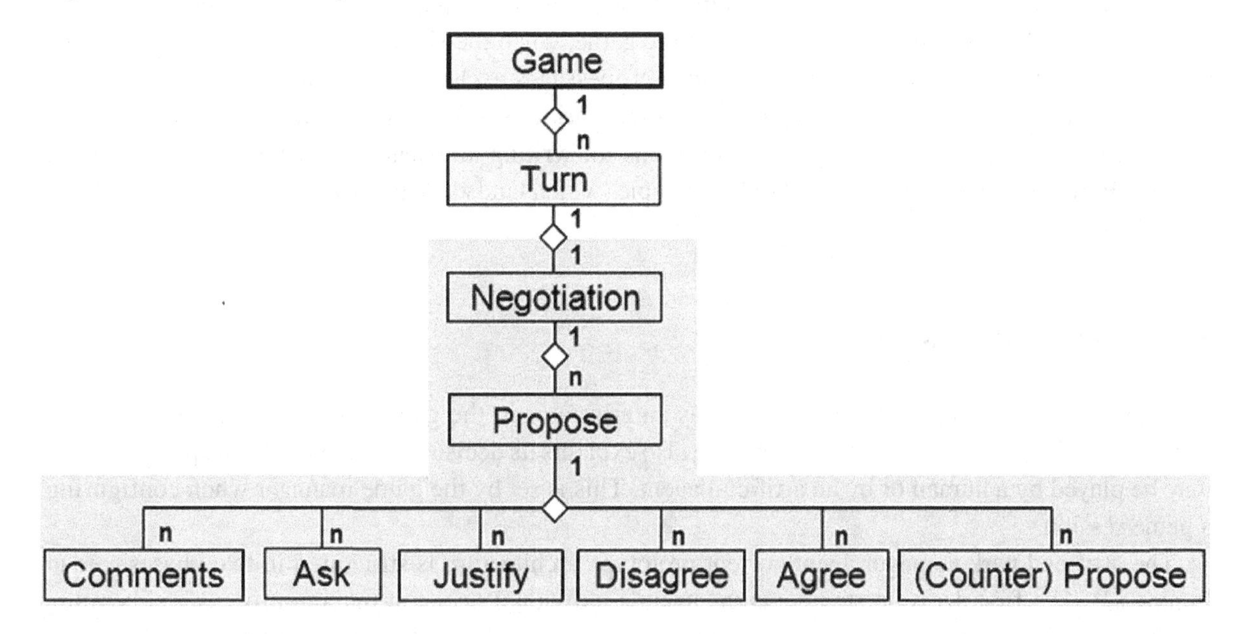

Figure 11. Example of threading structured by the rhetorical markers

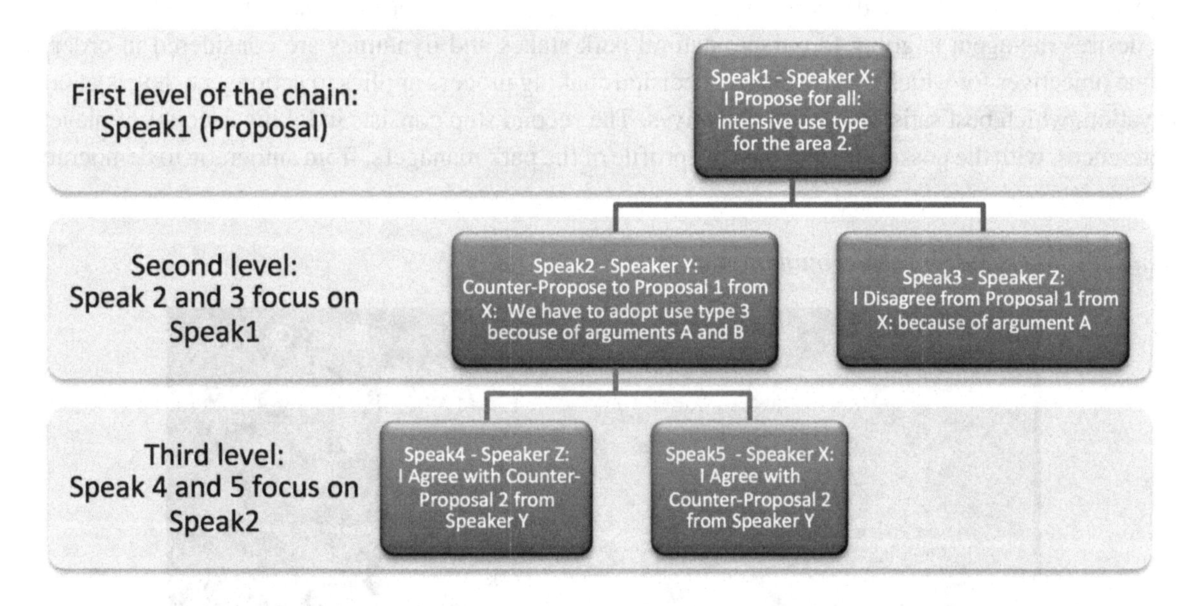

Current Interface

The outputs of the design phase were: interaction diagrams, class diagrams, class and entity relationship model for the database. We then created a first fast prototype in order to evaluate the appearance and usage, before implementing current interface. We have tried to balance a support for some structure of the text of the dialogue and also sufficient fluidity.

We focus here on the interface corresponding to step 3 of the game, i.e., negotiation between players (see at Figure 12). It is indeed a central part of the game, when the shared knowledge is jointly negotiated and built. It includes an area for the history of messages exchanged, which could be presented in chronological order or hierarchical order. It also offers various ways of selecting messages (by intention, topic, player). Another area below contains options for writing and sending messages with rhetorical markers for intention (e.g., propose), the chosen topic (which landscape unit), and the unique or multiple recipients.

PARK MANAGER AGENT

As explained earlier, the park manager acts as an arbitrator in the game, making a final decision for conservation levels for each landscape unit. He also explains its decision to all players. The park manager may be played by a human or by an artificial agent. This is set by the game manager when configuring a game session.

The artificial park manager agent current prototype architecture is structured in two phases (see at Figure 13). The first decision step concerns agent's individual decision-making process, deliberating about conservation types for each landscape unit. Broadly speaking, the park manager agent builds its preference preorder over allowed levels of conservation. An argumentation-based framework has been implemented to support the decision making. The key idea is to use the argumentation system to select the desires the agent is going to pursue: natural park stakes and dynamics are considered in order to define objectives for which to aim. Hence, decision-making process applies to actions, i.e. levels of conservation, which best satisfy selected objectives. The second step consists in taking account of players' preferences, with the possibility to adjust the profile of the park managers, from autocratic to democratic,

Figure 12. Interface for the negotiation step

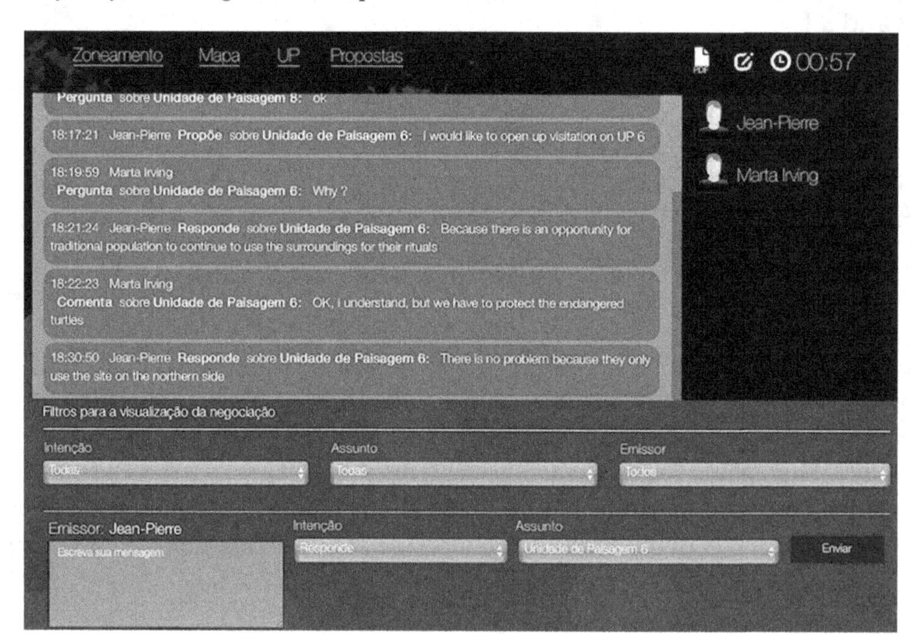

Figure 13. Park manager agent 2 steps decision architecture

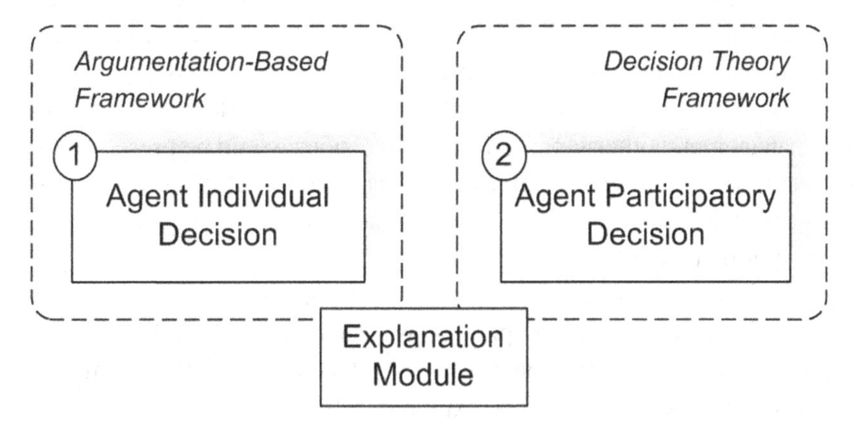

and therefore the influence of players' votes. Therefore, an original method for combining players votes and the park manager vote, based on an influence function, has been designed.

Objectives

Participatory management aims to emphasize the role of local actors in managing protected areas. Participatory management aims to emphasize the role of local actors in managing protected areas. However, the park manager is the ultimate arbiter of all policy on devolved matters. He acts like an expert who decides on validity of collective concerted management policies. Moreover, he is not a completely fair and objective arbiter: he still brings his personal opinions and preferences in the debate. Therefore, we aimed at developing an artificial agent modeling the following behaviors:

- **Personal Preferential Profile:** Park manager decision-making process is supposed to be influenced by its sensibility to natural park stakes and conflicts. In decision theory terms, we can affirm that park manager's preferential profile could be intended as a preference relation over conservation policies. One of the key issues is to understand that we cannot define a strict bijection between preferential profile and preference relation. Agent's preference relation is partially dependent on natural park resources and realities. Moreover, this relation is not likely to be an order or a preorder. Hence, our agent must be able to generate its preference relation according with its preferential profile. We distinguish two preferential profiles:
    - **Preservationist:** Aims to preserve ecosystems and the natural environment.
    - **Socio-Conservationist:** Generally accepts the notion of sustainable yield – that man can make indirect use of resources from a natural environment on a regular basis without compromising the long term viability of the ecosystem.
- **Taking into account Stakeholders' Decisions:** A participatory decision-making leader seeks to involve stakeholders in the process, rather than taking autocratic decisions. However, the question of how much influence stakeholders are given may vary on manager's preferences and beliefs. Hence, our objective is to model the whole spectrum of participation, from autocratic decisions to fully democratic ones. To do so, we want the park manager agent to generate a preference preorder over conservation policies. This is because it should be able to calculate the distance between any

two conservation policies. This way, we can merge stakeholders' preference preorders with manager's one to establish one participatory final decision. Autocratic/democratic manager attitude will be modeled by an additional parameter during the merge process.

- **Expert Decision:** The park manager's final decision must consider legal constraints related to environmental management; otherwise, non-viable decisions would be presented to the players, thus invalidating the game's learning objectives. These constraints are directly injected in the cognitive process of the agent. Hence, the agent will determine a dynamic preference preorder over allowed levels of conservation (according to its preferential profile).
- **Explaining Final Decision:** In order to favor the learning cycle, the park manager agent must be able to explain its final decision to the players. We can consider that the players could eventually argue about its decision; the agent should then defend its purposes using some kind of argumentative reasoning. Even if such cases will be explored in future work, it is our concern to conceive a cognitive architecture which provides a good basis for managing these situations.

Architecture Overview

Let us now present an architecture overview of the park manager agent. As depicted at Figure 13, the architecture is structured in two phases. We believe that sequential decision-making mechanisms can model complex cognitive behaviors along with enhanced explanation capabilities.

The first decision step concerns agent's individual decision-making process: the agent deliberates about the types of conservation for each landscape unit. Broadly speaking, park manager agent builds its preference preorder over allowed levels of conservation. An argumentation-based framework is implemented to support the decision making.

The second step consists in taking account of players' preferences. The result of the execution is the final park manager decision, influenced by the different stakeholder's preferences.

Individual Decision

Recently, argumentation has been gaining increasing attention in the multi-agent community. Dung's seminal work (Dung, 1995) proposed formal proof that argumentation systems can handle epistemic reasoning under open-world assumptions, usually modeled by non-monotonic logics. Argumentation thus became an established approach for reasoning with inconsistent knowledge, based on the construction and the interaction between arguments. More recently, some research has considered argumentation systems capabilities to model practical reasoning, aimed at reasoning about what to do (Hulstijn & Torre, 2003) (Amgoud & Kaci, 2004) (Rahwan & Amgoud, 2006). Indeed, argumentative deliberation provides a mean for an agent to choose or discard a desire as an intention. Another important motivation for choosing argumentation as a foundation is the capacity of explaining the decisions made. We believe that argumentation "tracking" represents an effective choice for accurate explanations. Conflicts between arguments can be reported to the players, following agent's reasoning cycle, thus enhancing user comprehension.

From this starting position, we have developed an artificial agent on the basis of Rahwan and Amgoud's work (2006). The key idea is to use an argumentation system to select the desires the agent is going to pursue: natural park stakes and dynamics are considered in order to define objectives for which to aim. Hence, decision-making process applies to actions, i.e. levels of conservation, which best satisfy selected objectives. In order to deal with arguments and knowledge representation, we use first-order logic. Various inference rules were formulated with the objective of providing various types of reasoning capability. Note that the agent's knowledge base is not updated during execution, since it is not directly exposed to social interactions. Knowledge base and inference rules consistency-checking methods are, therefore, not necessary.

The reasoning cycle of the manager agent includes 3 successive phases:

1. **Reasoning about Desires:** In order to construct a set of consistent desires (defended by arguments), by managing conflicts between arguments;
2. **Reasoning about Decisions:** In order to construct a set of gain vectors (gain for each desire) for each action (type of conservation);
3. **Multicriteria Aggregation:** In order to select actions (types of conservation) that best satisfy selected desires, by applying an aggregation operator on gain vectors.

For example, a simple rule for generating desires from beliefs (i.e. natural park stakes), used during phase 1, is:

Fire → Avoid_Fires, 4

where Fire (fire danger in the park) is a belief in agent's knowledge base and Avoid_Fires is the desire that is generated from the belief. The value 4 represents the intensity of the generated desire.

Examples of rules for selecting actions (types of conservation) from desires, used during phase 2, are:

Primitive → Avoid_Fires, 0.4

Intangible → Avoid_Fires, 0.8

where Primitive and Intangible represent levels of conservation and values 0.4 and 0.8 respectively represent their utility in order to satisfy the corresponding desire (Avoid_Fires). The gain of an action A for a desire D is defined and computed as the product of the utility of A for satisfying D with the intensity of D.

During phase 3, the aggregation of gain vectors for each action leads to a single value, measure of its performance. The ordering of the performance of actions (types of conservations) leads to a preference ordering (for the 3 best values), ex such as:

Extensive > Primitive > Intangible

Further details about the architecture may be found in (Sordoni, 2008).

Participatory Decision

In the real world, a park manager can be more or less participatory, i.e., more or less open to taking into account stakeholders' opinions in his final decision. We propose a method, fitted into the social-choice framework, in which the manager participatory attitude is a model parameter. In a real case scenario, a decision-maker would examine each stakeholder's preferences in order to reach the compromise that best reflects its participatory attitude. Our idea is to represent this behavior by weighting each player's vote according to the manager's point of view.

This concept is illustrated at Figure 14. The process is structured in two phases. Firstly, the manager agent injects its own preferences into the players' choices by means of an influence function describing agent's participatory attitude. Stronger influence translates into more autocratic managers. Secondly, modified players' choices are synthesized, using an aggregation function, i.e. Condorcet voting method. The result of the execution will be the agent participatory decision.

For example, let the following be players' choices, where > is a preference relation (a > b means "a is preferred to b") and A = {Intangible, Primitive, Extensive} the candidates' set. Players' choices are converted into numeric vectors specifying the candidates' rank (each column corresponds to each of the candidates) for each vote. Suppose that preferences of the 1st player are:

player_1: Intangible > Primitive > Extensive

Intangible is the most preferred by player_1, and thus has the highest rank (3), then Primitive is in 2nd position (2), and Extensive as the last one (1). The corresponding vector of 1st player's preferences is thus:

$v_1 = (3, 2,1)$

Suppose now that the 2nd player preferences are:

player_2: Extensive > Primitive > Intangible

Figure 14. Park manager agent participatory decision architecture

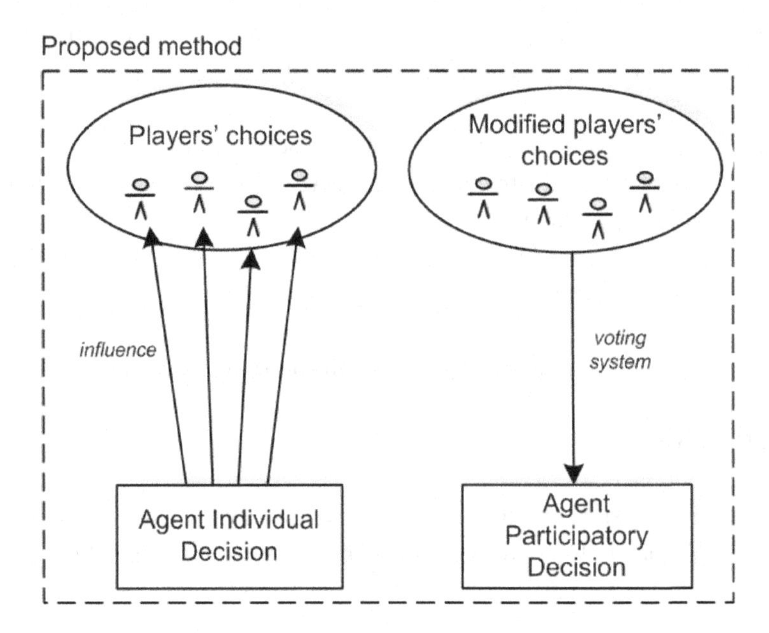

Intangible is the least preferred by $player_2$, and thus has the lowest rank (1), Primitive is in 2nd position (2), and Extensive has the highest rank (3). The corresponding vector of 2nd player's preferences is thus:

$v_2 = (1, 2,3)$

Suppose that the preferences of the 3rd player are:

$player_3$: Primitive > Extensive > Intangible

This leads to the corresponding vector:

$v_3 = (1, 3,2)$

Suppose now that the manager individual decision (as produced by previous individual decision module) is:

manager: Extensive > Primitive > Intangible

$v_M = (1, 2,3)$

Let the influence function be defined as follows:

$$f(x,y) = \begin{cases} x & if\ x = y \\ x * 1 / |x - y| & otherwise \end{cases}$$

Modified player's vectors will be:

$mv_1 = (f(v_1(1), v_M(1)), f(v_1(2), v_M(2)), f(v_1(3), v_M(3)))$

$= (f(3, 1), f(2, 2), f(1, 3))$

$= (3*1/|3-1|, 2, 1*1/|1-3|)$

$= (1.5, 2, 0.5)$

$mv_2 = (1, 2,3)$

$mv_3 = (1, 3,2)$

In order to compute the manager participatory decision, we apply the Choquet integral C_{μ} (Choquet, 1953) choosing a symmetric capacity measure $\mu(S) = |S|^2/|A|^2$, where A is the candidates set.

C_{μ}(Intangible) = 1.05

C_{μ}(Primitive) = 2.12

C_{μ}(Extensive) = 1.27

Primitive has the highest value (and Intangible the lowest), thus the final decision will be:

manager_participatory: Primitive > Extensive > Intangible

From Arguments to Explanations

As stated above, the chain of arguments are interesting material for explanation of the decisions. In one of our tested scenarios, the manager individual decision is the following:

manager: Intangible > Recuperation

Arguments for Intangible are:

Endangered_Species & Tropical_Forest → Maximal_Protection

Intangible → Maximal_Protection

Arguments for Recuperation are:

Fire & Agriculture_Activities → Recover_deteriorated_zone

Recuperation → Recover_deteriorated_zone

Note that these traces of arguments represent a basis of explanation of how/why the decision has been made. Meanwhile, making it upto a complete explanation facility remains as a future work.

Implementation Framework

The architecture presented in this paper has been implemented in the Jason multi-agent platform (Bordini et al., 2007). Besides interpreting the original AgentSpeak(L) language, thus disposing of logic programming capabilities, Jason also features extensibility by user-defined internal actions, written in Java. Hence, it has been possible to easily implement aggregation methods.

The resulting architecture (Sordini et al., 2010) has been implemented and tested offline and its outputs (decision and arguments) have been validated by our project domain experts. In the current architecture of the artificial park manager, only static information about the park and about the final votes of the players are considered. We are considering exploring how to introduce dynamicity in the decision model, taking into account the dynamics of negotiation among the players (the evolution of player's proposed votes during negotiation).

VIABILITY EXPERT AGENT

Technocracy vs. Relativism

Regarding environmental governance and public policies, in our case management of protected areas for biodiversity conservation, we believe that there is some fundamental complementarity, as well as some tension, between the needs for participation of social actors involved (stakeholders) and the needs for a minimal kind of technical expertise to help at evaluate management proposals and decisions. Let's illustrate this as follows. On the one hand, a pertinent decision taken without any consultation may not be well accepted by stakeholders because seen as too technocratic and autocratic. (For this reason, we designed an artificial park manager with an adjustable level of incorporation of players proposals). On the other hand, a completely democratic and consensual decision could be very inappropriate regarding the future of the protected area. Furthermore, it is important to note that the technical tools for evaluation and decision making are usually only in the hands of the technical expert(s) or/and the decision maker(s). The potential limits of participation is therefore that the discussion between members of the management council may reach some limits because of the difficulty to analyze and compare (commensurate) the pros and cons of their respective perspectives and proposals. This may lead to blockage and frustration. On a more epistemological and political perspective, we believe that pure participation without any means for minimal grounded technical expertise raises the issue of the incommensurability of proposals (in other words, relativism), thus leaving unclear on what ground a decision could be finally taken (persuasion, trading mutual support for respective objectives, lobbying, coalition formation, intimidation, and in extreme cases force…).

What we propose therefore is to provide the stakeholders/participation with some access to technical tools to self-evaluate and also to self-relate and compare their viewpoints and proposals. Obviously, it is a very ambitious wish and it touches upon some fundamental educational issues. But we believe that with computer-supported recent advances, there are some ways towards empowering stakeholders with more technical expertise, and not let stay this technical expertise only in the hands of the decision makers.

Assistant and Expert Agents

We envision technical assistance of two kinds. One kind is based on (shallow) decision theory in order to provide players with information about their relative and global positioning within the collective decision process (e.g., relations of dominance and equity properties).

A simplified preliminary version of this kind has already been implemented and tested, in the form of assistant agents to assist players through the game. The basic initial function of these agents is to present and explain each step of the game. During the negotiation step, assistant agents may also provide some helpful information to participants, in order to improve their analysis concerning the negotiation. For instance, different modules inform a player about computed indicators such as affinity (compatibility)

Figure 15. One of the modules (affinity) of the assistant agent

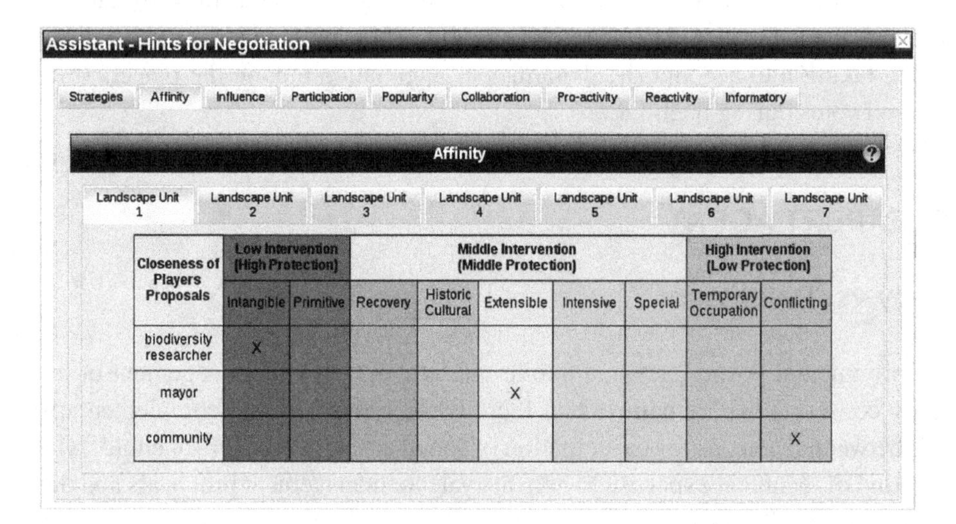

Closeness of Players Proposals	Low Intervention (High Protection)		Middle Intervention (Middle Protection)					High Intervention (Low Protection)	
	Intangible	Primitive	Recovery	Historic Cultural	Extensible	Intensive	Special	Temporary Occupation	Conflicting
biodiversity researcher	X								
mayor					X				
community									X

with other players proposals (see at Figure 15), participation, influence, as simple indicators of current position of the player within the other players proposals and the collective decision process. It is important to emphasize that the user has total control over his assistant, enabling or disabling it at anytime. Moreover, since we have decided to favor a bottom-up approach, we decided to avoid intrusive assistant agents through the game. We believe that intrusive assistant agents could interfere in the players' cognitive processes. This is why our assistant agents do not suggest decisions to players.

Viability Theory

A more ambitious kind of technical expert agent is based on viability theory (Aubin, 1991). It is a mathematical formalism which, as we will see, allows to identify the policies that can retain or restore desirable properties of a dynamical system (biological, economical…) subject to constraints. It has been applied to various domains, including to environment, as for instance for the modeling of lake eutrophication and as a way to analyze viability and resilience properties (Martin, 2004. In our case, the underlying hypothesis is that negotiation between stakeholders could more easily focus on the constraints that each player may want to define (what he considers to be the desirable properties of the park to be viable, e.g., considering the survival of an endangered species, or the sustainability of an economical model of park visitation), rather than on predefined objectives for management decisions. The objective is therefore to provide each player with some technical evaluation of the impact of the constraints he defines as well as the way he could enforce them, in other words, the feasibility of the constraints and objectives that a player himself defines and wants to negotiate.

Viability theory considers a dynamical system subject to constraints in the state space with the assumption that its evolution may be influenced by a control (which may vary in time). Rather than defining criteria for optimality of the system, viability theory considers constraints on the possible values of the variables (e.g., intervals of values), which should be satisfied for the system to be and remain viable. The methods and the tools of viability theory allow to find conditions (states and decisions) for which such operational constraints will be always satisfied and thus progress in a viable and sustainable way. The viability kernel is defined as the set of states of the system for which always exists (at least) a control function which maintains the system within the constraints. The primary method is to determine the viability kernel and algorithms have been developed to compute it (Saint-Pierre, 1994). Figure 16 shows an example of viability kernel. The dark blue part of the figure represents the viability kernel Viab(K) for the set of constraints K. x_0, y_0 and y_1 are initial states verifying the constraints. The green/light trajectory starting from x_0 and directed towards the right corresponds to a viable evolution. All other red/dark trajectories correspond to non viable evolutions.

Another useful study is to analyze the possible trajectories corresponding to a given control strategy in order to see which ones will stay or not within the viability kernel (in other words, analyze if control strategies are viable or not). Figure 17 displays an example of two trajectories. The red/dark trajectory which exits towards the right is obtained with an "empty" strategy with a constant control. The green/light spiral-shape trajectory corresponds to a strategy where the control is variable and adjusted (points correspond to the adjustments of the control values) in order to stay within the viability kernel. Note that it is also possible from current implementation of viability theory to automatically generate a viable control (Aubin et al., 2011). Last, note that there are some other interesting concepts of the viability theory,

Figure 16. Example of viability kernel

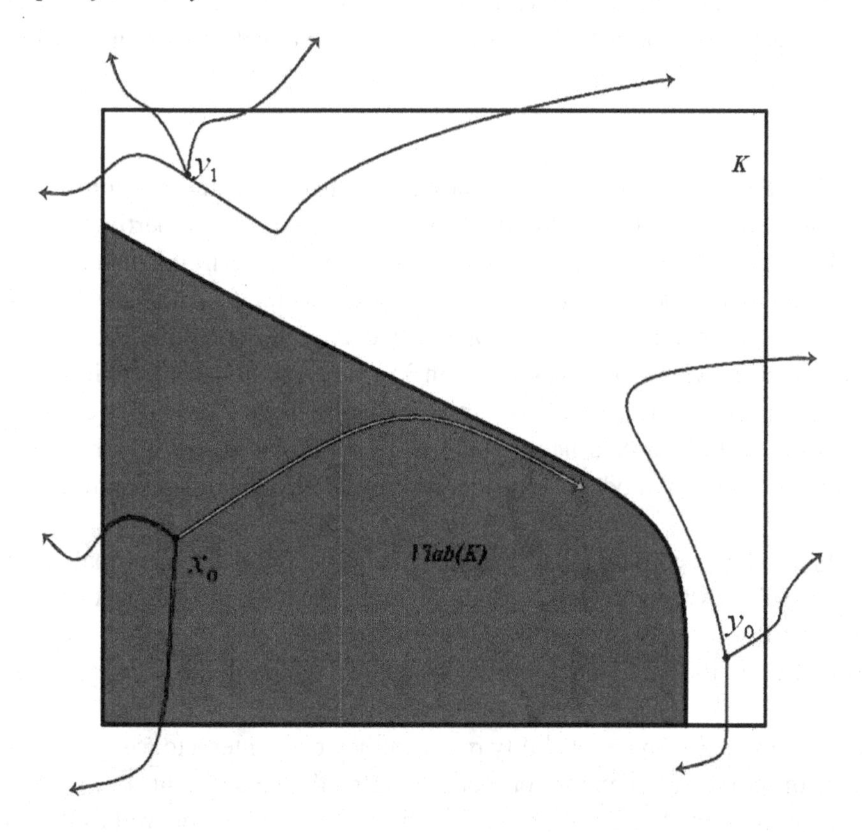

Figure 17. Example of viable and non viable trajectories

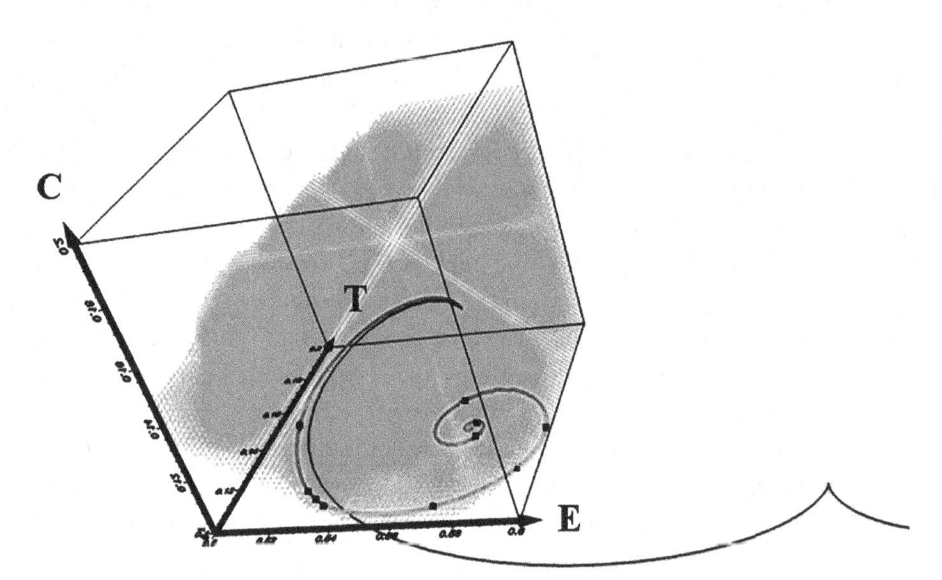

such as the capture basin (all states for which it exists a control function to reach a given target in some temporal horizon), but we are not (yet) using them as a basis for analysis within our SimParc project.

Park Viability

In summary, viability theory helps at identifying the policies that can retain or restore desirable properties of a dynamical system. It turns out easier to ask the players for desirable properties than for optimization objectives, which are generally not unique and may be unknown for environmental issues. With viability theory, players have to define desirable properties as constraints on the park state variables.

To illustrate the potential of the viability approach, we have chosen a simplified model representing the park environmental and visitation dynamics, with 3 variables: tourists (T), environmental quality (E) and animals (A). It captures the idea of a park sub-area (landscape unit) where tourists come to observe animals (e.g., birds or turtles) in their natural habitat. In addition to these 3 variables, the model also includes a set of parameters: 10 model parameters and one control parameter, introduced as follows:

- Animals (A) develop according to a population evolution law with a maximal growth rate γ_A.
- The reference value of the carrying capacity σ_A represents the environment capacity to host animals (the better is the environment, the higher its capacity).
- The threshold E_0 is the minimal environment quality value, under which animal reproduction stops.
- Animals may be killed with a probability σ_T when they encounter a tourist.
- The number of tourists (T) depends on both the attractiveness (quality) of the environment (E) and the number of animals with respective coefficients ω_E and ω_A, as well as the number of tourists (they are mutually exclusive with coefficient ω_T – tourists do not like the visitation site to be overcrowded).

- The quality of the environment (E) is described by a classical logistic equation, with a growth rate γ_E and a maximal value K.
- The environment is damaged by tourists with a rate α.
- Finally, u is the control parameter which represents the investment rate to restore the environment.

The model is defined as follows:

$$A' = A * (\gamma_A(\sigma_A(E - E_0) - A) - \sigma_T T)$$

$$T' = T * (\omega_E E + \omega_A A - \omega_T T)$$

$$E' = E * (\gamma_E(1 - \frac{E}{K}) - \alpha T) + u(K - E)$$

We initialize the parameters with the following default values, shown at Table 1.

Viability Expert Agent

We have designed and implemented a prototype viability expert agent based on a viability analysis (Wei et al., 2012). During a game session, each player proposes his set of desirable constraints for each landscape unit (on the three variables A, E and T, for instance, a desired maximum number of tourists). A default association between conservation types (Intangible, Primitive,…) and constraints (intervals of values for variables A, E and T) is proposed to the players and can be adjusted by them. A subset of this table is presented at Table 2.

Once a player has entered his desired constraints, the expert agent can compute and visualize the viability kernel corresponding to the player constraints. A player can then use these results as a technical basis for negotiating with other players. If the desired constraints are too strong, the expert agent will

Table 1. Default values for the model parameters

γ_A	σ_A	E_0	σ_T	ω_E	ω_A	ω_T	γ_E	K	α	u
1	1	10	0.001	0.1	3	0.1	10	100	0.001	[0 0.5]

Table 2. Default correspondence between conservation types and constraints over variables

Type / *Variable*	Intangible	Primitive	Cultural	Intensive	Occupation
T (Tourists)	0	< 200	[400 1500]	[1500 3000]	[0 1000]
E (Environment)	100%	> 90%	> 70%	> 60%	> 70%
A (Animals)	100%	> 95%	> 85%	> 80%	> 85%

compute an empty viability kernel, which means that the park is not viable with the constraints chosen. The expert agent can then propose a relaxation of the constraints. Also the expert agent can help players at analyzing and comparing more than one viability kernel/set of constraints, in order to help players at comparing their strategies and their viabilities.

Suppose that a first player at some point of the game specifies the following constraints (shown at Table 3). The visualization of the corresponding viability kernel by the viability expert agent is shown (in green) at Figure 18(a).

Table 3. Set of constraints specified by the 1st player

A_1	E_1	T_1
[60 80]	[60 100]	[1000 2500]

Figure 18: Examples of viability kernels for the park visitation model: individual and intersections

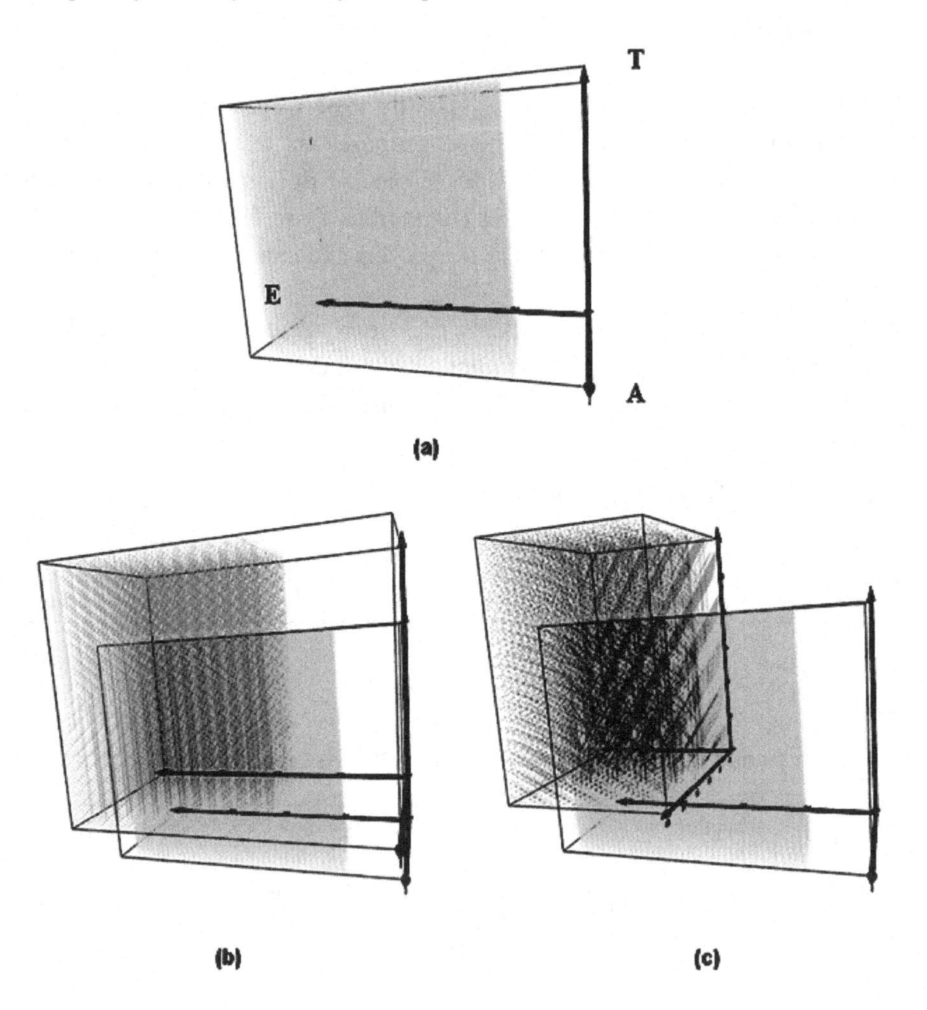

Table 4. Alternative set of constraints specified by the 1st player

A_2	E_2	T_2
[64 90]	[60 100]	[1400 3000]

Suppose now that this player wants to compare this situation with an alternative (slightly different) set of constraints, shown at Table 4. Figure 18(b) shows the comparison (and the intersection) of the first viability kernel (in green/right, corresponding to the first set of constraints shown at Table 3) with the second kernel (in blue/left, corresponding to constraints at Table 4).

Last, suppose that another (second) player is exploring another set of constraints, shown at Table 5, and is currently negotiating with the first player. Figure 18(c) shows the comparison of the first viability kernel (in green/right, corresponding to the first player's first set of constraints at Table 3) with the viability kernel (in red/left) corresponding to the second player's set of constraints (Table 5). This illustrates the capacity of the viability expert agent to help a player to analyze one viability kernel corresponding to a set of constraints that he himself decided, but also to compare with alternative kernels/constraints explored by himself or proposed by other players during the negotiation. Therefore, this provides the players with a basic way to quantify and analyze the degree of feasibility and viability of proposals. Instead of just comparing the constraint sets, the viability expert compares the viability kernels, which are based on the link between the dynamics and the constraints. Small changes in constraint sets can have a broad range of impacts depending on the dynamics. For example, authorizing lower values for the environment quality would not change the viability kernel from Figure 18(a), since the dynamics cannot guarantee the constraint on animals (A) with relatively high constraints on visitation levels (T) and low levels of environment quality (E). In this case, large changes in environmental constraints have no effect. On the contrary, for higher values of environmental quality (E), small changes can diminish terribly the size of the viability kernel. The main interest of the viability agent, through the computation of the viability kernel, is to show the coupling between the system dynamics and the constraints specified by the player. Furthermore, the comparison of viability kernels may help the player to identify constraints (intervals of values) which are irrelevant because of the system dynamics. For instance, in Figure 18(a), states where value of E is too small are irrelevant because they are unviable, thus it is useless to authorize them.

Once computed and analyzed a viability kernel, a player can also ask the viability expert agent to compute and visualize a possible evolution of the model, by specifying a starting state and the time duration (number of time steps for computing the evolution of the model). Figure 19 shows an example of two trajectories corresponding to two different values of the control parameter u (degree of investment for restoring the environment). The two trajectories start from the same point/state x within the viability

Table 5. Set of constraints specified by the 2nd player

A_3	E_3	T_3
[64 90]	[80 100]	[1500 3000]

Figure 19. Examples of viable and non viable trajectories for the park visitation model

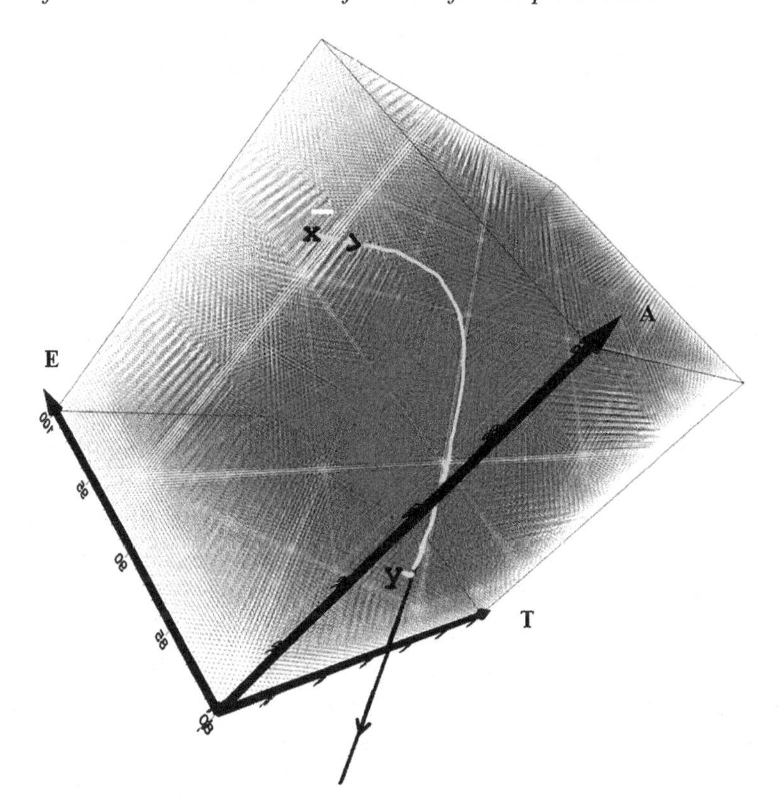

kernel. The green/light trajectory corresponds to a viable trajectory, as it stays within the viability kernel (stops at point/state y). The black/dark trajectory initially follows the green/light trajectory (until point/state y) but exits downwards from the viability kernel and thus corresponds to an unviable trajectory. This is because in the second case no viable control (no viable value of investment) has been applied.

In summary, we see that the viability expert agent helps the players to visualize and analyze and also to compare their constraints and their control strategies. This technical evaluation helps them to evaluate the viability of their proposals (constraints as well as control strategies), compare them, and provides them with objective arguments for the negotiation. In order to help the players at handling the informations, an interface for the viability expert agent has been built. It is shown at Figure 20. With this interface, players can:

- Select among predefined models;
- Define constraints (by entering values intervals);
- Select and execute an algorithm to compute the viability kernel;
- Analyze the results;
- Study the trajectories, i.e. the evolutions.

A session of test of the viability expert agent is depicted at Figure 21. Different sessions have been conducted without and with a viability expert agent in order to analyze the benefits of its use (as well as the possible complexity of its use). During the tests conducted, it turned out that it was easier for players

Figure 20. Viability expert agent user interface

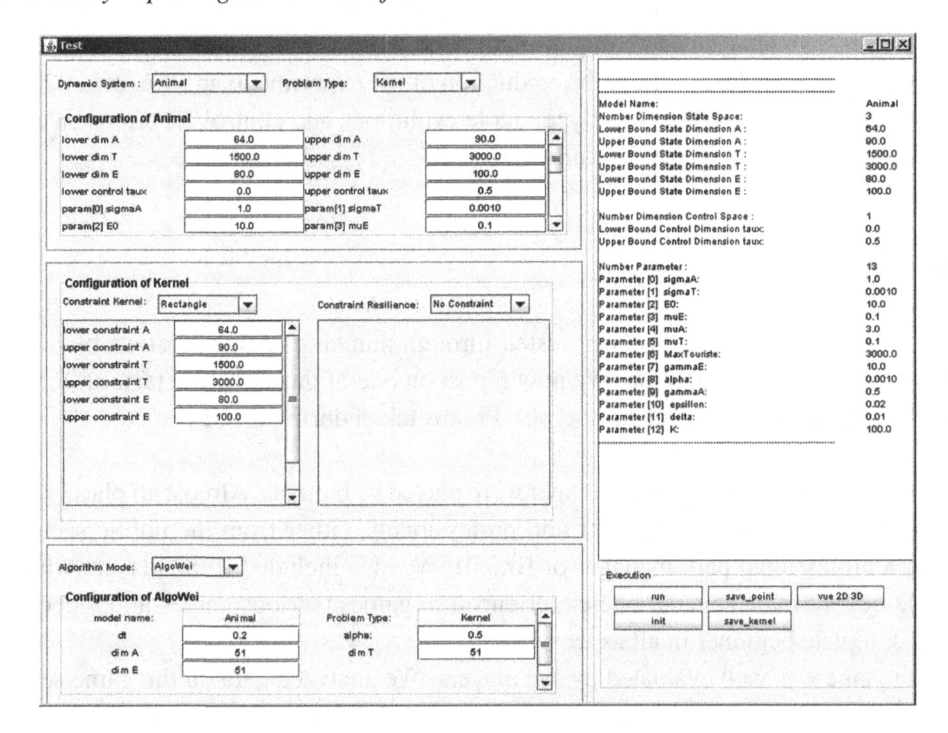

Figure 21. Viability expert test session

to negotiate about constraints than about the types of conservation. It seems that through the viability expert agent, players can analyze and compare their respective desired constraints and their associated viability analysis (see more details about the evaluation of the experiments in (Wei et al., 2012)). Moreover, the expert agent helps players to analyze viable evolutions and control functions. We believe that these results, although preliminary, are promising.

EVALUATION

The SimParc computer prototype has been tested through numerous game sessions by domain expert players, with debriefing and evaluation. We now report on one of the extensive test sessions and evaluations (with the second version of the prototype). Photos taken during the session are shown at Figures 22 and 23.

The 9 roles of the game and the park manager were played by humans. Almost all players were experts in park management (researchers, students and professionals, either from the public sector or NGOs, also including a professional park manager of Brazil). We also included some players which were not knowledgeable in park management: one experienced in games (serious games and video games) and another one a complete beginner in all aspects.

Overall, the game was well evaluated by the players. We analyzed data on the game sessions (written questionnaires, recorded debriefing, etc.). Two aspects of the game were positively evaluated by the participants of the game session: the structure, (script, rules and tasks) and contents (scenes, conflicts, environmental management). Through successful integration of structure and content, SimParc was evaluated as a game that reached the goal of creating, in fact, a virtual arena of management. Although the game does not constitute a tool for decision-making directly applicable to real parks, but only a support for epistemic and pedagogical goals, it was highlighted as a positive aspect the closeness of SimParc to

Figure 22. SimParc game session

Figure 23. SimParc game session

a virtual scenario corresponding to the reality of park management, making it more attractive for those people working directly in real parks.

In the analysis of the test, two key aspects to future improvements of the game play were highlighted: information access and interaction. Considering the issue of information access, the evaluation identified the need for improving the conditions for access to content, form, data quality, amount of information available, among others. Regarding the issue of interaction, were mainly considered the resources and tools available to help players negotiate. The information issue is certainly one of the key issues in the SimParc game. For example, it was considered essential that the proposed zoning of each other possible player could always be viewed with decision changes shown in real time, in order for each player to see how others players are defending their interests. The interaction between the players seems also a key element in improving SimParc. Considering that the game requires a continuous and dynamic interaction between players, it has been highlighted the importance of the use of flexible systems with additional features such as hyperlinks to send messages directly to a player and use of tools that allow the creation of parallel trading rooms.

Aiming to investigate whether SimParc is approaching its epistemic and pedagogical goals, participants of the tests were asked about what would be the main goal of the game. The responses were related to the following issues: management practice involving negotiation between different social actors; experiencing different roles to facilitate practicing dialogue and negotiation; illustration of the dynamics of conflict; learning environmental expertise and park management; and dissemination of the importance of biodiversity conservation. In the interpretation of the players about if the game had reached such objectives, the players felt that yes, the game was a great exercise for negotiation, stimulating active interaction and interest of players, further encouraged by the possible exchange of roles. Participants also reported that the main knowledge acquired after the game was related to the territorial zoning process of parks, specially for players who did not have advanced knowledge about biodiversity conservation. Players who are professionals about biodiversity conservation explained that they improved their knowledge about the specificities of each type of zoning commonly used in park management. It was also mentioned that the game could be considered as an exercise on process and techniques of negotiation, although the game does not suggest any technique to the players.

Another point that was mentioned refers to the recognition of the diversity of interests in the management of a park. Even though most players knew many of the conflicts illustrated by the game (tension between development demands and conservation issues, distinct perception of nature, etc.), they mentioned that it was possible to improve their analysis based on different roles and groups of social actors that the game illustrates. Last, the fact that the game does not need to be installed as a program on computers, which means greater mobility for applications and larger dissemination of this game, was recognized as an important point.

An interesting finding after the game sessions was also that all players learned and took benefit of the game. The experts explored and refined strategies for negotiation and management, whereas the beginner player took benefit of the game as a more general educational experience about biodiversity conservation and associated issues. In other words, the game appeared to be tolerant to the actual level of expertise of players, an aspect which had not been planned ahead.

A summary of the different test sessions and workshops is available at: http://www-desir.lip6.fr/~briot/simparc/evenements-en.html

PROSPECTS

We believe that our project, although prospective, has already produced some interesting prototype artifacts, methodological findings and results. Meanwhile, it also opens the way for future works. We will now mention some of them.

Game Server Prototype

We are currently using the third version (generation) of the SimParc Web server. It has already been tested and updated. We are currently organizing new large scale test sessions in order to evaluate it.

Artificial Players

Artificial players represent an ongoing research based on previous experiences on virtual players in a computer-supported role-playing game, such as JogoMan-ViP (Adamatti et al., 2007). The objective is to possibly replace some of the human players by artificial agents players. The two main motivations are: (1) the possible absence of sufficient number of human players for a game session; and (2) the need for testing in a systematic way specific configurations of players' profiles, and thus broader possibilities of game experiences. Note that our existing manager agent architecture provides a basis for artificial players decision. We could complement it with the addition of negotiation and interaction modules. The argumentation capabilities could also be used to generate and control the negotiation process. One complementary direction is to use automated analysis of recorded traces of interaction between human players in order to infer models of artificial players. In some previous work (Guyot et al., 2006), genetic programming had been used as a technique to infer interaction models, but alternative induction and machine learning techniques, e.g., inductive logic programming could also be used.

It is important to note that, as opposed to artificial players in the JogoMan role-playing game (Adamatti et al., 2007), where interaction protocol between players (human or artificial) is simplified and fixed, in our SimParc role-playing game, interactions are not a priori restricted and include arbitrary text. Meanwhile, it is possible to exploit rethorical markers and the choice of preferences by human players to provide artificial players with some useful guidance information. Also, as pointed out above, arguments generated by the actual manager agent architecture provide a good basis for generating negotiation messages, see for instance the work of Amgoud & Vesic (2012) on the role of argumentation in negotiation dialogues. Furthermore, remember that we do not intend the game to correspond exactly to the situation of an actual park, but to help players at exploring and learning (in an epistemic way) about the participatory process.

Assistant Agents

As noted above, current implementation of assistant expert agents is preliminary. Expert agents based on decision theory analysis methods may inform a player about his relative as well as global positioning within the collective decision process (e.g., relations of dominance and equity properties).

Viability Expert

The viability expert is one of the most ambitious direction. This is because it is challenging to combine the inner formalism of viability theory (as well as the complex computations involved) with an intuitive assistance for the human player. The current prototype is therefore more of a proof of concept than a definitive system. Meanwhile, as discussed in the section on the viability expert agent, the preliminary tests that we have conducted have been very positively received by the participants (although the model used was very basic). Also, we need to explore and combine various formal models of the ecosociosystem dynamics to propose a little more realistic evaluation. Therefore, we plan to identify cases of usage conflicts (e.g., between tourism and conservation of an endemic species) and model the dynamics of the system (in an individual-based/multi-agent model or/and in an aggregated model).

Prospects are also to offer more possibilities to the players, such as computing the resilience of the park (i.e., its resistance to perturbation, regarding the issues at stake), as well as suggesting modifications of a given policy in order to improve the resilience. Last, we have recently developed a new implementation framework for viability analysis more flexible and efficient (Alvarez et al., 2016), by representing viability sets with kd-tree data structures (Rouquier et al., 2015), and we will use it for the next version of the viability assistant.

CONCLUSION

As proposed in the article, the key idea that has inspired us the SimParc project, game and prototype, is to explore methodologies which may help to consolidate democratic forums of decision in cases of protection of nature (Irving, 2006). In this sense, the game intends to be a pedagogical tool to explore and test practices of "good governance" (Graham et al., 2003), one of the major issues for biological biodiversity conventions. Although this is an innovative proposal, with wide application in the present context, the experience has shown that quick and simple solutions to modeling the complexity of this process can become a great risk of loss of meaning of the game. Considering that the game could be (and already has been) played by some professional park managers, it is important to reflect how far the game, that is fun and educational, should be closer to reality and what are the necessary representations and abstractions to achieve the required goals. For example, how the process of negotiating social pacts and democratic management of protected areas can be promoted without losing the focus on respect to real problems and operational by the tax legislation and guidelines for management? Similarly, how to balance technical and scientific expertise in the social participation in the management of biodiversity conservation strategies? Although more evaluation is needed, we believe the feedback gained from the different game session tests we conducted is encouraging for the future. We are welcoming any feedback and input from similar or related projects.

More information about SimParc project (papers, tests, workshops…) is available at: http://www-desir.lip6.fr/~briot/simparc/

ACKNOWLEDGMENT

The authors would like to thank to all the past members of SimParc research team: Altair Sancho, Davis Sansolo, Diana Adamatti, Felipe Martins, Ivan Bursztyn, Lucas Dias, Paul Guyot, Sophie Martin, Vinícius Sebba Patto and Wei Wei, for their contributions to the project. We also thank all the participants of the various game sessions for their participation.

This research has been funded by various programs, such as: ARCUS Program (France), CNRS & Cemagref Ingénierie Ecologique Program (France), MCT/CNPq/CT-INFO Grandes Desafios Program (Brazil). Some additional individual support has been provided through doctorship or fellowship programs by: French Ministry of Research (France), AlBan (Europe), CAPES, CNPq and FAPERJ (Brazil).

REFERENCES

Adamatti, D., Sichman, S., & Coelho, H. (2007). Virtual players: From manual to semi-autonomous RPG. In F. Barros, C. Frydman, N. Giambiasi and B. Ziegler (Eds.), *Proceedings of the AISCMS'07 International Modeling and Simulation Multiconference (IMSM'07)*. The Society for Modeling Simulation International (SCS).

Alvarez, I., Reuillon, R., & de Aldama, R. (2016). Viabilitree: A kd-tree Framework for Viability-based Decision. Research report, LIP6, Paris. https://hal.archives-ouvertes.fr/hal-01319738

Amgoud, L., & Kaci, S. (2005). On the generation of bipolar goals in argumentation-based negotiation. In *Argumentation in Multi-Agent Systems – Proceedings of the First International Workshop (ArgMAS'04) – Expanded and Invited Contributions*, (LNCS), (vol. 3366). Heidelberg, Germany: Springer Verlag. doi:10.1007/978-3-540-32261-0_13

Amgoud, L., & Vesic, S. (2012). A formal analysis of the role of argumentation in negotiation dialogues. *Journal of Logic and Computation*, *5*(22), 957–978. doi:10.1093/logcom/exr037

Annett, J., & Duncan, K. D. (1967). Task analysis and training design. *Journal of Occupational Psychology*, *41*, 211–221.

Aubin, J.-P. (1991). *Viability theory*. Basel, Switzerland: Birkhäuser.

Aubin, J.-P., Bayen, A. M., & Saint-Pierre, P. (2011). *Viability Theory: New Directions*. Heidelberg, Germany: Springer Verlag. doi:10.1007/978-3-642-16684-6

Aureliano, V., Silva, B., & Barbosa, S. (2006). Extreme Designing: Binding Sketching to an Interaction Model in a Streamlined HCI Design Approach.*Proceedings of the VII Simpósio Brasileiro sobre Fatores Humanos em Sistemas Computacionais (IHC'06)*.

Barreteau, O. (2003a). Our Companion Modelling Approach. *Journal of Artificial Societies and Social Simulation*, *6*(2), Article 1.

Barreteau, O. (2003b). The joint use of role-playing games and models regarding negotiation processes: Characterization of associations. *Journal of Artificial Societies and Social Simulation*, *6*(2), Article 3.

Bordini, R. H., Hübner, J. F., & Wooldridge, M. (2007). *Programming multi-agent systems in AgentSpeak using Jason*. Chichester, UK: Wiley.

Brazil. (2000). *Lei N° 9.985, que regulamenta o Art. 225, parágrafo 1º, incisos I, II, III, VII da Constituição Federal, institui o Sistema Nacional de Unidades de Conservação da Natureza e dá outras providências*. Brazil.

Brazil. (2002). *Decreto No. 4340 que regulamenta os artigos da Lei No. 9985 que institui o Sistema Nacional de Unidades de Conservação – SNUC*. Brazil.

Briot, J.-P., & Guyot, P., & Irving, M. (2007). *Participatory simulation for collective management of protected areas for biodiversity conservation and social inclusion*. Academic Press.

Choquet, G. (1953). Theory of capacities. *Journal of Fourier Institute*, *5*, 131–295. doi:10.5802/aif.53

Crookall, D. (2010). Serious Games, Debriefing, and Simulation/Gaming as a Discipline. *Simulation & Gaming*, *41*(6), 898–920. doi:10.1177/1046878110390784

de Souza, C. S. (2005). *The Semiotic Engineering of Human-Computer Interaction*. Boston, MA: MIT Press.

Ekman, P. (2007). *Emotions Revealed – Recognizing Faces and Feelings to Improve Communication and Emotional Life* (2nd ed.). New York, NY: Holt, Henry & Company, Inc.

Finin, T., Fritzon, R., McKay, D., & McEntire, R. (1994). KQML as an Agent Communication language. In *Proceedings of the Third International Conference on Information and Knowledge Management (CIKM'94)*. ACM Press. doi:10.1145/191246.191322

FIPA. (2002). *FIPA Agent Communication specifications*. Foundation for Intelligent Physical Agents. Retrieved from http://www.fipa.org/repository/aclspecs.html

Fuks, H., Pimentel, M., & Lucena, C. J. P. (2006). R-U-Typing-2-Me? Evolving a chat tool to increase understanding in learning activities. *Computer-Supported Collaborative Learning*, *1*(1), 117–142. doi:10.1007/s11412-006-6845-3

Gerosa, M. A., Pimentel, M., Fuks, H., & Lucena, C. J. P. (2006). Development of Groupware Based on the 3C Collaboration Model and Component Technology. In Y. A. Dimitriadis, I. Zigurs, & E. Gómez-Sánchez (Eds.), *Groupware: Design, Implementation, and Use (LNCS)*, (Vol. 4154, pp. 302–309). Springer. doi:10.1007/11853862_24

Graham, J., Amos, B., & Plumptre, T. (2003). Governance principles for protected areas in the 21st century. Durban, South Africa: IUCN (International Union for Conservation of Nature).

Guyot, P., Drogoul, A., & Honiden, S. (2006). Power and negotiation: Lessons from agent-based participatory simulations. In P. Stone, & G. Weiss (Eds.), *Proceedings of the 5th International Joint Conference on Autonomous Agents and Multiagent Systems (AAMAS'06)*. doi:10.1145/1160633.1160636

Hulstijn, J., & van der Torre, L. (2003). Combining goal generation and planning in an argumentation framework. In *Proceedings of the 15th Belgium-Netherlands Conference on Artificial Intelligence (BNAIC'2003)*.

Irving, M. A. (2006). *Áreas Protegidas e Inclusão Social: Construindo Novos Significados*. Rio de Janeiro, Brazil: Aquarius.

Kirschner, P. A., Shum, J. B., & Carr, S. C. (Eds.). (2003). *Visualizing Argumentation: Software Tools for Collaborative and Educational Sense-Making*. Heidelberg, Germany: Springer. doi:10.1007/978-1-4471-0037-9

Le Page, C., Becub, N., Bommel, P., & Bousquet, F. (2012). Participatory Agent-Based Simulation for Renewable Resource Management: The Role of the Cormas Simulation Platform to Nurture a Community of Practice. *Journal of Artificial Societies and Social Simulation*, *15*(1), 10. doi:10.18564/jasss.1928

Martin, S. (2004). The cost of restoration as a way of defining resilience: a viability approach applied to a model of lake eutrophication. *Conservation Ecology*, *9*(2), Article 8.

Michael, D., & Chen, S. (2006). *Serious Games Games that Educate, Train and Inform*. Thomson Course Technology.

Putnam, L. L., & Roloff, M. E. (Eds.). (1992). Communication and Negotiation. Newbury Park, CA: Sage. doi:10.4135/9781483325880

Rahwan, I., & Amgoud, L. (2006). An Argumentation-based Approach for Practical Reasoning. In *Proceedings of the 5th International Joint Conference on Autonomous Agents & Multi Agent Systems (AAMAS'2006)*. New York, NY: ACM Press. doi:10.1145/1160633.1160696

Raiffa, H. (1982). *The Art & Science of Negotiation*. Cambridge, MA: Harvard University Press.

Rao, A., & Georgeff, M. (1991). Modelling Rational Agents within a BDI-Architecture. In *Proceedings of the 2nd International Conference on Principles of Knowledge Representation and Reasoning*, (pp. 473–484).

Rouquier, J.-B., Alvarez, I., Reuillon, R., & Wuillemin, P.-H. (2015). A kd-tree algorithm to discover the boundary of a black box hypervolume or how to peel potatoes by recursively cutting them in halves. *Annals of Mathematics and Artificial Intelligence*, *75*(3), 335–350.

Saint-Pierre, P. (1994). Approximation of the viability kernel. *Applied Mathematics & Optimization*, *29*(2), 187–209. doi:10.1007/BF01204182

Searle, J. R. (1969). *Speech Acts: An Essay in the Philosophy of Language*. Cambridge, UK: Cambridge University Press. doi:10.1017/CBO9781139173438

Sordoni, A. (2008). *Conception et implantation d'un agent artificial dans le cadre du projet SimParc*. Rapport de stage de Master 2ème année, Master Intelligence Artificielle et Décision (IAD), Université Pierre et Marie Curie – Laboratoire d'Informatique de Paris 6, Paris, France.

Sordoni, A., Briot, J.-P., Alvarez, I., Vasconcelos, E., Irving, M. A., & Melo, G. (2010). Design of a participatory decision making agent architecture based on argumentation and influence function – Application to a serious game about biodiversity conservation. *RAIRO – An International Journal on Operations Research*, *44*(4), 269–284.

Vasconcelos, E., Briot, J.-P., Irving, M. A., Barbosa, S., & Furtado, V. (2009a). A user interface to support dialogue and negotiation in participatory simulations. In N. David & J. S. Sichman (Eds.), Multi-Agent-Based Simulation IX, (LNAI) (vol. 5269, pp. 127–140). Springer-Verlag. doi:10.1007/978-3-642-01991-3_10

Vasconcelos, E., Melo, G., Briot, J.-P., Patto, V. S., Sordoni, A., Irving, M. A., & Lucena, C. et al. (2009b). A serious game for exploring and training in participatory management of national parks for biodiversity conservation: Design and experience. In *Proceedings of the VIII Brazilian Symposium on Games and Digital Entertainment (SBGAMES'09)*. doi:10.1109/SBGAMES.2009.19

Wall, J. A., & Blum, M. W. (1991). Negotiations. *Journal of Management*, *17*(2), 273–303. doi:10.1177/014920639101700203

Wei, W., Alvarez, I., & Martin, S. (2013). Sustainability analysis: Viability concepts to consider transient and asymptotical dynamics in socio-ecological tourism-based systems. *Ecological Modelling*, *251*, 103–113. doi:10.1016/j.ecolmodel.2012.10.009

Wei, W., Alvarez, I., Martin, S., Briot, J.-P., Irving, M. A., & Melo, G. (2012). Integration of viability models in a serious game for the management of protected areas. In A. Palma dos Reis, P.S.P. Wang, & A.P. Abraham (Eds.), *Proceedings of the IADIS Intelligent Systems and Agents Conference (ISA'2012)*.

KEY TERMS AND DEFINITIONS

Agent: An autonomous software entity being able to take decision and action (including communication with other agents).

Assistant Agent: An artificial agent dedicated to help and assist a human user.

Expert Agent: An artificial agent to assist game players with technical expertise.

Multi-Agent System: An organization of interacting agents.

National Parks: A type of protected areas where there cannot be direct use of resources.

Park Manager Agent: An artificial agent which models the decision behavior of the manager of a national park.

Protected Area: A geographically delimited area where public policies of biodiversity conservation are conducted.

Role-Playing Game: A game where players play a certain role, partially reproducing a certain situation.

Serious Game: A role-playing game intended for training and exploring participatory behaviors which incorporates computer support inspired by video games.

Stakeholder: A social actor who is concerned by the public policy. Examples of stakeholders for a national parks are: traditional population, environmental ONG, tourism operator.

Viability: A mathematical formalism for modeling the policies that can retain or restore desirable properties of a dynamical system (biological, economical…).

Chapter 14
Using Probability Distributions in Parameters of Variables at Agent-Based Simulations:
A Case Study for the TB Bacillus Growth Curve

Marcilene Fonseca de Moraes
Universidade Federal do Rio Grande, Brazil

Albano Oliveira de Borba
Universidade Federal do Rio Grande, Brazil

Diana Francisca Adamatti
Universidade Federal do Rio Grande, Brazil

Adriano Velasque Werhli
Universidade Federal do Rio Grande, Brazil

Andrea von Groll
Universidade Federal do Rio Grande, Brazil

ABSTRACT

Even treatable and preventable with medication, tuberculosis (TB) continues to infect and cause deaths globally, especially in the poorest countries and in most vulnerable parts of the rich countries. Given this situation, the study of the growth curve of Mycobacterium tuberculosis, which causes tuberculosis, can be a strong ally against TB. This study models the growth curve of Mycobacterium tuberculosis using simulation based agents, aiming to simulate the curve with the minimum possible error when compared to in vitro results. To implement this model, the agents represent the bacteria in their habitat and how they interact with each other and the environment. Some parameters of the agents are modelled with probability distributions.

DOI: 10.4018/978-1-5225-1756-6.ch014

INTRODUCTION

Tuberculosis (TB) is a major public health problem, affecting predominantly low and middle-income countries, developing among immigrants, poorest and vulnerable parts of high-income countries (Lönnroth et al., 2015). According to Burgos and Pym (2002)*Mycobacterium tuberculosis*, which causes tuberculosis, is one of the most successful bacterial pathogens in the humanity history.

A report published in 2015 by the World Health Organization (WHO) estimates that in the year 2014 were 9.6 million new cases of tuberculosis (TB) and 1.5 million deaths, and together HIV virus, tuberculosis presents one of the biggest causes of deaths from infectious diseases.

Therefore, the study of *Mycobacterium tuberculosis* growth curve becomes extremely important, because this study will test behavior hypotheses in cases of environmental stress (Voskuil et al., 2004), verify bacillus drug reactions and help to develop new ones (Andries, 2005). For these objectives, growth curve are important not only to determine its dynamic but also sampling bacteria in different phase of growth to determine its behavior variation during the growth. However, the tuberculosis bacillus has a very slow rate of population growth, form clumps to grow and requires enrichment medium. Because this behavior, in vitro experiments are very costly and need the maximum of tools in order to rational design of the studies involving the growth curve and sampling of the bacteria. According to Rebonatto (2000), computer simulations methods have been shown to be effective in situations involving high costs and risks. Computer simulations allow the study of various problems more effectively, as possible, in most cases, the viewing behavior, and specific details of the study object.

Multi-agent systems, a field of artificial intelligence, enables, by means of their tools, to simulate behavioral rules of a system computationally. According to Garcia and Sichman (2005), "Agents are computer characters that act according to the program set, directly or indirectly, by a user. They can act alone or in communities, trainees multi-agent systems ".

Many measurable phenomena present in nature have probability distributions similar to some probabilistic models. Often, these models are used to represent the probability density function of random variables. Probabilistic models are useful in many real situations, to make the variable predictions study and assist in decision support. It is believed that the main variables that model the Mycobacterium tuberculosis bacillus growth curve also resemble a probabilistic model.

The main goal of this work is to model the tuberculosis growth curve, using agent-based simulations, where the values of some variables are drawn from probability distributions, thus making the system developed more similar to real systems.

TUBERCULOSIS

Tuberculosis (TB) is a contagious infectious disease transmitted by *Mycobacterium tuberculosis*. It was discovered on March 24, 1882 by the German scientist Robert Koch, and so the tuberculosis bacillus, *Mycobacterium tuberculosis*, is also known as Koch's bacillus.

Mycobacterium tuberculosis is an intracellular pathogen that can affect various animal species, although humans are the main hosts. It grows with more success in tissues that contain high levels of oxygen, such as the lungs (Lawn & Zumla, 2011).

TB is an ancient disease that continues to be one of the major health problems worldwide. During the humanity history, TB received different names and was often associated with high rates of infection and mortality (Groll, 2010).

Passion and Gontijo (2007) reported that although preventable and treatable with medication, tuberculosis has been presented with an intensity of contamination, with repercussions on health and mortality rates.

In a report published in 2015 by the World Health Organization (WHO), in the 2014, worldwide, there were 9.6 million new cases of tuberculosis (TB) and 1.5 million deaths, and together HIV, tuberculosis represents one of the major causes of death from infectious diseases.

GROWTH CURVE

When Mycobacterium tuberculosis is inoculated in a medium that contains all the nutrients necessary for its survival, the bacteria tend to duplicate. Initially, it adjusts to the new environment (lag phase) until it can begin the process of division regularly (exponential phase/ log phase). When growth become limited, the cells stop dividing (stationary phase), until finally they die for the saturation of environment (Todar, 2013). These phases as shown in Figure 1.

Figure 1. Growth curve and its phases
Source: Todar, 2013

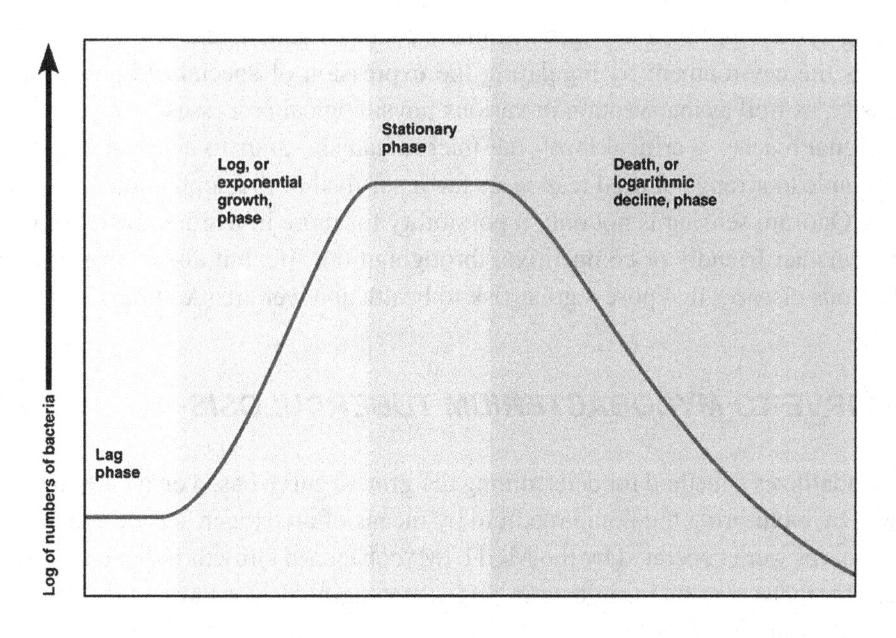

1. **Latency Phase:** Immediately after inoculation, the population temporarily remains unchanged. Although there is no apparent cell division, the bacteria can be grown in mass or volume, synthesizing enzymes, proteins and RNA.
2. **Exponential Phase:** After the adaptation, they rapid growth and consume the maximum amount of nutrients from the environment, also in this stage release wastes and chemical signaling molecules in the environment.
3. **Stationary Phase:** After increasing population growth, the bacteria begin to decrease due to saturation of available nutrients, waste accumulation in the environment and the lack of space. In this stage, they enter dormancy stage where saving energy to increase the time survival.
4. **Decrease Phase (Death):** The last phase of the curve is the death, where they begin to die for lack of nutrients.

QUORUM SENSING

For many years, researchers believed that bacteria existed as individual cells, they acted as independent cell populations and reproduced under favorable conditions. However, in recent decades, several studies have shown that these microorganisms can communicate through signs (Antunes, 2003).

The quorum sensing mechanism is based on the production and diffusion of small signal molecules that may be detected by bacteria. This process occurs when there is awareness of high cell density, allowing the entire population initiate an action, once the critical concentration has been reached (Whitehead et al., 2001).

This signaling system has been identified in many bacterial genera and allows the coordination of behavior towards the environment by regulating the expression of specialized genes, in response to population density, as well as intervention in various physiological processes.

When this signal reaches a critical level, the microorganisms start to act as a single multicellular organism, being able to arrange unified responses favor survival of the population (Rumjanek, 2004).

The study of Quorum sensing is not only a possibility for more in-depth knowledge of bodies with which we live, whether friendly or competitive, throughout our life, but also a promising way for the control of infectious diseases that pose a great risk to health and welfare (Antunes, 2003).

GROWTH CURVE TO *MYCOBACTERIUM TUBERCULOSIS*

Groll (2010) standardizes a method for determining the growth curve based on a system that obtains the bacterial growth by monitoring the liquid medium by means of an oxygen sensor which emits fluorescence. Growth curves were generated by the MGIT (Mycobacteria Growth Indicator Tube).

The data are for the strains that originate in different geographical regions and they are resistant to different drugs, according to Table 1.

For each of the strains, Groll (2010) made two experiments with the same solubility. The monitoring was conducted for 25 days and every time the equipment bacterial growth medium, the growth expressed as Growth Units (GU). Figure 2 expresses the average of experiments for each strain.

Table 1. Used strains and their sources

Strains Identification	Source
GC 01-2522	Georgia
GC 02-2761	Bangladesh
GC 03-0850	Georgia
GC 03-2922	Georgia
H37Rv	ATCC

Figure 2. Real curves in Groll (2010) experiment

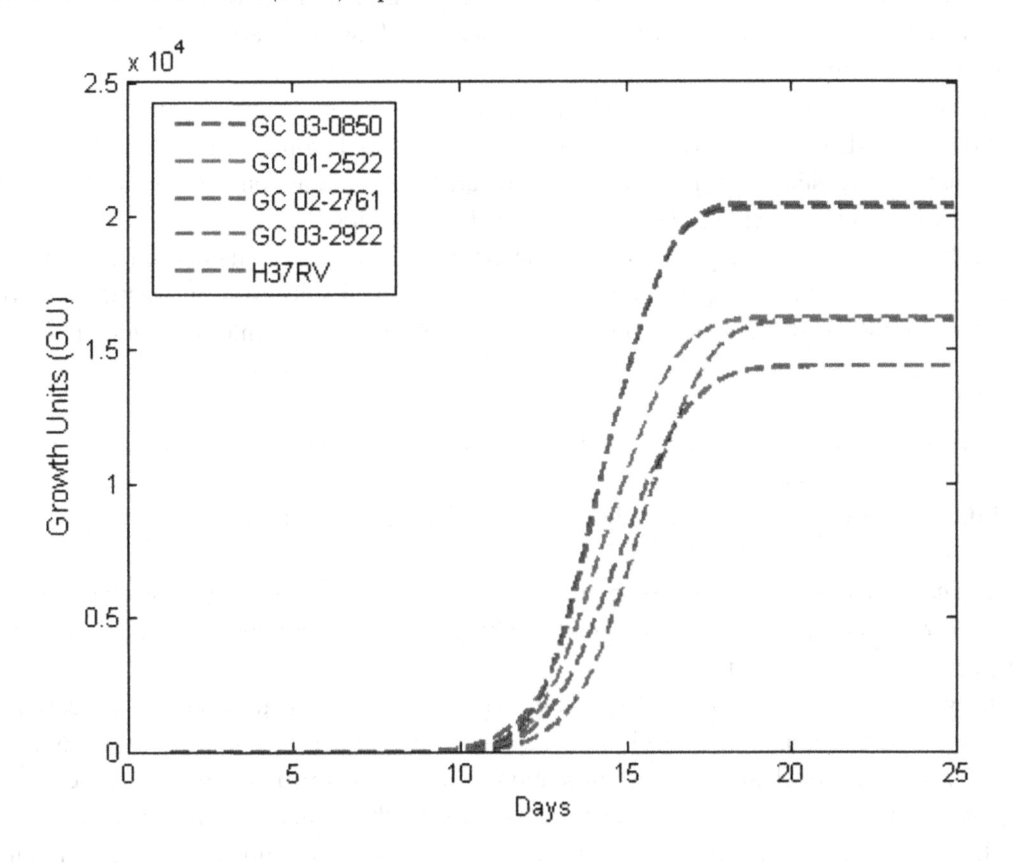

*For a more accurate representation of this figure, please see the electronic version.

However, this method does not allow to know how long the bacteria remained in the stationary phase or the rate of population decline in the death phase. This is because the measurement comes from the metabolic activity of the bacteria, and therefore can only be monitored if there is population growth.

AGENT-BASED SYSTEMS

The conventional simulation is one of the more viable tools available to project, plan, control and evaluate new alternatives/changes to real world. To use the computational simulation, we need softwares to represent the functions of the real world (Rebonatto, 2000).

According to Azevedo (2007), the use of computational simulations is very appropriate to describe real systems, because the simulation tries to reproduce a real situation artificially, where hypothesis could be verify without risks.

One type of simulation is the agent-based simulation. In this type of simulation is possible recreate a population of a real system, where each individual of this population is represented by an agent and each agent has a set of specific rules to define its behavior, its interactions with other agents and the environment where it is inserted (Collela, Klopper, & Resnick, 2001).

In agent-based systems, a real phenomenon is decomposed in a set of elements and their interactions. Each element is modeled as an agent and a general model is the results of all interactions between the agents. Strack (1984) apud Adamatti (2011) reports that the simulation could be divided in three steps:

1. **Modelling Step:** Build the phenomenon model;
2. **Experiment Step:** Apply variation in the built model, changing parameters that influence in the resolution process; and
3. **Validation Step:** Compare the simulated data with real data, to analyses the results.

To simulate computationally a real problem, it is necessary a rigorous study to be able to abstract all variables and relationships that define the model. This study is usually done through observation and analysis of the real phenomenon.

Capturing all of the components of the simulation model is not an easy task and the higher the number of variables and more detailed the model to be simulated, the greater the computational work.

After defining the model and done the simulations, the results are compared with those observed in natural phenomena, in order to evaluate their equivalence, ie, the similarity with the reality.

The choice of agent-based simulation to model of Mycobacterium tuberculosis growth curve was done because the need for integration of different behaviors among the agents, the dynamism that this kind of modeling can enable, such as the interaction between agents, the interaction between agents and environment, and the flexibility to do modifications and extensions in the model.

PROBABILISTIC MODELS

Probabilistic models are useful to represent real situations, or to describe a random experiment. Simulation studies try to play in a controlled environment what is happening in a real environment (Bussab & Morettin, 2005).

The frequency distributions constructed from observations can be represented by mathematical formulas. These formulas are used for the idealization of the actual data and they are called theoretical distributions.

Given a random studied variable, the goal is to know the probability density function itself. In probabilities theory, there are many probabilistic models that describe the probability distribution of several variables.

Among the most important model is the normal distribution, being applicable in various phenomena and constantly used for the theoretical development of statistical inference. To Devore and Silva (2006), the normal distribution is a key model in probability and statistical inference.

The Normal Model

It is one of the most important probability distributions and applicable in various phenomena and constantly used for the theoretical development of statistical inference.

- **Definition:** It is said that the random variable X has a normal distribution with μ parameters (mean of X) and σ (standard deviation of X), if its density function is given by Equation 1:

$$f\left(x;\mu,\sigma^2\right) = \frac{1}{\sigma\sqrt{2\pi}} e^{\frac{-\left(x-\mu\right)^2}{2\sigma^2}}, -\infty < x < \infty \quad (1)$$

- **Graphic:** see Figure 3.

- **Momentum:**
  - Calculating the average of a probability distribution, it is obtained the average value that we would expect to have if we could repeat indefinitely the evidence. It represents the probable value of a random variable, so it is often called expected value or hope.
  - The average of a normal distribution is given by:

$$\boldsymbol{E}\left(\boldsymbol{x}\right) = ¼ \quad (2)$$

  - The variance is a measure of how the probability distribution is dispersed around the mean. A large variance reflects considerable dispersion, while a smaller variance less variability translates with relatively close to the average values. The variance of a normal distribution is represented by Equation 3.

$$\mathrm{Var(x)} = \sigma^2 \quad (3)$$

- **Cumulative Distribution Function:** See Equation 4.

Figure 3. Probability density function for a normal variable with mean x e standard deviation σ

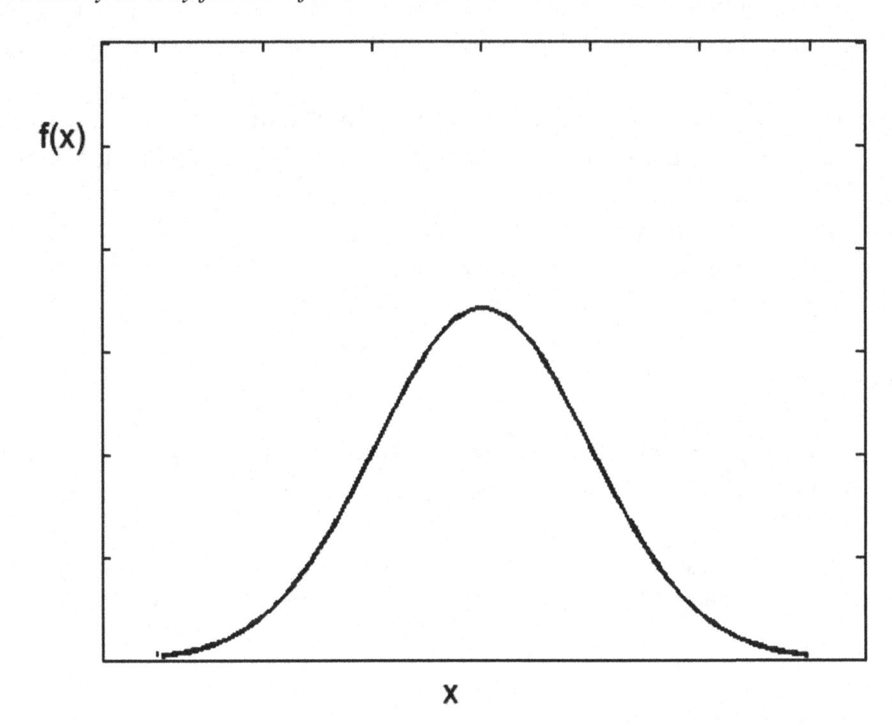

$$P\left(X \leq x\right) = \int_{-\infty}^{x} f\left(x\right) dx = \frac{1}{\sqrt{2\pi\sigma^2}} \int_{-\infty}^{x} exp \frac{1}{2\pi\sigma^2} \left(x - \mu\right)^2 dx \tag{4}$$

The integral of Equation 4 cannot be calculated analytically and the indicated probability can only be obtained approximately by numerical integration. Therefore, it uses the transformation of Equation 5.

$$Z = \frac{x - \mu}{\sigma} \tag{5}$$

The calculation of the reduced variable Z is a transformation of the actual values in coded values. Once calculated the reduced variable Z, refers to normal standardized table to identify the accumulated probability for Z left, i.e., the probability of occurrence values less than or equal to the value of Z consulted.

Sensitivity to Changes in Model Parameters

The normal distribution is unimodal, symmetric about its mean, and tends increasingly to the horizontal axis as it moves away from the average (see Figure 3).

- The distribution N (0,1) to N(3,1) changes the central tendency, but the variability is constant;
- The distribution N(6, 0.25) to N(6,4) changes variability, but the central tendency is constant; and
- Among other distributions combinations of Figure 4, the central trends change and variability.

Figure 4. Normal Distribution with variation in x and σ

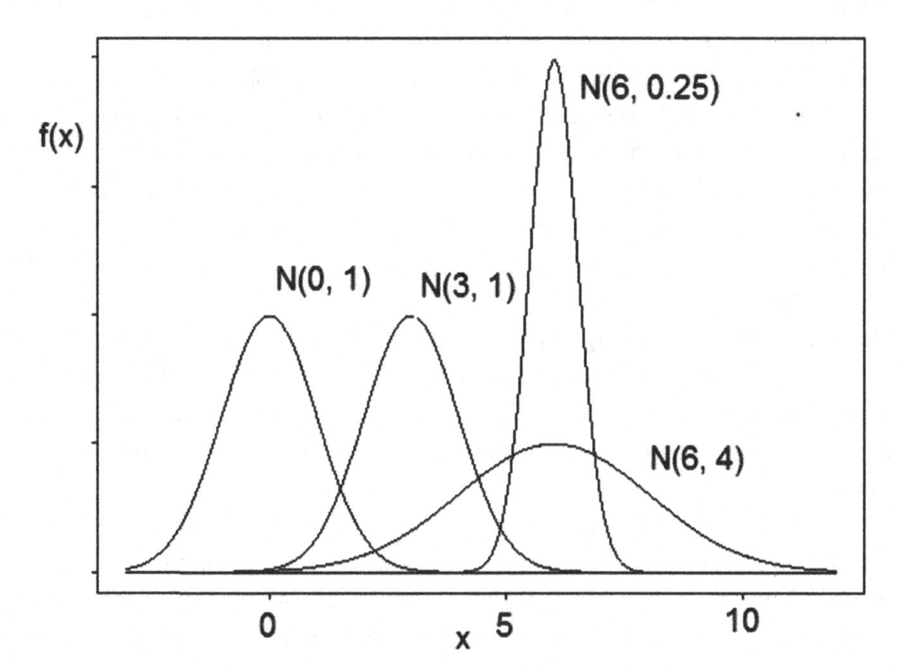

RELATED WORKS

A variety of mathematical models and statistical techniques are used in order to simulate the behavior of random variables. Probabilistic models are useful to represent real situations, or to describe a random experiment.

In order to find a probability density function that more accurately describe the hydraulic conductivity, Mesquita et al. (2002) found the adjustment of the normal, lognormal, gamma and beta. According to the authors the saturated hydraulic conductivity (Ksat) is one of the most important variables for water and solutes motion studies in soil. Therefore, their determination in laboratory and field, produces results with high dispersion, which may indicate that this variable does not have symmetrical distribution. The experiment was to determine the saturated hydraulic conductivity in samples of three soils with different textures of the center west region of São Paulo, and submit the results to statistical tests to identify the most appropriate asymmetric distribution to represent them. After conducted the tests, they concluded that the Ksat showed no normal distribution and set the lognormal, gamma, beta, and the lognormal probability density functions were the most suitable to describe the variable data in study.

Other work that finds the adhesiveness of density functions probability in sample data is Cargnelutti Son, Matzenauer and Trinity (2004). In this work, the authors checked the fit of the series of average global solar radiation data of 22 municipalities of Rio Grande do Sul State (Brazil) to, normal probability, lognormal, gamma, Gumbel and Weibull distributions functions. According to the authors, knowledge of the behavior of precipitation, temperature and relative humidity, evaporation, wind direction and speed, solar radiation, occurrence of dew, fog, hail, frost and snow, is an important tool in making decisions activities related to agribusiness, and the study of solar radiation is essential. The data used in the tests were generated by 47 years of daily observation (1956-2003). To verify compliance of the data set, the authors have made the Kolmogorov-Smirnov test, comparing the empirical probability of the set of sampling data with the probabilities generated by the test distribution.

Cargnelutti Son, Matzenauer, and Trinity (2004) treat data from global average ten-day solar radiation and they use functions of normal, lognormal, gamma, Gumbel and Weibull probability distributions. They showed better adjustment to the normal probability distribution function.

With recent computational advances it became possible to analyze social and economic systems through simulation studies, including the agent-based modeling. Terano (2011) says that the simulation method based on agents is very important, since it can produce results without assumptions unlike conventional approaches.

Werlang (2013) describes a model to growth curves of Mycobacterium tuberculosis based multi-agent systems. For the author, the study of this bacterium is very important as it allows the study of characteristics and the development of new drugs. The performance of experimental tests with this bacillus are slow, taking at least three weeks to show some result, and often fail because of contamination or dehydration means.

In the model of Werlang (2013), agents that represent bacteria have individual characteristics, and these are extremely important for agents to represent their roles in the environment and to interacting with it, similarly as to Mycobacterium tuberculosis bacteria would interact in their natural living environment. The author reports that the results were very satisfactory and the curves found reached a very close similarity to the real curves. Finally, the author also points out that the model is useful as it enables the testing of hypotheses in some hours in opposition to those carried out in vitro which would take days.

In the literature, there are no studies that use simulation based on agents and probability distributions. However, according Porcellis, Moraes, and Bertin (2015) to choose the model (equations or agent-based) depends on how detailed you want to be the system to be simulated. The authors point out that it is possible to work with these two techniques in an associated way, to have better results

GROWTH MODEL CURVE OF *MYCOBACTERIUM TUBERCULOSIS*

When the growth curves are obtained by MGIT, the information that these curves present are the product of several factors of population dynamics. Therefore, it is not possible to extract isolated information, such as: how much they consume or how much fail to consume after reaching environmental saturation, or the proportion of signaling molecules are required to enact the saturation.

Considering these circumstances, it is necessary to infer how many are the variables that affect the population growth, using just the observation of the results.

To simulate the population dynamics we used NetLogo programming environment. The agents based model implemented simulates an environment where agents represent Mycobacterium tuberculosis.

The model agents have specific rules of behavior, which are modeled as variables of the agents. These rules are essential for them to represent their role in the environment and how they interact.

The simulation has the time division: the tick. Every tick, the agents perform one or more actions. These actions are modeled by functions set out in the model, as follows: feed, continue, signals and reproduce.

As many measurable phenomena present in our everyday lives tend to be distributed according to some probabilistic theoretical models, there is a possibility that the main variables that growth curve of Mycobacterium tuberculosis model can also be distributed as normal model.

How many random variables biological fit a normal distribution (Callegari-Jacques, 2003), i.e., the central values are more frequent and extreme rarer (very low values as infrequent as very high). It was assumed that the model variables also are distributed normally. The variables of the agents are:

- **Reproducing Tick:** Time that the bacterium takes to adapt to the environment, whereas this threshold is not reached, it cannot reproduce.
- **Power:** Indicates how healthy the agent is, it accumulates energy every time it consumes nutrients.
- **Consumption:** Nutrient amount that the agent consumes the environment every tick.
- Vital functions: They indicate the amount of energy that will be subtracted from the variable energy each tick.
- **Reproducing Power:** In the proposed model, each agent will start the process of division when healthy, i.e., when it reaches the amount of energy required to reproduce.
- **Reduces Energy:** Amount of energy that the agent spends to reproduce.
- **Bacteria Sensor:** When bacteria realize saturation situation, it releases a signaling molecule, indicating the population that the environment is full and enters reduced power state.
- **Signals Sensor:** Even though bacteria has not released the signaling molecule, it can go into reduced consumption, for detecting a signal threshold in the environment.
- **Reduced Consumption:** This variable defines how much will fall in nutrient consumption value when the agent go into reduced power state.
- **Vital Functions in Reduced Consumption:** They indicate the amount of energy that will be spent every tick when it is in reduced power state.

Agents are inserted in an environment that is shared by all. In this environment, they find the nutrients needed to survive and deposit the waste from their metabolism.

Each space in the environment is called *patch.* In each patch, there is a number of nutrients and waste. Nutrients are used by agents throughout the simulation to keep their vital functions active and accumulate energy. The waste is deposited by the agents after the metabolization of nutrients.

As *in vitro* experiments, that are performed in the laboratory, where a number of bacteria are inoculated into a container, the model initialization is done with a number of agents in the environment.

Agents receive different values for each variable. The set of possible values have different probabilities of occurrence, a characteristic of normal distribution. When an agent is generated, its variables receive simulated values of the distribution used.

After the start of the simulation, the agents begin to move in the environment to search for nutrients. However, in the beginning, they only main is survival, which they are adapting to the environment, and therefore unable to reproduce. This same behavior is observed in vitro experiments that the bacteria need a certain time to adjust their metabolic functions to the new *habitat.*

Later this adaptation time, the agent starts to perform normally all its functions, including reproduction. It consumes nutrients and then transform them into energy, which will be deposited in its reservation to indicate how healthy it is and to keep active vital functions.

Each agent of the model has a different time to reproduce. This time indicates the amount of energy that the agent should have available in its energy reservation to perform the reproduction function. In this way, the energy is a limiting factor for reproduction.

There is an amount of energy to maintain the agent alive at each cycle. If it consume less than necessary to maintain its metabolism, it will start spending power of its reserve in order to survive and the end of this reserve will take its death.

When there is accumulation of waste in patches, the agents have difficulty absorbing nutrients, because it becomes more difficult to agents survive in a very saturated environment.

Another aspect in the curve modeling is the bacteria sensor. This variable determines how many agents in the environment will reach the saturation situation. When this threshold is reached, the agents release a signal molecule, called *quorum sensing,* that warn others that the environment is full. According to Whitehead et al. (2001), this process occurs when there is awareness of high cell density, allowing the entire population initiate an action, once the critical concentration was achieved.

The main action in the proposed growth model is the decision to reduce consumption. Once the agent perceives the situation of saturation, and release a signal molecule, it enters reduced power state, which consumes less nutrients, and it generates less waste and reproduce less.

Reduced consumption is a boolean variable that can only receive two values, true or false. Once the real consumption is reduced, the bacteria reduces the quantity of nutrients that will absorb the environment by tick and also decreases proportionally to the amount of energy required to keep it alive.

In the model initialization, all agents have false value for reduced consumption and it just gets true value when the agent detects a significant number of agents or signals in the environment. This information is stored into the variables, sensor signals or sensor bacteria.

Reducing consumption of nutrients and energy to maintain vital functions agent aim to make it grow less, and so can survive longer in the environment.

Figure 5 shows the agent's life cycle, clarifying the actions and decisions they must take every tick.

To better understand the real model and the agent-based model developed, we present the main functions performed by the agents as well as the actions and decisions that they take every tick.

Figure 5. Flowchart of agents lifecycle
Source: Werlang, 2013

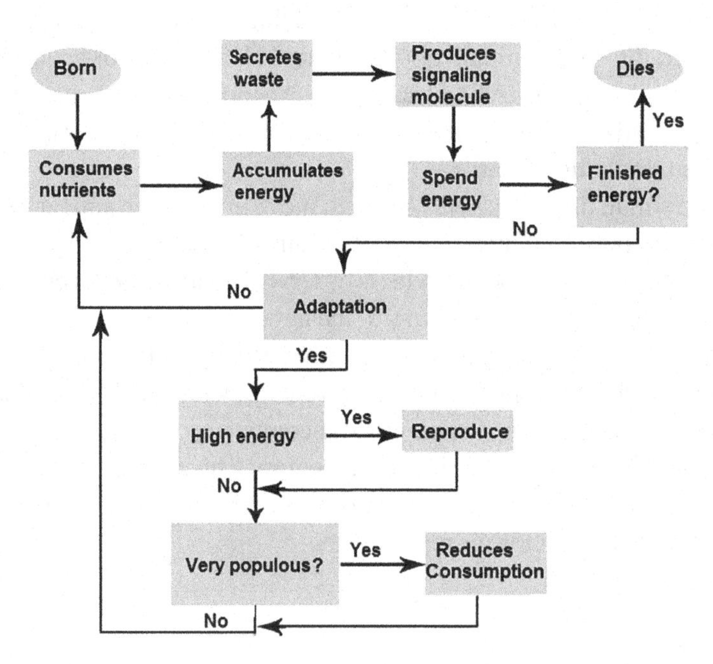

- **Function Responsible for Metabolizing Agents (Feed):** If the patch where the agent is located have nutrients, it will absorb the same amount of nutrients established by consumption, and the energy generated will be determined by consumption and saturation of the patch. After feeding, the agent secrets the waste generated by metabolism. Regardless of the patch contain (or not) nutrients, the agent takes energy to stay alive. If after the function execution feeds, the agent's power is less than or equal to 1, then it dies.
- **Function Responsible for Moving Agents (Movement):** If the agent keep alive, its energy is regulated and it is questioned whether the number of signals in the environment are no longer sufficient for it to reduce its consumption. If the signal threshold has been reached, then the agent enters reduced power state, resetting its consumption rate and vital functions with lower values. After, the agent looks for a neighboring patch that is free and has available nutrients, going to this patch.
- **Function Responsible for The Reproduction of Agents (Reproduce):** If the agent has reached the stage adaptation then it is ready to reproduce. If the agent has all the prerequisites for reproduction, which are: to have passed the stage of adaptation, have greater than or equal energy that defined the reproduce power, then it will play a similar agent to it. The similarity of agents is set when initializing the new agent variables are done based on the player agent variables, setting with normal distribution with average equal to the variables of the "father" and a very low standard deviation. Each time the agent reproduce, it loses energy at a rate given by the variable reduces energy.
- **Function Responsible for Resolving The Agents of The Signaling Step (Signals):** If the number of agents in the environment is greater than or equal to the variable bacteria sensor that sets the maximum number of agents to define environmental saturation, then the agent releases a signaling molecule and enters reduced power state.

Model Implementation

After defining the population dynamics, the next step was to implement the model in NetLogo. Through user intervention, it is possible configure some environmental parameters, such as the initial number of agents and the amount of available nutrients. Figure 6 shows the model interface created and its variables.

Each of the simulated strains have different times to start its phases. Thus, each receiving different values for the variables. To not create five simulations for each of them, a device in which the five simulations can be run on the same code. Just choose the strain to be simulated and start it in Cepa_name button.

The results obtained by Groll (2010) represent the growth curves of Mycobacterium tuberculosis in only two stages, the adaptation and the exponential phase. In order to facilitate comparisons between real growth and simulated, a key has been created that enables the user to simulate curves with the same stages of growth, i.e., it is detected that the curve is not growing, the simulation is terminated.

Values Interpretation

The curves from the MGIT express the result in Units Growth (GU). Therefore, an arbitrary measure of equivalence becomes necessary GU relates to the number of simulation agents. In the model developed was used an agent to represent every two GUs.

Figure 6. Interface of developed model in NetLogo

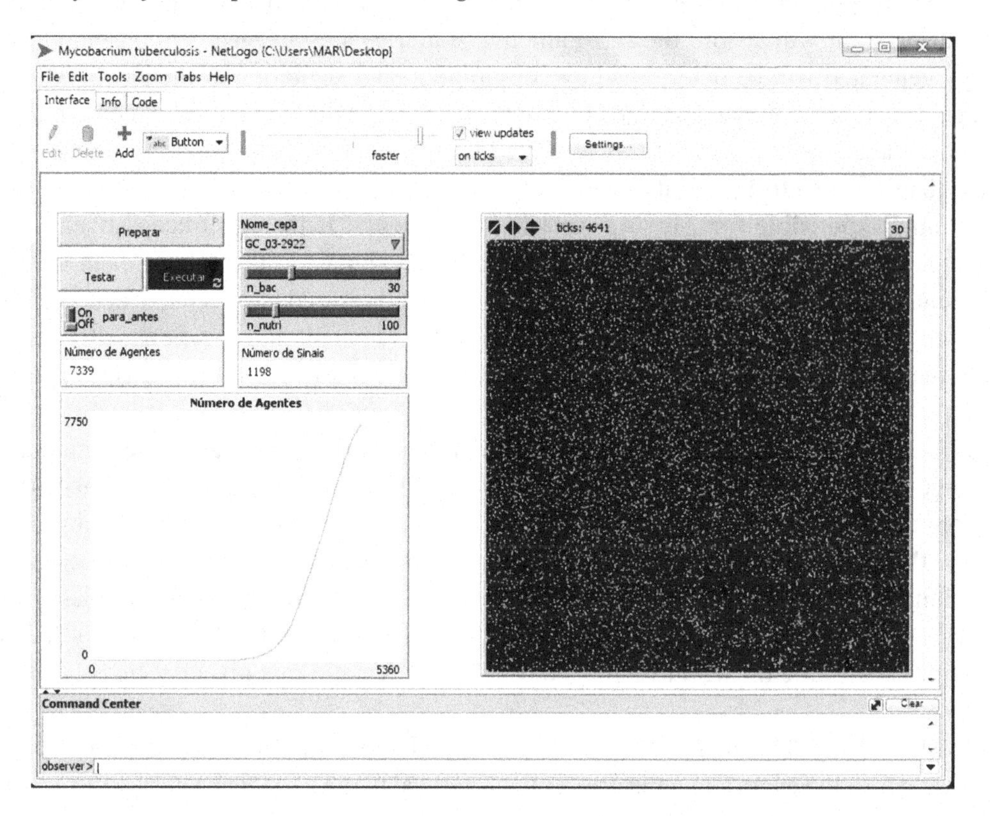

Likewise, it can represent the GU for agents using the same equivalence. Figure 7 is the same as Figure 2, except with different units and growth time. See that equivalence was 20 000 GUs 10 000 agents.

Using the same idea, an equivalence was also obtained for the growing time. As the experimental curve expresses the results by the day, and as seen in the simulation proposed model is temporal division ticks, the equivalence of 260 ticks for each day simulation was used.

RESULTS

This section presents the results obtained after the implementation of the proposed model. Initially, it presents the parameterization of the environment. In a second step, it shows the curves generated by the model, and their validation with the real data (in vitro) and the curves generated in the work of Werlang (2013).

Environmental Parameterization

After modeling the growth curve of Mycobacterium tuberculosis, the next step was to calibrate the characteristics and parameters of the model variables, seeking a curve similar to a bacterial growth. This search becomes necessary because, depending on the characteristics of the environment and the agent variables, this similarity is not always found.

Figure 7. Real growths represented by agents

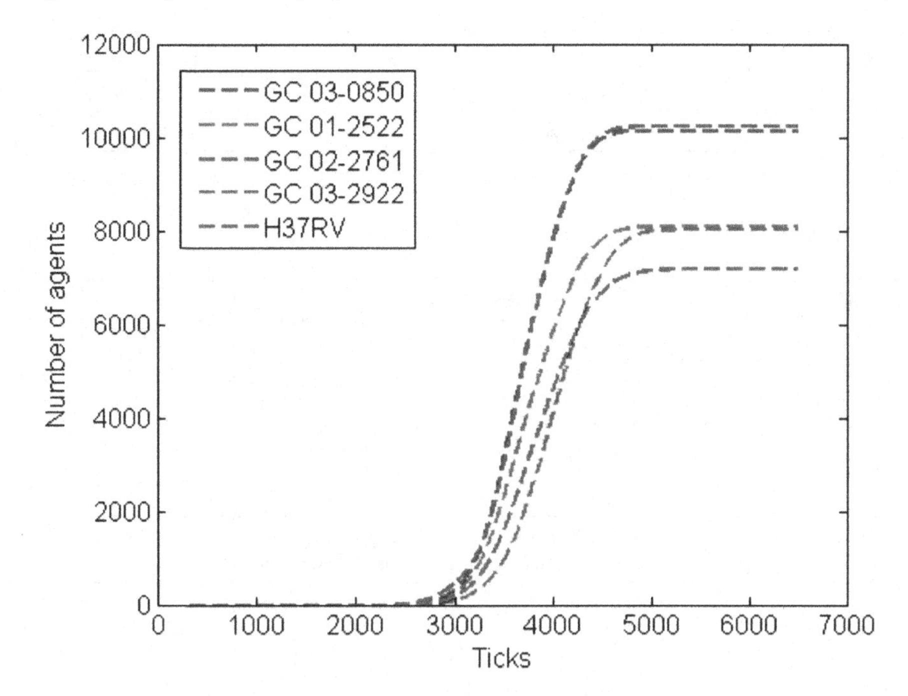

**For a more accurate representation of this figure, please see the electronic version.*

After discovering the curve with the same growth patterns of real curves, some parameters and variables have been set. There are some parameters that are common to the five strains, according to Tables 2 and 3.

Considering that the created environment is used to simulate five populations, each referring to a strain, it becomes necessary variations in the characteristics of variable agents, to permit representation of all.

Due to the fact that the proposed model is composed of several variables, each one can receive an infinite number of values in its initialization, and it is not feasible to work with all. Given these circumstances, a study was done to select what would be the best to use them as objective function.

For each of the variables various values were tested for the parameters. Some have shown to be more sensitive to change than others, and more susceptible to variation to be expected.

Among the possible choices, we have selected those that had both greater impact in behavior of the curve and predictability to the variation of its parameters. Table 4 shows that the variables and initialization values for each of the strain.

It is important to note that the tick_reproduce is the only variable in the model that we can estimate basing on the observation of real data. With the standardized method by Groll (2010), it is possible to determine how long the bacteria are in the environment to start the reproduction phase. Therefore, the time for adaptation should be done for the conversion from days to ticks.

Table 2. Common parameters for all strains of agents

Parameter	Description	
n_bac	Initial number of agentes	30 agents
n_nutri	Nutrients in each *patch*	100 units

Table 3. Common parameters in agent variables for all strains

Variable	Description	Mean	Standard Deviation
Energy	Amount of initial energy of agent	100	7
Consume	Amount of nutrients that agent consume in each tick.	1	0.07
sensor_signal	Even the bacterium has not released the signaling molecule, it can enter low power consumption because a signal detection threshold in the environment.	1000	100
consume_red	Define how much will fall in nutrient consumption value when the agent enters into reduced power state.	0.45	0.0315
reduce_energy	Amount of energy that the agent spends to reproduce.	0.7	0.049

Table 4. Parameters of variables those are different from each strain

	GC 03-0850		GC 02-2761		GC 01-2522		GC 03-2922		H37Rv	
Variable	Mean	Standard deviation	Mean	Standard deviation	Mean	Standard deviation	Mean	Standard deviation	Mean	Standard deviation
tick_reproduce	2400	72	2350	70.5	2516	75.48	2611	78.33	2505	75.15
func_vitals	0.55	0.0385	0.55	0.0385	0.50	0.0315	0.45	0.0315	0.5	0.35
energy_reproduce	200	14	200	14	225	15.75	230	16.1	235	16.45
sensor_bac	2000	200	1950	195	3000	300	2600	260	2600	260

Simulation Results for Each Strain

Figure 8 shows the results generated for all strains. Generated values were obtained by the averaging of 10 (ten) simulations for each of the strains. The gray shaded area represents the standard deviation.

In all simulations, CG 02-2761 strain presented the highest variation, as can be seen in Figure 8. The highest standard deviation is an indicator of greater flexibility with respect to possible curves to be generated, noting that the higher standard deviation the lowest the robustness level (stability) of the model.

In Table 5 are represented numerically some moments of each simulated curve. The adaptation time indicates how much time, measured in ticks the curve took to begin to increase; maximum growth refers to the tick corresponding to the highest growth in the curve; growth refers to the number of registered ticks to the population to stop growing; and the population is the maximum number of detected agents in the environment.

Figure 8. Simulate Curves of all strain

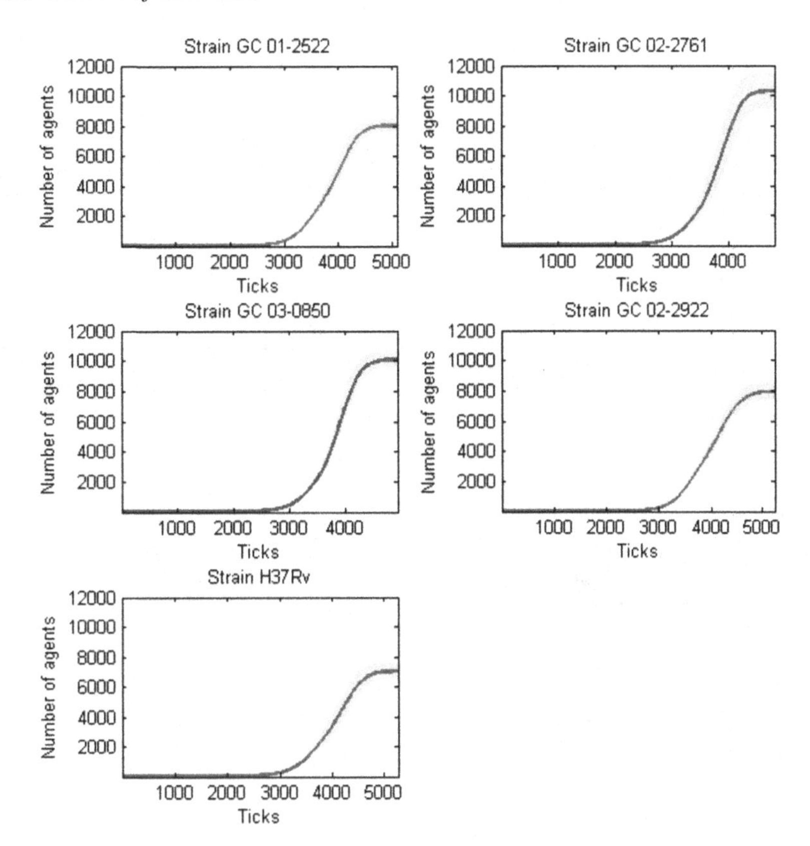

For each simulation, the values of the moments were recorded and stored. The results presented are for the average of 10 (ten) simulations.

Model Validation

The proposed model has been tested in order to be validated. The graphic and numeric results presented in the previous section were compared with real data obtained by Groll (2010). To do the comparison it was necessary to convert the simulated data. The ticks were converted to days and the agents number for Growth Units. Remember that each day is equals to 260 ticks, and each agent represents two Growth Units.

The graphs of Figure 9 are composed from two growth curves:

Table 5. Some moments of each simulated curve

Moments	GC 01-2522	GC 02-2761	GC 03-0850	GC 03-2922	H37RV
Adaptation	2387	2229	2300	2507	2381
Maximum growth	2476	2276	2366	2532	2439
Growth	4904	4632	4766	5058	5151
Population	8022	10286	10081	7981	7024

Figure 9. Comparison between real and simulated curves

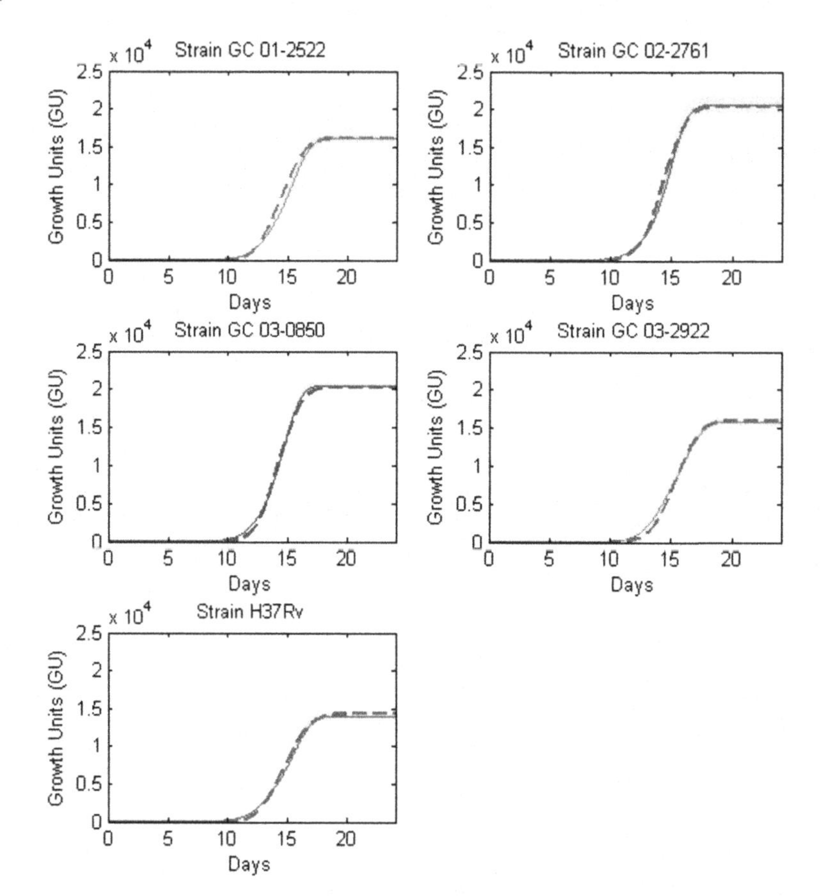

- The dotted line represents the real curve data; and
- Continuous curve is the simulated data curve.

The continuous line was obtained from the average of ten simulations, and the gray area represents the standard deviation.

Tables from 6 to 10 represent a comparison of moments in the simulated time curves (which have previously explained and shown in isolate way) with the real curves.

Table 6. Time of adaptation, maximum grown, growth, and population to GC 01-2522

GC 01-2522			
Momentum	**Real**	**Simulated**	**Error**
Adaptation	2535	2387	6.20%
Maximum growth	2556	2476	3.23%
Growth	5070	4904	3.38%
Population	8090	8022	0.85%

Table 7. Time of adaptation, maximum grown, growth, and population to GC 02-2761

GC 02-2761			
Momentum	**Real**	**Simulated**	**Error**
Adaptation	2384	2229	6.95%
Maximum growth	2425	2276	6.54%
Growth	4820	4632	4.06%
Population	10217	10286	0.67%

Table 8. Time of adaptation, maximum grown, growth, and population to GC 03-0850

GC 03-0850			
Momentum	**Real**	**Simulated**	**Error**
Adaptation	2425	2300	5.43%
Maximum growth	2460	2366	3.97%
Growth	4919	4766	3.21%
Population	10139	10081	0.57%

Table 9. Time of adaptation, maximum grown, growth, and population to GC 03-2922

GC 03-2922			
Momentum	**Real**	**Simulated**	**Error**
Adaptation	2620	2507	4.50%
Maximum growth	2655	2532	4.86%
Growth	5220	5058	3.20%
Population	8014	7981	0.41%

Table 10. Time of adaptation, maximum grown, growth, and population to H37Rv

H37Rv			
Momentum	**Real**	**Simulated**	**Error**
Adaptation	2524	2381	6.00%
Maximum growth	2556	2439	4.80%
Growth	5299	5151	2.87%
Population	7179	7024	2.20%

To do the numerical analysis, it was necessary to perform data conversion. However, now in and inverse way, because we choose to represent the data for ticks and agents. It is observed that were converted days for ticks and growth units for the number of agents.

Various parameters settings were tested. These adjustments allowed the behavior of the simulated growth curves to be reproduced with a relatively similarity to the real growth curves.

By analyzing the obtained data, it is clear that some represent better the beginning of growth; other the growth medium; and most of them can represent faithfully the end of growth. In this way, we cannot conclude that got better representation. However, the simulation of GC 01-2522 strain is the strain that worse represents the real model.

Making an analysis of the tables, the simulated strain with lower error rate was the CG 03-850, with mean error equal 3.29%, and the bigger error was the CG 02-2761, with mean error equal 4.55%. The CG 01-2522 strain, which apparently is more difference between real and simulated curves, not obtained the highest average error. As these means are for the moment curve, it is believed that should be given more reliability to the plotted data, as representing a whole and not just parts of the curve.

COMPARATION WITH WERLANG (2013) RESULTS

Werlang (2013) also created a model based on multi-agent systems to represent the Mycobacterium tuberculosis growth curves. However, the agents of his model behave the same way, ignoring the individuality of each agent (he did not use probabilities distribution).

Figure 10 shows the growth curves of each of the strains. We presented in the same graph the real growth curves obtained by Groll (2010), the simulated presented by Werlang (2013) and our simulated curves.

GC 01-2522 strain was the most distanced from the real growth curve in the model developed in our work. The other strains had a closer behavior to the real curves compared the representations of simulated curves obtained by Werlang (2013).

We believed that the results of our model have been more approximate to the real growth curve, because we used the probability distribution for possible values of the variables. In this way, we have inserted a diffuse behavior for each agent, expressing their individuality.

However, one can not affirm with certainty the reasons for the better representation, since the calibration parameters were made in different ways and by different researchers.

CONCLUSION

Tuberculosis is one of the oldest diseases, with wide geographical distribution, constituting a serious public health problem worldwide. The study of growth curve of Mycobacterium tuberculosis, which causes tuberculosis, enables the understanding of various behaviors of bacilli, such as response to different chemical agents and environment conditions. In spite of the behavior of bacillus face to different conditions still no predict by in silicum models. Modeling growth curve is a tool allows better designing studies involving growth curve from normal condition of growth and maximizing the analyses.

This work had as main objective the development of a growth curve model of Mycobacterium tuberculosis, able to reproduce the real curves given as input. The proposal was inspired by the work of

Figure 10. Comparison of real and simulated curves

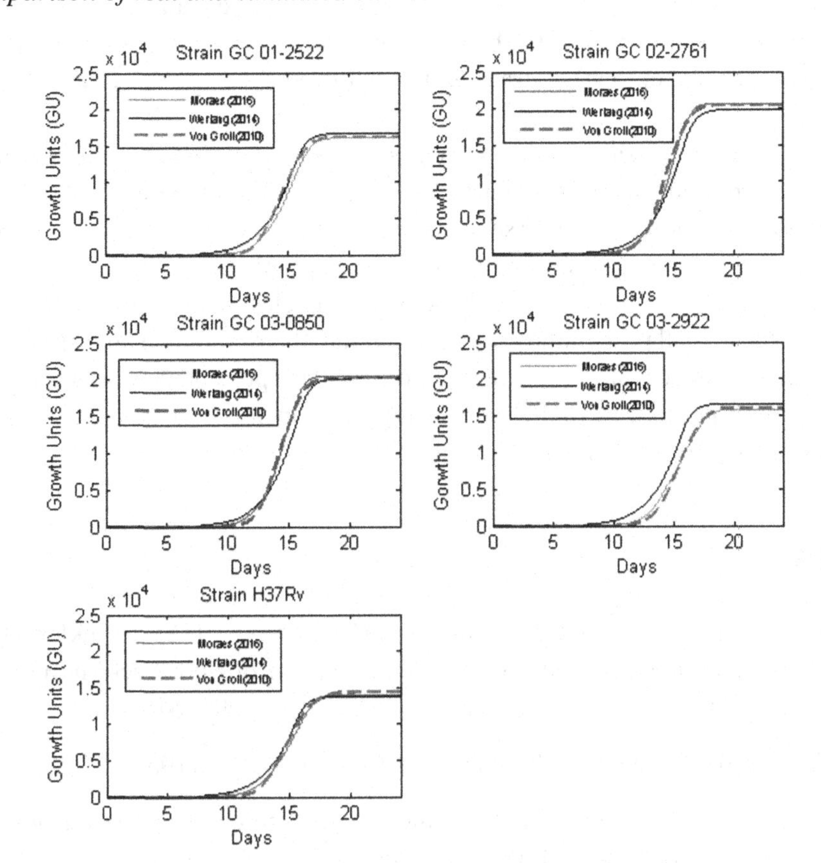

Werlang (2013), which created a model based on multi-agent capable of representing the real curves of Mycobacterium tuberculosis with a certain degree of similarity and Groll (2010), which has standardized a method to measure growth population bacillus when placed in a medium. The results obtained by Groll (2010) enabled to verify the model with more fidelity to reproduce the real curves.

The proposed growth curve was modeled using simulation based agents. Wooldridge (2009) says that multi-agent systems are a powerful and flexible tool for modeling environmental/social systems, because this type of system allows to analyze the behavior of each individual, rather than an average behavior of them.

The behavioral rules of the agents were established according to the knowledge gained by experts and bibliographic references area.

To make the proposed model more similar to real growth model, we proposed that possible values of the variables were stochastically distributed, thus, treating the individuality of each agent. We have as hypotheses estimate the best possible probability density function of the variables, with the final product an identical curve, or minimum error when compared to real data.

After conducting studies and research, we found that the normal distribution is one of the most used models to represent random variables. Therefore, the model variables also used normal distribution to select their values.

After finishing the first simulation model, we can observe that the parameter adjustments of certain variables that generate greater impact on the behavior of the curve. Each of these variables was adjusted in order to represent each one of the five strains to be simulated: GC01-2522, CG02-2761 CG03-0850, CG03-2922 and H37Rv.

The results were satisfactory, since the growth curve modeled revealed a similar real curve. Thus, the model was effective in reproducing growth curves of Mycobacterium tuberculosis.

Every search task and optimization have several components. The most important is the search space, where are considered all the possibilities for solving a given problem.

As the calibration of the parameters of variables in this model were made manually, the search space must be reduced, because if all possibilities were considered, there would be an infinite number of possible values to be taken into account for the number of variables presented. In this way, there are many search methods, and we propose as future work using some methods, as genetic algorithms, to seek optimal (or approximately optimal) solutions for the values of variables.

REFERENCES

Andries, K., Verhasselt, P., Guillemont, J., Göhlmann, H. W., Neefs, J. M., Winkler, H., & Jarlier, V. et al. (2005). A diarylquinoline drug active on the ATP synthase of Mycobacterium tuberculosis. *Science, 307*(5707), 223–227. doi:10.1126/science.1106753 PMID:15591164

Antunes, L. C. M. (2003). A linguagem das bactérias. *Ciência hoje, 33*(193).

Burgos, M., & Pym, A. (2002). Molecular epidemiology of tuberculosis. *The European Respiratory Journal, 20*(36), 54–65. doi:10.1183/09031936.02.00400702 PMID:12168748

Bussab, W., & Morettin, P. (2005). *Estatística Básica*. São Paulo: Saraiva.

Callegari-jacques, S. M. (2003). *Bioestatística: princípios e aplicações*. Porto Alegre: Artmed.

Cargnelutti Filho, A., Matzenauer, R., & Trindade, J. d. (2004). Ajustes de funções de distribuição de probabilidade à radiação solar global no Estado do Rio Grande do Sul. *Revista Agropecuária Brasileira, 39*(12), 1157–1166.

Colella, V. S., Klopfer, E., & Resnick, M. (2001). *Adventures in modeling: exploring complex, dynamic system with StarLogo*. Teachers college Press.

Devore, J. L., & da Silva, J. P. N. (2006). *Probabilidade e estatística: para engenharia e ciências*. São Paulo: Pioneira Thomson Learning.

Garcia, A., & Sichman, J. (2005). Agentes e Sistemas Multiagentes. In *Sistemas Inteligentes Fundamentos e Aplicações* (pp. 269–304). São Paulo: Manole.

Groll, A., Martin, A., Stehr, M., Singh, M., Portaels, F., Silva, P. E. A., & Palomino, J. C. (2010). *Fitness of Mycobacterium tuberculosis strains of the W-Beijing and Non-W-Beijing genotype*. Academic Press.

Lawn, S. D., & Zumla, A. I. (2011). Tuberculosis. *The Lancet Journal, 378*(9785), 57–72. doi:10.1016/S0140-6736(10)62173-3 PMID:21420161

Lönnroth, K, & Migliori, G. B., & Abubakar. (2015). Towards tuberculosis elimination: An action framework for low-incidence countries. *The European Respiratory Journal*, *45*(4), 928–952. PMID:25792630

Mesquita, M. G., Moraes, S. O., & Corrente, J. E. (2002). More adequate probability distributions to represent the saturated soil hydraulic conductivity. *Scientia Agrícola*, *59*(4), 789–793. doi:10.1590/S0103-90162002000400025

NetLogo. (2015). Retrieved from https://ccl.northwestern.edu/netlogo/

Paixão, L. M. M., & Gontijo, E. E. (2007). Perfil de casos de tuberculose notificados e fatores associados ao abandono. *Revista de Saude Publica*, *41*(2), 205–213. doi:10.1590/S0034-89102007000200006 PMID:17384794

Porcellis, D., Moraes, M., & Bertin, C. (2015) Modelo Baseado em Equações versus Modelo Baseado em Agentes: uma abordagem usando sistema predador-presa. In *IX Workshop-Escola de Sistemas de Agentes, Seus Ambientes e Aplicacões* (pp. 100–111). Niterói.

Rebonatto, M. (2000). *Simulação paralela de eventos discretos com uso de memória compartilhada distribuída, Dissertação (Mestrado em Ciência da Computação). Programa de Pós-Graduação em Computação*. Porto Alegre: UFRGS.

Rumjanek, N. G., Fonseca, M. C. C., & Xavier, G. R. (2004). Quorum sensing em sistemas agrícolas. In *Revista Biotecnologia Ciência*. Belo Horizonte: Desenvolvimento.

Todar, K. (2013). The growth of bacterial populations. In *Todar's Online Textbook ofBacteriology*. Author.

Voskuil, M. I., Visconti, K. C., & Schoolnik, G. K. (2004). *Mycobacterium tuberculosis* gene expression during adaptation to stationary phase and low-oxygen dormancy. *Tuberculosis (Edinburgh, Scotland)*, *84*(3-4), 218–227. doi:10.1016/j.tube.2004.02.003 PMID:15207491

Werlang, P. (2013). *Simulação da curva de crescimento do Mycobacterium tuberculosis utilizando sistemas multiagentes. Dissertação (Mestrado em Modelagem Computacional). Programa de Pós-Graduação em Modelagem Computacional*. Rio Grande: FURG.

Werlang, P. S. (2013). *Multi-Agent-Based Simulation of Mycobacterium tuberculosis growth*. In MABS 2013 - 14th International Workshop on Multi-Agent-Based Simulation, Saint Paul, MN. doi:10.1007/978-3-642-54783-6_9

Whitehead, N. A., Barnard, A. M. L., Slater, H., Simpson, N. J. L., & Salmond, G. P. C. (2001). Quorum-sensing in Gram-negative bacteria. *FEMS Microbiology Reviews*, *25*(4), 365–404. doi:10.1111/j.1574-6976.2001.tb00583.x PMID:11524130

Wooldridge, M. (2009). *An Introduction to Multiagent Systems*. John Wiley & Sons.

World Health Organization. (2015). *Global tuberculosis report 2015*. WHO/HTM/TB/2015.10. Geneva: World Health Organization.

Compilation of References

Abbott, R. (2006). Emergence explained: Abstractions - Getting epiphenomena to do real work. *Complexity*, *12*(1), 13–26. doi:10.1002/cplx.20146

Abreu, C. G., Coelho, C. G. C., & Ralha, C. G. (2015). MASE-BDI: Agents with Practical Reasoning for Land Use and Cover Change Simulation. In *Proceedings of the 6th Workshop of Applied Computing for the Management of the Environment and Natural Resources*. XXXV Congress of the Brazilian Computer Society.

Abreu, C. G., Coelho, C. G. C., Ralha, C. G., & Macchiavello, B. (2014). A Model and Simulation Framework for Exploring Potential Impacts of Land Use Policies: The Brazilian Cerrado Case. In *Proceedings of the 47th Hawaii International Conference on System Sciences (HICSS)*, (pp. 847-856). doi:10.1109/HICSS.2014.113

Abreu, P. C., Bergesch, M., Proença, L. A., Garcia, C. A. E., & Odebrecht, C. (2010). Short-and long-term chlorophyll a variability in the shallow microtidal Patos Lagoon estuary, southern Brazil. *Estuaries and Coasts*, *33*(2), 554–569. doi:10.1007/s12237-009-9181-9

Adamatti, D., Sichman, S., & Coelho, H. (2007). Virtual players: From manual to semi-autonomous RPG. In F. Barros, C. Frydman, N. Giambiasi and B. Ziegler (Eds.), *Proceedings of the AISCMS'07 International Modeling and Simulation Multiconference (IMSM'07)*. The Society for Modeling Simulation International (SCS).

Adamatzky, A. (1997). Automatic programming of cellular automata: Identification approach. *Kybernetes*, *26*(2-3), 126–135. doi:10.1108/03684929710163074

Adami, C. (1998). *Introduction to Artificial Life*. Berlin: Springer. doi:10.1007/978-1-4612-1650-6

Adams, B. M., Bauman, L. E., Bohnhoff, W. J., Dalbey, K. R., Ebeida, M. S., Eddy, J. P., . . . Wildey, T. M. (2014). *Dakota, A Multilevel Parallel Object-Oriented Framework for Design Optimization, Parameter Estimation, Uncertainty Quantification, and Sensitivity Analysis: Version 6.0 User's Manual*. Sandia Technical Report SAND2014-4633.

Agarwal, C., Green, G. M., Grove, J. M., Evans, T. P., & Schweik, C. M. (2002). *A Review and Assessment of Land-use Change Models: Dynamics of Space, Time, and Human Choice*. Gen. Tech. Rep. NE-297. Newton Square, PA: U.S. Department of Agriculture, Forest Service, Northeastern Research Station.

Agency for Toxic Substances and Disease Registry (ATSDR). (2016). *Public Health Assessment Guidance Manual, 2005*. Available from http://www.atsdr.cdc.gov/HAC/PHAmanual/ch2.html

Aguiar, M. S. (1998). *Análise Formal da Complexidade de Algoritmos Genéticos (Dissertação de Mestrado em Ciência da Computação)*. Porto Alegre, RS, Brazil: PPGC/UFRGS.

Aguiar, M. S., & Toscani, L. V. (1997). Algoritmos Genéticos. In *I Workshop sobre métodos formais e qualidade de software* (pp. 78-87). Porto Alegre/RS.

Alberts, B., Bray, D., Lewis, J., Raff, M., Roberts, K., & Watson, J. D. (2002). Universal mechanisms of animal development. In *Molecular Biology of the Cell.* New York: Garland Publishing. Retrieved from http://www.ncbi.nlm.nih.gov/books/NBK26825/

Aldrich, C. (2009). The Complete Guide to Simulations and Serious Games. John Wiley & Sons.

Ali, M., & Uzmi, Z. (2004, May). CSN: a network protocol for serving dynamic queries in large-scale wireless sensor networks. In Proceedings of Communication Networks and Services. (pp. 165–174). doi:10.1109/DNSR.2004.1344725

Alicea, B. & Gordon. (2014). Toy models for macroevolutionary patterns and trends. *BioSystems, 122*, 25-37.

Alicea, B. (2016). *DevoWorm Project.* Retrieved from http://devoworm.weebly.com

Alicea, B., McGrew, S., Gordon, R., Larson, S., Warrington, T., & Watts, M. (2014). *DevoWorm: differentiation waves and computation in* C. elegans *embryogenesis.* Retrieved from http://www.biorxiv.org/content/early/2014/10/03/009993

Ali, M., & Langendoen, K. (2007). A Case for Peer-to-Peer Network Overlays in Sensor Networks. In *International Workshop on Wireless Sensor Network Architecture* (pp. 56–61).

Allan, J. D., & Benke, A. C. (2005). Overview and prospects. In Rivers of North America. Elsevier.

Allan, R. (2011). *Survey of Agent Based Modelling and Simulation Tools. Version 1.1.* Available at http://www.grids.ac.uk/Complex/ABMS/

Alliance, Z. (2005). *ZigBee Specification 1.1* (Tech. Rep.).

Alto, L. T., Havton, L. A., Conner, J. M., Hollis, E. R. II, Blesch, A., & Tuszynski, M. H. (2009). Chemotropic guidance facilitates axonal regeneration and synapse formation after spinal cord injury. *Nature Neuroscience, 12*(9), 1106–U1108. doi:10.1038/nn.2365 PMID:19648914

Altschul, S. F., Madden, T. L., Schäffer, A. A., Zhang, J., Zhang, Z., Miller, W., & Lipman, D. J. (1997). Gapped blast and psi-blast: A new generation of protein database search programs. *Nucleic Acids Research, 25*(17), 3389–3402. doi:10.1093/nar/25.17.3389 PMID:9254694

Alvarez, I., Reuillon, R., & de Aldama, R. (2016). Viabilitree: A kd-tree Framework for Viability-based Decision. Research report, LIP6, Paris. https://hal.archives-ouvertes.fr/hal-01319738

Amgoud, L., & Kaci, S. (2005). On the generation of bipolar goals in argumentation-based negotiation. In *Argumentation in Multi-Agent Systems – Proceedings of the First International Workshop (ArgMAS'04) – Expanded and Invited Contributions,* (LNCS), (vol. 3366). Heidelberg, Germany: Springer Verlag. doi:10.1007/978-3-540-32261-0_13

Amgoud, L., & Vesic, S. (2012). A formal analysis of the role of argumentation in negotiation dialogues. *Journal of Logic and Computation, 5*(22), 957–978. doi:10.1093/logcom/exr037

Amigoni, F., & Schiaffonati, V. (2007). Multiagent-based simulation in biology. In *Model-based reasoning in science, technology, and medicine* (pp. 179–191). Springer Berlin Heidelberg. doi:10.1007/978-3-540-71986-1_10

Anderson, A. R. A., Chaplain, M. A. J., & Rejniak, K. A. (Eds.). (2007). *Single-cell-based models in biology and medicine.* Birkähuser Basel; doi:10.1007/978-3-7643-8123-3

Anderson, K., & Silver, J. M. (1998). Modulation of Anger and Aggression. *Seminars in Clinical Neuropsychiatry, 3*, 232–242. PMID:10085211

Andrade, S. F. R. (2015). *Risk assessment to health applied to nickel mining activity in southern Bahia* (Unpublished doctoral thesis). State University of Santa Cruz.

Andreeva, A., Howorth, D., Chandonia, J. M., Brenner, S. E., Hubbard, T. J., Chothia, C., & Murzin, A. G. (2008). Data growth and its impact on the scop database: New developments. *Nucleic Acids Research*, *36*(suppl 1), D419–D425. doi:10.1093/nar/gkm993 PMID:18000004

Andries, K., Verhasselt, P., Guillemont, J., Göhlmann, H. W., Neefs, J. M., Winkler, H., & Jarlier, V. et al. (2005). A diarylquinoline drug active on the ATP synthase of Mycobacterium tuberculosis. *Science*, *307*(5707), 223–227. doi:10.1126/science.1106753 PMID:15591164

Anfinsen, C. (1973). Principles that govern the folding of protein chains. *Science*, *181*(4096), 223–230. doi:10.1126/science.181.4096.223 PMID:4124164

Anfinsen, C., Haber, E., Sela, M., & White, F. H. J. (1961). The kinetics of formation of native ribonuclease during oxidation of the reduced polypeptide chain. *Proceedings of the National Academy of Sciences of the United States of America*, *47*(9), 1309–1314. doi:10.1073/pnas.47.9.1309 PMID:13683522

An, G., Mi, Q., Dutta-Moscato, J., & Vodovotz, Y. (2009). Agent-based models in translational systems biology. *Wiley Interdisciplinary Reviews: Systems Biology and Medicine*, *1*(2), 159–171. doi:10.1002/wsbm.45 PMID:20835989

Annett, J., & Duncan, K. D. (1967). Task analysis and training design. *Journal of Occupational Psychology*, *41*, 211–221.

Ansel, J., Kamil, S., Veeramachaneni, K., Ragan-Kelley, J., Bosboom, J., O'Reilly, U. M., & Amarasinghe, S. (2014). OpenTuner: An Extensible Framework for Program Autotuning. In *Proceedings of the 23rd International Conference on Parallel Architectures and Compilation*, (pp. 303–316). New York, NY: ACM. doi:10.1145/2628071.2628092

Antunes, L. C. M. (2003). A linguagem das bactérias. *Ciência hoje, 33*(193).

Ashby, W. R. (1968). Some consequences of Bremermann's limit for information-processing systems. In H. L. Oestreicher & D. L. Moore (Eds.), *Problems in Bionics* (pp. 69–76). New York: Gordon and Breach Science Publishers, Inc.

Aubin, J.-P. (1991). *Viability theory*. Basel, Switzerland: Birkhäuser.

Aubin, J.-P., Bayen, A. M., & Saint-Pierre, P. (2011). *Viability Theory: New Directions*. Heidelberg, Germany: Springer Verlag. doi:10.1007/978-3-642-16684-6

Aureliano, V., Silva, B., & Barbosa, S. (2006). Extreme Designing: Binding Sketching to an Interaction Model in a Streamlined HCI Design Approach.*Proceedings of the VII Simpósio Brasileiro sobre Fatores Humanos em Sistemas Computacionais (IHC'06).*

Axelrod, R. (1997). Advancing the art of simulation in the social sciences. In R. Conte, R. Hegselmann, & P. Terna (Eds.), *Simulating social phenomena*. Berlin: Springer-Verlag.

Balci, O. (1998). Verification, validation, and accreditation. In *Proceedings of the 30th Conference on Winter Simulation*. IEEE Computer Society Press. doi:10.1109/WSC.1998.744897

Ballet, P., Zemirline, A., & Marcé, L. (2004). The Biodyn language and simulators, application to an immune response and E. coli and phage interaction. In *Proceeding of the spring school on Modelling and simulation of biological processes in the context of genomics.*

Bandini, S., Manzoni, S., & Vizzari, G. (2009). *Crowd behaviour modelling: from cellular automata to multi-agent systems.*Taylor and Francis Group, LLC.

Bandini, S., & Mauri, G. (1999). Multilayered cellular automata. *Theoretical Computer Science*, *217*(1), 99–113. doi:10.1016/S0304-3975(98)00152-2

Banitz, T., Gras, A., & Ginovart, M. (2015). Individual-based modeling of soil organic matter in NetLogo: Transparent, user-friendly, and open. *Environmental Modelling & Software*, *71*, 39–45. doi:10.1016/j.envsoft.2015.05.007

Baraniak, P. R., & McDevitt, T. C. (2010). Stem cell paracrine actions and tissue regeneration. *Regenerative Medicine*, *5*(1), 121–143. doi:10.2217/rme.09.74 PMID:20017699

Barreteau, O. (2003a). Our Companion Modelling Approach. *Journal of Artificial Societies and Social Simulation*, *6*(2), Article 1.

Barreteau, O. (2003b). The joint use of role-playing games and models regarding negotiation processes: Characterization of associations. *Journal of Artificial Societies and Social Simulation*, *6*(2), Article 3.

Bateson, P., Barker, D., Clutton-Brock, T., Deb, D., DUdine, B., Foley, R. A., & Sultan, S. E. et al. (2004). Developmental plasticity and human health. *Nature*, *430*(6998), 419–421. doi:10.1038/nature02725 PMID:15269759

Bates, P. A., Kelley, L. A., MacCallum, R. M., & Sternberg, M. J. (2001). Enhancement of protein modeling by human intervention in applying the automatic programs 3d-jigsaw and 3d-pssm. *Proteins: Struct., Funct. Bioinf.*, *45*(S5), 39–46.

Batten, D. F. (2001). Complex landscapes of spatial interaction. *The Annals of Regional Science*, *35*(1), 81–111. doi:10.1007/s001680000032

Batty, M. (2003). Geocomputation using cellular automata. In R. J. Abrahart, S. Openshaw, L. M. See, & C. R. C. Press (Eds.), Geocomputation (pp. 95–126). Academic Press.

Baxevanis, A. D., & Ouellette, B. F. (2004). *Bioinformatics: a practical guide to the analysis of genes and proteins*. John Wiley & Sons.

Bellifemine, F. B., Caire, G., & Greenwood, D. (2016). *JAVA Agent Development Framework* (Version 4.3.3) [Software]. Available from http://jade.tilab.com/

Bellifemine, F. L., Caire, G., & Greenwood, D. (2007). *Developing multi-agent systems with jade*. John Wiley & Sons. doi:10.1002/9780470058411

Bellifemine, F., Caire, G., Poggi, A., & Rimassa, G. (2008). JADE: A Software Framework for Developing Multi-agent Applications. Lessons Learned. *Information and Software Technology*, *50*(1-2), 10–21. doi:10.1016/j.infsof.2007.10.008

Bellifemine, F., Poggi, A., & Rimassa, G. (2001). Developing multi-agent systems with JADE. *Software, Practice & Experience*, *31*(2), 103–128. doi:10.1002/1097-024X(200102)31:2<103::AID-SPE358>3.0.CO;2-O

Beloussov, L. V. (2015). Morphogenetic fields: History and relations to other concepts. In D. Fels, M. Cifra, & F. Scholkmann (Eds.), *Fields of the Cell* (pp. 271–282). Kerala, India: Research Signpost.

Beloussov, L. V., Opitz, J. M., & Gilbert, S. F. (1997). Life of Alexander G. Gurwitsch and his relevant contribution to the theory of morphogenetic fields. *The International Journal of Developmental Biology*, *41*(6), 771–779. PMID:9449452

Berman, H. M., Westbrook, J., Feng, Z., Gilliland, G., Bhat, T., Weissig, H., & Bourne, P. E. et al. (2000). The protein data bank. *Nucleic Acids Research*, *28*(1), 235–242. doi:10.1093/nar/28.1.235 PMID:10592235

Bermúdez, J. L. (2001). Normativity and Rationality in Delusional Psychiatric Disorders. *Mind & Language*, *16*(5), 493–457. doi:10.1111/1468-0017.00179

Bijelic, A., Molitor, C., Mauracher, S. G., Al-Oweini, R., Kortz, U., & Rompel, A. (2015). Hen egg-white lysozyme crystallisation: Protein stacking and structure stability enhanced by a tellurium(VI)-centred polyoxotungstate. *ChemBioChem*, *16*(2), 233–241. doi:10.1002/cbic.201402597 PMID:25521080

Boas, F. E., & Harbury, P. B. (2007). Potential energy functions for protein design. *Current Opinion in Structural Biology, 17*(2), 199–204. doi:10.1016/j.sbi.2007.03.006 PMID:17387014

Boeckmann, B., Bairoch, A., Apweiler, R., Blatter, M.-C., Estreicher, A., & Gasteiger, E. et al.. (2003). The SWISS-PROT protein knowledgebase and its supplement TrEMBL in 2003. *Nucleic Acids Research, 31*(1), 365–370. doi:10.1093/nar/gkg095 PMID:12520024

Boghanim, H. C., Kim, J. R., Dinsdale, R. M., Guwy, A. J., & Premier, G. C. (2013). *Analysis of the dynamic performance of a microbial fuel cell using a system identification approach. Journal of Power Sources.*

Bollen, L., & van Joolingen, W. R. (2013). Simsketch: Multiagent simulations based on learner-created sketches for early science education. *IEEE Transactions on Learning Technologies, 6*(3), 208–216. doi:10.1109/TLT.2013.9

Bonabeau, E. (2002). Agent-based modeling: Methods and techniques for simulating human systems. *Proceedings of the National Academy of Sciences of the United States of America, 99*(Suppl 3), 7280–7287. doi:10.1073/pnas.082080899 PMID:12011407

Bordini, R. H., Hübner, J. F., & Wooldridge, M. (2007). *Programming multi-agent systems in AgentSpeak using Jason.* Chichester, UK: Wiley.

Borgia, E. (2014). The Internet of Things vision: Key features, applications and open issues. *Computer Communications, 54*, 1–31. doi:10.1016/j.comcom.2014.09.008

Bornholdt, S. (1998). Genetic algorithm dynamics on a rugged landscape. *Physical Review E: Statistical Physics, Plasmas, Fluids, and Related Interdisciplinary Topics, 57*(4), 3853–3860. doi:10.1103/PhysRevE.57.3853

Bortolussi, L., Dal Palu, A., Dovier, A., & Fogolari, F. (2004). Protein folding simulation in CCP. In Proceedings of bioconcur2004.

Bortolussi, L., Dovier, A., & Fogolari, F. (2005). Multi-agent simulation of protein folding. In *Proceedings of the first international workshop on multi- agent systems for medicine, computational biology, and bioinformatics.*

Bortolussi, L., Dovier, A., & Fogolari, F. (2007). Agent-based protein structure prediction. *Multiagent and Grid Systems, 3*(2), 183–197. doi:10.3233/MGS-2007-3204

Boucher, D., Elias, P., Lininger, K., May-Tobin, C., Roquemore, S., & Saxon, E. (2011). *The root of the problem. What's driving tropical deforestation today? Technical Report. Tropical Forest and Climate Initiative.* Union of Concerned Scientists.

Bourgine, P., & Lesne, A. (Eds.). (2010). *Morphogenesis: Origins of Patterns and Shapes.* Berlin: Springer Science & Business Media.

Bower, J.M. (2005). Looking for Newton: realistic modeling in modern biology. *Brains, Minds and Media, 1*(2).

Bradbury, R. H. (2002). Futures, prediction and other foolishness. In M. A. Janssen (Ed.), *Complexity and ecosystem management: The theory and practice of multi-agent systems* (pp. 48–62). Cheltenham, UK: Edward Elgar.

Branden, C., & Tooze, J. (1998). *Introduction to protein structure* (2nd ed.). New York: Garlang Publishing Inc.

Bratman, M. (1987). *Intentions, Plans and Practical Reason.* Harvard University Press.

Brazil. (2000). *Lei N° 9.985, que regulamenta o Art. 225, parágrafo 1º, incisos I, II, III, VII da Constituição Federal, institui o Sistema Nacional de Unidades de Conservação da Natureza e dá outras providências.* Brazil.

Brazil. (2002). *Decreto No. 4340 que regulamenta os artigos da Lei No. 9985 que institui o Sistema Nacional de Unidades de Conservação – SNUC*. Brazil.

Bremermann, H. J. (1967). Quantum noise and information. In *Proceedings of the Berkeley Symposium on Mathematical Statistics and Probability*. University of California Press.

Bresciani, P., Perini, A., Giorgini, P., Giunchiglia, F., & Mylopoulos, J. (2004). Tropos: An Agent-Oriented Software Development Methodology. *Autonomous Agents and Multi-Agent Systems*, *8*(3), 203–236. doi:10.1023/B:AGNT.0000018806.20944.ef

Briot, J.-P., & Guyot, P., & Irving, M. (2007). *Participatory simulation for collective management of protected areas for biodiversity conservation and social inclusion*. Academic Press.

Brooks, R., Bruccoleri, R., Olafson, B., States, D., Swaminathan, S., & Karplus, M. (1983). Charmm: A program for macromolecular energy, minimization, and dynamics calculations. *Journal of Computational Chemistry*, *4*(2), 187–217. doi:10.1002/jcc.540040211

Brown, D. G., Walker, R., Manson, S., & Seto, K. (2004). Modeling Land Use and Land Cover Change. In Land Change Science: Observing, Monitoring and Understanding Trajectories of Change on the Earth's Surface, (pp. 395–409). Dordrecht: Springer Netherlands.

Bruine de Bruin, W., Parker, A. M., & Fischhoff, B. (2007). Individual differences in adult decision-making competence. *Journal of Personality and Social Psychology*, *92*(5), 938–956. doi:10.1037/0022-3514.92.5.938 PMID:17484614

Bruno, S., Collino, F., Tetta, C., & Camussi, G. (2013). Dissecting paracrine effectors for mesenchymal stem cells. In Mesenchymal Stem Cells: Basics and Clinical Application I. Academic Press.

Bryant, S. H., & Altschul, S. (1995). Statistics of sequence-structure threading. *Current Opinion in Structural Biology*, *5*(2), 236–244. doi:10.1016/0959-440X(95)80082-4 PMID:7648327

Bubak, M., & Czerwiński, P. (1999). Traffic simulation using cellular automata and continuous models. *Computer Physics Communications*, *121*, 395–398. doi:10.1016/S0010-4655(99)00363-X

Buchan, D. W., Minneci, F., Nugent, T. C., Bryson, K., & Jones, D. T. (2013). Scalable web services for the psipred protein analysis workbench. *Nucleic Acids Research*, *41*(W1), W349–W357. doi:10.1093/nar/gkt381 PMID:23748958

Bujnicki, J., Elofsson, A., Fischer, D., & Rychlewski, L. (2001a). Structure prediction meta server. *Bioinformatics (Oxford, England)*, *17*(8), 750–751. doi:10.1093/bioinformatics/17.8.750 PMID:11524381

Bujnicki, J., Elofsson, A., Fischer, D., & Rychlewski, L. (2001b). Livebench-2: large-scale automated evaluation of protein structure prediction servers. *Proteins: Struct., Funct. Bioinf.*, *45*(S5), 184–191.

Burgos, M., & Pym, A. (2002). Molecular epidemiology of tuberculosis. *The European Respiratory Journal*, *20*(36), 54–65. doi:10.1183/09031936.02.00400702 PMID:12168748

Burguillo, J. C. (2013). Playing with complexity: From cellular evolutionary algorithms with coalitions to self-organizing maps. *Computers & Mathematics with Applications (Oxford, England)*, *66*(2), 201–212. doi:10.1016/j.camwa.2013.01.020

Bussab, W., & Morettin, P. (2005). *Estatística Básica*. São Paulo: Saraiva.

Buzsaki, G. (2010). Neural syntax: Cell assemblies, synapsembles, and readers. *Neuron*, 68. PMID:21040841

Byrne, H., & Drasdo, D. (2008). *Individual-based and continuum models ofgrowing cell populations: a comparison. Journal of Mathematical Biology*.

Bystroff, C., Thorsson, V., & Baker, D. (2000). Hmmstr: A hidden markov model for local sequence-structure correlations in proteins. *Journal of Molecular Biology*, *301*(1), 173–190. doi:10.1006/jmbi.2000.3837 PMID:10926500

Calder, M. A. (n.d.). *Modelling of a Microbial Fuel Cell* (Master thesis). Norwegian University of Science and Technology.

Callegari-jacques, S. M. (2003). *Bioestatística: princípios e aplicações*. Porto Alegre: Artmed.

Campeotto, F., Dovier, A., & Pontelli, E. (2013). Protein structure prediction on gpu: a declarative approach in a multi-agent framework. In *Parallel processing (icpp), 2013 42nd international conference on* (pp. 474–479). doi:10.1109/ICPP.2013.57

Campeotto, F., Dovier, A., & Pontelli, E. (2015). A declarative concurrent system for protein structure prediction on gpu. *Journal of Experimental & Theoretical Artificial Intelligence*, *27*(5), 503–541. doi:10.1080/0952813X.2014.993503

Cancedda, P., & Caire, G. (2008). *Creating ontologies by means of the bean-ontology class. JADE 4.0*. Telecom Italia S.A.

Cancedda, P., & Caire, G. (2010). *Application-defined content languages and ontologies. JADE 4.0. Ontology bean-Generator for JADE (Version 3.5)* [Software]. Telecom Italia S.A.

Cannata, N., Corradini, F., Merelli, E., Omicini, A., & Ricci, A. (2005). An agent-oriented conceptual framework for Systems Biology. In E. Merelli, P. P. González Pérez & A. Omicini (Eds.), Transactions on Computational Systems Biology III (Vol. 3737, pp. 105–122). Springer. 8 doi:10.1007/11599128_8

Canolty, R. T., Cadieu, C. F., Koepsell, K., Ganguly, K., Knight, R. T., & Carmena, J. M. (2012). Detecting event-related changes of multivariate phase coupling in dynamic brain networks. *Journal of Neurophysiology*, *107*(7), 2020–2031. doi:10.1152/jn.00610.2011 PMID:22236706

Canutescu, A., Shelenkov, A., & Dunbrack, R. Jr. (2001). A graph-theory algorithm for rapid protein side chain prediction. *Proteins*, *12*(9), 2001–2014. doi:10.1110/ps.03154503 PMID:12930999

Cargnelutti Filho, A., Matzenauer, R., & Trindade, J. d. (2004). Ajustes de funções de distribuição de probabilidade à radiação solar global no Estado do Rio Grande do Sul. *Revista Agropecuária Brasileira*, *39*(12), 1157–1166.

Carlsson, M., & Mildner, P. (2012). Sicstus prologâthe first 25 years. *Theory and Practice of Logic Programming*, *12*(1-2), 35–66. doi:10.1017/S1471068411000482

Carneiro, T. G. S., de Andrade, P. R., Câmara, G., Monteiro, A. M. V., & Pereira, R. R. (2013). TerraME: An extensible toolbox for modeling nature society interactions. *Environmental Modelling & Software*, *46*, 104–117. doi:10.1016/j.envsoft.2013.03.002

Carriero, N., & Gelernter, D. (1989). Linda in context. *Communications of the ACM*, *32*(4), 444–458. doi:10.1145/63334.63337

Carroll, J. B. (1993). *Human cognitive abilities: A survey of factor-analytic studies*. Cambridge, UK: Cambridge University Press. doi:10.1017/CBO9780511571312

Carruthers, P. (2006). *The Architecture of the Mind: Massive Modularity and the Flexibility of Thought*. Oxford University Press. doi:10.1093/acprof:oso/9780199207077.001.0001

Castle, C. J. E., & Crooks, A. T. (2006). *Principles and Concepts of Agent-Based Modelling for Developing Geospatial Simulations. Technical Report, UCL Centre For Advanced Spatial Analysis*. London: UCL.

Cavalcanti, R. B., & Joly, C. A. (2002). Biodiversity and Conservation Priorities in the Cerrado Region Biodiversity and Conservation Priorities in the Cerrado Region. In The Cerrados of Brazil: Ecology and Natural History of a Neotropical Savana, (pp. 351–367). New York: Columbia University Press.

Cavanagh, J., Fairbrother, W., Palmer, A. III, & Skelton, N. (1995). *Protein nmr spectroscopy: principles and practice* (1st ed.). New York: Academic Press.

Cazala, J. (2015). *Perceptron*. Retrieved from https://github.com/cazala/synaptic/wiki/Architect

Chang, C. C., & Lin, C. J. (2011). LIBSVM: A library for support vector machines. *ACM Transactions on Intelligent Systems and Technology*, *2*(3), 27. doi:10.1145/1961189.1961199

Chivian, D., Robertson, T., Bonneau, R., & Baker, D. (2003). Ab initio methods. *Methods of Biochemical Analysis*, *44*, 547–557. PMID:12647404

Chopard, B. (1990). A cellular automata model of large-scale moving objects. *Journal of Physics. A, Mathematical and General*, *23*(10), 1671–1687. doi:10.1088/0305-4470/23/10/010

Choquet, G. (1953). Theory of capacities. *Journal of Fourier Institute*, *5*, 131–295. doi:10.5802/aif.53

Chou, K. C. (2004). Structural bioinformatics and its impact to biomedical science. *Current Medicinal Chemistry*, *11*(16), 2105–2134. doi:10.2174/0929867043364667 PMID:15279552

Chowdhury, R. R. (2006). Driving forces of tropical deforestation: The role of remote sensing and spatial models. *Singapore Journal of Tropical Geography*, *27*(1), 82–101. doi:10.1111/j.1467-9493.2006.00241.x

Cickovski, T. M., Huang, C., Chaturvedi, R., Glimm, T., Hentschel, H. G. E., Alber, M. S., Glazier, J. A., Newman, S. A. & Izaguirre, J. A. (2005). A framework for three-dimensional simulation of morphogenesis. *IEEE/ACM Transactions on Computational Biology and Bioinformatics*, *2*(4), 273–288. doi: 10.1109/TCBB.2005.46

Cipriano, R. (2008). On the hybridization of constraint programming and local search techniques: Models and software tools. In *Logic programming* (pp. 803–804). Springer Berlin Heidelberg. doi:10.1007/978-3-540-89982-2_81

Cock, P. J. A., Antao, T., Chang, J. T., Chapman, B. A., Cox, C. J., Dalke, A., & de Hoon, M. J. L. et al. (2009). Biopython: Freely available Python tools for computational molecular biology and bioinformatics. *Bioinformatics (Oxford, England)*, *25*(11), 1422–1423. doi:10.1093/bioinformatics/btp163 PMID:19304878

Codd, E. F., Codd, S. B., & Salley, C. T. (1993). Providing OLAP (on-line analytical processing) to user-analysts: An IT mandate. *Codd and Date, 32*.

Coelho, C. G. C., Abreu, C. G., & Ramos, R. M. (2016). MASE-BDI: Agent-based simulator for environmental land change with efficient and parallel auto-tuning. *Applied Intelligence*, 1–19. doi:10.1007/s10489-016-0797-8

Colella, V. S., Klopfer, E., & Resnick, M. (2001). *Adventures in modeling: exploring complex, dynamic system with StarLogo*. Teachers college Press.

Cook, M. (2004). Universality in elementary cellular automata. *Complex Systems*, *15*(1), 1–40.

Cooper, J., & Osborne, J. (2013). Connecting models to data in multiscale multicellular tissue simulations. *Procedia Computer Science*, *18*, 712–721. doi:10.1016/j.procs.2013.05.235

Cornell, W., Cieplak, P., Bayly, C., Gould, I., Merz, K. Jr, Ferguson, D., & Kollman, P. et al. (1995). A second generation force field for the simulation of proteins, nucleic acids, and organic molecules. *Journal of the American Chemical Society*, *117*(19), 5179–5197. doi:10.1021/ja00124a002

Costa, A. C. R. (2014). *On the Bases of an Architectural Style for Agent Societies: Concept and Core Operational Structure*. Available at: www.ResearchGate.net

Costa, A. C. R. (2009). DIMURO, Graçaliz Pereira. A Minimal Dynamical MAS Organization Model. In *Multiagent Systems - Semantics and Dynamics of Organization Models* (pp. 419–445). Hershey, PA: IGI Global. doi:10.4018/978-1-60566-256-5.ch017

Costa, A. C. R. (2016). Situated Ideological Systems: A Formal Concept, a Computational Notation, some Applications. *Axiomathes*.

Costanza, R., & Ruth, M. (1998). Using Dynamic Modeling to Scope Environmental Problems and Build Consensus. *Environmental Management*, *22*(2), 183–195. doi:10.1007/s002679900095 PMID:9465128

Creighton, T. E. (1990). Protein folding. *The Biochemical Journal*, *270*(1), 1–16. doi:10.1042/bj2700001 PMID:2204340

Crescenzi, P., Goldman, D., Papadimitriou, C., Piccolboni, A., & Yannakakis, M. (1998). On the complexity of protein folding. *Journal of Computational Biology*, *5*(3), 423–465. doi:10.1089/cmb.1998.5.423 PMID:9773342

Crookall, D. (2010). Serious Games, Debriefing, and Simulation/Gaming as a Discipline. *Simulation & Gaming*, *41*(6), 898–920. doi:10.1177/1046878110390784

Dada, J. O., & Mendes, P. (2011). Multi-scale modelling and simulation in systems biology. *Integrative Biology*, *3*(2), 86–96. doi:10.1039/c0ib00075b PMID:21212881

Dascalu, M., Stefan, G., Zafiu, A., & Plavitu, A. (2011). Applications of multilevel cellular automata in epidemiology. In *Proceedings of the 13th WSEAS international conference on Automatic control, modelling & simulation*. World Scientific and Engineering Academy and Society (WSEAS).

Datta, A., & Thomas, H. (1999). The cube data model: A conceptual model and algebra for on-line analytical processing in data warehouses. *Decision Support Systems*, *27*(3), 289–301. doi:10.1016/S0167-9236(99)00052-4

de Koster, C. G., & Lindenmayer, A. (1987). Discrete and continuous models for heterocyst differentiation in growing filaments of blue-green bacteria. *Acta Biotheoretica*, *36*(4), 249–273. doi:10.1007/BF02329786

De Mori, G. M., Micheletti, C., & Colombo, G. (2004). All-atom folding simulations of the villin headpiece from stochastically selected coarse-grained structures. *The Journal of Physical Chemistry B*, *108*(33), 12267–12270. doi:10.1021/jp0477699

de Souza, C. S. (2005). *The Semiotic Engineering of Human-Computer Interaction*. Boston, MA: MIT Press.

Deffuant, G., Weisbuch, G., Amblard, F., & Faure, T. (2003). Simple is beautiful . . . and necessary. *Journal of Artificial Societies and Social Simulation*, *6*(1).

Deisboeck, T. S., Wang, Z., Macklin, P., & Cristini, V. (2011). Multiscale cancer modeling. *Annual Review of Biomedical Engineering*, *13*(1), 127–155. doi:10.1146/annurev-bioeng-071910-124729 PMID:21529163

Dessbesell, G. J. (2015). *Simulação de Controle adaptativo de tráfego urbano através de sistema multiagentes e com base em dados reais. Dissertação (Mestrado em Sistemas e Processos Industriais)*. Santa Cruz do Sul/RS.

Deutsch, A., Maini, P., & Dormann, S. (Eds.). (2007). *Cellular Automaton Modeling of Biological Pattern Formation: Characterization, Applications, and Analysis*. Birkhäuser Boston.

Devore, J. L., & da Silva, J. P. N. (2006). *Probabilidade e estatística: para engenharia e ciências*. São Paulo: Pioneira Thomson Learning.

Dill, K., Bromberg, S., Yue, K., Fiebig, K., Yee, D., Thomas, P., & Chan, H. (1995). Principles of protein folding: A perspective from simple exact models. *Protein Science*, *4*(4), 561–602. doi:10.1002/pro.5560040401 PMID:7613459

Dimakopoulos, V. V., & Pitoura, E. (2003). A Peer-to-Peer Approach to Resource Discovery in Multi-agent Systems. In *Cooperative information agents* (Vol. 2782, pp. 62–77). Springer. doi:10.1007/978-3-540-45217-1_5

Dittmer, J., & Leyh, B. (2014). Paracrine effects of stem cells in wound healing and cancer progression[Review]. *International Journal of Oncology*, *44*(6), 1789–1798. PMID:24728412

Dlodlo, N., & Kalezhi, J. (2015, May). The internet of things in agriculture for sustainable rural development. In *International Conference on Emerging Trends in Networks and Computer Communications (ETNCC)* (pp. 13–18). doi:10.1109/ETNCC.2015.7184801

Dobrescu, R. & V.I. Purcarea (2011). Emergence, self-organization and morphogenesis in biological structures. *Journal of Medicine and Life, 4*(1), 82-90.

Dorn, M., Silva, M. B., Buriol, L. S., & Lamb, L. C. (2014). Three-dimensional protein structure prediction: Methods and computational strategies. *Computational Biology and Chemistry*, *53*, 251–276. doi:10.1016/j.compbiolchem.2014.10.001 PMID:25462334

DOro, S., Galluccio, L., Morabito, G., & Palazzo, S. (2015). Exploiting Object Group Localization in the Internet of Things: Performance Analysis. *IEEE Transactions on Vehicular Technology*, *64*(8), 3645–3656. doi:10.1109/TVT.2014.2356231

Drachman, D. A. (2005). Do we have brain to spare? *Neurology*, *64*(12), 2004–2005. doi:10.1212/01.WNL.0000166914.38327.BB PMID:15985565

Drasdo, D., & Forgacs, G. (2000). Modeling the Interplay of Generic and Genetic Mechanisms in Cleavage, Blastulation, and Gastrulation. *Developmental Dynamics*, *219*(2), 182–191. doi:10.1002/1097-0177(200010)219:2<182::AID-DVDY1040>3.3.CO;2-1 PMID:11002338

Dunker, A., Lawson, J., Brown, C., Williams, R., Romero, P., Oh, J., & Obradovic, Z. et al. (2001). Intrinsically disordered protein. *Journal of Molecular Graphics & Modelling*, *19*(1), 26–59. doi:10.1016/S1093-3263(00)00138-8 PMID:11381529

Dunker, A., Silman, I., Uversky, V., & Sussman, J. (2008). Function and structure of inherently disordered proteins. *Current Opinion in Structural Biology*, *18*(6), 756–764. doi:10.1016/j.sbi.2008.10.002 PMID:18952168

Dunn, A. (2010). Hierarchical cellular automata methods. In A. G. Hoekstra, J. Kroc, & P. M. A. Sloot (Eds.), *Simulating Complex Systems by Cellular Automata* (pp. 59–80). doi:10.1007/978-3-642-12203-3_4

Dunn, A. G., & Majer, J. D. (2007). Simulating weed propagation via hierarchical, patch-based cellular automata. *Lecture Notes in Computer Science*, *4487*, 762–769. doi:10.1007/978-3-540-72584-8_101

Duraccio, V., Falcone, D., Silvestri, A., & Di Bona, G. (2007). Use of Simulation for the Prevention of Environmental Problems. In *Proceedings of the 2007 summer computer simulation conference* (pp. 863–866). San Diego, CA: Society for Computer Simulation International.

Dyo, V., Ellwood, S. A., Macdonald, D. W., Markham, A., Trigoni, N., Wohlers, R., & Yousef, K. (2012, September). WILDSENSING: Design and Deployment of a Sustainable Sensor Network for Wildlife Monitoring. *ACM Trans. Sen. Netw., 8*(4), 29:1–29:33.

Dzemydiene, D., & Dzindzalieta, R. (2010). Development of architecture of embedded decision support systems for risk evaluation of transportation of dangerous goods. *Technological and Economic Development of Economy*, *16*(4), 654–671. doi:10.3846/tede.2010.40

Edmonds, B., & Moss, S. (2004). *From KISS to KIDS - An "anti-simplistic" modelling approach.Joint Workshop on Multi-Agent and Multi-Agent-Based Simulation*, New York, NY.

Ekman, P. (2007). *Emotions Revealed – Recognizing Faces and Feelings to Improve Communication and Emotional Life* (2nd ed.). New York, NY: Holt, Henry & Company, Inc.

Elmenreich, W., & Fehervari, I. (2011). Evolving self-organizing cellular automata based on neural network genotypes. *Lecture Notes in Computer Science*, *6557*, 16–25. doi:10.1007/978-3-642-19167-1_2

Environmental Company of the State São Paulo (CETESB). (2001). *Manual management of contaminated areas*. São Paulo, Brazil: Author.

Epstein, J. M. (1999). Agent-based computational models and generative social science. *Complexity*, *4*(5), 41–60. doi:10.1002/(SICI)1099-0526(199905/06)4:5<41::AID-CPLX9>3.0.CO;2-F

Epstein, J. M., & Axtell, R. L. (1996). *Growing Artificial Societies. Social science from the bottom up*. Washington, DC: Brookings Institution Press.

Esser, D. S., Leveau, J. H. J., & Meyer, K. M. (2015). *Modelling microbial growth and dynamics. In Applied Microbiology Biotechnology*. Springer-Verlag Berlin Heidelberg.

European Union - EU. (2003). *Technical Guidance Document (TGD) Part I. EUR20418 EN/1*. European Commission-Joint Research Center, Directive 98/8EC, 14 JRC Publication N°: JRC23785, EUR 20418 EN, 2003.

Evans, J. St. B. T. (2003). In two minds: Dual process accountsof reasoning. *Trends in Cognitive Sciences*, *7*(10), 454–459. doi:10.1016/j.tics.2003.08.012 PMID:14550493

Evans, J. St. B. T. (2006). The heuristic-analytic theory of reasoning: Extension and evaluation. *Psychonomic Bulletin & Review*, *13*(3), 378–395. doi:10.3758/BF03193858 PMID:17048720

Evans, J. St. B. T. (2007). On the resolution of conflict in dual-process theories of reasoning. *Thinking & Reasoning*, *13*(4), 321–329. doi:10.1080/13546780601008825

Evans, J. St. B. T. (2008). Dual-processing accounts of reasoning, judgment and social cognition. *Annual Review of Psychology*, *59*(1), 255–278. doi:10.1146/annurev.psych.59.103006.093629 PMID:18154502

Faceli, K., Carvalho, A., Lorena, A., & Gama, J. (2011). *Inteligência Artificial – uma abordagem de aprendizado de máquina*. Rio de Janeiro: LTC.

Feig, M., Rotkiewicz, P., Kolinski, A., Skolnick, J., & Brooks, C. (2000). Accurate reconstruction of all-atom protein representations from side-chain- based low-resolution models. *Proteins*, *41*(1), 86–97. doi:10.1002/1097-0134(20001001)41:1<86::AID-PROT110>3.0.CO;2-Y PMID:10944396

Fernández-López, M., Gómez-Pérez, A., & Juristo, N. (1997). *Methontology: from ontological art towards ontological engineering.AAAI Symposium on Ontological Engineering*, Stanford, CA.

Ferrando, N., Gosálvez, M. A., Cerdá, J., Gadea, R., & Sato, K. (2011). Octree-based, GPU implementation of a continuous cellular automaton for the simulation of complex, evolving surfaces. *Computer Physics Communications*, *182*(3), 628–640. doi:10.1016/j.cpc.2010.11.004

Ferrando, N., Gosálvez, M. A., & Colóm, R. J. (2012). Evolutionary continuous cellular automaton for the simulation of wet etching of quartz. *Journal of Micromechanics and Microengineering*, *22*(2), 025021. doi:10.1088/0960-1317/22/2/025021

Ferreira, C. (2001). Gene expression programming: A new adaptive algorithm for solving problems. *Complex Systems*, *13*(2), 87–129.

Fingelkurts, A., & Fingelkurts, A. (2004). Making complexity simpler: Multivariability and metastability in the brain. *The International Journal of Neuroscience*, *114*(7), 843–862. doi:10.1080/00207450490450046 PMID:15204050

Fingelkurts, A., Fingelkurts, A., & Kähkönen, S. (2005). Functional connectivity in the brain, is it an elusive concept? *Neuroscience and Biobehavioral Reviews*, *28*(8), 827–836. doi:10.1016/j.neubiorev.2004.10.009 PMID:15642624

Finin, T., Fritzon, R., McKay, D., & McEntire, R. (1994). KQML as an Agent Communication language. In *Proceedings of the Third International Conference on Information and Knowledge Management (CIKM'94).* ACM Press. doi:10.1145/191246.191322

FIPA. (2002). *FIPA Agent Communication specifications.* Foundation for Intelligent Physical Agents. Retrieved from http://www.fipa.org/repository/aclspecs.html

Fischer, D. (2000). Hybrid fold recognition: Combining sequence derived properties with evolutionary information. *Pacific Symposium on Biocomputing. Pacific Symposium on Biocomputing*, *5*, 119–130. PMID:10902162

Fischer, D. (2006). Servers for protein structure prediction. *Current Opinion in Structural Biology*, *16*(2), 178–182. doi:10.1016/j.sbi.2006.03.004 PMID:16546376

Fischer, D., & Eisenberg, D. (1996). Protein fold recognition using sequence- derived predictions. *Protein Science*, *5*(5), 947–955. doi:10.1002/pro.5560050516 PMID:8732766

Fischer, D., Elofsson, A., Rychlewski, L., Pazos, F., Valencia, A., Rost, B., & Dunbrack, R. L. et al. (2001). Cafasp2: the second critical assessment of fully automated structure prediction methods. *Proteins: Struct., Funct. Bioinf.*, *45*(S5), 171–183.

Floudas, C. (2007). Computational methods in protein structure prediction. *Biotechnology and Bioengineering*, *97*(2), 207–213. doi:10.1002/bit.21411 PMID:17455371

Fodor, J. (1983). *The Modularity of Mind.* Cambridge, MA: MIT Press.

Fogolari, F., Pieri, L., Dovier, A., Bortolussi, L., Giugliarelli, G., Corazza, A., & Viglino, P. et al. (2007). Scoring predictive models using a reduced representation of proteins: Model and energy definition. *BMC Structural Biology*, *7*(1), 1–17. doi:10.1186/1472-6807-7-15 PMID:17378941

Forrest, S., & Mitchell, M. (2014). Relative Building-Block Fitness and the Building Block Hypothesis. In *Foundations of Genetic Algorithms II* (pp. 109–126). San Mateo, CA: Morgan Kaufmann.

Foundation for Intelligent Physical Agentes (FIPA). (2002). Communicative Act Library Specification. Document number SC00037J - FIPA TC Communication. Geneva, Switzerland: Author.

Franklin, S. (1995). *Artificial minds.* Cambridge, MA: MIT Press.

Fredkin, E. (1992). Finite nature. *Progress in Atomic Physics Neutrinos and Gravitation*, *72*, 345–354.

Fuks, H., Pimentel, M., & Lucena, C. J. P. (2006). R-U-Typing-2-Me? Evolving a chat tool to increase understanding in learning activities. *Computer-Supported Collaborative Learning*, *1*(1), 117–142. doi:10.1007/s11412-006-6845-3

Fukui, M., & Ishibashi, Y. (1996). Traffic flow in 1D cellular automaton model including cars moving with high speed. *Journal of the Physical Society of Japan*, *65*(6), 1868–1870. doi:10.1143/JPSJ.65.1868

Galán, R., Ermentrout, G., & Urban, N. (2006). Predicting synchronized neural assemblies from experimentally estimated phase-resetting curves. *Neurocomputing*, 1–2.

Galluccio, L. (2011, June). On the potentials of object group localization in the Internet of Things. In *IEEE International Symposium on a World of Wireless, Mobile and Multimedia Networks* (pp. 1–9). doi:10.1109/WoWMoM.2011.5986489

Gammack, D. (2015). Using NetLogo as a tool to encourage scientific thinking across disciplines. *Journal of Teaching and Learning with Technology*, *4*(1), 22–39. doi:10.14434/jotlt.v4n1.12946

Ganesan, P., Bawa, M., & Garcia-Molina, H. (2004). Online Balancing of Range-partitioned Data with Applications to Peer-to-peer Systems. In *Proceedings of the Thirtieth International Conference on Very Large Data Bases* (Vol. 30, pp. 444–455). doi:10.1016/B978-012088469-8.50041-3

Garcia, A., & Sichman, J. (2005). Agentes e Sistemas Multiagentes. In *Sistemas Inteligentes Fundamentos e Aplicações* (pp. 269–304). São Paulo: Manole.

Gardner, M. (1970). Mathematical Games: The fantastic combinations of John Conway's new solitaire game "life". *Scientific American, 223*(4), 120-123.

Garro, A., Terracina, G., & Ursino, D. (2004). A multi-agent system for supporting the prediction of protein structures. *Integr. Comput. Aid. E.*, *11*(3), 259–280.

Gary, P. J., Izquierdo, L. R., & Gotts, N. M. (2005). The ghost in the model (and other effects of floating point arithmetic). *Journal of Artificial Societies and Social Simulation*, *8*(1). Retrieved from http://jasss.soc.surrey.ac.uk/8/1/5.html

Geist, H. J., & Lambin, E. F. (2001). What Drives Tropical Deforestation? A meta-analysis of proximate and underlying causes of deforestation based on subnational case study evidence. LUCC International Project Office.

Geist, H.J., & Lambin, E.F. (2002). Proximate Causes and Underlying Driving Forces of Tropical Deforestation. *BioScience, 52*(2),143-150.

Gelfand, A. (2013). The biology of interacting things: The intuitive power of agent-based models. *Biomedical Computation Review*, *1*, 20–27.

Gerosa, M. A., Pimentel, M., Fuks, H., & Lucena, C. J. P. (2006). Development of Groupware Based on the 3C Collaboration Model and Component Technology. In Y. A. Dimitriadis, I. Zigurs, & E. Gómez-Sánchez (Eds.), *Groupware: Design, Implementation, and Use (LNCS),* (Vol. 4154, pp. 302–309). Springer. doi:10.1007/11853862_24

Ghobara, Smith, Schoefs, Vinayak, Gebeshuber, & Gordon. (2016). On light and diatoms: A photonics and photobiology review. *Advances in Optics and Photonics*. Unpublished.

Gibas, C., & Jambeck, P. (2001). *Developing bioinformatics computer skills*. Sebastopol, CA: O'Reilly Media, Inc.

Gibson, M. A., & Bruck, J. (2000). Efficient exact stochastic simulation of chemical systems with many species and many channels. *The Journal of Physical Chemistry A*, *104*(9), 1876–1889. doi:10.1021/jp993732q

Gillespie, D. T. (1977). Exact stochastic simulation of coupled chemical reactions. *Journal of Physical Chemistry*, *81*(25), 2340–2361. doi:10.1021/j100540a008

Ginalski, K., Elofsson, A., Fischer, D., & Rychlewski, L. (2003). 3d-jury: A simple approach to improve protein structure predictions. *Bioinformatics (Oxford, England)*, *19*(8), 1015–1018. doi:10.1093/bioinformatics/btg124 PMID:12761065

Ginovart, M. (2014). Discovering the power of individual-based modelling in teaching and learning: The study of a predator-prey system. *Journal of Science Education and Technology*, *23*(4), 496–513. doi:10.1007/s10956-013-9480-6

Gkiolmas, A., Karamanos, K., Chalkidis, A., Skordoulis, C., & Papaconstantinou, M. D. S. (2013). Using simulation of Netlogo as a tool for introducing Greek high-school students to eco-systemic thinking. *Advances in Systems Science and Application*, *13*(3), 275–297.

Gluckman, P. D., Hanson, M. A., & Low, F. M. (2011). The role of developmental plasticity and epigenetics in human health. *Birth Defects Research Part C-Embryo Today-Reviews*, *93*(1), 12–18. doi:10.1002/bdrc.20198 PMID:21425438

Goldberg, D. (1989). *Genetic algorithms in search, optimization, and machine learning*. Addison-Wesley. *Artificial Intelligence*.

Golomb, D., Hansel, D., Shraiman, B., & Sompolinsky, H. (1992). Clustering in globally coupled phase oscillators. *Physical Review A*., *45*(6), 3516–3530. doi:10.1103/PhysRevA.45.3516 PMID:9907399

González Pérez, P. P., Omicini, A., & Sbaraglia, M. (2013). A biochemically-inspired coordination-based model for simulating intracellular signalling pathways. *Journal of Simulation*, *7*(3), 216–226. doi:10.1057/jos.2012.28

González Pérez, P., Cárdenas, M., Camacho, D., Franyuti, A., Rosas, O., & Lagúnez-Otero, J. (2003). Cellulat: An agent-based intracellular signalling model. *Bio Systems*, *68*(2–3), 171–185. doi:10.1016/S0303-2647(02)00094-1 PMID:12595116

Gordon, N. K., & Gordon, R. (2016b). The organelle of differentiation in embryos: the cell state splitter [invited review]. Theoretical Biology and Medical Modelling, 13.

Gordon, R. (1980). Monte Carlo methods for cooperative Ising models. In Cooperative Phenomena in Biology. New York: Pergamon Press. doi:10.1016/B978-0-08-023186-0.50010-X

Gordon, D., Marshall, S., & Mayo, S. (1999). Energy functions for protein design. *Current Opinion in Structural Biology*, *9*(4), 509–513. doi:10.1016/S0959-440X(99)80072-4 PMID:10449371

Gordon, N. K., & Gordon, R. (2016a). *Embryogenesis Explained*. Singapore: World Scientific Publishing. doi:10.1142/8152

Gordon, R. (1966). On stochastic growth and form. *Proceedings of the National Academy of Sciences of the United States of America*, *56*(5), 1497–1504. doi:10.1073/pnas.56.5.1497 PMID:5230309

Gordon, R. (1970). On Monte Carlo algebra. *Journal of Applied Probability*, *7*(02), 373–387. doi:10.1017/S002190020003494X

Gordon, R. (1999). *The Hierarchical Genome and Differentiation Waves: Novel Unification of Development, Genetics and Evolution*. London: World Scientific & Imperial College Press. doi:10.1142/2755

Gordon, R. (2015). Walking the tightrope: the dilemmas of hierarchical instabilities in Turing's morphogenesis. In S. B. Cooper & A. Hodges (Eds.), *The Once and Future Turing: Computing the World* (pp. 150–164). Cambridge, UK: Cambridge University Press.

Gordon, R., Goel, N. S., Steinberg, M. S., & Wiseman, L. L. (1972). A rheological mechanism sufficient to explain the kinetics of cell sorting. *Journal of Theoretical Biology*, *37*(1), 43–73. doi:10.1016/0022-5193(72)90114-2 PMID:4652421

Gordon, R., Goel, N. S., Steinberg, M. S., & Wiseman, L. L. (1975). A rheological mechanism sufficient to explain the kinetics of cell sorting. In G. D. Mostow (Ed.), *Mathematical Models for Cell Rearrangement* (pp. 196–230). New Haven, CT: Yale University Press.

Gordon, R., & Melvin, C. A. (2003). Reverse engineering the embryo: A graduate course in developmental biology for engineering students at the University of Manitoba, Canada. *The International Journal of Developmental Biology*, *47*(2/3), 183–187. PMID:12705668

Gordon, R., & Rangayyan, R. M. (1984a). Feature enhancement of film mammograms using fixed and adaptive neighborhoods. *Applied Optics*, *23*(4), 560–564. doi:10.1364/AO.23.000560 PMID:18204600

Gordon, R., & Rangayyan, R. M. (1984b). Correction: Feature enhancement of film mammograms using fixed and adaptive neighborhoods. *Applied Optics*, *23*(13), 2055. doi:10.1364/AO.23.002055 PMID:20424727

Gordon, R., & Stone, R. (2016in press). Cybernetic embryo. In R. Gordon & J. Seckbach (Eds.), *Biocommunication*. London: World Scientific Publishing. doi:10.1142/q0013

Gorochowski, T. E., Matyjaszkiewicz, A., Todd, T., Oak, N., Kowalska, K., Reid, S., & di Bernardo, M. et al. (2012). 08). BSim: An agent-based tool for modeling bacterial populations in systems and synthetic biology. *PLoS ONE*, *7*(8), 1–9. doi:10.1371/journal.pone.0042790 PMID:22936991

Gosálvez, M. A., Xing, Y., Sato, K., & Nieminen, R. M. (2009). Discrete and continuous cellular automata for the simulation of propagating surfaces. *Sensors and Actuators. A, Physical*, *155*(1), 98–112. doi:10.1016/j.sna.2009.08.012

Government of Canada. (2004). Canadian Handbook on Health Impact Assessment: The Multidisciplinary Team (vols. 1-3). Ottawa, Canada: Author.

Government of Canada. Health Canada's Social Media Tools. (2012). Guidance *Documents related to Human Health Risk Assessment.* Revised 2012, ISBN: 978-1-100-17671-0, Cat.: H128-1/11-632E-PDF. Retrieved from http://www.hc-sc.gc.ca/ewh-semt/contamsite/docs/index-eng.php#a1

Government of Netherland. National Institute for Public Health and the Environment (RIVM). (2015). *Risks of Chemicals - Proast.* Retrieved from http://www.rivm.nl/en

Governo do Distrito Federal, G. D. F. (2009). *Plano Diretor de Ordenamento Territorial do Distrito Federal: documento técnico. Technical Report.* Brasília, DF, Brazil: Secretaria de Habitação, Regularização e Desenvolvimento Urbano - SEDHAB.

Goyal, A., Szarzynska, B., & Fankhauser, C. (2013). Phototropism: At the crossroads of light-signaling pathways. *Trends in Plant Science*, *18*(7), 393–401. doi:10.1016/j.tplants.2013.03.002 PMID:23562459

Graham, J., Amos, B., & Plumptre, T. (2003). Governance principles for protected areas in the 21st century. Durban, South Africa: IUCN (International Union for Conservation of Nature).

Gravan, C. P., & Lohoz-Beltra, R. (2004). Evolving morphogenetic fieds in the zebra skin pattern based on Turing's morphogen hypothesis. *International Journal of Applied Mathematics and Computer Science*, *14*(3), 351–361.

Gray, J. R., Chabris, C. F., & Braver, T. S. (2003). Neural mechanisms of general fluid intelligence. *Nature Neuroscience*, *6*(3), 316–322. doi:10.1038/nn1014 PMID:12592404

Greenfield, S. A. (2002). Mind, brain and consciousness. *The British Journal of Psychiatry*, *181*, 91–93. PMID:12151275

Greenfield, S. A., & Collins, T. F. T. (2005). A neuroscientific approach to consciousness. *Progress in Brain Research Journal*, *150*, 11–23. doi:10.1016/S0079-6123(05)50002-5 PMID:16186012

Grimm, V., Berger, U., Bastiansen, F., Eliassen, S., Ginot, V., Giske, J., & Huse, G. (2006). A standard protocol for describing individual-based and agent-based models. *Ecological Modelling*, *198*(1-2), 115–126. doi:10.1016/j.ecolmodel.2006.04.023

Grimm, V., Berger, U., DeAngelis, D. L., Polhill, J. G., Giske, J., & Railsback, S. F. (2010). The ODD protocol: A review and first update. *Ecological Modelling*, *221*(23), 2760–2768. doi:10.1016/j.ecolmodel.2010.08.019

Grimm, V., & Railsback, S. F. (2005). *Individual-based Modeling and Ecology.* Princeton, NJ: Princeton University Press. doi:10.1515/9781400850624

Groll, A., Martin, A., Stehr, M., Singh, M., Portaels, F., Silva, P. E. A., & Palomino, J. C. (2010). *Fitness of Mycobacterium tuberculosis strains of the W-Beijing and Non-W-Beijing genotype.* Academic Press.

Grover, M. (2011). Parallels between Gluconeogenesis and Synchronous Machines. *International Journal on Computer Science and Engineering*, *3*(1), 185–191.

Gruner, S. (2010). *Mobile agents systems and cellular automata. Autonomous, Agent Multi-agent System Journal.*

Guan, Q., & Clarke, K. C. (2010). A general-purpose parallel raster processing programming library test application using a geographic cellular automata model. *International Journal of Geographical Information Science*, *24*(5), 695–722. doi:10.1080/13658810902984228

Guex, N., & Peitsch, M. C. (1997). SWISS-MODEL and the Swiss-Pdb Viewer: An environment for comparative protein modeling. *Electrophoresis*, *18*(15), 2714–2723. doi:10.1002/elps.1150181505 PMID:9504803

Gummadi, K. P., Saroiu, S., & Gribble, S. D. (2002). King: Estimating Latency Between Arbitrary Internet End Hosts. In *Proceedings of the Second ACM SIGCOMM Workshop on Internet measurment* (pp. 5–18). doi:10.1145/637201.637203

Gunasekaran, K., Tsai, C., Kumar, S., Zanuy, D., & Nussinov, R. (2003). Extended disordered proteins: Targeting function with less scaffold. *Trends in Biochemical Sciences*, *28*(2), 81–85. doi:10.1016/S0968-0004(03)00003-3 PMID:12575995

Guyeux, C., Côte, N. M.-L., Bahi, J. M., & Bienia, W. (2014). Is protein folding problem really a np-complete one? first investigations. *Journal of Bioinformatics and Computational Biology*, *12*(01), 1350017–1350041. doi:10.1142/S0219720013500170 PMID:24467756

Guyot, P., Drogoul, A., & Honiden, S. (2006). Power and negotiation: Lessons from agent-based participatory simulations. In P. Stone, & G. Weiss (Eds.), *Proceedings of the 5th International Joint Conference on Autonomous Agents and Multiagent Systems (AAMAS'06)*. doi:10.1145/1160633.1160636

Haider, W., ur Rehman, A., & Durrani, N. M. (2013, Sept). Towards decision support model for ubiquitous agriculture. In *International Conference on Digital Information Management (ICDIM)* (pp. 308–313). doi:10.1109/ICDIM.2013.6693987

Halbach, M., & Hoffmann, R. (2005). Optimal behavior of a moving creature in the cellular automata model. *Lecture Notes in Computer Science*, *3606*, 129–140. doi:10.1007/11535294_11

Halgren, T. A. (1995). Potential energy functions. *Current Opinion in Structural Biology*, *5*(2), 205–210. doi:10.1016/0959-440X(95)80077-8 PMID:7648322

Hall, M., Frank, E., Holmes, G., Pfahringer, B., Reutemann, P., & Witten, I. H. (2009). The WEKA data mining software: an update. Waikato Environment for Knowledge Analysis (Version 3.7.5) [Software]. *ACM SIGKDD Explorations Newsletter, 11*(1), 10-18.

Hao, M., & Scheraga, H. (1999). Designing potential energy functions for protein folding. *Current Opinion in Structural Biology*, *9*(2), 184–188. doi:10.1016/S0959-440X(99)80026-8 PMID:10322206

Hart, J. K., & Martinez, K. (2015). Toward an environmental Internet of Things. *Earth and Space Science*, *2*(5), 194–200. doi:10.1002/2014EA000044

Haykin, S. (1999). Neural Networks. A comprehensive Foundation (2nd ed.). Prentice-Hall.

Haykin, S.O. (2011). *Neural Networks and Learning Machines*. Pearson Education.

Heinen, M., Sá, C., Silveira, F., Cesconetto, C., & Sohn, G. (2013). *Controle Inteligente de Semáforos Utilizando Redes Neurais Artificiais com Funções de Base Radial*. Frederico Westphalen/RS.

Hentenryck, P. V., & Michel, L. (2009). *Constraint-based local search*. The MIT Press.

Hochberger, C., Hoffmann, R., & Waldschmidt, S. (1999). CDL++ for the description of moving objects in cellular automata. *Lecture Notes in Computer Science*, *1662*, 428–435. doi:10.1007/3-540-48387-X_44

Hocking, A. M., & Gibran, N. S. (2010). Mesenchymal stem cells: Paracrine signaling and differentiation during cutaneous wound repair. *Experimental Cell Research*, *316*(14), 2213–2219. doi:10.1016/j.yexcr.2010.05.009 PMID:20471978

Hodges, A. (2014). Alan Turing: The Enigma. New York: Simon & Schuster. doi:10.1515/9781400865123

Hoehme, S., & Drasdo, D. (2010). A cell-based simulation software for multi-cellular systems. *Bioinformatics (Oxford, England)*, *26*(20), 2641–2642. doi:10.1093/bioinformatics/btq437 PMID:20709692

Hoffman, D. D. (2009). The interface theory of perception: Natural selection drives true perception to swift extinction. In S. J. Dickinson, A. Leonardis, B. Schiele, & M. J. Tarr (Eds.), *Object Categorization: Computer and Human Vision Perspectives* (pp. 148–165). Cambridge, UK: Cambridge University Press. doi:10.1017/CBO9780511635465.009

Hohm, T., & Zitzler, E. (2009). Multicellular pattern formation. *IEEE Engineering in Medicine and Biology Magazine*, *28*(4), 52–57. doi:10.1109/MEMB.2009.932905 PMID:19622425

Holcombe, M., Adra, S., Bicak, M., Chin, S., Coakley, S., Graham, A. I., & Worth, D. et al. (2012). Modelling complex biological systems using an agent-based approach. *Integrative Biology*, *4*(1), 53–64. doi:10.1039/C1IB00042J PMID:22052476

Holland, J. H. (1992). Adaptation in Natural and Artificial Systems: An Introductory Analysis with Applications to Biology, Control, and Artificial Intelligence. Cambridge, MA: MIT Press.

Hoppensteadt, F., & Izhikevich, E. (1997). *Weakly Connected Neural Networks*. Berlin: Springer. doi:10.1007/978-1-4612-1828-9

Houser, N. (1992). *The Essential Peirce: Selected Philosophical Writings (1867-1893)* (Vol. 1). Indiana University Press.

Hovmoller, T., & Ohlson, T. (2002). Conformation of amino acids in protein. *Acta Crystallographica*, *58*(5), 768–776. PMID:11976487

Howarth, R. W., Swaney, D. P., Butler, T. J., & Marino, R. (2000). Climatic control on eutrophication of the Hudson River Estuary. *Ecosystems (New York, N.Y.)*, *3*(2), 210–215. doi:10.1007/s100210000020

Hsee, C. K., & Hastie, R. (2006). *Decision and Experience: Why Don't We Choose What Makes Us Happy? Trends in Cognitive Sciences*. Retrieved from http://ssrn.com/abstract=929914

Hsu, C. W., & Lin, C. J. (2002). A comparison of methods for multiclass support vector machines. *Neural Networks. IEEE Transactions on*, *13*(2), 415–425.

Huang, H., Feng, C., Pan X., Wu, H., Ren, Y., Wu, C., & Wei, C. (2013). Thiocyanate Oxidation by Coculture from a Coke Wastewater Treatment Plant. *Journal of Biomaterials and Nanobiotechnology, 4*(2A).

Huang, J. H., Chen, Y. Y., Huang, Y. T., Lin, P. Y., Chen, Y. C., Lin, Y. F., & Chen, L. J. et al. (2010, June). Rapid Prototyping for Wildlife and Ecological Monitoring. *IEEE Systems Journal*, *4*(2), 198–209. doi:10.1109/JSYST.2010.2047294

Huber, A. B., Kolodkin, A. L., Ginty, D. D., & Cloutier, J. F. (2003). Signaling at the growth cone: Ligand-receptor complexes and the control of axon growth and guidance. *Annual Review of Neuroscience*, *26*(1), 509–563. doi:10.1146/annurev.neuro.26.010302.081139 PMID:12677003

Huberman, B. A., & Glance, N. S. (1993). Evolutionary games and computer simulations. *Proceedings of the National Academy of Sciences of the United States of America*, *90*(16), 7716–7718. doi:10.1073/pnas.90.16.7716 PMID:8356075

Hulstijn, J., & van der Torre, L. (2003). Combining goal generation and planning in an argumentation framework. In *Proceedings of the 15th Belgium-Netherlands Conference on Artificial Intelligence (BNAIC'2003)*.

IEEE 802.15. (2005). *IEEE Standard for Information Technology - Telecommunications and Information Exchange Between Systems - Local and Metropolitan Area Networks - Specific Requirements Part 15.4: Wireless Medium Access Control (MAC) and Physical Layer (PHY) Specifications for Low-Rate Wireless Personal Area Networks (LR-WPANs)* (Tech. Rep.).

Ilachinski, A. (2001). *Cellular Automata: A Discrete Universe.* Retrieved from http://www.worldscientific.com/worldscibooks/10.1142/4702

Ilhan Akbas, M., Brust, M. R., Turgut, D., & Ribeiro, C. H. (2015). A preferential attachment model for primate social networks. *Computer Networks, 76*, 207–226. doi:10.1016/j.comnet.2014.11.009

Instituto Brasileiro de Geografia e Estatística (IBGE). (2008). Pesquisa Nacional de Saneamento Básico. Rio de Janeiro: IBGE.

International Agency for Research on Cancer (IARC/WHO). (2016). *List of Classifications. Agents Classified by the IARC.* Retrieved from http://monographs.iarc.fr/ENG/Classification/index.php

Irving, M. A. (2006). *Áreas Protegidas e Inclusão Social: Construindo Novos Significados*. Rio de Janeiro, Brazil: Aquarius.

Izaguirre, J. A., Chaturvedi, R., Huang, C., Cickovski, T., Coffland, J., Thomas, G., & Glazier, J. A. (2004). Compucell, a multi-model framework for simulation of morphogenesis. *Bioinformatics (Oxford, England), 20*(7), 1129–1137. doi:10.1093/bioinformatics/bth050 PMID:14764549

Jacob, F. (1977). Evolution and tinkering. *Science, 196*(4295), 1161–1166. doi:10.1126/science.860134 PMID:860134

Jacobson, M. J., & Wilensky, U. (2006). Complex systems in education: Scientific and educational importance and implications for the learning science. *Journal of the Learning Sciences, 15*(1), 11–34. doi:10.1207/s15327809jls1501_4

Jaskowski, W., Blazewicz, J., Lukasiak, P., Milostan, M., & Krasnogor, N. (2007). 3d-judge–a metaserver approach to protein structure prediction. *Found. Comput. Decis. Sci., 32*(1), 3–14.

Jassby, A. D. (2008). *Phytoplankton in the Upper San Francisco Estuary: Recent biomass trends, their causes and their trophic significance.* Retrieved from http://repositories.cdlib.org/cgi/viewcontent.cgi?article51103&context5jmie/sfew

Jenkins, S. (2015). *Tools for critical thinking in biology.* New York, NY: Oxford University Press.

Jennings, N. R., Sycara, K., & Wooldridge, M. (1998). A roadmap of agent research and development. *Autonomous Agents and Multi-Agent Systems, 1*(1), 7–38. doi:10.1023/A:1010090405266

Jiang, D., Lu, M. X., Li, F. S., Zhou, Y. Y., & Gu, Q. B. (2011). Review on current application of decision support systems for contaminated site management. *Huanjing Kexue yu Jishu, 34*(3), 170-174.

Jiang, R., & Zhang, Y. (2013, September). Research of Agricultural Information Service Platform Based on Internet of Things. In *International Symposium on Distributed Computing and Applications to Business Engineering and Science.*

Jiang, T., Cui, Q., Shi, G., & Ma, S. (2003). Protein folding simulations of the hydrophobich-hydrophilic model by combining tabu search with genetic algorithms. *The Journal of Chemical Physics, 119*(8), 4592–4596. doi:10.1063/1.1592796

Jimenez, R., Osmani, F., & Knutsson, B. (2011, August). Sub-Second lookups on a Large-Scale Kademlia-Based overlay. In *11th IEEE Conference on Peer-to-Peer Computing*. doi:10.1109/P2P.2011.6038665

Jin, H., & Kim, I. C. (2004). Plan-based coordination of a multi-agent system for protein structure prediction. In *International Conference on AI, Simulation, and Planning in High Autonomy Systems* (pp. 224–232). Springer Berlin Heidelberg.

Johnson, S. (2010). *Where Good Ideas Come From.* Penguin Publishing Group.

Jones, D. T. (1999). Genthreader: An efficient and reliable protein fold recognition method for genomic sequences. *Journal of Molecular Biology*, *287*(4), 797–815. doi:10.1006/jmbi.1999.2583 PMID:10191147

Jones, D., Taylor, W., & Thornton, J. (1992). A new approach to protein fold recognition. *Nature*, *358*(6381), 86–89. doi:10.1038/358086a0 PMID:1614539

Jorgensen, W., & Tirado-Rives, J. (2005). Potential energy functions for atomic- level simulations of water and organic and biomolecular systems. *Proceedings of the National Academy of Sciences of the United States of America*, *102*(19), 6665–6670. doi:10.1073/pnas.0408037102 PMID:15870211

Kaandorp, J. A., Botman, D., Tamulonis, C., & Dries, R. (2012). Multi-scale Modeling of Gene Regulation of Morphogenesis. In *How the World Computes: Turing Centenary conference and 8th conference on computability in Europe* (LNCS), (vol. 7318, pp. 355-362). Berlin: Springer. doi:10.1007/978-3-642-30870-3_36

Kahneman, D. (2000). A psychological point of view: Violations of rational rules as a diagnostic of mental processes[Commentary on Stanovich and West]. *Behavioral and Brain Sciences*, *23*(5), 681–683. doi:10.1017/S0140525X00403432

Kahneman, D. (2011). *Thinking, fast and slow*. New York: Farrar, Straus and Giroux.

Kaji, M., & Inohara, T. (2014). Numerical analysis focused on each agent's moving on refuge under congestion circumstances by using cellular automata. In *2014 IEEE International Conference on Systems, Man and Cybernetics*. doi:10.1109/SMC.2014.6974374

Kambayashi, Y., & Harada, Y. (2007). A Resource Discovery Method Based on Multi-agents in P2P Systems. In Agent and Multi-Agent Systems: Technologies and Applications (Vol. 4496, pp. 364–374). Springer. doi:10.1007/978-3-540-72830-6_38

Kang, S., Kahan, S., & Momeni, B. (2014). Simulating microbial community patterning using biocellion. In Engineering and Analyzing Multicellular Systems (pp. 233-253). Springer. doi:10.1007/978-1-4939-0554-6_16

Kang, S., Kahan, S., McDermott, J., Flann, N., & Shmulevich, I. (2014). Biocellion: Accelerating computer simulation of multicellular biological system models. *Bioinformatics (Oxford, England)*, *30*(21), 3101–3108. doi:10.1093/bioinformatics/btu498 PMID:25064572

Karplus, K., Karchin, R., Barrett, C., Tu, S., Cline, M., Diekhans, M., & Hughey, R. et al. (2001). What is the value added by human intervention in protein structure prediction? *Proteins: Struct., Funct. Bioinf.*, *45*(S5), 86–91.

Katkovnik, V., Egiazarian, K., & Astola, J. (2006). *Local Approximation Techniques in Signal and Image Processing*. SPIE Press. doi:10.1117/3.660178

Kaukalias, T., & Chatzimisios, P. (2016). Internet of Things (IoT). In Encyclopedia of Information Science and Technology, (3rd ed.; pp. 7623–7632).

Kelley, L. A., MacCallum, R. M., & Sternberg, M. J. (2000). Enhanced genome annotation using structural profiles in the program 3d-pssm. *Journal of Molecular Biology*, *299*(2), 501–522. doi:10.1006/jmbi.2000.3741 PMID:10860755

Kier, L. B., Bonchev, D., & Buck, G. A. (2005). Modeling biochemical networks: A cellular-automata approach. *Chemistry & Biodiversity*, *2*(2), 233–243. doi:10.1002/cbdv.200590006 PMID:17191976

Kiester, A. R., & Sahr, K. (2008). Planar and spherical hierarchical, multi-resolution cellular automata. *Computers, Environment and Urban Systems*, *32*(3), 204–213. doi:10.1016/j.compenvurbsys.2008.03.001

Kim, D. E., Chivian, D., & Baker, D. (2004). Protein structure prediction and analysis using the Robetta server. *Nucleic Acids Research*, *32*(S2), W526–W531. doi:10.1093/nar/gkh468 PMID:15215442

Kirkpatrick, S., Gelatt, C., & Vecchi, M. (1983). Optimization by simulated annealing. *Science*, *220*(4598), 671–680. doi:10.1126/science.220.4598.671 PMID:17813860

Kirschner, P. A., Shum, J. B., & Carr, S. C. (Eds.). (2003). *Visualizing Argumentation: Software Tools for Collaborative and Educational Sense-Making*. Heidelberg, Germany: Springer. doi:10.1007/978-1-4471-0037-9

Kitakawa, A. (2004). On a segregation intensity parameter in continuous-velocity lattice gas cellular automata for immiscible binary fluid. *Chemical Engineering Science*, *59*(14), 3007–3012. doi:10.1016/j.ces.2004.04.032

Kitakawa, A. (2005). Simulation of break-up behavior of immiscible droplet under shear field by means of continuous-velocity lattice gas cellular automata. *Chemical Engineering Science*, *60*(20), 5612–5619. doi:10.1016/j.ces.2005.05.011

Kitano, H. (2002). Systems Biology: A brief overview. *Science*, *295*(5560), 1662–1664. doi:10.1126/science.1069492 PMID:11872829

Klaczynski, P. A., & Lavallee, K. L. (2005). Domain-specific identity, epistemic regulation, and intellectual ability as predictors of belief-based reasoning: A dual-process perspective. *Journal of Experimental Child Psychology*, *92*(1), 1–24. doi:10.1016/j.jecp.2005.05.001 PMID:16005013

Klausberger, T., Magill, P. J., Marton, L. F., Roberts, J. D., Cobden, P. M., Buzsaki, G., & Somogyi, P. (2003). Brain-state- and cell-type-specific firing of hippocampal interneurons in vivo. *Nature*, *421*(6925), 844–848. doi:10.1038/nature01374 PMID:12594513

Klein, J. (2014). *Breve* [computer software]. Available from http://www.spiderland.org/breve/

Klein, J., & Spector, L. (2009). 3D Multi-Agent Simulations in the Breve Simulation Environment. In M. Komosinski & A. Adamatzky (Eds.), *Artificial Life Models in Software* (pp. 79–106). Springer. doi:10.1007/978-1-84882-285-6_4

Klink, C. A., & Machado, R. B. (2005). A conservação do Cerrado brasileiro. *Megadiversidade*, *1*(1), 147–155.

Klink, C. A., & Moreira, A. G. (2002). Past and Current Human Occupation, and Land Use. In R. J. Marquis & P. S. Oliveira (Eds.), *The Cerrados of Brazil: Ecology and Natural History of a Neotropical Savana*. New York: Columbia University Press. doi:10.7312/oliv12042-004

Knight, R. D. (2009). *Física: Uma Abordagem Estratégica* (Vol. 4). Bookman.

Koestler, A. (1967). *The Ghost in the Machine*. New York: Macmillan.

Kohl, P., & Noble, D. (2009). Systems biology and the virtual physiological human. *Molecular Systems Biology*, 5. PMID:19638973

Kokis, J., Macpherson, R., Toplak, M., West, R. F., & Stanovich, K. E. (2002). Heuristic and analytic processing: Age trends and associations with cognitive ability and cognitive styles. *Journal of Experimental Child Psychology*, *83*(1), 26–52. doi:10.1016/S0022-0965(02)00121-2 PMID:12379417

Kolinski, A., & Skolnick, J. (2004). Reduced models of proteins and their applications. *Polymer*, *45*(2), 511–524. doi:10.1016/j.polymer.2003.10.064

Kopell, N., Whittington, M. A., & Kramer, M. A. (2011). Neuronal assembly dynamics in the beta1 frequency range permits short-term memory. *Proceedings of the National Academy of Sciences of the United States of America*, *108*(9), 37793784. doi:10.1073/pnas.1019676108 PMID:21321198

Koppensteiner, W. A., & Sippl, M. J. (1995). Knowledge-based potentials-back to the roots. *Biochemistry, 63*, 247–252. PMID:9526121

Krajzewicz, D., Eedmann, J., Behrisch, M., & Bieker, L. (2012). Recent Development and Applications of SUMO - Simulation of Urban MObility. *International Journal On Advances in Systems and Measurements, 3-4*(5), 128–138.

Kridi, D., Carvalho, C., & Gomes, D. (2014). A Predictive Algorithm for Mitigate Swarming Bees Through Proactive Monitoring via Wireless Sensor Networks. In *Proceedings of the 11th ACM symposium on Performance evaluation of wireless ad hoc, sensor, & ubiquitous networks* (pp. 41–47). New York, NY: ACM. doi:10.1145/2653481.2653482

Kryshtafovych, A., Fidelis, K., & Moult, J. (2014). Casp10 results compared to those of previous casp experiments. *Proteins: Struct., Funct. Bioinf., 82*(S2), 164–174.

Kubera, Y., Mathieu, P., & Picault, S. (2008). Interaction-oriented agent simulations: From theory to implementation. In *Proceedings of the 18th European Conference on Artificial Intelligence* (pp. 383-387). IOS Press.

Kurowski, M. A., & Bujnicki, J. M. (2003). Genesilico protein structure prediction meta-server. *Nucleic Acids Research, 31*(13), 3305–3307. doi:10.1093/nar/gkg557 PMID:12824313

Lalli, C., & Parsons, T. R. (1997). *Biological Oceanography: An Introduction: An Introduction.* Butterworth-Heinemann.

Lambin, E. F., & Geist, H. J. (2001). Global land-use and land-cover change: What have we learned so far? *Global Change Newsletter, 29*(46), 27–30.

Lambin, E. F., Geist, H. J., & Lepers, E. (2003). Dynamics of Land-Use and Land-Cover Change in Tropical. *Annual Review of Environment and Resources, 28*(1), 205–241. doi:10.1146/annurev.energy.28.050302.105459

Lashkari, Y., Metral, M., & Maes, P. (1994). Collaborative interface agents. In *Proceedings of the 12th national conference on artificial intelligence* (p. 444-449). Elsevier.

Laskowiski, R., Watson, J., & Thornton, J. (2005a). Profunc: A server for predicting protein functions from 3d structure. *Nucleic Acids Research, 33*, 89–93. doi:10.1093/nar/gki414

Laskowiski, R., Watson, J., & Thornton, J. (2005b). Protein function prediction using local 3d templates. *Journal of Molecular Biology, 351*(3), 614–626. doi:10.1016/j.jmb.2005.05.067 PMID:16019027

Laspidou, C. S., Kungolos, A., & Samaras, P. (2010). *Cellular-automatta and individual-based approaches for the modelling of biofilm structures: Pros and cons. Desalination Journal.*

Lau, K., & Dill, K. (1989). A lattice statistical mechanics model of the conformation and sequence spaces of proteins. *Macromolecules, 22*(10), 3986–3997. doi:10.1021/ma00200a030

Lawn, S. D., & Zumla, A. I. (2011). Tuberculosis. *The Lancet Journal, 378*(9785), 57–72. doi:10.1016/S0140-6736(10)62173-3 PMID:21420161

Lazaridis, T., & Karplus, M. (2000). Effective energy functions for protein structure prediction. *Current Opinion in Structural Biology, 10*(2), 139–145. doi:10.1016/S0959-440X(00)00063-4 PMID:10753811

Le Page, C., Bousquet, F., Bakam, I., Bah, A., & Baron, C. (2000). CORMAS: A multiagent simulation toolkit to model natural and social dynamics at multiple scales. In Workshop The ecology of scales, Wageningen, The Netherlands.

Le Page, C., Becub, N., Bommel, P., & Bousquet, F. (2012). Participatory Agent-Based Simulation for Renewable Resource Management: The Role of the Cormas Simulation Platform to Nurture a Community of Practice. *Journal of Artificial Societies and Social Simulation, 15*(1), 10. doi:10.18564/jasss.1928

Leavitt, D. (2006). *The Man Who Knew Too Much: Alan Turing and the Invention of the Computer.* New York: W. W. Norton.

Lehninger, A., Nelson, D., & Cox, M. (2005). *Principles of biochemistry* (4th ed.). New York, USA: W.H. Freeman.

Lesk, A. M. (2002). *Introduction to bioinformatics* (1st ed.). New York, NY: Oxford University Press Inc.

Lesk, A. M. (2010). *Introduction to protein science: architecture, function, and genomics.* New York: Oxford University Press.

Letcher, R. A. K., Jakeman, A. J., Barreteau, O., Borsuk, M. E., ElSawah, S., Hamilton, S. H., & Voinov, A. A. et al. (2013). Selecting among five common modelling approaches for integrated environmental assessment and management. *Environmental Modelling & Software*, *47*, 159–181. doi:10.1016/j.envsoft.2013.05.005

Levin, M. (2012). Morphogenetic fields in embryogenesis, regeneration, and cancer: non-local control of complex patterning. *Biosystems, 109*(3), 243-261.

Levin, M. (2011). The wisdom of the body: Future techniques and approaches to morphogenetic fields in regenerative medicine, developmental biology and cancer. *Regenerative Medicine*, *6*(6), 667–673. doi:10.2217/rme.11.69 PMID:22050517

Levinthal, C. (1968). Are there pathways for protein folding? *Journal de Chimie Physique*, *65*(1), 44–45.

Liljas, A., Liljas, L., Piskur, J., Lindblom, G., Nissen, P., & Kjeldgaard, M. (2009). *Textbook of structural biology.* Singapore: World Scientific. doi:10.1142/6620

Linard, C., Ponçon, N., Fontenille, D., & Lambin, E. F. (2009). A multi-agent simulation to assess the risk of malaria re-emergence in southern France. *Ecological Modelling*, *220*(2), 160–174. doi:10.1016/j.ecolmodel.2008.09.001

Linden, R. (2008). *Algoritmos Genéticos.* Rio de Janeiro: RJ Brasport.

Lioy, P. J. (1990). Assessing total human exposure to contaminants. *A multidisciplinary approach. Environmental Science & Technology*, *24*(7), 938–945. doi:10.1021/es00077a001

Lipinski-Paes, T., & De Souza, O. N. (2014). Masters: A general sequence- based multiagent system for protein tertiary structure prediction. *Electronic Notes in Theoretical Computer Science*, *306*, 45–59. doi:10.1016/j.entcs.2014.06.014

Liu, Y., He, Y., Li, M., Wang, J., Liu, K., Mo, L., & Li, X.-Y. (2011, April). Does wireless sensor network scale? A measurement study on GreenOrbs. In INFOCOM, 2011 Proceedings IEEE (pp. 873–881).

Liu, Y., Cheng, H. D., Huang, J. H., Zhang, Y. T., & Tang, X. L. (2012). An effective approach of lesion segmentation within the breast ultrasound image based on the cellular automata principle. *Journal of Digital Imaging*, *25*(5), 580–590. doi:10.1007/s10278-011-9450-6 PMID:22237810

Li, Y., Gosálvez, M. A., Pal, P., Sato, K., & Xing, Y. (2015). Particle swarm optimization-based continuous cellular automaton for the simulation of deep reactive ion etching. *Journal of Micromechanics and Microengineering*, *25*(5), 055023. doi:10.1088/0960-1317/25/5/055023

Lobo, D., & Levin, M. (2015). Inferring regulatory networks from experimental morphological phenotypes: A computational method reverse-engineers planarian regeneration. *PLoS Computational Biology*, *11*(6), e1004295. doi:10.1371/journal.pcbi.1004295 PMID:26042810

Lodish, H., Berk, A., Matsudaira, P., Kaiser, C. A., Krieger, M., & Scott, M. (1990). Molecular cell biology (5th ed.). New York: Scientific American Books, W.H. Freeman.

Logan, B. E. (2008). *Microbial Fuel Cells.* John Wiley & Sons.

Lönnroth, K, & Migliori, G. B., & Abubakar. (2015). Towards tuberculosis elimination: An action framework for low-incidence countries. *The European Respiratory Journal*, *45*(4), 928–952. PMID:25792630

Lopes da Silva, F. (1991). Neural mechanisms underlying brain waves: from neural membranes to networks. *Electroencephalography and Clinical Neurophysiology*, *79*(2), 81-93. 10.1016/0013-4694(91)90044-5

Lopes dos Santos, V., Conde-Ocazionez, S., Nicolelis, M., Ribeiro, S. T., & Tort, A. B. L. (2011). Neuronal assembly detection and cell membership specification by principal component analysis. *PLoS ONE*, *6*(6), 20996. doi:10.1371/journal.pone.0020996 PMID:21698248

Lotka, A. J. (1925). *Elements of Physical Biology*. Williams & Wilkins Company.

Low, F. M., Gluckman, P. D., & Hanson, M. A. (2012). Developmental plasticity, epigenetics and human health. *Evolutionary Biology*, *39*(4), 650–665. doi:10.1007/s11692-011-9157-0

Lucas, L. V., Thompson, J. K., & Brown, L. R. (2009). Why are diverse relationships observed between phytoplankton biomass and transport time. *Limnology and Oceanography*, *54*(1), 381–390. doi:10.4319/lo.2009.54.1.0381

Luke, S., Cioffi-Revilla, C., Panait, L., Sullivan, K., & Balan, G. (2005). MASON: A multiagent simulation environment. *Simulation*, *81*(7), 517–527. doi:10.1177/0037549705058073

Lund, O., Frimand, K., Gorodkin, J., Bohr, H., Bohr, J., Hansen, J., & Brunak, S. (1997). Protein distance constraints predicted by neural networks and probability density functions. *Protein Engineering*, *10*(11), 1241–1248. doi:10.1093/protein/10.11.1241 PMID:9514112

Lundstrom, J., Rychlewski, L., Bujnicki, J., & Elofsson, A. (2001). Pcons: A neural-network based consensus predictor that improves fold recognition. *Protein Science*, *10*(11), 2354–2362. doi:10.1110/ps.08501 PMID:11604541

Lynch, S. C., & Ferguson, J. (2014). Reasoning about complexity - software models as external representations. In *Proceedings of the 25th Workshop of The Psychology of Programming Interest Group*.

Macal, C. M., & North, M. J. (2010). Tutorial on agent-based modelling and simulation. *Journal of Simulation*, *4*(3), 151–162. doi:10.1057/jos.2010.3

MacKerell, A. D. (2004). Empirical force fields for biological macromolecules: Overview and issues. *Journal of Computational Chemistry*, *25*(13), 1584–1604. doi:10.1002/jcc.20082 PMID:15264253

Macklin, P., Edgerton, M. E., Thompson, A. M., & Cristini, V. (2012). Patient-calibrated agent-based modelling of ductal carcinoma in situ (DCIS): From microscopic measurements to macroscopic predictions of clinical progression. *Journal of Theoretical Biology*, *301*, 122–140. doi:10.1016/j.jtbi.2012.02.002 PMID:22342935

Mancuso, S., & Viola, A. (2015). *Brilliant Green: The Surprising History and Science of Plant Intelligence*. Island Press.

Manzanares-Lopez, P., Muñoz Gea, J. P., Malgosa-Sanahuja, J., & Sanchez-Aarnoutse, J. C. (2011, May). An Efficient Distributed Discovery Service for EPCglobal Network in Nested Package Scenarios. *Journal of Network and Computer Applications*, *34*(3), 925–937. doi:10.1016/j.jnca.2010.04.018

Maria, A. (1997) Introduction to modeling and simulation.*Proceedings of the 1997 Winter Simulation Conference.*

Markus, M., Böhm, D., & Schmick, M. (1999). Simulation of vessel morphogenesis using cellular automata. *Mathematical Biosciences*, *156*(1-2), 191–206. doi:10.1016/S0025-5564(98)10066-4 PMID:10204393

Martin, S. (2004). The cost of restoration as a way of defining resilience: a viability approach applied to a model of lake eutrophication. *Conservation Ecology*, *9*(2), Article 8.

Martínez, G. J., Adamatzky, A., & Alonso-Sanz, R. (2013). Designing complex dynamics in cellular automata with memory. *International Journal of Bifurcation and Chaos in Applied Sciences and Engineering*, *23*(10), 1330035. doi:10.1142/S0218127413300358

Martí-Renom, M., Stuart, A., Fiser, A., Sanchez, A., Mello, F., & Sali, A. (2000). Comparative protein structure modeling of genes and genomes. *Annual Review of Biophysics and Biomolecular Structure*, *29*(16), 291–325. doi:10.1146/annurev.biophys.29.1.291 PMID:10940251

McAlpine, C. A., Etter, A., Fearnside, P. M., Seabrook, L., & Laurance, W. F. (2009). Increasing world consumption of beef as a driver of regional and global change: A call for policy action based on evidence from Queensland (Australia), Colombia and Brazil. *Global Environmental Change*, *19*(1), 21–33. doi:10.1016/j.gloenvcha.2008.10.008

McGuinness, D. L., & Van Harmelen, F. (2004). OWL web ontology language overview. *W3C Recommendation, 10*(10), 2004.

McRee, D. (1999). *Practical protein crystallography* (1st ed.). London: Academic press.

Medler, D. A. (1998). A brief history of connectionism. *Neural Computing Surveys*, *1*, 18–72.

Meinig, C., Stalin, S. E., Nakamura, A. I., Gonzalez, F., & Milburn, H. B. (2005, Sept). Technology developments in real-time tsunami measuring, monitoring and forecasting. In *OCEANS. Proceedings of MTS/IEEE* (Vol. 2, pp. 1673–1679). doi:10.1109/OCEANS.2005.1639996

Meir, E., Perry, J., Stal, D., Maruca, S., & Klopfer, E. (2005). How effective are simulated molecular-level experiments for teaching diffusion and osmosis? *Cell Biology Education*, *4*(3), 235–248. doi:10.1187/cbe.04-09-0049 PMID:16220144

Melo, F. B. (2005). *Proposição de Medidas Favorecedoras à Acessibilidade e Mobilidade de Pedestres em áreas Urbanas. Estudo de Caso: O Centro de Fortaleza. Dissertação (Mestrado em Engenharia de Transportes).* Fortaleza, CE: Universidade Federal do Ceará.

Merelli, E., Armano, G., Cannata, N., Corradini, F., dInverno, M., Doms, A., & Luck, M. et al. (2007). Agents in bioinformatics, computational and systems biology. *Briefings in Bioinformatics*, *8*(1), 45–59. doi:10.1093/bib/bbl014 PMID:16772270

Merks, R. M. H., Guravage, M., Inzé, D., & Beemstaer, G. T. S. (2001). VirtualLeaf: An open-source framework for cell-based modelling of plant tissue growth and development. *Plant Physiology*, *155*(2), 656–666. doi:10.1104/pp.110.167619 PMID:21148415

Mesquita, M. G., Moraes, S. O., & Corrente, J. E. (2002). More adequate probability distributions to represent the saturated soil hydraulic conductivity. *Scientia Agrícola*, *59*(4), 789–793. doi:10.1590/S0103-90162002000400025

Metcalf & Eddy. (2003). *Wastewater Engineering treatment Disposal Reuse.* New York: McGraw - Hill Book.

Michael, D., & Chen, S. (2006). *Serious Games Games that Educate, Train and Inform.* Thomson Course Technology.

Micheletti, C., Seno, F., & Maritan, A. (2000). Recurrent oligomers in proteins: an optimal scheme reconciling accurate and concise backbone representations in automated folding and design studies. *Proteins: Struct., Funct. Bioinf.*, *40*(4), 662–674.

Milano, M., & Roli, A. (2004). MAGMA: A multiagent architecture for metaheuristics. *IEEE Transactions on Systems, Man, and Cybernetics. Part B, Cybernetics*, *34*(2), 925–941. doi:10.1109/TSMCB.2003.818432 PMID:15376840

Ministério de Minas e Energia. (2015). *Resenha Energética Brasileira – Exercício de 2014*. Retrieved from http://www.mme.gov.br/web/guest/publicacoes-e-indicadores Acessed in 15/03/2015

Moczek, A. P., Sultan, S., Foster, S., Ledon-Rettig, C., Dworkin, I., Nijhout, H. F., . . . Pfennig, D. W. (2011). The role of developmental plasticity in evolutionary innovation. *Proc. R. Soc. B-Biol. Sci., 278*(1719), 2705-2713. doi:10.1098/rspb.2011.0971

Modi, P. J., Shen, W.-M., Tambe, M., & Yokoo, M. (2005). Adopt: Asynchronous distributed constraint optimization with quality guarantees. *Artificial Intelligence, 161*(1), 149–180. doi:10.1016/j.artint.2004.09.003

Mofrad, M. H., Sadeghi, S., Rezvanian, A., & Meybodi, M. R. (2015). Cellular edge detection: Combining cellular automata and cellular learning automata. *AEÜ. International Journal of Electronics and Communications, 69*(9), 1282–1290. doi:10.1016/j.aeue.2015.05.010

Momany, F., McGuire, R., Burgess, A., & Scheraga, H. (1975). Energy parameters in polypeptides vii, geometric parameters, partial charges, non- bonded interactions, hydrogen bond interactions and intrinsic torsional potentials for naturally occurring amino acids. *Journal of Physical Chemistry, 79*(22), 2361–2381. doi:10.1021/j100589a006

Monetti, R. A., & Albano, E. V. (1997). On the emergence of large-scale complex behavior in the dynamics of a society of living individuals: The Stochastic Game of Life. *Journal of Theoretical Biology, 187*(2), 183–194. doi:10.1006/jtbi.1997.0424 PMID:9405136

Monod, J. (1949). The growth of bacterial cultures. *Annual Review of Microbiology, 3*(1), 371–394. doi:10.1146/annurev.mi.03.100149.002103

Montagna, S., Omicini, A., & Pianini, D. (2016). Extending the Gillespie's stochastic simulation algorithm for integrating discrete-event and multi-agent based simulation. In B. Gaudou & J. S. Sichman (Eds.), Multi-Agent Based Simulation XVI (Vol. 9568, pp. 3–18). Springer. 1 doi:10.1007/978-3-319-31447-1_1

Montagna, S., Pianini, D., & Viroli, M. (2012). A model for Drosophila Melanogaster development from a single cell to stripe pattern formation. In D. Shin, C.-C. Hung & J. Hong (Eds.), *27th annual ACM Symposium on Applied Computing* (SAC 2012) (pp. 1406–1412). Riva del Garda, Italy: ACM. doi:10.1145/2245276.2231999

Montagna, S., Donati, N., & Omicini, A. (2010). An Agent-based Model for the Pattern Formation in Drosophila Melanogaster. In *Proceedings of the 12th International Conference on the Synthesis and Simulation of Living Systems*. The MIT Press.

Montagna, S., Ricci, A., & Omicini, A. (2008). A&A for modelling and engineering simulations in Systems Biology. *International Journal of Agent-Oriented Software Engineering, 2*(2), 222–245. doi:10.1504/IJAOSE.2008.017316

Montresor, A., & Jelasity, M. (2009, September). PeerSim: A scalable P2P simulator. In *Proceedings of the 9th Int. Conference on Peer-to-Peer(P2P'09)* (pp. 99–100).

Morigi, M., Imberti, B., Zoja, C., Corna, D., Tomasoni, S., Abbate, M., & Remuzzi, G. et al. (2004). Mesenchymal stem cells are renotropic, helping to repair the kidney and improve function in acute renal failure. *Journal of the American Society of Nephrology, 15*(7), 1794–1804. doi:10.1097/01.ASN.0000128974.07460.34 PMID:15213267

Morozova, N., & Shubin, M. (2012). *The Geometry of Morphogenesis and the Morphogenetic Field Concept*. Retrieved from http://arxiv.org/abs/1205.1158v1

Morton, D. C., Defries, R. S., Shimabukuro, Y. E., Anderson, L. O., Arai, E., Espirito-santo, F. B., & Espirito-santo, B. (2006). Cropland expansion changes deforestation dynamics in the southern Brazilian Amazon. *Proceedings of the National Academy of Sciences of the United States of America, 103*(39), 14637–14641. doi:10.1073/pnas.0606377103 PMID:16973742

Moult, J., Fidelis, K., Kryshtafovych, A., Schwede, T., & Tramontano, A. (2014). Critical assessment of methods of protein structure prediction (CASP)-round x. *Proteins: Struct., Funct. Bioinf., 82*(S2), 1–6.

Moussa, N. (2005). A 2-dimensional cellular automaton for agents moving from origins to destinations. *International Journal of Modern Physics C, 16*(12), 1849–1860. doi:10.1142/S0129183105008370

Mulder, N. J., Apweiler, R., Attwood, T. K., Bairoch, A., Barrell, D., & Bateman, A. et al.. (2003). The interpro database, 2003 brings increased coverage and new features. *Nucleic Acids Research, 31*(1), 315–318. doi:10.1093/nar/gkg046 PMID:12520011

Muscalagiu, I., Popa, H. E., & Vidal, J. (2013). Clustered computing with netlogo for the evaluation of asynchronous search techniques. In *Intelligent software methodologies, tools and techniques (somet), 2013 IEEE 12th international conference on* (pp. 115–120). IEEE. doi:10.1109/SoMeT.2013.6645651

Muscalagiu, I., Iordan, A., Osaci, M., & Panoiu, M. (2012). Modeling and simulation of the protein folding problem in discsp-netlogo. *Global Journal on Technology, 2*, 229–234.

Muscalagiu, I., Popa, H. E., Panoiu, M., & Negru, V. (2013). Multi-agent systems applied in the modelling and simulation of the protein folding problem using distributed constraints. In *Multiagent system technologies* (pp. 346–360). Springer Berlin Heidelberg. doi:10.1007/978-3-642-40776-5_29

National Research Council (NRC). (1983). *Risk Assessment in the Federal Government: Managing the Process.* Washington, DC: The National Academies Press. doi: 10.17226/366

Naumowicz, T., Freeman, R., Heil, A., Calsyn, M., Hellmich, E., Brandle, A., & Schiller, J. (2008). Autonomous Monitoring of Vulnerable Habitats Using a Wireless Sensor Network. In *Proceedings of the workshop on real-world wireless sensor networks* (pp. 51–55). New York, NY: ACM. doi:10.1145/1435473.1435488

Nelder, J. A., & Mead, R. (1965). A Simplex Method for Function Minimization. *The Computer Journal, 7*(4), 308–313. doi:10.1093/comjnl/7.4.308

NetLogo. (2015). Retrieved from https://ccl.northwestern.edu/netlogo/

Neumaier, A. (1997). Molecular modeling of proteins and mathematical prediction of protein structure. *Society for Industrial and Applied Mathematics Rev., 39*, 407.

Nikolai, C., & Madey, G. (2009). Tools of the Trade: A Survey of Various Agent Based Modeling. *Journal of Artificial Societies and Social Simulation, 12*(2).

Nikolai, C., & Madey, G. (2009). Tools of the trade: A survey of various agent based modelling platforms. *Journal of Artificial Societies and Social Simulation, 12*(2), 2.

Nissen, S. (2003). *Implementation of a fast artificial neural network library (fann).* Report, Department of Computer Science University of Copenhagen (DIKU).

Nizam, A. and Shanmugham, B. (2013). Self-organizing Genetic Algorithm: A survey. *International Journal of Computer Applications, 65*(18), 0975-8887.

Noble, D. (2002). The rise of computational biology. *Nature Reviews. Molecular Cell Biology, 3*(6), 459–463. doi:10.1038/nrm810 PMID:12042768

North, M. J., Collier, N. T., Ozik, J., Tatara, E. R., Macal, C. M., Bragen, M., & Sydelko, P. (2013). Complex adaptive systems modeling with Repast Symphony. *Complex Adaptive Systems Modeling, 1*(1), 3. doi:10.1186/2194-3206-1-3

North, M. J., Collier, N. T., & Vos, J. R. (2006). Experiences Creating Three Implementations of the Repast Agent Modeling Toolkit. *ACM Transactions on Modeling and Computer Simulation*, *16*(1), 1–25. doi:10.1145/1122012.1122013

Noy, N. F., & McGuinness, D. L. (2001). *Ontology development 101: A guide to creating your first ontology*. Palo Alto, CA: Stanford University Press.

Nunes, G. (2013). *Estudo e análise da Dispersão de Poluentes: um estudo de caso para a cidade de Rio Grande/RS. Dissertação (Mestrado em Modelagem Computacional)*. Rio Grande, RS: Universidade Federal do Rio Grande.

Nwana, H. S. (1996). Software agents: An overview. *The Knowledge Engineering Review*, *11*(3), 205–244. doi:10.1017/S026988890000789X

Odebrecht, C., Abreu, P. C., & Carstensen, J. (2015). Retention time generates shortterm phytoplankton blooms in a shallow microtidal subtropical estuary. *Estuarine, Coastal and Shelf Science*, *162*, 35–44. doi:10.1016/j.ecss.2015.03.004

Odum, E. P. (1953). Fundamentals of Ecology. Cengage Learning.

Odum, E. P. (1964). The New Ecology. *Bioscience*, *14*(7), 14–16. doi:10.2307/1293228

Odum, H. T. (1994). *Ecological and General Systems - An Introduction to Systems Ecology*. Univ. Press of Colorado.

Odum, H. T., & Odum, B. (2003). Concepts and Methods of Ecological Engineering. *Ecological Engineering*, *20*(5), 339–361. doi:10.1016/j.ecoleng.2003.08.008

Olami, Z., Feder, H. J. S., & Christensen, K. (1992). Self-organized criticality in a continuous, nonconservative cellular automaton modeling earthquakes. *Physical Review Letters*, *68*(8), 1244–1247. doi:10.1103/PhysRevLett.68.1244 PMID:10046116

Olmedo, M. T. C., Pontius, R. G., Paegelow, M., & Mas, J.-F. (2015). Comparison of simulation models in terms of quantity and allocation of land. *Environmental Modelling & Software*, *69*, 214–221. doi:10.1016/j.envsoft.2015.03.003

Olsen, M. M., & Siegelmann, H. T. (2013). Multiscale agent-based model of tumor angiogenesis. *Procedia Computer Science*, *18*, 1016–1025. doi:10.1016/j.procs.2013.05.267

Omicini, A., & Zambonelli, F. (2004). MAS as complex systems: A view on the role of declarative approaches. In J. A. Leite, A. Omicini, L. Sterling & P. Torroni (Eds.), Declarative Agent Languages and Technologies (Vol. 2990, pp. 1–17). Springer. 1 doi:10.1007/978-3-540-25932-9_1

Opitz, J. M., & Neri, G. (2013). Historical perspective on developmental concepts and terminology. *American Journal of Medical Genetics. Part A*, *161*(11), 2711–2725. doi:10.1002/ajmg.a.36244 PMID:24123982

Osguthorpe, D. J. (2000). Ab initio protein folding. *Current Opinion in Structural Biology*, *10*(2), 146–152. doi:10.1016/S0959-440X(00)00067-1 PMID:10753815

Ouali, M., & King, R. D. (2000). Cascaded multiple classifiers for secondary structure prediction. *Protein Science*, *9*(06), 1162–1176. doi:10.1110/ps.9.6.1162 PMID:10892809

Paganelli, F., & Parlanti, D. (2012). A DHT-Based Discovery Service for the Internet of Things. *Journal of Computer Networks and Communications*, *2012*, 107041:1-107041:11.

Paixão, L. M. M., & Gontijo, E. E. (2007). Perfil de casos de tuberculose notificados e fatores associados ao abandono. *Revista de Saude Publica*, *41*(2), 205–213. doi:10.1590/S0034-89102007000200006 PMID:17384794

Palopoli, L., & Terracina, G. (2003). A framework for improving protein structure predictions by teamwork. In *Proceedings of the first asia-pacific bioinformatics conference on bioinformatics 2003* (vol. 19, pp. 163–171).

Park, B., Huang, E., & Levitt, M. (1997). Factors affecting the ability of energy functions to discriminate correct from incorrect folds. *Journal of Molecular Biology, 266*(4), 831–846. doi:10.1006/jmbi.1996.0809 PMID:9102472

Parker, D. C., Berger, T., & Manson, S. M. (2001). *Agent-Based Models of Land- Use and Land- Cover Change.* Technical Report No.6. Report and Review of an International Workshop. L. R. No.6, Irvine, CA.

Parker, A. M., & Fischhoff, B. (2005). Decision-making competence: External validation through an individual differences approach. *Journal of Behavioral Decision Making, 18*(1), 1–27. doi:10.1002/bdm.481

Pauling, L., & Corey, R. (1951). The pleated sheet, a new layer configuration of polypeptide chains. *Proceedings of the National Academy of Sciences of the United States of America, 37*(5), 251–256. doi:10.1073/pnas.37.5.251 PMID:14834147

Pauling, L., Corey, R., & Branson, H. (1951). The structure of proteins: Two hydrogen-bonded helical configurations of the polypeptide chain. *Proceedings of the National Academy of Sciences of the United States of America, 37*(4), 205–211. doi:10.1073/pnas.37.4.205 PMID:14816373

Pérez, P. P. G., Beltrán, H. I., Rojo-Domínguez, A., Eduardo, M., & Gutiérrez, S. (2009). Multi-agent systems applied in the modeling and simulation of biological problems: A case study in protein folding. *World Acad. Sci. Eng. Technol., 3*(10), 497–506.

Pezzulo, G., & Levin, M. (2015). Re-Membering the Body: Applications of computational neuroscience to the top-down control of regeneration of limbs and other complex organs. *Integrative Biology : Quantitative Biosciences from Nano to Macro, 7*(12), 1487–1517. doi:10.1039/C5IB00221D PMID:26571046

Pianini, D., Montagna, S., & Viroli, M. (2013). Chemical-oriented simulation of computational systems with Alchemist. *Journal of Simulation, 7*(S3), 202–215. doi:10.1057/jos.2012.27

Piccioreanu, C., Loosdrecht, M. C. M., & Heijnen, J. J. (n.d.). Mathematical modeling of biofilm structure with a hybrid differential-discrete cellular automaton approach. *Biotechnology and Bioengineering, 58*. PMID:10099266

Pidwirny, M. (2009). *Fundamentals of Physical Geography* (2nd ed.). Available at: http://www.physicalgeography.net

Pisarev, A., Poustelnikova, E., Samsonova, M. & Reinitz, J. (2009). FlyEx, the quantitative atlas on segmentation gene expression at cellular resolution. *Nucleic Acids Research, 37*(1), 560–566. doi: 10.1093/nar/gkn717

Pizarro, G. E., Teixeira, J., Sepúlveda, M., & Noguera, D. R. (2005). Bitwise implementation of a two-dimensional cellular automata biofilm model. *Journal of Computing in Civil Engineering, 19*(3), 258–268. doi:10.1061/(ASCE)0887-3801(2005)19:3(258)

Platt, J. (1988). *Sequetial minimal optimization: A fast algorithm for training support vector machines.* Technical Report MST-TR-98-14. Microsoft Research.

Pogson, M., Smallwood, R., Qwarnstrom, E., & Holcombe, M. (2006). Formal agent-based modelling of intracellular chemical interactions. *Bio Systems, 85*(1), 37–45. doi:10.1016/j.biosystems.2006.02.004 PMID:16581178

Pokahr, A., Braubach, L., & Lamersdorf, W. (2003). Jadex: Implementing a BDI Infrastructure for JADE Agents. *EXP, 3*(3), 76–85.

Pokala, N., & Handel, T. (2000). Review: Protein design - where we were, where we are, where were going. *Journal of Structural Biology, 134*(2-3), 269–281. doi:10.1006/jsbi.2001.4349 PMID:11551185

Pollock, J. L. (1995). *Cognitive carpentry: A blueprint for how to build a person.* Cambridge, MA: MIT Press.

Pontius, R. G., Boersma, W., Castella, J. C., Clarke, K., de Nijs, T., Dietzel, C., & Verburg, P. H. (2008). Comparing the input, output, and validation maps for several models of land change. *The Annals of Regional Science*, *42*(1), 11–37. doi:10.1007/s00168-007-0138-2

Porcellis, D., Moraes, M., & Bertin, C. (2015) Modelo Baseado em Equações versus Modelo Baseado em Agentes: uma abordagem usando sistema predador-presa. In *IX Workshop-Escola de Sistemas de Agentes, Seus Ambientes e Aplicacões* (pp. 100–111). Niterói.

Portegys, T. E. (2002). An abstraction of intercellular communication. In *Alife VIII Proceedings*.

Portegys, T., & Wiles, J. (2004). A robust game of life. In *The International Conference on Complex Systems (ICCS2004)*. Retrieved from http://www.necsi.edu/events/iccs/openconf/author/abstractbook.php

Potter, M. C. (1911). Electrical effects accompanying the decomposition of organic compounds. *Proceedings of the Royal Society of London. Series B, Containing Papers of a Biological Character*, *84*(571), 260–276. doi:10.1098/rspb.1911.0073

Prada, R., Prendinger, H., Yongyuth, P., Nakasoneb, A., & Kawtrakulc, A. (2015, February). AgriVillage: A Game to Foster Awareness of the Environmental Impact of Agriculture. *Comput. Entertain.*, *12*(2), 3:1–3:18.

Prain, V., & Waldrip, B. (2006). An exploratory study of teachers and students use of multimodal representations of concepts in primary science. *International Journal of Science Education*, *28*(15), 1843–1866. doi:10.1080/09500690600718294

Pritchard, D. W. (1967). Observations of circulation in coastal plain estuaries. American Association for the Advancement of Science.

Protégé Software. (2016). *Protégé is a core component of The National Center for Biomedical Ontology* (Version 4.3.3) [Software]. Stanford Center for Biomedical Informatics Research. Available from http://protege.stanford.edu/

Pruitt, K. D., Tatusova, T., & Maglott, D. R. (2005). Ncbi reference sequence (refseq): A curated non-redundant sequence database of genomes, transcripts, and proteins. *Nucleic Acids Research*, *33*(Database issueS1), D501–D504. doi:10.1093/nar/gki025 PMID:15608248

Putnam, L. L., & Roloff, M. E. (Eds.). (1992). Communication and Negotiation. Newbury Park, CA: Sage. doi:10.4135/9781483325880

Rahimnejad, M., Adhami, A., Darvari, S., Zirepour, A., & Oh, S. (2015). Microbial fuel cell as new technology for bioelectricity generation: A review. *Alexandria Engineering Journal*, *54*(3), 745–756. doi:10.1016/j.aej.2015.03.031

Rahwan, I., & Amgoud, L. (2006). An Argumentation-based Approach for Practical Reasoning. In *Proceedings of the 5th International Joint Conference on Autonomous Agents & Multi Agent Systems (AAMAS'2006)*. New York, NY: ACM Press. doi:10.1145/1160633.1160696

Raiffa, H. (1982). *The Art & Science of Negotiation*. Cambridge, MA: Harvard University Press.

Railsback, S. F., Lytinen, S. L., & Jackson, S. K. (2006). Agent-based Simulation Platforms: Review and Development Recommendations. *Simulation*, *82*(9), 609–623. doi:10.1177/0037549706073695

Ralha, C. G., Abreu, C. G., Coelho, C. G., Zaghetto, A., Macchiavello, B., & Machado, R. B. (2013). A multi-agent model system for land-use change simulation. *Environmental Modelling & Software*, *42*, 30–46. doi:10.1016/j.envsoft.2012.12.003

Ranhel, J. (2013). Neural Assemblies and Finite State Automata. In *BRICS-CCI-CBIC '13 Proceedings of the 2013 BRICS Congress on Computational Intelligence and 11th Brazilian Congress on Computational Intelligence* (pp. 28-33). IEEE Computer Society. doi:10.1109/BRICS-CCI-CBIC.2013.16

Rao, A., & Georgeff, M. (1991). Modelling Rational Agents within a BDI-Architecture. In *Proceedings of the 2nd International Conference on Principles of Knowledge Representation and Reasoning*, (pp. 473–484).

Rebelo, A., Ferra, I., Gonçalves, I., & Marques, A. M. (2014). A risk assessment model for water resources: Releases of dangerous and hazardous substances. *Journal of Environmental Management*, *140*, 51–59. doi:10.1016/j.jenvman.2014.02.025 PMID:24726965

Rebonatto, M. (2000). *Simulação paralela de eventos discretos com uso de memória compartilhada distribuída, Dissertação (Mestrado em Ciência da Computação). Programa de Pós-Graduação em Computação.* Porto Alegre: UFRGS.

Rendell, P. (2002). Turing universality of the game of life. In A. Adamatzky (Ed.), *Collision-Based Computing* (pp. 513–539). London: Springer London. doi:10.1007/978-1-4471-0129-1_18

Ren, F., Zhang, J., Wu, Y., He, T., Chen, C., & Lin, C. (2013, May). Attribute-Aware Data Aggregation Using Potential-Based Dynamic Routing in Wireless Sensor Networks. *IEEE Transactions on Parallel and Distributed Systems*, *24*(5), 881–892. doi:10.1109/TPDS.2012.209

Repenning, A., Smith, C., Owen, B., & Repenning, N. (2012). Agentcubes: Enabling 3d creativity by addressing cognitive and affective programming challenges. In *Proceedings of World Conference on Educational Multimedia, Hypermedia and Telecommunications* (pp. 2762-2771). Chesapeake, VA: AACE.

Repenning, A., & Sumner, T. (1995). Agentsheets: A medium for creating domain-oriented visual languages. *IEEE Computer*, *28*(3), 17–25. doi:10.1109/2.366152

Resnick, M. (1994). *Turtles, Termites, and Traffic Jams: Explorations in Massively Parallel Microworlds*. Cambridge, MA: MIT Press.

Retore, P., Santos, R., Marietto, M., & Sá, C. (2006). *Sistemas Multi-Agentes Reativos Modelando o Controle de Tráfego Urbano*. Uruguaiana/RS.

Reynolds, C. S. (2000). Hydroecology of river plankton: The role of variability in channel flow. *Hydrological Processes*, *14*(16-17), 3119–3132. doi:10.1002/1099-1085(200011/12)14:16/17<3119::AID-HYP137>3.0.CO;2-6

Reynolds, C. W. (1987, August). Flocks, Herds and Schools: A Distributed Behavioral Model. *SIGGRAPH Comput. Graph.*, *21*(4), 25–34. doi:10.1145/37402.37406

Rezende, S. O. (2003). *Sistemas Inteligentes: fundamentos e aplicações*. Editora Manole Ltda.

Richardson, J. S. (1981). The anatomy and taxonomy of protein structure. *Advances in Protein Chemistry*, *34*, 167–339. doi:10.1016/S0065-3233(08)60520-3 PMID:7020376

Rindfuss, R. R., Walsh, S. J., Turner, B. L., Fox, J., & Mishra, V. (2004). Developing a science of land change: Challenges and methodological issues. *Proceedings of the National Academy of Sciences of the United States of America*, *101*(39), 13976–13981. doi:10.1073/pnas.0401545101 PMID:15383671

Rinzel, J., & Ermentrout, B. (1998). Article. In C. Koch & I. Segev (Eds.), Methods in Neuronal Modelling (2nd ed.; p. 251). Cambridge, MA: MIT Press.

Rittmann, B. E., & McCarty, P. L. (2001). *Environmental Biotechnology: Principles and Applications*. New York: McGraw-Hill Book Co.

Rocha, V., & Brandão, A. A. F. (2015). Towards conscientious peers: Combining agents and peers for efficient and scalable video segment retrieval for VoD services. *Engineering Applications of Artificial Intelligence*, *45*, 180–191. doi:10.1016/j.engappai.2015.07.001

Rohl, C. A., Strauss, C. E., Misura, K. M., & Baker, D. (2004). Protein structure prediction using rosetta. *Methods in Enzymology*, *383*, 66–93. doi:10.1016/S0076-6879(04)83004-0 PMID:15063647

Rossi, F., Van Beek, P., & Walsh, T. (2006). *Handbook of constraint programming*. Elsevier.

Rouillard, A. D., & Holmes, J. W. (2014). Coupled agent-based and finite-element models for predicting scar structure following myocardial infarction. *Progress in Biophysics and Molecular Biology*, *115*(2–3), 235–243. doi:10.1016/j.pbiomolbio.2014.06.010 PMID:25009995

Rouquier, J.-B., Alvarez, I., Reuillon, R., & Wuillemin, P.-H. (2015). A kd-tree algorithm to discover the boundary of a black box hypervolume or how to peel potatoes by recursively cutting them in halves. *Annals of Mathematics and Artificial Intelligence*, *75*(3), 335–350.

Rumjanek, N. G., Fonseca, M. C. C., & Xavier, G. R. (2004). Quorum sensing em sistemas agrícolas. In *Revista Biotecnologia Ciência*. Belo Horizonte: Desenvolvimento.

Russell, S. J., & Norvig, P. (2010). *Artificial Intelligence: A Modern Approach* (3rd ed.). Upper Saddle River, NJ: Pearson Education.

Ruxton, G. D. (1996). Effects of the spatial and temporal ordering of events on the behaviour of a simple cellular automaton. *Ecological Modelling*, *84*(1-3), 311–314. doi:10.1016/0304-3800(94)00145-6

Ruxton, G. D., & Saravia, L. A. (1998). The need for biological realism in the updating of cellular automata models. *Ecological Modelling*, *107*(2-3), 105–112. doi:10.1016/S0304-3800(97)00179-8

Saint-Pierre, P. (1994). Approximation of the viability kernel. *Applied Mathematics & Optimization*, *29*(2), 187–209. doi:10.1007/BF01204182

Salamon, T. (2011). *Design of Agent-Based Models: Developing Computer Simulations for a Better Understanding of Social Processes. Academic Series*. Repin, Czech Republic: Bruckner Publishing.

Salgado, A. J., & Gimble, J. M. (2013). Secretome of mesenchymal stem/stromal cells in regenerative medicine Foreword. *Biochimie*, *95*(12), 2195. doi:10.1016/j.biochi.2013.10.013 PMID:24210144

Samuels, R. (2005). The complexity of cognition: Tractability arguments for massive modularity. In P. Carruthers, S. Laurence, & S. Stich (Eds.), *The innate mind* (pp. 107–121). Oxford, UK: Oxford University Press. doi:10.1093/acprof:oso/9780195179675.003.0007

Sánchez, R., & Sali, A. (1997). Advances in comparative protein-structure modeling. *Current Opinion in Structural Biology*, *7*(2), 206–214. doi:10.1016/S0959-440X(97)80027-9 PMID:9094331

Sano, E. E., Rosa, R., Brito, J. L. S., & Ferreira, L. G. (2010). *Mapeamento do Uso do Solo e Cobertura Vegetal–Bioma Cerrado: ano base 2002*. Report MMA/SBF.

Sano, E. E., Rosa, R., Brito, J. L. S., & Ferreira, L. G. (2008). Mapeamento Semi Detalhado do Uso da Terra do Bioma. *Pesquisa Agropecuaria Brasileira*, *43*(1), 153–156. doi:10.1590/S0100-204X2008000100020

Satoh, S. (2012). Computational identity between digital image inpainting and filling-in process at the blind spot. *Neural Computing & Applications*, *21*(4), 613–621. doi:10.1007/s00521-011-0646-y

Sayad, S. (2016). *Artificial Neural Network*. Retrieved from http://www.saedsayad.com/artificial_neural_network.htm

Scheef, E., & Fink, J. (2003). Fundamentals of protein structure. In P. Bourne & H. Weissig (Eds.), *Structural bioinformatics* (Vol. 44). Hoboken, NJ: John Wiley & Sons, Inc.

Schmidt, C., & Parashar, M. (2003, June). Flexible information discovery in decentralized distributed systems. In *IEEE International Symposium on High Performance Distributed Computing* (pp. 226–235). doi:10.1109/HPDC.2003.1210032

Schneider, W., & Shiffrin, R. M. (1977, January). Controlled and automatic human information processing: I. Detection, search, and attention. *Psychological Review*, *84*(1), 1–66. doi:10.1037/0033-295X.84.1.1

Schrödinger, L. L. C. (2015). *The PyMOL Molecular Graphics System, Version 1.8.* Available at http://pymol.org

Schroeder, B., & Harchol-Balter, M. (2006, February). Web Servers Under Overload: How Scheduling Can Help. *ACM Transactions on Internet Technology*, *6*(1), 20–52. doi:10.1145/1125274.1125276

Searle, J. R. (1969). *Speech Acts: An Essay in the Philosophy of Language*. Cambridge, UK: Cambridge University Press. doi:10.1017/CBO9781139173438

Seeliger, U., & Odebrecht, C. (1997). *Introduction and overview. Subtropical convergence environments: the coast and sea in the southwestern Atlantic*. New York: Springer. doi:10.1007/978-3-642-60467-6

Setty, Y. (2012). Multi-scale computational modeling of developmental biology. *Bioinformatics (Oxford, England)*, *28*(15), 2022–2028. doi:10.1093/bioinformatics/bts307 PMID:22628522

Shannon, P., Markiel, A., Ozier, O., Baliga, N. S., Wang, J. T., Ramage, D., & Ideker, T. et al. (2003). Cytoscape: A software environment for integrated models of biomolecular interaction networks. *Genome Research*, *13*(11), 2498–2504. doi:10.1101/gr.1239303 PMID:14597658

Shi, J., Blundell, T. L., & Mizuguchi, K. (2001). Fugue: Sequence-structure homology recognition using environment-specific substitution tables and structure-dependent gap penalties. *Journal of Molecular Biology*, *310*(1), 243–257. doi:10.1006/jmbi.2001.4762 PMID:11419950

Shin, D. G., Liu, L., Loew, M., & Schaff, J. (1998). Virtual cell: A general framework for simulating and visualizing cellular physiology. *Visual Database Systems*, *4*, 214–220.

Siew, N., Elofsson, A., Rychlewski, L., & Fischer, D. (2000). Maxsub: An automated measure for the assessment of protein structure prediction quality. *Bioinformatics (Oxford, England)*, *16*(9), 776–785. doi:10.1093/bioinformatics/16.9.776 PMID:11108700

Silva, E., & Manfredini, V. (2010). *Aplicação de Conceitos da Inteligência Artificial no Controle de Tráfego Rodoviário. Trabalho de Conclusão de Curso* (Graduação em Ciência da Computação). São Paulo/SP.

Simons, K. T., Kooperberg, C., Huang, E., & Baker, D. (1997). Assembly of protein tertiary structures from fragments with similar local sequences using simulated annealing and Bayesian scoring functions. *Journal of Molecular Biology*, *268*(1), 209–225. doi:10.1006/jmbi.1997.0959 PMID:9149153

Sinnott, E.W. (1961). *Cell and psych. The biology of purpose*. Academic Press.

Sinnott, E. W. (1960). *Plant Morphogenesis*. New York: McGraw-Hill Book Co. doi:10.5962/bhl.title.4649

Sinnott, E. W. (1962a). *The Biology of The Spirit*. New York: The Viking Press.

Sinnott, E. W. (1962b). *Matter, Mind and Man: The Biology of Human Nature*. New York: Atheneum.

Sinnott, E. W. (1966). *The Bridge of Life, From Matter to Spirit*. New York: Simon and Schuster.

Sivaramakrishna, R. (1998). Breast image registration using a textural transformation. *Medical Physics*, *25*(11), 2249. doi:10.1118/1.598426

Sivaramakrishna, R., & Gordon, R. (1997). Mammographic image registration using the Starbyte transformation. In P. G. McLaren & W. Kinsner (Eds.), *WESCSANEX'97* (pp. 144–149). Winnipeg, Canada: IEEE. doi:10.1109/WESCAN.1997.627128

Slepoy, A., Thompson, A. P., & Plimpton, S. J. (2008). A constant-time kinetic Monte Carlo algorithm for simulation of large biochemical reaction networks. *The Journal of Chemical Physics*, *128*(20), 205101. doi:10.1063/1.2919546 PMID:18513044

Sloman, A. (1993). The mind as a control system. In C. Hookway & D. Peterson (Eds.), *Philosophy and cognitive science* (pp. 69–110). Cambridge, UK: Cambridge University Press.

Sloman, A., & Chrisley, R. (2003). Virtual machines and consciousness. *Journal of Consciousness Studies*, *10*, 133–172.

Smajgl, A., Brown, D. G., Valbuena, D., & Huigen, M. G. A. (2011). Empirical characterisation of agent behaviours in socio-ecological systems. *Environmental Modelling & Software*, *26*(7), 837–844. doi:10.1016/j.envsoft.2011.02.011

Søballe, D. M., & Kimmel, B. L. (1987). A large-scale comparison of factors influencing phytoplankton abundance in rivers, lakes, and impoundments. *Ecology*, *68*(6), 1943–1954. doi:10.2307/1939885

Solé, R. V., Valverde, S., Rosas Casals, M., Kauffman, S. A., Farmer, D., & Eldredge, N. (2013). The evolutionary ecology of technological innovations. *Complexity*, *18*(4), 15–27. doi:10.1002/cplx.21436

Sordoni, A. (2008). *Conception et implantation d'un agent artificial dans le cadre du projet SimParc*. Rapport de stage de Master 2ème année, Master Intelligence Artificielle et Décision (IAD), Université Pierre et Marie Curie – Laboratoire d'Informatique de Paris 6, Paris, France.

Sordoni, A., Briot, J.-P., Alvarez, I., Vasconcelos, E., Irving, M. A., & Melo, G. (2010). Design of a participatory decision making agent architecture based on argumentation and influence function – Application to a serious game about biodiversity conservation. *RAIRO – An International Journal on Operations Research*, *44*(4), 269–284.

Spinoza, B. (1677). Ethica ordine geometrico demonstrata. In *English translation: "The Ethics*. Radford, VA: Wilder Publications.

Srinivasan, R., & Rose, G. D. (1995). Linus: a hierarchic procedure to predict the fold of a protein. *Proteins: Struct., Funct. Bioinf.*, *22*(2), 81–99.

Stanovich, K. E. (2004). Balance in psychological research: The dual process perspective. *Behavioral and Brain Sciences*, *27*(03), 357–358. doi:10.1017/S0140525X0453008X

Stanovich, K. E. (2008). Higher-order preferences and the Master Rationality Motive. *Thinking & Reasoning*, *14*(1), 111–127. doi:10.1080/13546780701384621

Stanovich, K. E., & West, R. F. (1997). Reasoning independently of prior belief and individual differences in actively open-minded thinking. *Journal of Educational Psychology*, *89*(2), 342–357. doi:10.1037/0022-0663.89.2.342

Stanovich, K. E., & West, R. F. (1999). Discrepancies between normative and descriptive models of decision making and the understanding/acceptance principle. *Cognitive Psychology*, *38*(3), 349–385. doi:10.1006/cogp.1998.0700 PMID:10328857

Stanovich, K. E., & West, R. F. (2000). Individual differences in reasoning: Implications for the rationality debate? *Behavioral and Brain Sciences*, *23*(5), 645–665. doi:10.1017/S0140525X00003435 PMID:11301544

Stanovich, K. E., & West, R. F. (2003). Evolutionary versus instrumental goals: How evolutionary psychology misconceives human rationality. In D. Over (Ed.), *Evolution and the psychology of thinking: The debate* (pp. 171–230). Hove, UK: Psychology Press.

Stanovich, K. E., & West, R. F. (2007). Natural myside bias is independent of cognitive ability. *Thinking & Reasoning*, *13*(3), 225–247. doi:10.1080/13546780600780796

Stanovich, K. E., & West, R. F. (2008). On the failure of intelligence to predict myside bias and one-sided bias. *Thinking & Reasoning*, *14*, 129–167. doi:10.1080/13546780701679764

Stanovich, K. E., & West, R. F. (2008). On the relative independence of thinking biases and cognitive ability. *Journal of Personality and Social Psychology*, *94*(4), 672–695. doi:10.1037/0022-3514.94.4.672 PMID:18361678

Stanovich, K. E., West, R. F., & Toplak, M. E. (2011). The complexity of developmental predictions from dual process models. *Developmental Review*, *31*(2-3), 103–118. doi:10.1016/j.dr.2011.07.003

Starruß, J., de Back, W., Brusch, L., & Deutsch, A. (2014). Morpheus: A user-friendly modeling environment for multiscale and multicellular systems biology. *Bioinformatics (Oxford, England)*, *30*(9), 1331–1332. doi:10.1093/bioinformatics/btt772 PMID:24443380

Steventon, B. & Arias, A.M. (2016). *Evo-engineering and the cellular and molecular origins of the vertebrate spinal cord*. doi:10.1101/068882

Stillinger, F. H., Head-Gordon, T., & Hirshfeld, C. L. (1993). Toy model for protein folding. *Physical Review E: Statistical Physics, Plasmas, Fluids, and Related Interdisciplinary Topics*, *48*(2), 1469–1477. doi:10.1103/PhysRevE.48.1469 PMID:9960736

Stoica, I., Morris, R., Karger, D., Kaashoek, M. F., & Balakrishnan, H. (2001). Chord: A Scalable Peer-to-peer Lookup Service for Internet Applications. In *Proceedings of the 2001 Conference on Applications, Technologies, Architectures, and Protocols for Computer Communications* (pp. 149–160). New York, NY: ACM. doi:10.1145/383059.383071

Stroeve, S. H., Blom, H. A. P., & Van der Park, M. N. J. (2003). Multi-agent situation awareness error evolution in accident risk modelling. In *Proceedings of the Fifth USA/Europe ATM R&D Seminar* (pp. 1-25).

Stroup, W. M., & Wilensky, U. (2014). On the embedded complementary of agent-based and aggregate reasoning in students developing understanding of dynamic systems. *Technology. Knowledge and Learning*, *19*(1-2), 19–52. doi:10.1007/s10758-014-9218-4

Sun, X., Rosin, P. L., & Martin, R. R. (2011). Fast rule identification and neighborhood selection for cellular automata. *IEEE Transactions on Systems, Man, and Cybernetics. Part B, Cybernetics*, *41*(3), 749–760. doi:10.1109/TSMCB.2010.2091271 PMID:21134817

Surkova, S., Kosman, D., Kozlov, K., Manu, , Myasnikova, E., Samsonova, A. A., & Reinitz, J. et al. (2008). Characterization of the Drosophila segment determination morphome. *Developmental Biology*, *313*(2), 844–862. doi:10.1016/j.ydbio.2007.10.037 PMID:18067886

Szigeti, B., Gleeson, P., Vella, M., Khayrulin, S., Palyanov, A., Hokanson, J., & Larson, S. et al. (2014). OpenWorm: An open-science approach to modelling *Caenorhabditis elegans*. *Frontiers in Computational Neuroscience*, *8*(137). PMID:25404913

Tabatabaee, V., Tiwari, A., & Hollingsworth, J. K. (2005). Parallel Parameter Tuning for Applications with Performance Variability. In *Proceedings of the 2005 ACM/IEEE Conference on Supercomputing*. Washington, DC: IEEE Computer Society. doi:10.1109/SC.2005.52

Tanaka, S. (2015). Simulation frameworks for morphogenetic problems. *Computation*, *3*(2), 197–221. doi:10.3390/computation3020197

Tang, Y., Parsons, S., & Skalar, E. (2006). Modeling human education data: From equation-based modeling to agent-based modeling. Hakodate.

Tang, J., Zhou, Z. B., Niu, J., & Wang, Q. (2014). An energy efficient hierarchical clustering index tree for facilitating time-correlated region queries in the Internet of Things. *Journal of Network and Computer Applications*, *40*, 1–11. doi:10.1016/j.jnca.2013.07.009

Tanin, E., Harwood, A., & Samet, H. (2007, April). Using a Distributed Quadtree Index in Peer-to-peer Networks. *The VLDB Journal*, *16*(2), 165–178. doi:10.1007/s00778-005-0001-y

Ţăpuş, C., Chung, I. H., & Hollingsworth, J. K. (2002). Active Harmony: Towards Automated Performance Tuning. In *Proceedings of the 2002 ACM/IEEE Conference on Supercomputing, SC '02*, (pp. 1-11). Los Alamitos, CA: IEEE Computer Society Press.

Technologies, E. (2015). *Integrated development environment for language C++*. Retrieved from https://www.embarcadero.com/br/products/cbuilder

Tessier-Lavigne, M., & Goodman, C. S. (1996). The molecular biology of axon guidance. *Science*, *274*(5290), 1123–1133. doi:10.1126/science.274.5290.1123 PMID:8895455

Tessier-Lavigne, M., Placzek, M., Lumsden, A. G. S., Dodd, J., & Jessell, T. M. (1988). Chemotropic guidance of developing axons in the mammalian central nervous system. *Nature*, *336*(6201), 775–778. doi:10.1038/336775a0 PMID:3205306

Tisue, S., & Wilensky, U. (2004). Netlogo: A simple environment for modeling complexity. In *International conference on complex systems* (vol. 21).

Todar, K. (2013). The growth of bacterial populations. In *Todar's Online Textbook of Bacteriology*. Author.

Tompa, P. (2002). Intrinsically unstructured proteins. *Trends in Biochemical Sciences*, *27*(10), 527–533. doi:10.1016/S0968-0004(02)02169-2 PMID:12368089

Tompa, P., & Csermely, P. (2004). The role of structural disorder in the function of RNA and protein chaperones. *The FASEB Journal*, *18*(11), 1169–1175. doi:10.1096/fj.04-1584rev PMID:15284216

Tompkins, N., Li, N., Girabawe, C., Heymann, M., Ermentrout, G. B., Epstein, I. R., & Fraden, S. (2014). Testing Turings theory of morphogenesis in chemical cells. *Proceedings of the National Academy of Sciences of the United States of America*, *111*(12), 4397–4402. doi:10.1073/pnas.1322005111 PMID:24616508

Toplak, M. E., & Stanovich, K. E. (2003). Associations between myside bias on an informal reasoning task and amount of post-secondary education. *Applied Cognitive Psychology*, *17*(7), 851–860. doi:10.1002/acp.915

Torrellas, G. S. (2004). A framework for multi-agent system engineering using ontology domain modelling for security architecture risk assessment in e-commerce security services. In *Network Computing and Applications, 2004. (NCA 2004). Proceedings. Third IEEE International Symposium on* (pp. 409-412). IEEE. doi:10.1109/NCA.2004.1347810

Tramontano, A., & Lesk, A. M. (2006). *Protein structure prediction* (1st ed.). Weinheim, Germany: John Wiley and Sons, Inc.

Tria, F., Loreto, V., Servedio, V. D. P., & Strogatz, S. H. (2014). The dynamics of correlated novelties. *Scientific Reports*, 4. PMID:25080941

Trott, O., & Olson, A. J. (2010). AutoDock Vina: Improving the speed and accuracy of docking with a new scoring function, efficient optimization and multithreading. *Journal of Computational Chemistry*, *31*, 455–461. PMID:19499576

Turing, A. M. (1952). The chemical basis of morphogenesis. *Philosophical Transactions of the Royal Society of London. Series B, Biological Sciences*, *237*(641), 37–72. doi:10.1098/rstb.1952.0012

Turky, M., Ahmad, S., & Yusoff, M. (2009). The Use of Genetic Algorithm for Traffic Light and Pedestrian Crossing Control. *International Journal of Computer Science and Network Security*, *9*(2).

Turner, B. L., II, Skole, D., Sanderson, S., Fischer, G., Fresco, L., & Leemans, R. (1995). *Land-use and land-cover change, science/research plan. Iiasa policy report*. International Geosphere-Biosphere Programme (IGBP) Report No. 35/HDP Report No. 7.

Tyldum, M. (2014). *The Imitation Game* [movie]. Retrieved from http://theimitationgamemovie.com

Tyler, S. E. B. (2014). The work surfaces of morphogenesis: The role of the morphogenetic field. *Biological Theory*, *9*(2), 194–208. doi:10.1007/s13752-014-0177-8

Uhrmacher, A. M., Degenring, D., & Zeigler, B. (2005). Discrete event multi-level models for Systems Biology. In C. Priami (Ed.), Transactions on Computational Systems Biology I (Vol. 3380, pp. 66–89). Springer. 6 doi:10.1007/978-3-540-32126-2_6

Uhrmacher, A. M., & Weyns, D. (Eds.). (2009). *Multi-agent systems: Simulation and applications* (1st ed.). Boca Raton, FL: CRC Press.

Ulam, S. (1962). On some mathematical problems connected with patterns of growth of figures.*Symposium in Applied Mathematics,14*, 215-224. doi:10.1090/psapm/014/9947

Unger, R., & Moult, J. (1993). Finding the lowest free energy conformation of a protein is an np-hard problem: Proof and implications. *Bulletin of Mathematical Biology*, *55*(6), 1183–1198. doi:10.1007/BF02460703 PMID:8281131

United States Environmental Protection Agency (USEPA). (1986). Guidelines for Carcinogen Risk Assessment. EPA/630/R-00/004. Author.

United States Environmental Protection Agency (USEPA). (1989). Risk Assessment Guidance for Superfund. Vol I: Human Health Evaluation Manual (Part A). Washington, DC: Author.

United States Environmental Protection Agency (USEPA). (1992). Guidelines for Exposure Assessment. EPA/600Z-92/001. Author.

United States Environmental Protection Agency (USEPA). (2004). Supplemental Guidance for Dermal Risk Assessment. Washington, DC: Author.

United States Environmental Protection Agency (USEPA/IRIS). (2016). *Integrated Risk Information System - USEPA-IRIS* [Data file]. Retrieved from http://www.epa.gov/IRIS/

Uversky, V. (2001). What does it mean to be natively unfolded? *European Journal of Biochemistry*, *269*(1), 2–12. doi:10.1046/j.0014-2956.2001.02649.x PMID:11784292

Valbuena, D., Verburg, P. H., & Bregt, A. K. (2008). A method to define a typology for agent-based analysis in regional land-use research. *Agriculture, Ecosystems & Environment*, *1281*(1-2), 27–36. doi:10.1016/j.agee.2008.04.015

van Nimwegen, E., Crutchfield, J. P., & Huynen, M. (1999). Neutral evolution of mutational robustness. *Proceedings of the National Academy of Sciences of the United States of America*, *96*(17), 9716–9720. doi:10.1073/pnas.96.17.9716 PMID:10449760

Vapnik, V. N., & Vapnik, V. (1998). *Statistical learning theory* (Vol. 1). New York: Wiley.

Vasconcelos, E., Briot, J.-P., Irving, M. A., Barbosa, S., & Furtado, V. (2009a). A user interface to support dialogue and negotiation in participatory simulations. In N. David & J. S. Sichman (Eds.), Multi-Agent-Based Simulation IX, (LNAI) (vol. 5269, pp. 127–140). Springer-Verlag. doi:10.1007/978-3-642-01991-3_10

Vasconcelos, E., Melo, G., Briot, J.-P., Patto, V. S., Sordoni, A., Irving, M. A., & Lucena, C. et al. (2009b). A serious game for exploring and training in participatory management of national parks for biodiversity conservation: Design and experience. In *Proceedings of the VIII Brazilian Symposium on Games and Digital Entertainment (SBGAMES'09)*. doi:10.1109/SBGAMES.2009.19

Vecchi, D., & Hernández, I. (2014). The epistemological resilience of the concept of morphogenetic field. In A. Minelli & T. Pradeu (Eds.), *Towards a Theory of Development*. doi:10.1093/acprof:oso/9780199671427.003.0005

Veldkamp, A., & Fresco, L. O. (1996). CLUE: A conceptual model to study the Conversion of Land Use and its Effects. *Ecological Modelling*, *85*(2-3), 253–170. doi:10.1016/0304-3800(94)00151-0

Veldkamp, A., & Lambin, E. F. (2001). Predicting land-use change. *Agriculture, Ecosystems & Environment*, *85*(1-3), 1–6. doi:10.1016/S0167-8809(01)00199-2

Verburg, P. H., Schot, P. P., Dijst, M. J., & Veldkamp, A. (2004). Land use change modelling: Current practice & research priorities. *GeoJournal*, *61*(4), 309–324. doi:10.1007/s10708-004-4946-y

von Neumann, J., & Burks, A. W. (1966). *Theory of Self-Reproducing Automata*. Urbana, IL: University of Illinois Press.

Voskuil, M. I., Visconti, K. C., & Schoolnik, G. K. (2004). *Mycobacterium tuberculosis* gene expression during adaptation to stationary phase and low-oxygen dormancy. *Tuberculosis (Edinburgh, Scotland)*, *84*(3-4), 218–227. doi:10.1016/j.tube.2004.02.003 PMID:15207491

Vuduc, R., Demmel, J. W., & Yelick, K. A. (2005). OSKI: A library of automatically tuned sparse matrix kernels. *Journal of Physics: Conference Series*, *16*(1), 521–530. doi:10.1088/1742-6596/16/1/071

Waddington, C. H., & Cowe, R. J. (1969). Computer simulation of a mulluscan pigmentation pattern. *Journal of Theoretical Biology*, *25*(2), 219–225. doi:10.1016/S0022-5193(69)80060-3 PMID:5383503

Wagner, A., & Rosen, W. (2014). Spaces of the possible: Universal Darwinism and the wall between technological and biological innovation. *Journal of the Royal Society, Interface*, *11*(97), 20131190. doi:10.1098/rsif.2013.1190 PMID:24850903

Wagner, G. P., & Altenberg, L. (1996). Complex Adaptations and the Evolution of Evolvability. *Evolution; International Journal of Organic Evolution*, *50*(3), 967–976. doi:10.2307/2410639

Walker, D. C., & Southgate, J. (2009). The virtual cell – a candidate co-ordinator for 'middle-out' modelling of biological systems. Briefings in Bioinformatics, 10(4), 450-461.

Walker, D. C., & Southgate, J. (2009). The virtual cell-a candidate co-ordinator for middle-out modelling of biological systems. *Briefings in Bioinformatics*, *10*(4), 450–461. doi:10.1093/bib/bbp010 PMID:19293250

Walker, D., Wood, S., Southgate, J., Holcombe, M., & Smallwood, R. (2006). An integrated agent-mathematical model of the effect of intercellular signalling via the epidermal growth factor receptor on cell proliferation. *Journal of Theoretical Biology*, *242*(3), 774–789. doi:10.1016/j.jtbi.2006.04.020 PMID:16765384

Wallace, A. C., Laskowski, R. A., & Thornton, J. M. (1996). LIGPLOT: A program to generate schematic diagrams of protein-ligand interactions. *Protein Engineering*, *8*(2), 127–134. doi:10.1093/protein/8.2.127 PMID:7630882

Wall, J. A., & Blum, M. W. (1991). Negotiations. *Journal of Management*, *17*(2), 273–303. doi:10.1177/014920639101700203

Wang, Z., & Cai, M. (2015). Reinforcement Learning Applied to Single Neuron. In *Computer Science and Artificial Intelligence*. Cornell University.

Wang, Z., Butner, J. D., Kerketta, R., Cristini, V., & Deisboeck, T. S. (2015). Simulating cancer growth with multiscale agent-based modeling. *Seminars in Cancer Biology*, *30*, 70–78. doi:10.1016/j.semcancer.2014.04.001 PMID:24793698

Waterman, T. (2009). *The Fundamentals of Landscape Architecture*. Lausanne: AVA Publishing.

Weimar, J. R. (2001). Coupling microscopic and macroscopic cellular automata. *Parallel Computing*, *27*(5), 601–611. doi:10.1016/S0167-8191(00)00080-6

Weiss, G. (Ed.). (1999). *Multiagent Systems: A Modern Approach to Distributed Artificial Intelligence*. Cambridge, MA: MIT Press.

Weisstein, E. W. (2016a). *Cellular Automaton*. Retrieved from http://mathworld.wolfram.com/CellularAutomaton.html

Weisstein, E. W. (2016b). *Moore Neighborhood*. Retrieved from http://mathworld.wolfram.com/MooreNeighborhood.html

Wei, W., Alvarez, I., & Martin, S. (2013). Sustainability analysis: Viability concepts to consider transient and asymptotical dynamics in socio-ecological tourism-based systems. *Ecological Modelling*, *251*, 103–113. doi:10.1016/j.ecolmodel.2012.10.009

Wei, W., Alvarez, I., Martin, S., Briot, J.-P., Irving, M. A., & Melo, G. (2012). Integration of viability models in a serious game for the management of protected areas. In A. Palma dos Reis, P.S.P. Wang, & A.P. Abraham (Eds.), *Proceedings of the IADIS Intelligent Systems and Agents Conference (ISA'2012)*.

Werlang, P. S. (2013). *Multi-Agent-Based Simulation of Mycobacterium tuberculosis growth*. In MABS 2013 - 14th International Workshop on Multi-Agent-Based Simulation, Saint Paul, MN. doi:10.1007/978-3-642-54783-6_9

Werlang, P. (2013). *Simulação da curva de crescimento do Mycobacterium tuberculosis utilizando sistemas multiagentes. Dissertação (Mestrado em Modelagem Computacional). Programa de Pós-Graduação em Modelagem Computacional*. Rio Grande: FURG.

West-Eberhard, M. J. (2002). *Developmental Plasticity and Evolution*. Oxford, UK: Oxford University Press.

Weston, J., & Watkins, C. (1998). *Multi-class support vector machines*. Technical Report CSD-TR-98-04. Department of Computer Science, Royal Holloway, University of London.

Weyns, D., Omicini, A., & Odell, J. J. (2007). Environment as a first class abstraction in multi-agent systems. *Autonomous Agents and Multi-Agent Systems*, *14*(1), 5–30. doi:10.1007/s10458-006-0012-0

Whitehead, N. A., Barnard, A. M. L., Slater, H., Simpson, N. J. L., & Salmond, G. P. C. (2001). Quorum-sensing in Gram-negative bacteria. *FEMS Microbiology Reviews*, *25*(4), 365–404. doi:10.1111/j.1574-6976.2001.tb00583.x PMID:11524130

White, T., & Pagurek, B. (1998). Towards multi-swarm problem solving in networks. In *3rd International Conference on Multi Agent Systems (ICMAS '98)* (pp. 333–340). doi:10.1109/ICMAS.1998.699217

Wikipedia. (2016a). *Lenna*. Retrieved from https://en.wikipedia.org/wiki/Lenna

Wikipedia. (2016b). *List of animals by number of neurons*. Retrieved from https://en.wikipedia.org/wiki/List_of_animals_by_number_of_neurons

Wilensky, U. (1999). *Netlogo*. Center for Connected Learning and Computer-Based Modeling. Available at http://ccl.northwestern.edu/netlogo/

Wilensky, U. (1999). *NetLogo. Center for Connected Learning and Computer-Based Modeling*. Northwestern University. Retrieved from http://ccl.northwestern.edu/netlogo/

Wilensky, U. (1999). *NetLogo*. Retrieved from http://ccl.northwestern.edu/netlogo/

Wilensky, U. (n.d.). *The NetLogo 5.3.1 User Manual*. Retrieved from https://ccl.northwestern.edu/netlogo/

Willmott, S., Dale, J., Burg, B., Charlton, P., & O'Brien, P. (2001). Agentcities: a worldwide open agent network. *Agentlink News, 8*.

Wolfram, S. (1984). Cellular automata as models of complexity. *Nature, 311*(5985), 419–424. doi:10.1038/311419a0

Wolfram, S. (2001). *A New Kind of Science*. Champaign, IL: Wolfram Media.

Wood, J. M. (2007). Understanding and Computing Cohen's Kappa: A Tutorial. *WebPsychEmpiricist*. Retrieved November 8, 2015, from http://wpe.info/vault/wood07/wood07ab.html

Wooldridge, M. (2009). *An introduction to multiagent systems* (2nd ed.). John Wiley & Sons.

Wooldridge, M. (2009). *An Introduction to MultiAgent Systems* (2nd ed.). Wiley Publishing.

Wooldridge, M. (2009). *An Introduction to Multiagent Systems*. John Wiley & Sons.

World Health Organization (WHO). (2010). Risk Assessment Toolkit: chemical hazards. WHO/IPCS.

World Health Organization (WHO/IPCS/ICHEM). (2016). *International Programmed on Chemical Safety - Chemical Safety Information from Intergovernmental Organizations - INCHEM* [Data file]. Retrieved from http://www.inchem.org/

World Health Organization. (2015). *Global tuberculosis report 2015*. WHO/HTM/TB/2015.10. Geneva: World Health Organization.

Wright, P., & Dyson, H. (1999). Intrinsically unstructured proteins: Re-assessing the protein structure-function paradigm. *Journal of Molecular Biology, 293*(2), 321–331. doi:10.1006/jmbi.1999.3110 PMID:10550212

Wu, R., Xu, Y., Li, L., Zha, J., & Li, R. (2014). Advanced Technologies in Ad Hoc and Sensor Networks. In *Proceedings of the 7th China Conference on Wireless Sensor Networks*. Berlin: Springer Berlin Heidelberg.

Wu, J. (2012). Landscape Ecology. In A. Hastings & L. J. Gross (Eds.), *Encyclopedia of Theoretical Ecology* (pp. 392–396). Berkeley, CA: California Press.

Wyczalkowski, M. A., Chen, Z., Filas, B. A., Varner, V. D., & Taber, L. A. (2012). Computational models for mechanics of morphogenesis. *Birth Defects Research Part C-Embryo Today-Reviews, 96*(2), 132–152. doi:10.1002/bdrc.21013 PMID:22692887

Xie, Z., & He, Z. (2014, June). The Study on the Development of Decision Support Systems in Response to Catastrophic Social Risks. In *Third International Conference on Computer Science and Service System*. Atlantis Press. doi:10.2991/csss-14.2014.108

XJ Technologies Company Ltd. (2012). *AnyLogic* [computer software]. Available from www.xjtek.com

Xu, D., & Zhang, Y. (2012). Ab initio protein structure assembly using continuous structure fragments and optimized knowledge-based force field. *Proteins: Struct., Funct. Bioinf., 80*(7), 1715–1735.

Xu, J., Li, M., Kim, D., & Xu, Y. (2003). Raptor: Optimal protein threading by linear programming. *Journal of Bioinformatics and Computational Biology, 1*(01), 95–117. doi:10.1142/S0219720003000186 PMID:15290783

Xu, Y., Xu, D., & Liang, J. (2007). *Computational methods for protein structure prediction and modeling*. Springer.

Yadgar, O. (2007). Emergent Ad Hoc Sensor Network Connectivity in Large-scale Disaster Zones. In *Proceedings of the 6th International Joint Conference on Autonomous Agents and Multiagent Systems* (pp. 269:1–269:2). New York, NY: ACM. doi:10.1145/1329125.1329450

Yokoo, M., Durfee, E. H., Ishida, T., & Kuwabara, K. (1998). The distributed constraint satisfaction problem: Formalization and algorithms. *IEEE Transactions on Knowledge and Data Engineering*, *10*(5), 673–685. doi:10.1109/69.729707

Yu, J., Liu, W., & Song, J. (2007, Dec). C2WSN: A Two-Tier Chord Overlay Serving for Efficient Queries in Large-Scale Wireless Sensor Networks. In *International Conference on Advanced Computing and Communications* (pp. 237–242). doi:10.1109/ADCOM.2007.25

Yu, L., Wang, N., & Meng, X. (2005, Sept). Real-time forest fire detection with wireless sensor networks. In *International Conference on Wireless Communications, Networking and Mobile Computing* (Vol. 2, pp. 1214–1217).

Zambonelli, F., & Omicini, A. (2004). Challenges and research directions in agent-oriented software engineering. *Autonomous Agents and Multi-Agent Systems*, *9*(3), 253–283. doi:10.1023/B:AGNT.0000038028.66672.1e

Zemla, A. (2003). LGA: A method for finding 3d similarities in protein structures. *Nucleic Acids Research*, *31*(13), 3370–3374. doi:10.1093/nar/gkg571 PMID:12824330

Zenonos, A., Stein, S., & Jennings, N. R. (2015). Coordinating Measurements for Air Pollution Monitoring in Participatory Sensing Settings. In *Proceedings of the 2015 International Conference on Autonomous Agents and Multiagent Systems* (pp. 493–501).

Zhang, C., Krishnamurthy, A., & Wang, R. Y. (2004, May). *SkipIndex: Towards a Scalable Peer-to-Peer Index Service for High Dimensional Data* (Vol. TR-703-04; Tech. Rep.). Department of Computer Science, Princeton University.

Zhang, C., Xiao, W., Tang, D., & Tang, J. (2011). P2P-based Multidimensional Indexing Methods: A Survey. *Journal of Systems and Software*, *84*(12), 2348–2362. doi:10.1016/j.jss.2011.07.027

Zhang, J., & Patel, V. L. (2006). Distributed cognition, representation, and affordance. *Pragmatics & Cognition*, *14*(2), 333–341. doi:10.1075/pc.14.2.12zha

Zhang, W., Yang, J., He, B., Walker, S. E., Zhang, H., Govindarajoo, B., & Zhang, Y. (2015). *Integration of QUARK and I-TASSER for Ab Initio Protein Structure Prediction in CASP11. Proteins: Struct., Funct. Bioinf.*

Zhang, Y. (2008). I-tasser server for protein 3d structure prediction. *BMC Bioinf.*, *9*(40), 1–8. doi:10.1093/bib/bbn041 PMID:18215316

Zhang, Y. (2009). I-tasser: Fully automated protein structure prediction in casp8. *Proteins*, *77*(S9), 100–113. doi:10.1002/prot.22588 PMID:19768687

Zhang, Y., Kihara, D., & Skolnick, J. (2002). Local energy landscape flattering: Parallel hyperbolic Monte Carlo sampling of protein folding. *Proteins*, *48*(2), 192–201. doi:10.1002/prot.10141 PMID:12112688

Zhang, Y., & Skolnick, J. (2004a). Scoring function for automated assessment of protein structure template quality. *Proteins*, *57*(4), 702–710. doi:10.1002/prot.20264 PMID:15476259

Zhang, Y., & Skolnick, J. (2004b). Tertiary structure predictions on a comprehensive benchmark of medium to large size proteins. *Biophysical Journal*, *87*(4), 2647–2655. doi:10.1529/biophysj.104.045385 PMID:15454459

Zhang, Y., & Skolnick, J. (2004c). Automated structure prediction of weakly homologous proteins on a genomic scale. *Proceedings of the National Academy of Sciences of the United States of America*, *101*(20), 7594–7599. doi:10.1073/pnas.0305695101 PMID:15126668

Zhang, Y., & Skolnick, J. (2004d). Spicker: A clustering approach to identify near-native protein folds. *Journal of Computational Chemistry*, *25*(6), 20–22. doi:10.1002/jcc.20011 PMID:15011258

Zhao, W., Liu, X. Y., Ma, S., Yuan, C. Y., & Wang, L. F. (2011, Aug). A Distributed RFID Discovery System: Architecture, Component and Application. In *IEEE International Conference on Computational Science and Engineering (CSE)*, (pp. 518–525). doi:10.1109/CSE.2011.93

Zhao, Y., Wei, H. L., & Billings, S. A. (2012). A new adaptive fast cellular automaton neighborhood detection and rule identification algorithm. *IEEE Transactions on Systems, Man, and Cybernetics. Part B, Cybernetics*, *42*(4), 1283–1287. doi:10.1109/TSMCB.2012.2185790 PMID:22695356

Zhong, Y., Lin, J., Du, Q., & Wang, L. (2015). Multi-agent simulated annealing algorithm based on differential perturbation for protein structure prediction problems. *International Journal of Computer Applications in Technology*, *51*(3), 164–172. doi:10.1504/IJCAT.2015.069330

Zhong, Y., Wang, L., Wang, C., & Zhang, H. (2012). Multi-agent simulated annealing algorithm based on differential evolution algorithm. *Int. J. Bio-Inspir. Com.*, *4*(4), 217–228.

About the Contributors

Diana F. Adamatti is Assistant Professor at Universidade Federal do Rio Grande - Brazil. She received her PhD from Escola Politécnica da Universidade de São Paulo, Brazil. She has a MSc from Universidade Federal do Rio Grande do Sul and BSc from Universidade de Caxias do Sul, both in Brazil. Her main areas of research Aritificial Intelligence, Social Simulation and Multi-Agent-Based Simulation.

* * *

Carolina G. Abreu is a Ph.D. student at the Computer Science Graduate Program of the University of Brasilia in the Artificial Intelligence area, more specifically using multi-agent systems for environmental simulation. Currently, works as a legal analyst - Area Support Specialist in the Superior Court of Justice (STJ). She has a specialization in Project Management of Information Technology. Graduated in Biological Sciences (Bachelor Degree) at the University of Brasilia (2009) and Computer Science from the Catholic University of Brasilia (2008). She has a master degree in Computer Science from the University of Brasilia (2012). She participated in a scholarship and volunteer in various research projects, mainly in the following themes: health informatics, biomedical engineering, 3D modeling, virtual education systems, multi-agent systems, simulations and environmental monitoring.

Paulo Cesar Abreu is a Professor at the Institute of Oceanography Federal University of Rio Grande FURG.

Marilton Aguiar has his PhD in Computer Science at Federal University of Rio Grande do Sul (UFRGS, Brazil). His research interests are related to hybrid intelligent systems applied to hydrological, ecological and bioinformatics problems, such as: measures for hydrological regionalization; route recommendation based on pheromones; multiagent based simulation of dispersion of pollutants; natural resources management; motifs discovery in genetic expressions; and, spot recognition in 2D-eletrophoresis. Currently, he is an adjunct professor at the Technological Development Center of the Federal University of Pelotas (UFPel, Brazil). He is also member of Brazilian Society of Computing (SBC). For more information, please see Marilton's profile on Academia.edu.

Bradly Alicea has a Masters degree from the University of Florida, a PhD from Michigan State University, and completed a Postdoc at the University of Illinois Urbana-Champaign. He has published in multiple academic fields, and in venues including Nature Reviews Neuroscience, Stem Cells and Development, Biosystems, and Proceedings of Artificial Life. With interests spanning the biological, computational, and social sciences, he currently operates an academic start-up called Orthogonal Research. Bradly is also the administrator of Synthetic Daisies blog, and has an active interest in open and virtual science. For more information, please see Bradly's profile on Academia.edu.

Isabelle Alvarez, a Researcher at LISC (IRSTEA) and LIP6 (UPMC), has obtained her PhD in Computer Science in 1992. Her doctoral work was dedicated to geometric-based explanations in expert systems. She generalized this kind of explanation to numerical classifiers. As a researcher at Cemagref/Irstea, she applies her work to agricultural engineering and environmental problems. To promote the use of model in these fields she develops both at Irstea and at LIP6 geometric methods to provide relevant information to users of viability-based classifiers. In particular, this work was applied in food process to discover innovative control strategies. Her main interest is now focused on decision in the viability framework, with focuses on links between individual and collective viability, and on links between robustness, resilience and risk.

Luis Antunes holds a PhD in Computer Science from University of Lisbon (2001). He is Professor of Computer Science and Researcher at Agents and Systems Modelling Group, member of Bio systems and Integrative Sciences Institute, a Research Unit of Faculdade de Ciências da Universidade de Lisboa. He has been a researcher in Artificial Intelligence since 1988 and published more than 70 refereed scientific papers. He was the founder and first director of the Group of Studies in Social Simulation (GUESS). Luis Antunes is on the Program Committee of some of the most important international conferences on Artificial Intelligence, Multi-Agent Systems and Social Simulation, such as AAMAS, ECAI, ESSA, WCSS and MABS.

Pascal Ballet is an associate professor in computer science. He develops software for the field of computational biology, especially at the cellular scale.

Albano Oliveira Borba is graduating in Computer Engineering at Universidade Federal do Rio Grande, Brazil.

Míriam Blank Born is a Graduate in Computer Science from the Catholic University of Pelotas (UCPel), Master's Graduate in Computational Modeling the Federal University of Rio Grande (Furg), substitute teacher and special doctoral student in Computer Science from the Federal University of Pelotas (UFPel).

Anarosa Alves Franco Brandão has her PhD in Informatics at PUC-Rio, Brazil. Her research interests are related to software engineering and knowledge engineering applied to multiagent systems and educational systems domains. Currently, she is an assistant professor at the Computing Engineering and Digital Systems Department of the University of Sao Paulo - Brazil. She is also member of IEEE, ACM and the Brazilian Society of Computing (SBC).

Jean-Pierre Briot is a Directeur de recherche (Research Professor) in Centre National de la Recherche Scientifique (CNRS), France. He is conducting research at Laboratoire d'Informatique de Paris 6 (LIP6), the joint computer science research laboratory of Université Pierre et Marie Curie (Paris 6) and CNRS, in Paris, France, and also, as Permanent Visiting Professor, at Computer Science Department (DI) of

Pontifícia Universidade Católica do Rio de Janeiro (PUC-Rio), Brazil. Jean-Pierre Briot has received his Doctorship/PhD in 1984 and his Habilitation à diriger des recherches in 1989, both in computer science and from Université Pierre et Marie Curie. He also also holds faculty degrees in: mathematics, music, musical acoustics, and japanese language. His general research interests are about the design and construction of adaptive cooperative distributed software, at the crossing of programming languages, software engineering, artificial intelligence and distributed systems. He has created or participated in some projects with significant impact, e.g., on: reflexive (self-descriptive and adaptive) software architectures (ObjVlisp), concurrent programming frameworks (ABCL/1 and Actalk), fault-tolerant multi-agent systems (DarX), as well as working on interdisciplinary projects in computer music (Formes) or environment management (SimParc). Currently, he is interested in adaptive hybrid architectures for IoT, collaborative systems and computer music, incorporating software adaptation and modular machine learning techniques. Jean-Pierre Briot has headed more than 20 research projects (ANR, Europe…), several in cooperation with companies or/and international. At LIP6, he has headed the "Objects and Agents for Simulation and Information Systems" (OASIS) research department, including more than 60 researchers, from 1996 to 2005. He has been the founding Director of CNRS Brasil, the permanent representation office of CNRS in Brazil, from 2010 to 2014. He has edited or co-edited 10 books or journals special issues and has authored or co-authored more than 150 scientific publications. He has advised or co-advised more than 20 PhD students and more than 15 master students. For more details, see http://www-desir.lip6.fr/~briot/cv/

Leonardo de Lima Corrêa received his BSc degree in Computer Science from the Federal University of Pelotas, Brazil (2014). He is currently a MSc student in the Institute of Informatics of the Federal University of Rio Grande do Sul, Porto Alegre, Brazil. His research interests include Bioinformatics, Structural Bioinformatics, Metaheuristics, Optimization algorithms and High Performance Computing.

Antônio Carlos da Rocha Costa is Ph.D. in Computer Science (UFRGS, 1993). Invited lecturer and thesis advisor at PGIE/UFRGS and PPGComp/FURG. Main area of interest: Epistemology of Artificial Intelligence, Foundations of Agent Societies, Formal Languages for Social Sciences.

Márcio Dorn is an Assistant Professor with the Informatics Institute at Federal University of Rio Grande do Sul (UFRGS), Porto Alegre, Brazil. He holds a Ph.D. in Computer Science from the Federal Universityof Rio Grande do Sul in Computer Science (2012), Brazil. His research interests include Bioinformatics, Structural Bioinformatics, Machine Learning and Meta-heuristics. Dr. Dorn is currently a CNPq (Brazilian National Research Council) Advanced Fellow.

Alexandra Fronville went to the university Pierre et Marie Curie in Paris, where she studied mathematics and obtained her PhD in 1986. After a post-doctoral position at INRIA, she works for a couple of years for the university of Brest. Her research focuses on viability theory and morphological analysis applied to biological shapes. She also teaches how to use computers and scientific software for young teachers.

Eder Mateus Gonçalves is PhD in Electrical Engineering working with intelligent systems, multiagent systems, manufacturing and Petri nets.

Andrea von Groll, PhD, is adjunct professor at Universidade Federal do Rio Grande (FURG).

Marta de Azevedo Irving is Full Professor at the Federal University of Rio de Janeiro (UFRJ), in Brazil. She has an interdisiplinar training, with a degree in Biology from the Federal University of Rio de Janeiro (1978) and a degree in Psychology from the State University of Rio de Janeiro (1981). She has a Master from the University of Southampton (UK) in 1983 with theme management of coastal ecosystems. She has a Doctorate in Science from the University of São Paulo (1991) in marine biology. She has been invited researcher at the School of Higher Studies in Social Sciences (EHESS) in Paris and the Department of Ecology and Management of Biodiversity Museum of Natural History in Paris (2004-2005), as well as in other institutions (e.g., University of Santiago de Compostela in Spain). She has specialized in a critical reflection about sustainable development, society-nature relationships, democratic governance, social inclusion, tourism and public policies. At UFRJ, she leads the research group on Governance, Biodiversity, Protected Areas and Social Inclusion (GAPIS), under which were defended several dissertations and doctoral theses. She also coordinates the Centre of Governance and Protected Areas since 2006, with interlocution with academic institutions and other segments of society. Several studies have also been carried out in his career, in partnership with the public administration and various national and international institutions. She has authored several scientific texts and didactic-pedagogic material. She is currently a researcher in the EICOS Post-Graduate Program in Social Psychology of Communities and Social Ecology (EICOS/IP/UFRJ), in the Post-Graduate Program in Public Policy, Strategies and Development (PPED/IE/ UFRJ) and in the CNPq National Institute of Science and Technology on Public Policies, Strategies and Development. She is a member of the editorial board and referee of several scientific journals and research management agencies. In 2012 she has been awarded by France Palmes Académiques as a Chevalier.

Diogo Ortiz Machado is a Chemical Engineering with participation in automation, chemical engineering and education projects. Today develop work as a Teacher in the Federal Institute of Technology of Rio Grande do Sul - IFRS in the fields of simulation, automation, control and instrumentation.

Nuno Trindade Magessi is currently finishing is PhD thesis in Cognitive Science at Universidade de Lisboa. He integrates the Agents and Systems Modelling Group, member of Bio systems and Integrative Sciences Institute, a Research Unit of Faculdade de Ciências da Universidade de Lisboa.

Sara Montagna is a contract researcher and lecturer at the Alma Mater Studiorum, the University of Bologna. She is working in between biology and computer science (BIO-ICT convergence): on the one hand she studies the adoption of innovative techniques devised in the computer science field, such as ABM, for modelling biological systems in Computational Biology and in mobile Health applications; on the other hand she proposes the adoption of the biological metaphor for engineering computational systems -- self-organisation applied to pervasive systems in particular.

Andrea Omicini is Full Professor at DISI, the Department of Computer Science and Engineering of the Alma Mater Studiorum–Università di Bologna, Italy. He holds a PhD in Computer & Electronic Engineering, and his main research interests include coordination models, multi-agent systems, intelligent systems, programming languages, autonomous systems, middleware, simulation, software engineering, pervasive systems, self-organisation. On those subjects, he published over 300 articles, edited a number of international books, guest-edited several special issues of international journals, and held many invited talks and tutorials at international conferences and schools.

Carlos José Pereira de Lucena completed his undergraduate studies in Economics and Mathematics between 1962 and 1965 at the Catholic University of Rio de Janeiro (PUC-Rio). While an undergraduate he worked as a computer programmer at the Computing Center of PUC-Rio. This Computing Center, founded in 1960, was the first computer laboratory in Brazil. In 1965, he was hired by the Mathematics Department of PUC-Rio to coordinate the area of computational mathematics. In 1968, together with a small group of colleagues he founded the first Computer Science Department in Brazil, and since 1977 the first in the rank of the CAPES (Ministry of Education). Later, Prof. Lucena received his Masters degree from the University of Waterloo (1969), Canada, and his PhD from the University of California in Los Angeles (1974). Since 1965, he has been the Chairman of the Computer Science Department, the Dean of the Center of Science and Technology and Vice-rector at PUC-Rio. He is a Full Professor of Computer Science since 1982 and has served in all the university's formal committees. Prof. Lucena is an Adjunct Professor of Computer Science and a Senior Research Associate of the Computer Systems Group at the University of Waterloo which he has visited on a regular basis since 1975. His service to the Brazilian academic community since 1972 includes: his participation in the Computer Science Committee and the Governing Board of the Brazilian Research Council (CNPq), coordination of the area of Computer Science of CAPES, Ministry of Education (from 1980 to 1984), and the coordination of the scientific cooperation agreement between Brazil and Germany in the area of Information Technology (CNPq/GMD). Recently Professor Lucena was the coordinator of the Computer Science Area of CAPES in 2005-2007. From 1995 to 1996, Professor Lucena represented the Brazilian academic community in the Steering Committee of the Project Internet in Brazil. He also represents the President's Council for Science and Technology (Information Technology) for the last eleven years. Prof. Lucena is the author of over 450 refereed papers and 19 books in the area of formal methods in software engineering, his primary area of research. He has also been a member of the program committees of more than 30 national and international conferences as well as member of the editorial board of some important journals in his area of research. Until 2009, he has supervised 37 PhD theses and 105 MSc dissertations. His former PhD students are faculty members in universities in Brazil and abroad (e.g. USA, Canada and England). Prof. Lucena has received, among others, the Alvaro Alberto Award in Informatics in 1987 (Ministry of Science and Technology), twice the National Award in Informatics (1988 and 1991) and the Great Cross of the National Order of the Scientific Merit in 1996. Prof. Lucena is a full member of the Brazilian Academy of Sciences and also fellow of TWAS – Academy of Sciences for the Developing World, since Nov. 2008. Recently, he has received the following awards and honors: Sixty Anniversary Award from PUC-Rio and the CAPES Medal "50 Years", in 2001; The Scientific Merit Award from the Brazilian Computing Society (SBC), in 2002 and 2010; Research Felow from the Fraunhofer Institute for Computer Architecture and Software Technology (Berlin), in 2003; IBM Faculty Award and IBM Eclipse Innovation Award, in 2004; The Scientific Merit Award Carlos Chagas Filho, in 2005; IBM Eclipse Innovation Award and Assespro Personality Award 30 years, both in 2006, and "ACM Distinguished Scientist" (in 2009).

Gustavo Mendes de Melo, PhD in Social Psychology of Communities and Social Ecology from Universidade Federal do Rio de Janeiro (UFRJ) (2012). Postdoc in Technology Support for Natural Resource Management at CIRAD (France) (2014-1016). Member of the Instituto Nacional de Ciência e Tecnologia INCT/PPED about Public Policies and Social Development. Research on participatory management of protected areas for biodiversity conservation and social inclusion.

Marcilene Fonseca de Moraes received her MSc in Computational Modelling from Universidade Federal do Rio Grande, Brazil. She has a graduation in Mathematics at Universidade Federal do Rio Grande, in Brazil.

Karine Pichavant is an assistant professor in animal physiology. Research on physiological mechanisms developed by animals to face with changes at several integration levels (cell, tissue, organ, organism).

Diego de Abreu Porcellis received his MsC from Universidade Federal de Rio Grande, Brazil. Currently, he is teacher at Instituto Federal Farroupilha, in Brazil.

Tom Portegys received his Ph.D. in Computer Science from Northwestern University in Evanston Illinois in 1986. He has worked and taught at Lucent Technologies, Illinois State University, Microsoft, Citrix Labs, and EY. He has a research background in artificial intelligence, machine learning, and computational biology.

Alain Pothet is a teacher of Biology in secondary level (1989-2009) Associated teacher of French Institute of Pedagogy (Ife, ENS LYIN) between 1999 to 2008 Inspecteur Pédagogique Régional in Créteil academy since 2009.

Célia Ghedini Ralha is an associate professor at the Computer Science Department, University of Brasília. Célia is an active researcher in the field of intelligent information systems, particularly using multi-agent systems. She received her Ph.D. degree in Computer Science from Leeds University, England, UK (1996) and her M.Sc. from the Aeronautics Institute of Technology (ITA), São José dos Campos, São Paulo, Brazil (1990). She has experience in Computer Science with research interest on the treatment of information and knowledge. She is the research group leader of InfoKnow – Computer Systems for Information and Knowledge Treatment that is registered in the CNPq. She has been a member of the Special Committee on Information Systems (CE-SI) of the Brazilian Computer Society (SBC) since 2009, being the CE-SI Coordinator (2013-2015) and Vice Coordinator (2010-2011). She is part of the editorial board of important journals, e.g., Environmental Modelling & Software (Elsevier, 5-Year JCR 4.359).

Sergio Fred Ribeiro Andrade has PhD in Environmental Sciences, MSc in Systems and Computing, specialist in Advanced Computer Science, graduated in Information Systems and in Administration at the State University of Santa Cruz, Bahia-Brazil. He is an assistant professor at the Department of Exact and Technological Sciences of State University of Santa Cruz, has experience in computing, analysis and information systems, corporate database, data warehousing, data mining and environmental geoinformatics with application of multivariate models, neural networks and support vector machine.

Jérémy Rivière has a PhD in computer science and has been an associate professor at the Université de Bretagne Occidentale since 2014. His research focuses on multi-agent systems (MAS) for modelling and simulating complex biological phenomena, as well as cooperative agents for resolving problems. He is also interested in interaction and cognition mechanisms for intelligent agents, and their application in complex systems, multi-agent systems and agent-based simulations.

Vladimir Rocha is a Computer Engineer and holds a Master's degree in Computer Science at University of São Paulo. Specialist in Peer to Peer and Cloud Computing tecnhologies, works as CTO at Infomobile and as Senior Consulting in companies as RedHat, Infraero.

Vincent Rodin is a full professor of computer science at the University of Western Brittany in Brest, France. His research interests include image processing, multi-agent systems, and biological processes simulation.

Alessandro Sordoni, PhD from Université de Montréal (2016). Research scientist at Maluuba.

Lilia Marta Brandão Soussa Modesto holds a degree in Data Processing Technologist from Salvador University and a MSc degree in Regional Development and Environment at the State University of Santa Cruz, Bahia-Brazil. Has is currently professor at the State University of Santa Cruz. Has experience in the area of Computer Science with emphasis on Database.

Weslen Schiavon de Souza is an eternal student and lover of computer technologies.

José Eurico Vasconcelos Filho, PhD in Computer Science from Pontifícia Universidade Católica do Rio de Janeiro - PUC-Rio (2010). Master in Applied Computer Science (Artificial Intelligence) from Universidade de Fortaleza (2006). Coordinator of the Laboratório de Inovação em TI and Professor in Software Engineering at Universidade de Fortaleza. Director of Science Popularization at Coordenadoria de Ciência Tecnologia e Inovação (CITINOVA) of Fortaleza City Prefecture. Works in research and development in the areas of mHealth (mobile health technologies), Human-Computer Interaction, Serious Games and gamification.

Index

G

H

I

L

M

N

O

P

R

S

T

V

W

Printed in the United States
By Bookmasters